The Surface Wettability Effect on Phase Change

Marco Marengo · Joel De Coninck
Editors

The Surface Wettability Effect on Phase Change

Springer

Editors
Marco Marengo
School of Computing, Engineering and Mathematics
University of Brighton
Brighton, UK

Joel De Coninck
Laboratory of Surfaces and Interfacial Physics
University of Mons
Mons, Belgium

ISBN 978-3-030-82991-9 ISBN 978-3-030-82992-6 (eBook)
https://doi.org/10.1007/978-3-030-82992-6

This Springer imprint is published by the registered company Springer Nature Switzerland AG
The registered company address is: Gewerbestrasse 11, 6330 Cham, Switzerland

Introduction

The transition between matter phases is a natural phenomenon which can be encountered at any length and time scales, from nucleation to astronomical events: A comet tail is formed by the evaporation of components due to the heating from a close star, the beautiful ice crystals are generated starting from solidification at nanoscales, transpiration in plant is also linked to a phase transition, and we could continue with an incredible number of processes.

Phase change is one of the most important topics in physical and chemical sciences, mechanical, chemical and materials engineering, and nanotechnology. Even in medicine and biology, phase change has a major role in many phenomena, such as breathing, drug release, sweating, wound exudation, ultrasound analysis, and so on. In the food and pharmaceutical industry, phase change is fundamental for many processes to prepare dry and wet compounds. The list of products, applications, and devices where the transition from one phase to another is of critical importance could be as long as an encyclopaedia of technologies.

A phase change is a physical process in which a substance goes from one phase to another. Usually, this transition occurs due to adding or removing heat at a specific temperature and pressure. This process can happen in the bulk of the substance, i.e., in a local homogeneous condition, or in vicinity or in contact with a third phase, e.g., for the evaporation of a drop on a surface, through the so-called heterogenous phase change.

In the case of phase change in contact with a surface, several parameters may have a role such as the surface roughness, absorption and adsorption coefficients, and last but not least surface wettability.

The science of surfaces and interfaces involves a very particular zone of a system at the intersection of several media. From a mathematical point of view, this zone is defined as an ideal surface, but physics teaches us that this zone has a certain thickness, of the order of a few nanometers at least, where very surprising properties are likely to occur. The science of surface wetting, which is how a liquid behaves in the vicinity of another liquid or a solid. We are all familiar with rain, especially during winter. When it rains, drops of water hit the windows and slide down the glass. Part of the science of wetting is how the drops deform and slide over a medium. In fact, these properties themselves result from all the atomic interactions: between the atoms of the solid, between the atoms or molecules of the liquid and finally more particularly between the atoms of the solid and the atoms of the liquid. Depending on the magnitude of these interactions, the liquid will form a film (strong interactions) or a drop (weaker interactions). This science of wetting starts with Young and Dupré at the beginning of the nineteenth century. It regained strength and vigor a few years ago thanks to biomimicry which provided an extraordinary source of inspiration. Indeed, nature has imagined examples of remarkable properties in this context. Whether for plants or insects, we can find organisms that have a very special behavior toward water. This is the case with superhydrophobic surfaces like Lotus leaf. These leaves cannot be wetted by water. Raindrops roll over these surfaces until they are released. Imagine a tee-shirt like this. You could walk around in the rain without getting wet! We're not there yet but… we like to think that this dream will one day come true. Along with these developments in wetting, the science of heat exchange has also developed.

This book deals with the phenomena linked to the effects of wettability on phase change, and it is based on a series of workshops "Surface Wettability Effects on Phase Change Phenomena" held between Mons in Belgium and Brighton in the United Kingdom from 2018 to 2020. The book is organized in 4 main chapters and 11 contributions from well-known researchers in the field.

There are different reasons to consider this topic of primary importance nowadays.

Usually, the main parameter for the comprehension of the nucleation processes on a surface is the presence of a cavity or a heterogeneity. The Classical Nucleation Theory (CNT) [1, 2, 3] provides an excellent scheme to define the kinetics of formation of a phase (see Chap. 1). However, when the phase transition starts in contact with a solid surface, the nucleation threshold is significantly lower than the one in the homogeneous case. This is even most pronounced with pre-existing cavities where the volume is very small and the presence of air bubbles decrease the value of the Gibbs energy cost. The pioneering work of Bankoff [4] is usually considered as the first reference to link the cavity radii to the superheats necessary to start the nucleation. In such direction, in case, for example, of a liquid to vapor transition, bubbles nucleate at the imperfections on the surface submerged in the liquid: these imperfections are scratches, pits, and grooves, which form on the surface during its preparation. Early works indicated that the superheats associated with heterogeneous nucleation are much lower than expected from an evaluation of the surface structure. The reason for lower inception superheat can be easily rationalized if we consider that the cavities generally trap air or other incondensable gases, which may forms bubbles whose radii are much larger than the solid heterogeneities. However, only recently the importance of surface wettability starts to be properly investigated. See Chaps. 2.1 and 2.2.

Especially for electronics cooling or microsystems, such as micro-evaporators, micro-refrigerators, and so on, due to the fact that the manufacturing processes can be very accurate and the surface are extremely well defined with ultrasmooth characteristics, the importance of wettability is evident. Where the CNT may predict superheat of 200K due to a surface roughness of the order of few nanometres, the experiments show much lower values, challenging the researchers in finding the reasons for such discrepancies [5, 6]. See Chaps. 3.2, 3.3, 3.4.

For such applications, also the condensation phenomena are very important, and therefore, also for vapor/liquid transitions the surface wettability is critical [7]. There are also new interesting results at nanoscales, which may contradict the standard results at macroscales [8]. See Chap. 3.1.

Also for heterogenous liquid/solid transitions, the surface wettability is of primary importance. Many papers have been recently written on the effect of contact angle on icing formation [9, 10, 11]. On superhydrophobic surfaces, significant water roll-off and reduced water/ice adhesion reveals new fundamental insights into the kinetics of ice nucleation. Nucleation for an intermediate supercooling temperature is controlled by the liquid/substrate interface, i.e., the contact angle affects the forming of single nuclei. High contact angles lead to both the reduction of the liquid/substrate interfacial area and an increase in nucleation activation energy, which can lead to a delayed freezing. See Chap. 3.5.

Finally, the numerical simulations and modeling of phase changes has recently made very significant progresses, especially regarding the possibility of exploring the broad time and length scales of the process, from nucleation to the formation of the second phase with microscale size. Various methods are considered, from Molecular Dynamics [12, 13] to Volume of Fluids CFD [14], passing through Diffuse Interface simulations [15] and Lattice Boltzmann method [16, 17]. See Chaps. 4.1, 4.2 and 4.3.

Starting with the inexplicable case of very small superheats for superhydrophobic surfaces about 15 years ago, we have done a long journey in research in our labs and together with an increasing number of scientists in the world.

Nowadays, we get more and more questions to optimize processes implemented in our society. Whether in terms of energy consumption, waste recovery, to improve health, etc., the needs are immense. In this context, the science of surfaces and interfaces is no exception. Thousands of researchers are working on improving their properties. All kinds of qualifiers can complement these properties here: durable, robust, inexpensive, intelligent, etc. This is a fascinating research topic.

This book is thought has a collection of the last results and a review of the main physical understanding of the physical processes of phase change on surface with various wettabilities.

We wish you all to discover the pleasure we have had ourselves doing this research.

Brighton, UK
Mons, Belgium
April 2021

Joel De Coninck
Marco Marengo

References

1. Volmer, M., Weber, A. Z., Chem. Phys., 1926, 119, 277–301.
2. Becker, R., Döring, W., Ann. Phys., 1935, 24 (719), 752.

3. Frenkel, J. J., Chem. Phys., 1939, 7 (7), 538–547.
4. Bankoff, S. G. (1958), Entrapment of gas in the spreading of a liquid over a rough surface. AIChE J., 4: 24–26.
5. Bourdon B, Bertrand E, Di Marco P, Marengo M, Rioboo R, De Coninck J. Wettability influence on the onset temperature of pool boiling: Experimental evidence onto ultra-smooth surfaces. Adv Colloid Interface Sci. 2015; 221:34–40.
6. Bradley Bon, Cheng-Kang Guan, James F. Klausner, Heterogeneous nucleation on ultra-smooth surfaces, Experimental Thermal and Fluid Science, Volume 35, Issue 5, 2011, pp. 746–752.
7. Hyeongyun Cha, Hamed Vahabi, Alex Wu, Shreyas Chavan, Moon-Kyung Kim, Soumyadip Sett, Stephen A. Bosch, Wei Wang, Arun K. Kota, and Nenad Miljkovic, Dropwise condensation on solid hydrophilic surfaces, Science, Volume 6(2), 2020.
8. Niu, D. and Tang, G. H. The effect of surface wettability on water vapor condensation in nanoscale. *Sci. Rep.* **6**, 19192; https://doi.org/10.1038/srep19192 (2016)
9. Schutzius, Thomas M.; Jung, Stefan; Maitra, Tanmoy; Eberle, Patric; Antonini, Carlo; Stamatopoulos, Christos; et al. (2016): Physics of Icing and Rational Design of Surfaces with Extraordinary Icephobicity. ACS Publications. Collection. https://doi.org/10.1021/la502586a
10. Alizadeh, Azar; Yamada, Masako; Li, Ri; Shang, Wen; Otta, Shourya; Zhong, Sheng; et al. (2016): Dynamics of Ice Nucleation on Water Repellent Surfaces. ACS Publications Collection.
11. Amirfazli, A., & Antonini, C. (2016). Fundamentals of Anti-Icing Surfaces. In R. Ras, & A. Marmur (Eds.), Non-wettable Surfaces: Theory, Preparation and Applications (pp. 319–346). RSC.
12. Kimura, Tatsuto, and Shigeo Maruyama. Molecular dynamics simulation of heterogeneous nucleation of a liquid droplet on a solid surface. *Microscale Thermophysical Engineering* 6, no. 1 (2002): 3–13.
13. Fujinaga, Takuya, and Yasushi Shibuta. "Molecular dynamics simulation of athermal heterogeneous nucleation of solidification." *Computational Materials Science* 164 (2019): 74–81.
14. Georgoulas, Anastasios, Manolia Andredaki, and Marco Marengo. "An enhanced VOF method coupled with heat transfer and phase change to characterise bubble detachment in saturated pool boiling." *Energies* 10, no. 3 (2017): 272.
15. E. J. Gelissen, C. W. M. van der Geld, M. W. Baltussen, J. G. M. Kuerten, Modeling of droplet impact on a heated solid surface with a diffuse interface model, International Journal of Multiphase Flow, Vol. 123, 2020.
16. Li, Qing, Q. J. Kang, M. M. Francois, Y. L. He, and K. H. Luo. "Lattice Boltzmann modeling of boiling heat transfer: The boiling curve and the effects of wettability." *International Journal of Heat and Mass Transfer* 85 (2015): 787–796.
17. Li, Qing, P. Zhou, and H. J. Yan. "Improved thermal lattice Boltzmann model for simulation of liquid-vapor phase change." *Physical Review E* 96, no. 6 (2017): 063303.

Contents

Chapter 1
Introduction

Joel De Coninck and Marco Marengo

The transition between matter phases is a natural phenomenon which can be encountered at any length and time scales, from nucleation to astronomical events: A comet tail is formed by the evaporation of components due to the heating from a close star, the beautiful ice crystals are generated starting from solidification at nanoscales, transpiration in plant is also linked to a phase transition, and we could continue with an incredible number of processes.

Phase change is one of the most important topics in physical and chemical sciences, mechanical, chemical and materials engineering, and nanotechnology. Even in medicine and biology, phase change has a major role in many phenomena, such as breathing, drug release, sweating, wound exudation, ultrasound analysis, and so on. In the food and pharmaceutical industry, phase change is fundamental for many processes to prepare many dry and wet compounds. The list of products, applications, and devices where the transition from one phase to another is of critical importance could be as long as an encyclopaedia of technologies.

A phase change is a physical process in which a substance goes from one phase to another. Usually, this transition occurs due to adding or removing heat at a particular temperature and pressure. This process can happen in the bulk of the substance, i.e., in a local homogeneous condition, or in vicinity or in contact with a third phase, e.g., for the evaporation of a drop on a surface, so-called heterogenous phase change.

In the case of phase change in contact with a surface, several parameters may have a role such as the surface roughness, absorption and adsorption processes, and last but not least surface wettability.

The science of surfaces and interfaces involves a very particular zone of a system at the intersection of several media. From a mathematical point of view, this zone is defined as a functional z(x, y), but physics teaches us that this zone has a certain

J. De Coninck · M. Marengo (✉)
School of Architecture, Technology and Engineering, University of Brighton, Brighton BN2 4GJ, UK
e-mail: M.Marengo@brighton.ac.uk

M. Marengo and J. De Coninck (eds.), *The Surface Wettability Effect on Phase Change*,
https://doi.org/10.1007/978-3-030-82992-6_1

thickness, of the order of a few nanometers at least, where very surprising properties are likely to occur. The science of surfaces and interfaces is largely inspired by wetting, which is how a liquid behaves in the vicinity of another liquid or a solid. We are all familiar with rain, especially during winter. When it rains, drops of water hit the windows and slide down the glass. Part of the science of wetting is how the drops form, deform and slide over a medium. In fact, these properties themselves result from all the atomic interactions in presence: between the atoms of the solid, between the atoms or molecules of the liquid and finally more particularly between the atoms of the solid and the atoms of the liquid. Depending on the magnitude of these interactions, the liquid will form a film (strong interactions) or a drop (weaker interactions). This science of wetting starts with Young and Dupré at the beginning of the nineteenth century. It regained strength and vigor a few years ago thanks to biomimicry which provided an extraordinary source of inspiration. Indeed, nature has imagined examples of remarkable properties in this context. Whether for plants or insects, we can find organisms that have a very special behavior toward water. This is the case with superhydrophobic surfaces like Lotus leaf. These leaves cannot be wetted by water. Raindrops roll over these surfaces until they are released. Imagine a tee-shirt like this. You could walk around in the rain without getting wet! We're not there yet but… we like to think that this dream will one day come true. Along with these developments in wetting, the science of heat exchange has also developed.

This book deals with the phenomena linked to the effects of wettability on phase change, and it is based on a series of workshops "Surface Wettability Effects on Phase Change Phenomena" held between Mons in Belgium and Brighton in the United Kingdom from 2018 to 2020. The book is organized in 4 main chapters and 11 contributions from well-known researchers in the field.

There are different reasons to consider this topic of primary importance nowadays.

Usually, the main parameter for the comprehension of the nucleation processes on a surface is the presence of a cavity or a heterogeneity. The Classical Nucleation Theory (CNT) [1–3] provides an excellent scheme to define the kinetics of formation of a phase (See this chapter). However, when the phase transition starts in contact with a solid surface, the nucleation threshold is significantly lower than the one in the homogeneous case. This is even most pronounced with pre-existing cavities where the volume is very small and the presence of air bubbles decrease the value of the Gibbs energy cost. The pioneering work of Bankoff [4] is usually considered as the first reference to link the cavity radii to the superheats necessary to start the nucleation. In such direction, in case, for example of a liquid to vapor transition, bubbles nucleate at the imperfections on the surface submerged in the liquid: these imperfections are scratches, pits, and grooves, which form on the surface during its preparation. Early works indicated that the superheats associated with heterogeneous nucleation are much lower than expected from an evaluation of the surface structure. The reason for lower inception superheat can be easily rationalized if we consider that the cavities generally trap air or other incondensable gases, which may forms bubbles whose radii are much larger than the solid heterogeneities. However, only recently the importance of surface wettability starts to be properly investigated. See Chaps. 2.1 and 2.2.

Especially for electronics cooling or microsystems, such as micro-evaporators, micro-refrigerators, and so on, due to the fact that the manufacturing processes can be very accurate and the surface are extremely well defined with ultrasmooth characteristics, the importance of wettability is evident. Where the CNT may predict superheat of 200 K due to a surface roughness of the order of few nanometres, the experiments show much lower values, challenging the researchers in finding the reasons for such discrepancies [5, 6]. See Chaps. 3.2, 3.3 and 3.4.

For such applications, also the condensation phenomena are very important, and therefore, also for vapor/liquid transitions the surface wettability is critical [7]. There are also new interesting results at nanoscales, which may contradict the standard results at macroscales [8]. See Chap. 3.1.

Also for heterogenous liquid/solid transitions, the surface wettability is of primary importance. Many papers have been recently written on the effect of contact angle on icing formation [9–11]. On superhydrophobic surfaces, significant water roll-off and reduced water/ice adhesion reveals new fundamental insights into the kinetics of ice nucleation. Nucleation for an intermediate supercooling temperature is controlled by the liquid/substrate interface, i.e., the contact angle affects the forming of single nuclei. High contact angles lead to both the reduction of the liquid/substrate interfacial area and an increase in nucleation activation energy, which can lead to a delayed freezing. See Chap. 3.5.

Finally, the numerical simulations and modeling of phase changes has recently made very significant progresses, especially regarding the possibility of exploring the broad time and length scales of the process, from nucleation to the formation of the second phase with microscale size. Various methods are considered, from Molecular Dynamics [12, 13] to Volume of Fluids CFD [14], passing through Diffuse Interface simulations [15] and Lattice Boltzmann method [16, 17]. See Chaps. 4.1, 4.2 and 4.3.

Starting with the inexplicable case of very small superheats for superhydrophobic surfaces about 15 years ago, we have done a long journey in research in our labs and with an increasing number of scientists in the world.

Nowadays, we get more and more questions to optimize processes implemented in our society. Whether in terms of energy consumption, waste recovery, to improve health, etc., the needs are immense. In this context, the science of surfaces and interfaces is no exception. Thousands of researchers are working on improving their properties. All kinds of qualifiers can complement these properties here: durable, robust, inexpensive, intelligent, etc. This is a fascinating research topic.

This book is thought has a collection of the last results and a review of the main physical understanding of the physical processes of phase change on surface with various wettabilities.

We wish you all to discover the pleasure we have had ourselves doing this research.

Brighton/Mons, 15th April 2021

Joel De Coninck

Marco Marengo

References

1. Volmer, M., Weber, A. Z., Chem. Phys., 1926, 119, 277–301.
2. Becker, R., Döring, W., Ann. Phys., 1935, 24 (719), 752.
3. Frenkel, J. J., Chem. Phys., 1939, 7 (7), 538–547.
4. Bankoff, S.G. (1958), Entrapment of gas in the spreading of a liquid over a rough surface. AIChE J., 4: 24-26.
5. Bourdon B, Bertrand E, Di Marco P, Marengo M, Rioboo R, De Coninck J. Wettability influence on the onset temperature of pool boiling: Experimental evidence onto ultra-smooth surfaces. Adv Colloid Interface Sci. 2015; 221:34-40
6. Bradley Bon, Cheng-Kang Guan, James F. Klausner, Heterogeneous nucleation on ultra-smooth surfaces, Experimental Thermal and Fluid Science, Volume 35, Issue 5, 2011, Pages 746-752.
7. Hyeongyun Cha, Hamed Vahabi, Alex Wu, Shreyas Chavan, Moon-Kyung Kim, Soumyadip Sett, Stephen A. Bosch, Wei Wang, Arun K. Kota, and Nenad Miljkovic, Dropwise condensation on solid hydrophilic surfaces, Science, Volume 6(2), 2020
8. Niu, D. and Tang, G. H. The effect of surface wettability on water vapor condensation in nanoscale. *Sci. Rep.* **6**, 19192; doi: https://doi.org/10.1038/srep19192 (2016)
9. Schutzius, Thomas M.; Jung, Stefan; Maitra, Tanmoy; Eberle, Patric; Antonini, Carlo; Stamatopoulos, Christos; et al. (2016): Physics of Icing and Rational Design of Surfaces with Extraordinary Icephobicity. ACS Publications. Collection. https://doi.org/https://doi.org/10.1021/la502586a
10. Alizadeh, Azar; Yamada, Masako; Li, Ri; Shang, Wen; Otta, Shourya; Zhong, Sheng; et al. (2016): Dynamics of Ice Nucleation on Water Repellent Surfaces. ACS Publications Collection.
11. Amirfazli, A., & Antonini, C. (2016). Fundamentals of Anti-Icing Surfaces. In R. Ras, & A. Marmur (Eds.), Non-wettable Surfaces: Theory, Preparation and Applications (pp. 319-346). RSC.
12. Kimura, Tatsuto, and Shigeo Maruyama. Molecular dynamics simulation of heterogeneous nucleation of a liquid droplet on a solid surface. *Microscale Thermophysical Engineering* 6, no. 1 (2002): 3-13.
13. Fujinaga, Takuya, and Yasushi Shibuta. "Molecular dynamics simulation of athermal heterogeneous nucleation of solidification." *Computational Materials Science* 164 (2019): 74-81.
14. Georgoulas, Anastasios, Manolia Andredaki, and Marco Marengo. "An enhanced VOF method coupled with heat transfer and phase change to characterise bubble detachment in saturated pool boiling." *Energies* 10, no. 3 (2017): 272.
15. E.J. Gelissen, C.W.M. van der Geld, M.W. Baltussen, J.G.M. Kuerten, Modeling of droplet impact on a heated solid surface with a diffuse interface model, International Journal of Multiphase Flow, Vol. 123, 2020.
16. Li, Qing, Q. J. Kang, M. M. Francois, Y. L. He, and K. H. Luo. "Lattice Boltzmann modeling of boiling heat transfer: The boiling curve and the effects of wettability." *International Journal of Heat and Mass Transfer* 85 (2015): 787–796.
17. Li, Qing, P. Zhou, and H. J. Yan. "Improved thermal lattice Boltzmann model for simulation of liquid-vapor phase change." *Physical Review E* 96, no. 6 (2017): 063303.

Chapter 2
An Introduction to Wettability and Wetting Phenomena

Joël De Coninck

Abstract In this chapter, we discuss the foundations of wetting: Young–Dupré and Furmidge equations. We review the ideas to clarify these fundamental aspects, both statically and dynamically. We describe the different theories of wetting dynamics, emphasizing what links them. A global theory seems to us a close objective. Finally, we describe some potential applications related to this science of wetting and present how wetting can influence phase transition phenomena. Hopes are high, but the task is not over.

Abbreviations

θ^0	Equilibrium contact angle
θ_t	Dynamic contact angle
θ_m	Microscopic contact angle
θ_{app}	Apparent contact angle
σ_L	Liquid/vapor surface tension
σ_{SV}	Solid–vapor surface tension
σ_{SL}	Solid–liquid surface tension
U_{CL}	Speed of the contact line
f_{CL}	Friction at the contact line
CL	Contact line
MKT	Molecular kinetic theory
TPZ	Three phases zone

J. De Coninck (✉)
Laboratoire de Physique des Surfaces et des Interfaces, Université de Mons, 19, AV. Maistriau, 7000 Mons, Belgium
e-mail: joel.deconinck@umons.ac.be

M. Marengo and J. De Coninck (eds.), *The Surface Wettability Effect on Phase Change*,
https://doi.org/10.1007/978-3-030-82992-6_2

2.1 Introduction

Wetting of surfaces is dealing with the behavior of a liquid (simple or complex) in contact with a substrate, either at rest or in motion. The subject of this book relates more specifically to the behavior of this liquid when it is in conditions to change phases, to become vapor for example. But before going into these aspects, it is essential to understand what happens under simpler conditions, where the liquid is stable.

Several configurations where liquid wetting is at play can be easily considered such as, for example, the case of the liquid film covering all or part of the solid surface, the drop placed on a flat or fiber-shaped substrate, or on a bundle of fibers as for filter masks in this period of COVID-19 or even interstitial liquid between grains of various material that we find, for example, in the walls of mud houses or in sandcastles. The situations are very numerous.

The first scientific formulation of this problem is due to Thomas Young in 1805 in a famous essay [1]. In it, the author explains the mechanism of competition between the cohesion within the liquid and the adhesion of the liquid to the solid support in the case of a drop. It is this competition that gives rise to the contact angle of the drop of liquid placed on the solid support. These considerations materialized in the expression of a now very famous equation, known as Young–Dupré, which establishes a mathematical link between the equilibrium contact angle θ^0 and the surface tensions in presence σ_L, σ_{SV}, σ_{SL} at the three-phase zone (TPZ), meeting zone of the three adjacent phases: vapor (V), liquid (L) and solid (S):

$$\sigma_L \cdot \cos(\theta^0) = \sigma_{SV} - \sigma_{SL} \tag{2.1}$$

The underlying hypotheses for the validity of this equation are relative to the nature of the liquid (it has to be pure) and to the quality of the substrate (perfectly flat and without heterogeneities). The liquid drop is also supposed to be in equilibrium with the vapor phase and the substrate. Many textbooks have already described the experimental techniques to measure contact angles, I will not repeat that here. The profile of the drop is the solution of the Laplace equation which balances the pressure difference on either side of the interface by a local curvature which is written as follows:

$$p - p_{atm} = \sigma_L \cdot \left(\frac{1}{R_1} + \frac{1}{R_2}\right) = \sigma_L \cdot \left(\frac{z''}{(1 + z'^2)^{3/2}} + \frac{z'}{r \cdot (1 + z'^2)^{1/2}}\right) \tag{2.2}$$

where R_1 and R_2 are the radii of curvatures of the profile, $z(x, y)$ denotes the shape of the drop versus the coordinates x and y and r is the radial coordinate. Since in the absence of gravity the hydrostatic pressure becomes constant, we can conclude that the mean curvature $\frac{1}{R_1} + \frac{1}{R_2}$ is, therefore, also constant, what corresponds to one and only one solution: the spherical cap. Of course, in the presence of gravity, the equation of the profile of the drop becomes a partial differential equation of second order and the constraints (constant volume of liquid or not ...) with the particular edge conditions which vary from problem to problem make the resolution difficult. It was not until 200 years after Young that a solver of such equations appeared in the finite-element mathematical software "surface evolver" [2]. These Young–Dupré–Laplace equations constitute the first pillar of the science of surfaces and interfaces.

The thermodynamic approach I have described so far has been sufficient to develop the physics of wetting for a long time. Technological applications are numerous. They concern very varied fields such as construction, aeronautics, automotive, coating, cosmetics, pharmacy, the medical world, ... The list is really long. All my students know this well. There are always measurements to be performed in the laboratory for any manufacturer who wants to improve products using contact angles measurements.

In 1987 [3], another more fundamental approach appeared, linked to statistical physics and its microscopic description of phenomena. Can we describe the Young–Dupré equation as a consequence of the interactions present at the atomic scale? Simple and attractive idea but which has required some mathematical efforts. More recently still, several teams of researchers [4–6] have endeavored to prove the validity of these laws from mechanical arguments such as the balance of forces involved. Oddly enough, this is the kind of argument used to convince cohorts of first-year students, but these recent works have shown that the forces at the contact line are quite surprising and that they do not match exactly the naive description that is conventionally provided at the university.

These considerations are part of the equilibrium aspect of wetting. When the liquid drop is stable on the solid surface, we can study its characteristics. They will be presented in Sect. 2.2 of this chapter for simple liquids on substrates without defects. The case of complex liquids, mixtures, or with the addition of various objects such as nanoparticles or surfactants, will not be discussed here. I refer the reader interested in this subject to the excellent book [7].

When the substrates become heterogeneous, the situation is slightly more complicated. Yet this is the case with all surfaces of everyday life. These heterogeneities can be of chemical origin, another set of atoms or molecules on the surface, or of physical origin, a hole or a bump in the surface. This case was studied in particular by Furmidge in 1962 [8] who described the phenomenon of pinning of the contact line by a mathematical relation that still generates much ink today. This equation constitutes in my opinion the second pillar of the science of surfaces and interfaces. This point will be developed in Sect. 2.3.

The case of moving contact lines will be developed in Sect. 2.4. This concerns the dynamics of wetting. Despite the respectable age of this science, the dynamics of wetting remains to this day a subject of continuing controversy. Several theories are still opposed today. I will, therefore, present the most important ones and explain some recent progress. To the fundamental question of knowing whether today we can predict the wetting dynamics for any real system from its easily measurable physical properties such as viscosity, surface tension, and equilibrium contact angle, the answer is no. We must actually measure it and apply some model of one sort or another to try to understand it. This is a frustrating limitation. Nowadays that we are about to go to Mars, we still cannot predict how a simple drop of water will spread out on a solid surface. The root of the problem is our lack of understanding of the relative importance of dissipation channels appearing in any contact-line dynamics. For simplicity, let us focus for a moment on the spreading of a single drop. As the shape of the drop changes over time, the corresponding surface energy will change accordingly. Since the system is isolated, part of the energy must therefore dissipate somewhere. There are at least two families of dissipation channels mentioned in the literature, one related to flows during the process and the other related to the triple line. We will review them in Sect. 2.4 and compare these theories with experimental or numerical observations.

Section 2.5 will be devoted to some applications. I believe it is always important for a student to realize how useful science can be. This is the purpose of this part.

Finally, we will introduce some considerations about how wetting can possibly be related to phase changes, which is after all the main focus of this book. This will be developed in Sect. 2.6.

2.2 Equilibrium

The Young–Dupré Eq. (2.1) was validated very early on by thermodynamics since it is not difficult to show that it leads to a minimum of the surface energy (Fig. 2.1).

Nevertheless, the quantities σ_{SL} and σ_{SV} cannot be fixed by experiment. Since the left side of the equation can be determined quite easily by measurement, the equation is often used to actually determine the difference $\sigma_{SV} - \sigma_{SL}$. For a large class of interfaces, this equation has also been validated from microscopic considerations using statistical mechanics [9–12]. More precisely, considering phase B brought into contact with a wall S in coexistence with phase A, it has been shown that the shape of a fixed volume of B is given by Winterbottom's construction [13]. This general construction does not presuppose a priori that the corresponding interfacial tension σ_{AB} is isotropic as is the case, for example, for solid crystals. However, for liquids, σ_{AB} becomes isotropic and this construction can easily be shown to give Young's equation.

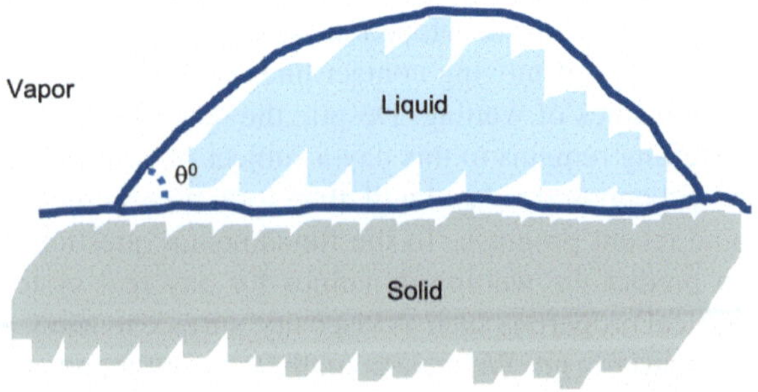

Fig. 2.1 A liquid sessile drop on top of a solid surface with an equilibrium contact angle θ^0

From a mechanical point of view, at the limit of the continuum, the interfaces are modeled as 2D surfaces and the contact zone where the three phases meet, that is to say, the intersection of a liquid/fluid interface with the solid/liquid interface, becomes a single line. The mechanical interpretation of Young's equation, understood as the balance of forces on the liquid and solid atoms present in the TPZ, is, however, not very clear at the atomic scale, where the triple line will have a thickness of a few nanometers. It is, therefore, quite naturally that a recent work has been devoted to the validity of Young's equation at this microscopic scale [14, 15]. This fundamental question has also been examined through high-quality experimental work such as the measurement of forces on carbon fibers and nanocones [16, 17]. Several recent articles using molecular dynamics or density functional theory have also been devoted to this equation at the nanometric scale. In particular, their object was to understand the mechanical forces exerted by the liquid on the solid in the TPZ [5, 18, 19]. What emerges from all these works is that the force exerted on the liquid due to the presence of the solid at the level of the contact line is given by $\sigma_L \cdot \cos\theta^0$ and $\sigma_L \cdot \sin\theta^0$ for, respectively, the tangential component and the normal component of this force. These results thus confirmed the validity of Young's equation up to the width of the TPZ, i.e. a few nanometers.

What is, however, much less obvious is the case of the tangential force exerted by the liquid on the solid at the level of the contact line. The results so far show that this force is greater than the work of adhesion expressed in $\sigma_L(1 + \cos - \theta^0)$. By introducing a small random roughness into the solid surface to destroy the layering of the liquid in close proximity to the substrate, it was confirmed that the corresponding tangential force is compatible with the adhesion energy $\sigma_L(1 + \cos \theta^0)$. This was recently revisited in [20] and led to even more open questions.

The TPZ mystery is not yet fully resolved despite the efforts made during all these years!

2.3 Pinning/Depinning

Except for a piece of glass or a silicon wafer, surfaces do always present heterogeneities. These defects can be either of chemical nature or of geometric nature such as, for example, with bumps and holes. When we revisit Young–Dupré's considerations in the presence of heterogeneities, things get a little more complicated. Indeed, it is then necessary to introduce the concept of advancing and receding contact angles. Consider a small piece of the contact line. The sum of forces parallel to the solid surface defines the so-called local spreading coefficient $\sigma_{SV} - \sigma_{SL} - \sigma_L \cdot \cos \theta_t = \sigma_L (\cos \theta^0 - \cos \theta_t)$, where θ_t is the local contact angle. θ_t is a macroscopic local quantity with a rather smooth variation at the macroscopic scale. The contact angle, before averaging, follows the heterogeneity of the solid surface energies $\sigma_{SV} - \sigma_{SL}$ and may vary in a certain range. If the local contact angle θ_t falls in this range, then the piece of contact line will undergo positive and negative spreading coefficients and thus will be pinned as illustrated in Fig. 2.2.

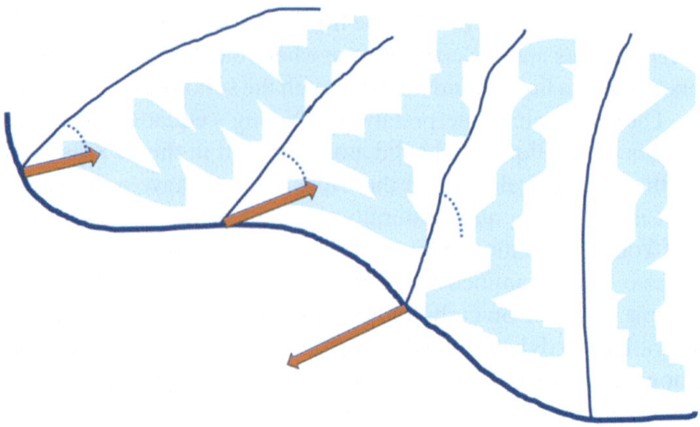

Fig. 2.2 A pinned interface. The arrows indicate the local spreading coefficients

Otherwise, it will move. In particular, when as a result of the force imbalance θ becomes larger than θ_{adv}, the line unpins and then advances, i.e., it wets a previously dry surface area. Conversely, when θ becomes smaller than θ_{rec}, the line unpins and recedes, i.e., it dewets a previously wet surface area. This picture is of course simplified since it deals with metastability through equilibrium macroscopic notions only. Nevertheless, it is enough to explain the details of the underlying phenomenon.

The advancing and receding angles θ_{adv} and θ_{rec} can be measured experimentally as illustrated in Fig. 2.3. We can easily recognize the tilting plate and the Johnson-Dettre methods which are by now standard.

But the basic mechanism should be valid and should imply the following scenario: for a drop on a horizontal substrate, the contact line is a circle. Suppose the contact angle is θ^0 with $\theta_{rec} < \theta^0 < \theta_{adv}$. Now tilt the substrate by a small angle α. The contact angle along the contact line becomes a function of azimuth, $\theta = \theta(\phi)$, oscillating around θ^0 and therefore satisfying $\theta_{rec} < \theta(\phi) < \theta_{adv}$ for all ϕ. The contact line is

Fig. 2.3 Cartoon representing the two classical ways to measure the advancing (adv) and the receding (rec) contact angles

pinned everywhere and remains circular. Upon increasing α, depending upon θ^0, the maximum of $\theta(\phi)$ will reach θ_{adv} or the minimum will reach θ_{rec} and a corresponding piece of the contact line will move by a finite amount, not yet the roll off. The remaining piece holds the drop. Upon increasing α further, eventually, the remaining piece will be unable to hold the drop, with the minimum contact angle at θ_{rec} and the maximum at θ_{adv}: the drop will then roll off. Such a scenario with three different transitions has been experimentally observed in [21]. If $\theta^0 = \theta_{adv}$ or θ_{rec}, of course, the first stage is skipped, and the circle is deformed as soon as $\alpha > 0$. The importance of the deposition history has been stressed in [22–24]. The variety of possible processes and motions makes the prediction of the final static contact angle challenging. To the best of my knowledge, there is no general correlation between hysteresis and roughness features for a given surface. When the corresponding substrate is tilted by a small angle α, the drop usually deforms its shape but remains pinned on the substrate. It is only when the tilt angle α becomes large enough, above the value α_c, that the drop starts to slide. It was proposed by Furmidge (Eq. (5) in [8]), that

$$\mathrm{mg} \sin \alpha_c = 2 \cdot C \cdot w \cdot \sigma_L (\cos \theta_{rec} - \cos \theta_{adv}), \tag{2.3}$$

where m is the mass of the drop, g the gravity constant, and w the width of the drop in the direction perpendicular to inclination. The dimensionless retention-force factor C is close to 2 more precisely $\pi/2 \leq C \leq 2$, depending on the physical situation. Several studies have been devoted to this equation through experiments and numerical or theoretical calculations, see [7] and references therein. Mostly, all these studies differ by their hypotheses concerning the shape of the contact line or different conditions for the experiments. This type of research is really fundamental since if we can design surfaces to control pinning, this would allow to transport droplets at will. This would have tremendous applications.

The hysteresis ΔH is defined as $\theta_{adv} - \theta_{rec}$. As seen before, it is of great interest to study wetting phenomena. Moreover, it should also be pointed out that dynamical effects can play a significant role here. Indeed, some part of the measuring liquid can penetrate pores or anfractuosities of the solid surface leading to dynamical effects. It is my opinion that such contact angles θ_{adv} and θ_{rec} should be measured with great care to characterize surfaces that are used in scientific publications. On the contrary, to avoid these pinning effects we can for instance vibrate the solid surface to overpass the energy barriers associated with the heterogeneities, see for instance [25].

I cannot close this paragraph without mentioning pinning–depinning associated with a moving contact line. This leads to the stick/slip motion which is nowadays a source of intense research. This dynamics on its own has been the subject of many reviews such as [24]. One practical example here is related to the deposition of nanoparticles to design specific structures or patterns on solid surfaces, the so-called 'coffee-stain effect' [26].

2.4 Dynamics of Wetting

Let us now consider the case of a moving contact line.

If the contact line is moving with respect to the solid surface, the corresponding liquid/vapor interface is changing with time and there is, therefore, some dissipation as pointed out above. If we denote by $x_{CL}(t)$ the position of the contact line versus time and by θ_t the contact angle, it is clear that

$$\frac{dx_{CL}}{dt} = f\left(\cos\left(\theta^0\right) - \cos(\theta_t)\right) \tag{2.4}$$

With $f(0) = 0$. The problem is in the nature of the function f. As pointed out by de Gennes [27], there are several ways to dissipate energy in the problem. This may happen at the contact line or in the vicinity of the contact line. More recently, it has been pointed out that this could be due to the presence of defects that will pin for a certain time the contact line. Last but not least, the mechanism of interface creation has also been developed. All these models lead to different functionals f(.).

Before briefly presenting the main models, let me highlight the basic problem.

A contact angle depends on the scale we use to measure it. Usually, experimental observations are at a scale greater than one micron. This is the so-called "apparent" or "macroscopic" contact angle θ_{app}. At the molecular level, things are a little more complicated since they are subject to thermal fluctuations and variations in density. It is therefore necessary to introduce the concept of "microscopic" contact angle θ_m [28], determined directly at the level of the contact line. In 2014, Chen et al. [29] observed significant mesoscopic structure using tapping AFM. It was a convex nanocurvature-shaped like a shoe tip at the advancing contact line. The foot of this protuberance amounted to around 20 nm on ultra-flat substrates. This profile and the corresponding contact angle, θ_m, varied systematically with the contact line velocity U_{CL}. Two related problems are, therefore, present in the study of mobile contact lines: (i) what happens around the contact line in terms of position, flow and therefore also θ_{app}, and (ii) the connection with the microscopic part of the contact line (of the size of the nanometer), where we measure θ_m.

That being said, let us now quickly see what the main theories consist of. More details can be found in the reviews [30, 31].

The first one is the molecular-kinetic theory (MKT) introduced by Blake and Haynes [32] and later revised to include the local effects of viscosity and solid–liquid interactions [33, 34]. At the molecular level, dynamic wetting is described in the model as a stress-modified, thermally activated process. The basic idea is that the leading edge of the liquid moves stochastically by way of individual molecular displacements between discrete interaction sites on the surface of the solid. The driving force for this process is the localized shear stress as indicated above. Per unit length of the contact line, this force is indeed $F_{CL} = \sigma_L \cdot (\cos\theta^0 - \cos\theta_t)$ as demonstrated by MD simulations [35]. Displacement of the contact line is the result of the collective thermal motion of the liquid atoms within the TPZ due to this capillary driving force. The derivation of the equation linking θ to U_{CL} is based on an effective model reducing the displacement of the CL to only two possibilities, either forward with a step $+\lambda$ or backward with a step $-\lambda$. The corresponding probabilities to jump are given by a classical exponential function involving the energy barrier (Kramers). This leads to the following relationship:

$$U_{CL} = 2\kappa^0\lambda\sinh\left[\gamma_L\left(\cos\theta^0 - \cos\theta_t\right)/2nk_BT\right]. \tag{2.5}$$

Here, κ^0 and λ are the characteristic frequency and length of the thermal displacements within the TPZ; n is the number of solid–liquid interaction sites per unit area of the solid surface. If the interaction sites are distributed uniformly, then $\lambda = 1/\sqrt{n}$. Whenever the argument of the sinh(.) function is small:

$$U_{CL} = \frac{\kappa^0\lambda}{nk_BT}\cdot\gamma_L\left(\cos\theta^0 - \cos\theta_t\right) \tag{2.6}$$

which we denote as

$$f_{CL}\cdot U_{CL} = \gamma_L\left(\cos\theta^0 - \cos\theta_t\right) \tag{2.7}$$

where $f_{CL} = nk_BT/\kappa^0\lambda$ is a friction per unit length. Many experimental data can be fitted by this straight-line but as is well known, it is not because a theory can fit the data that the theory explains correctly the mechanisms involved in the phenomenon. Let me add that this model allows a particularly good description of the spreading of liquid metals where the viscosity is small and the surface tension is large [36].

Let me now introduce the second approach.

The hydrodynamic model of the velocity dependence of the apparent dynamic contact angle was introduced by Voinov [37]. It was later extended to two-liquid systems by Cox [38] and is often known as the Cox–Voinov law. This model is relative to the macroscopic scale and where the viscosity of the second fluid phase is negligible. Complementary to that, the domain of validity of the de Gennes approach (based on lubrication theory, disjoining pressure, and the spreading parameter $S = \sigma_L - \sigma_{SV} + \sigma_{SL}$) has been extended. Still from a purely hydrodynamic point of view, several mechanisms have been studied to solve the classical paradox of the divergence of the force to be applied to the moving contact line along a solid substrate. In particular, besides using a classical Navier slip length, a new approach has been proposed in [39, 40], based only on the Kelvin effect (shift of saturation conditions due to strong interface curvature), without any other microscale effect. All these theories define their own length scales, the values of which have to be validated by comparison with experimental data. Importantly, at larger scales all theories match with a universal viscous bending regime, generally described by the Cox–Voinov relationship and expressing the contact angle as a function of the velocity:

$$\left(\theta_{app}\right)^3 = (\theta_m)^3 + 9Ca_{CL}\cdot\ln(L/L_m), \quad for\ \theta_{app} < 3\pi/4. \tag{2.8}$$

Here, $Ca_{CL} = U_{CL}\eta_L/\sigma_L$ is the capillary number based on U_{CL} and η_L is the shear viscosity of the liquid. The quantities L and L_m are appropriately chosen macroscopic and molecular length scales. The former represents the outer region where θ_{app} is measured, while the latter defines the inner region where surface effects, such as slip, are dominant. The analysis leading to Eq. (2.8) shows that the only important information to emerge from the inner region is θ_{m} [41, 42]. In solving the hydrodynamic problem, the slope of the interface in the outer region is matched to that of the inner region, potentially leading to an inflection point.

Typically $\ln(L/L_m)$ is of order 10. In practice, when fitting experimental data, $\ln(L/L_m)$ is treated as an unknown parameter; and while Voinov, Cox, and successive authors recognized that θ_m might well depend on U_{CL}, this angle is usually assumed to be invariant and set equal to the equilibrium contact angle θ^0. Accordingly, the variation in θ_{app} is ascribed only to bending of the meniscus by viscous shear-stress on a mesoscale greater than L_m. Despite these simplifications, Eq. (2.8) has proved effective in accounting for the observed behavior of the dynamic contact angle, especially for viscous liquids such as silicone oils on well-wetted solids. The model has proved less successful for two-liquid systems and for partially wetting liquids of low viscosity with large equilibrium contact angles. In these cases, it sometimes yields strange non-physical values of L/L_m [43, 44].

This review would not be complete without mentioning two other very interesting approaches.

The third class of models is due to Perrin et al. [45]. These authors propose a model that combines the hydrodynamic description of the liquid flow at large scales and a thermal activation process due to the presence of defects on the solid surface. They introduced contact-line friction via a Langevin approach in a formal way to construct a multiscale model of a contact line moving across a surface with pinning sites. They solve the Langevin equation using 4 free parameters: the equilibrium contact angle and the defect size, spacing, and amplitude. The parameters are determined by fitting their model to experimental data of contact angle versus the capillary number. None of the fitting parameters is the contact-line friction f_{CL} which is taken to be a linear function of viscosity and estimated using the lubrication approximation. They also use experimental data to support their conclusions.

The fourth one is due to Shikhmurzaev [46]. This model is, in fact, an extension of the hydrodynamic concepts, applied in the three-phase zone. The model is based on the notion that liquid elements at the liquid–vapor interface eventually traverse the three-phase zone to become an element at the solid–liquid interface. The surface properties of these elements have therefore to change to new values. Within the model it is assumed that Young's equation is also applicable to dynamic contact angles (which has been proved by MD simulations). Since the dynamic contact angle is different from the equilibrium contact angle, Shikhmurzaev concluded that the various surface tensions must vary accordingly. Since the surface tensions away from the three-phase zone are identical to the equilibrium ones, it follows that close to the three-phase zone, gradients of surface tensions exist during flow. It is also assumed that this surface tension relaxation along the interfaces is of a macroscopic scale, so that the influence of the contact line is also experienced outside the three-phase zone. Analyzing the hydrodynamic equations under these conditions, Shikhmurzaev identifies three regions of interest, similar to the regions identified by Cox. Matching the solutions for the different regions, and depending on the capillary and Reynolds numbers, expressions are found for the contact angle dependence on contact-line velocity. Here too, some experimental data support this approach.

Which theory is right, which is wrong?

This is of course a difficult question. Most probably, contact-line dynamics will deal with a combination of these different channels of dissipation either in time or in space or both. A large part of the experimental data does not allow to distinguish clearly between these theories. The reason is rather simple. The comparison between experimental data and theories so far is based on parametric fitting. It is very common that a set of data can be fitted equally well by different theories. This is due to the short interval in time or in speeds over which the acquisition has been considered. The interested reader can find a detailed analysis of this problem in [44].

Some progress have been achieved thanks to large-scale molecular dynamics simulations. More details about this technique can be found in another chap. 10 of this book. This is an interesting tool in physics. We just have to model the system at the atomistic scale (nm) describing how the atoms interact with each other. These interactions with some boundary conditions will provide the properties of the considered materials. Here we have to consider liquid and solid atoms which means that we will have to consider three types of interactions: liquid–liquid, solid–solid, and liquid–solid. A simple model often considered in the literature is given by fixing the liquid–liquid and solid–solid couplings to 1. The dynamics of spreading can thus be studied using a liquid drop on top of a solid surface varying the amplitude of the liquid–solid interaction or in other words the liquid–solid affinity. If this coupling is small, we may expect to have a large contact angle at equilibrium. However, if the affinity is large, we should have a very small contact angle at equilibrium. What comes out of all this work is that the dynamics of Lennard–Jones (LJ) liquid interfaces for large equilibrium contact angles can be fairly well described by MKT [47]. However, these liquids do present a small viscosity of the order of 0:2 cP and also a small surface tension of the order of 3 mN/m. For these reasons, all the results are compatible with the linear regime described in Eq. (2.7). Recently [48, 49], more specific geometries for LJ liquids were considered to study such dynamics of interfaces in the nonlinear regime and the results reveal again that MKT describes very well the dynamics of such interfaces.

In some very recent work [50], large-scale molecular dynamics (MD) and a coarse-grained model of liquid water have been used to explore the steady displacement of a liquid bridge under a constant applied force between two plane-parallel solid plates over a wide range of velocities (Fig. 2.4).

This liquid has a larger viscosity and a larger surface tension mimicking the values of real water. The results confirm that the velocity-dependence of the microscopic contact angle is well described by MKT, even into the nonlinear regime, while the Cox–Voinov hydrodynamic model accounts for the viscous bending, provided one uses, as input, the true microscopic angle rather than the equilibrium angle, as is

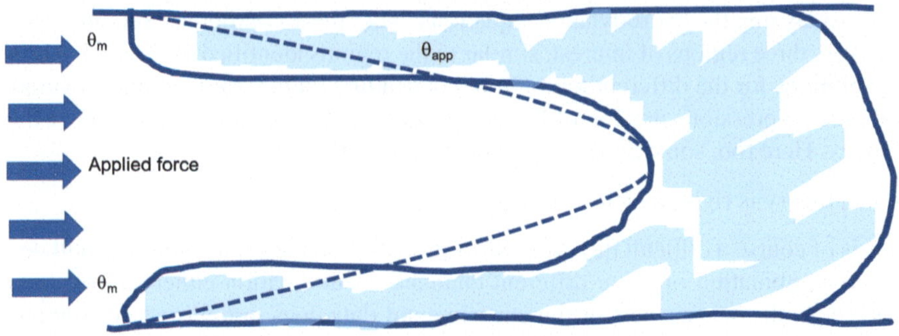

Fig. 2.4 Cartoon of a liquid bridge between two parallel plates under the influence of a large and constant external force. θ_m is the microscopic contact angle and θ_{app} is defined through a circular fit involving the top and bottom contact points and the middle point of the receding profile

usually done. The overall result is that the utility of Eq. (2.8) first introduced by Petrov and Petrov [51] to describe the dynamics of wetting has been confirmed.

Is there some hope to get a unique theory?

Well, I am rather optimistic. Some significant progress have been achieved in this direction recently. Brochard and de Gennes [52] have pointed out that for small angles the dynamics are more likely to be controlled by viscous dissipation, whereas, for large angles, contact-line friction would be the governing channel of dissipation. To overcome the inconsistency between these two models, de Ruijter et al. [53]and Petrov and Petrov [51] argued that both channels of dissipation should coexist and therefore combined these two approaches, though in different ways.

For the first piece of work, it is assumed that there is a unique contact angle describing the moving contact line. The absence of bending presupposes, therefore, a low viscosity liquid. For small enough speed U_{CL}, a linear approximation holds, and the contact angle appears as a solution of an ordinary differential equation which reveals two regimes versus time. The first regime is dominated by the friction theory and the second one by the hydrodynamic theory with a cross-over between the two in agreement thus with Brochard-de Gennes view. This has been validated by experiments in [54] and particularly for the fiber geometry in [55].

In the second one, Petrov and Petrov considered that the microscopic contact angle appearing in the HD approach is the dynamic contact angle described by MKT leading to an equation with 4 parameters to be fitted with the data. The validity of this approach has been established recently using MD for water-like systems (see above and in [50]).

Now the central question is related to our ability to predict the dynamics of wetting. Is that possible?

We have seen that each theory is involving parameters. These parameters K0, λ, f_{CL}, ln(L/Lm),... can be obtained by fitting the data. Is there a way to relate these values to measurable quantities such as viscosity, surface tension, equilibrium contact angles…? This would be great and would open the way for many practical optimizations.

At that stage, I believe that this requires still some work.

Let me explain.

In three recent papers [56–58], the contact-line fluctuations of a sessile drop were studied using large-scale MD simulations for two liquid systems (L-J liquid and water-like liquid). Combining these results to a stochastic Langevin approach, we were able to study the link between friction and fluctuations of the contact line at equilibrium for a meniscus of liquid between two parallel plates separated by a certain distance over some width L_y. More precisely, the contact-line position $x_{CL}(t)$ could be modeled as an over-damped one-dimensional Langevin oscillator confined around its equilibrium position x_{eq} by a harmonic potential $V(r) = 0.5k(x_{CL}(t) - x_{eq})^2$, where k is the stiffness of the oscillator. This leads to the equation

$$L_y \cdot f_{CL} \cdot \frac{dx_{CL}(t)}{dt} = -k\left(x_{CL}(t) - x_{eq}\right) + f \tag{2.9}$$

where f is a random force acting on the contact line, L_y is the length of the considered contact line, and f_{CL} is still the coefficient of contact-line friction (friction divided by contact-line length). The nature of this friction has already been described above. The random force f is due to the random fluctuations of the contact angle θ_t with respect to its equilibrium value θ^0, which induce a very fast variation in the capillary force. This force is uncorrelated at very short times and has a zero mean $f_t = 0$. This leads to a relation between the stiffness k and the temporal evolution of the signal $x_{CL}(t)$, which allows us to compute:

$$k = \frac{k_B T}{\sigma^2}, \tag{2.10}$$

where $\sigma^2 = x_{CL}^2(t_0)_{t0}$ is the variance of this fraction of contact line in terms of positions. Furthermore, the contact-line friction f_{CL} may be determined from the time decay of the self-correlation function $\langle x_{CL}(t_0 + t)x_{CL}(t_0)\rangle_{t0}$:

$$\langle x_{CL}(t_0 + t)x_{CL}(t_0)\rangle_{t0} = \sigma^2 \cdot e^{-b.t} \tag{2.11}$$

The parameter $b = k/f_{CL} \cdot L_y$ determines thus the rate of decay of the correlation. The larger the value of b, the more rapidly the system becomes uncorrelated. Combining this result with the previous expression for f_{CL} using MKT, we arrive at

$$f_{CL} = \frac{nk_B T}{\kappa^0 \lambda} = \frac{k_B \cdot T}{b \cdot \sigma^2 \cdot L_y}. \tag{2.12}$$

It is plausible that $\kappa^0 \sim b$ introducing a physical meaning in terms of interfaces for this parameter but it also reveals a more intriguing and unsolved analogy to determine l, the jump length.

The approach presented above was successfully used to determine the coefficient of contact-line friction of a simulated liquid bridge at equilibrium between two solid surfaces for a range of affinities between the liquid and the solid surface, i.e., a range of equilibrium contact angles. What comes out of this work is that the resulting frictions were identical to those found in MD simulations of spreading liquid drops and obtained by fitting the contact angle data to the linear approximation (see Eq. (2.7)).

We are thus able to predict the full dynamics of wetting within this linear regime simply by studying the fluctuations of the contact line at equilibrium! This is amazing. It is also remarkable that these results illustrate the idea that there is just one unique phenomenon of friction dissipation at the contact line.

This is in agreement with the model proposed by Blake and De Coninck [34] for the friction which should be proportional to the viscosity of the liquid and exponentially dependent on the work of adhesion at equilibrium $g_L{\cdot}(1 + \cos\theta^0)$. The validity of this expression has been tested successfully in many published papers (see [30] for references).

Of course, we have been considering the friction f_{CL} up to now. This parameter can be related to the jump frequency through the relation $f_{CL} = nk_BT/\kappa^0\lambda$. If we can determine f_{CL} as explained above, is there a way to extract properly the jump frequency or the associated b (see Eq. (2.11))? This is still an open question but no doubt that the answer to that question would allow huge progress in dynamic wetting.

The study on dynamics of receding contact angles in [51] may also provide a new way to understand the logarithmic term in the Cox–Voinov model. Indeed, it has been shown that this factor $\ln(L/L_m)$ is related to a dynamic wetting transition. For each θ^0, there is a critical contact-line velocity $U_{CL} = U_{crit}$ and contact angle $\theta_m = \theta_{crit}$ at which $\theta_{app} \to 0$ and the receding meniscus begins to deposit a thin liquid film on the solid plates. The associated contact line then continues to move at $U_{CL} \approx U_{crit}$, with a constant θ_m. Fixing $\theta_{app} = 0$, $\theta_m = \theta_{crit}$ and $U_{CL} = U_{crit}$ in the Voinov equation enables us to obtain the value of L/L_m directly. This is independent of θ^0 and may be used with the Voinov equation to determine θ_{app} from θ_m as a function of U_{CL}. These predicted values are in excellent agreement with those obtained from the simulations. This result seems to indicate that it is possible to have an independent determination of L/L_m. If this could be achieved experimentally, then we would be able to determine θ_m from θ_{app} and so greatly improve our understanding of dynamic wetting.

This review would not be complete without addressing the case of complete wetting, when the affinity between the liquid and the solid is so strong ($\theta^0 = 0$) that a film of this liquid spreads out in front of the droplet with a precursor film of molecular thickness. During the spreading process we observe, not only experimentally [59] but also via MD simulations [60, 61], a change in the curvature of the LV profile characterized by the presence of an inflection point at x_{ip}. This inflection point allows us to decompose the profile into two parts: a spherical cap for $x < x_{ip}$ and a precursor film for $x > x_{ip}$. We can then fit the spherical cap with a circle out of which we get the apparent location of the contact line according to the spherical cap fitting, rc, from the intersection of the fitted circle with the plate. Also, the tangent to this fitted circle at the intersection with the solid defines the apparent contact angle θapp. To the best of our knowledge, this central part has never been considered in detail before in MD simulations. Finally, we define the location of the 'precursor' film xp as the location of the LV profile obtained from the liquid layer closer to the plate. The evolution of x_{ap} is compatible with a power law fit $t^{0.08 +- 0.01}$ with an exponent similar to the classic Tanner's law [27] $t^{0:1}$ while the precursor film length $l = xp - xc$ behaves as $l^2 = 2 \cdot D \cdot t$ where D is a diffusion coefficient which can be related to a Hamaker constant describing the amplitude of interaction between the solid and the liquid.

2.5 Applications

Many applications are based on the use of an external stimuli or field to change the wettability of a solid surface (see Fig. 2.5). Clearly, each possibility deserves a review on its own. The most important source of inspiration here is provided by biomimicry. Living examples are really fascinating in many aspects.

The most active related field of research nowadays is the design of superhydrophobic surfaces by controlling the topography of the substrate at several scales [62]. The robustness of such construction is today the key problem. Such surfaces are also considered for anti-icing, self-cleaning, drag reduction, condensation, boiling, … The applications are numerous.

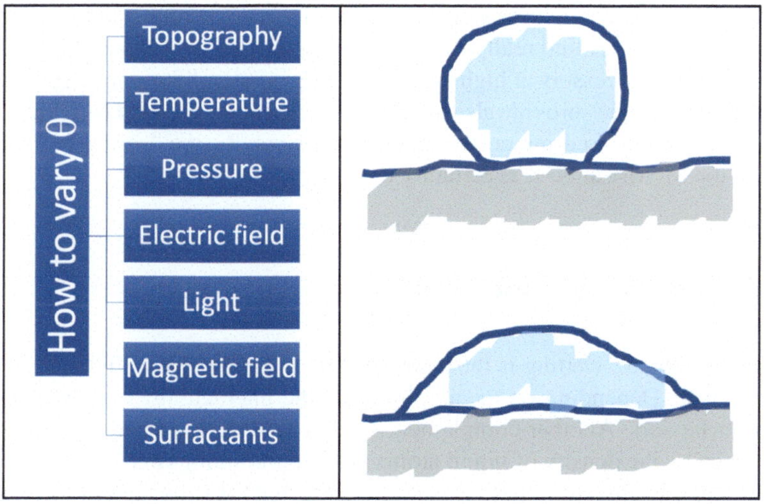

Fig. 2.5 Cartoon describing the most important possibilities to change the wettability

The case of the electric field is also well-known not only for electrowetting [63] but also for plenty of other applications such as smart glasses. In electrowetting, the idea is that by applying an electric field on the solid surface, its surface will change due probably to the migration of charges. This will then affect its wettability. By changing versus time the position where we apply the electric field, we can then move a drop at will. Some other applications have been considered such as for condensation [64] or to design smart lenses for cameras [65] for instance.

The case of temperature itself is fascinating [66] since it is well known that heating up a hydrophilic surface will make it even more hydrophilic. However, the hydrophobic case is much less obvious. Most hydrophobic surfaces do not change by increasing the temperature of the surface. Some others become even more hydrophobic. Those dealing with surface treatments by grafting a molecule of one sort or another on top of a solid surface are well aware that temperature is key to control the quality of the grafting. This is of utmost importance when we deal with the design of biosensors which is another field of applications for wetting [67].

There is also another method based on surfactants, see for instance [7]. Those are complex molecules that present two parts: one which likes water generally and another one which likes oil (and therefore dislikes water). These objects can preferentially be adsorbed by the solid modifying its wettability or can design an encapsulating layer for oil droplets. This is the basis of detergency.

Today, for the young generation, one of the most important problems will be the filtration of water [68]. Due to its crucial interest for humanity, the number of publications dealing with that topic is growing exponentially fast and all of them refer to wetting.

As can be seen, the science of wetting has many technological applications.

Concerning the dynamic aspect of wetting, let me mention fields of applications such as coating. I find really remarkable that nowadays it is possible to design industrially nanoprocessors at high speed using immersion lithography [69].

Spatial technologies provide also an important example. The best case I have seen is provided by cusps to drink coffee in absence of gravity, the "zero-g coffee cup", visible in the MoMA at N-Y. It certainly deserves the detour.

2.6 Phases Changes and Wetting

As described before, wetting is the study of how a liquid behaves in the vicinity of a solid surface. Depending on the amplitude of the interactions, we have observed different behaviors. Another complementary way to describe these properties is of course given by the density of liquid atoms close to the solid. This has been detailed by molecular dynamics simulations [47] revealing that for strong coupling, there are some layers of atoms in the liquid phase while for small couplings there are less atoms per unit of volume than in the bulk. These results are described in Fig. 2.6.

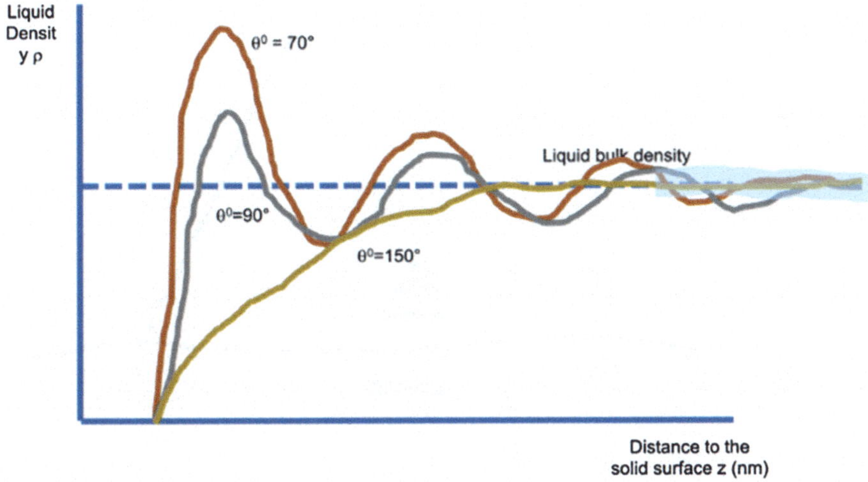

Fig. 2.6 Schematics of the liquid density in close vicinity to the solid surface (the height z is expressed in nm)

Wetting is thus intimately related to the density of liquid close to the solid. Another aspect is related to pinning–depinning. We have seen that a liquid–air interface can be pinned for a certain time by introducing heterogeneities. By combining both aspects, we do observe very interesting properties.

Consider first slip.

Many studies have been devoted to confined fluids and more recently to their dynamics. The notion of slip is there very important. Slip is the mechanism by which a moving substrate at a constant speed U_x will have a layer of liquid molecules in contact with the solid but moving at a different speed $\overline{U_x}$. The no-slip boundary conditions, very common in hydrodynamics, correspond to the case $\overline{U_x} = U_x$. This slip can be characterized by the so-called slip length, l_{slip}, which is defined as an extrapolated distance with respect to the solid surface where the tangential relative velocity component of the liquid becomes zero. The relationship between slip and wettability has been derived in [70]. More recently [71], studying wetting dynamics via large scale MD simulations, it has been shown that

$$l_{slip} = \delta \frac{\eta_L}{f_{CL}} \tag{2.14}$$

where d is the width or twice the standard deviation s of the TPZ and f_{CL} is given by Eq. (2.7). This links directly the properties of slip to the friction of the CL, that is to say, dynamics of wetting (Fig. 2.7).

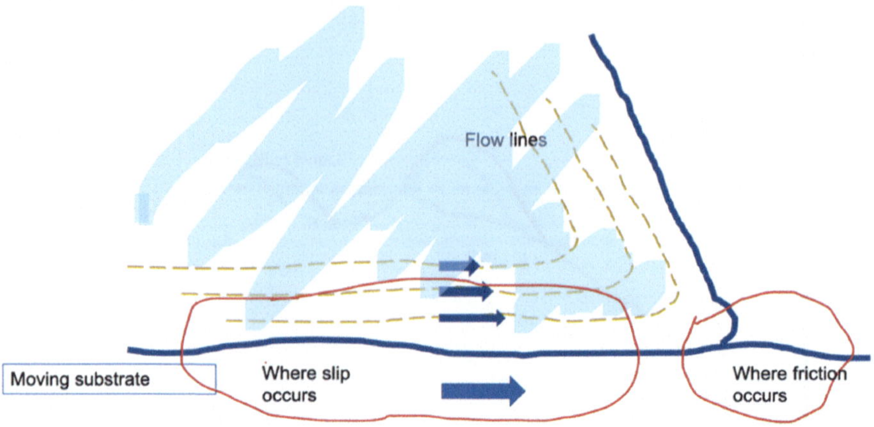

Fig. 2.7 Friction and slip for moving interfaces

Second case, boiling.

It has been shown experimentally [72] that for a very flat surface (glass), hydrophobicity enhances the onset of boiling, that is to say, the temperature at which the first vapor bubble appears. The effect is really important. For water, it can be of the order of 20 °C which is huge in terms of energy (nearly 20%). In these experiments, hydrophobicity was achieved by grafting a monolayer of molecules on the glass surface. By adding thus 1 nm of hydrophobic material on the surface, boiling was significantly modified to the point that a difference of 20 °C was observed for the onset of boiling. How is that possible? We believe this is due to the change of density of water close to the hydrophobic surface. If we now introduce a hydrophobic patch on this surface, we have the case studied in detail in [73]. In this paper, the authors do study the pinning–depinning phenomenon for the vapor bubble occurring during boiling. Furmidge has studied how gravity can force a drop to detach from some heterogeneity. Here these authors study the same problem but instead of gravity, they consider how the heat flux will modify the vapor volume until depinning occurs. Again, wetting is involved.

Third case, condensation.

Air is always humid to some extent. It contains water molecules and even nano or micro-droplets of such molecules. If these objects are adsorbed by the solid surface, they will remain on the surface (usually colder than the atmosphere) if the impact speed is not too high. These droplets continue to grow until they coalesce forming larger and larger objects. This growth will be subject to pinning–depinning of course. To remove such droplets from the surface, we have to refer to Furmidge's equation and use gravity to clean the surface and make it operational again for condensation. This is the reason why heterogeneous surfaces harvest more water than homogeneous ones [74].

Even more.

Many other domains will profit from progress in wetting. I have in mind not only boiling and condensation but also anti-icing, drag reduction, biosensor and biotechnology, insulation, microfluidic, micromachines, ... As illustrated in Fig. 2.8, the future looks promising (see, for instance [75, 76]).

I am convinced that we are just at the beginning of new important discoveries in these fields. We start to understand the basics of these phenomena. Soon, this understanding will allow optimization. I am for instance really impressed by the applications of wetting in forensic sciences [77, 78], another fascinating subject. Things are, however, changing with time. Previously, entire teams of physicists or chemists could study their own such kind of topics. Nowadays, we need to have multidisciplinary teams combining engineers, chemists, biologists, and physicists to design new smart materials with tremendous possible properties.

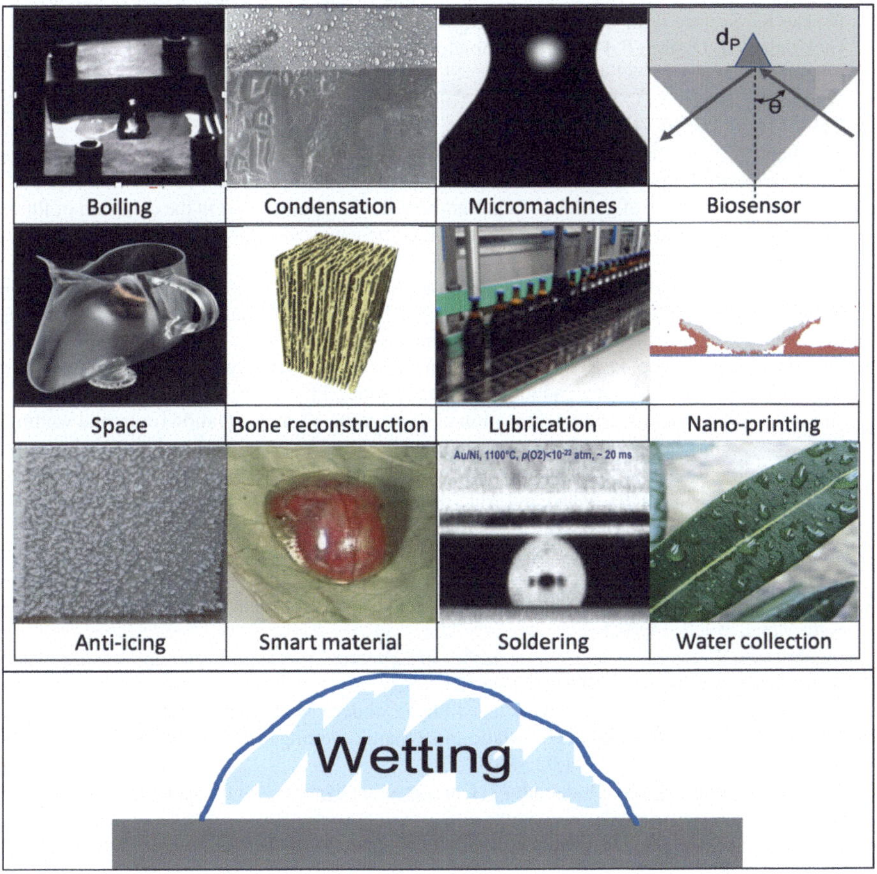

Fig. 2.8 Some R&D domains in which I have been working intimately related to wetting

Young researchers will have to face big changes. But I am rather optimistic. After all, this is what human beings have been doing since they exist.

Acknowledgements At that stage, I have to thank a lot of people and especially all my collaborators. Without them, nothing would have been possible. They will find here the expression of my most sincere gratitude. Let me here take the pleasure to also thank my old friend Terry Blake who helped me to make this adventure in wetting a real pleasure. He is certainly my best student in Molecular Dynamics. I would like to express here all my deepest gratitude for the fantastic "battles" we had together in front of an old blackboard.... A very special thanks also to Marco who has introduced me to this magic world of boiling.

References

1. Young T., An essay on the cohesion of fluids. Philos Trans R Soc London 1805;95:65–87.
2. Brakke K., Surface Evolver Manual (Susquehanna University, Selinsgrove, PA, 2013).
3. De Coninck J, Dunlop F. Partial to complete wetting: a microscopic derivation of the Young relation. J Stat Phys 1987;47:827–49.
4. Seveno, D., Blake, T.D., De Coninck, J., Young's equation at the nanoscale, Phys. Rev. Lett., 2013, 111(9), 096101-1-4.
5. Marchand A, Weijs JH, Snoeijer JH, Andreotti B. Why is surface tension a force parallel to the interface? Am J Phys 2011;79:999.
6. Fernandez-Toledano, J.-C., Blake, T.D., Lambert, P., De Coninck, J., On the cohesion of fluids and their adhesion to solids: Young's equation at the nanoscale, Adv. Colloid Interfac., 2017, 245, 102–107.
7. Butt, H.-J., Graft, K. and Kappl, M., Physics and Chemistry of Interfaces, Wiley, 2003.
8. Furmidge, C. G. L., Studies at phase interfaces. I. The sliding of liquid drops on solid surfaces and a theory for spray retention, J. Colloid Science 17, 309 (1962).
9. De Coninck, J., Dunlop, F., Rivasseau,V., On the microscopic validity of the Wulff construction and of the generalized Young equation, Commun. Math. Phys., 1989, 121 (3), 401–419.
10. Abraham, D.B., Ko, L.-F., Exact derivation of the modified Young equation for partial wetting, Phys. Rev. Lett. 1989, 63, 275.
11. Miracle-Sole, M. and Ruiz, J., On the Wulff construction : a problem of equivalence of statistical ensembles, in Micro, Meso and Macroscopic Approaches in Physics, A. Verbeure ed., PlenumPress, New York, 1994.
12. Bodineau, T., Ioffe, D., Velenik, Y. Winterbottom Construction for Finite Range Ferromagnetic Models: An L_1-Approach. J Stat Phys , 2001, 105, 93–131.
13. Winterbottom, W. L., Equilibrium shape of a small particle in contact with a foreign substrate, Acta Metall. 1967, 15, 303–310.
14. Ingebrigtsen T, Toxvaerd S., Contact angles of Lennard-Jones liquids and droplets on planar surfaces. J Phys Chem C 2007;111:8518–23.
15. Makkonen L. Young's equation revisited. J Phys Condens Matter 2016;28:135001.
16. Barber AH, Cohen SR, Wagner HD. Static and dynamic wetting measurements of single carbon nanotubes. Phys Rev Lett 2004;92:186103.
17. Delmas M, Monthioux M, Ondarçuhu T. Contact angle hysteresis at the nanometer scale. Phys Rev Lett 2011;106:136102.
18. Seveno D, Blake TD, De Coninck J. Young's equation at the nanoscale. Phys Rev Lett 2013;111:1–4.
19. Fan, J.C., De Coninck, J., Wu, H.A., Wang, F. Ch., Microscopic Origin of Capillary Force Balance at Contact Line, Phys. Rev. Lett., 2020, 124, 125502–1–6.

20. Berejnov, V. and Thorne, R.E., Effect of transient pinning on stability of drops sitting on an inclined plane, Phys. Rev. E 2007, 75, 066308.
21. Collet, P., De Coninck, J., Dunlop, F. and Regnard, A., Dynamics of the contact line: contact angle hysteresis, Phys. Rev. Lett. 1997, 79, 3704.
22. Chou, T.-H., Hong, S.-J., Sheng, Y.-J. and Tsao, H.-K., Drops sitting on a tilted plate: receding and advancing pinning, Langmuir, 2012, 28, 5158.
23. White, J. A., Santos, M. J., Rodríguez-Valverde, M. A. and Velasco, S., Numerical study of the most stable contact angle of drops on tilted surfaces, Langmuir, 2015, 31, 5326.
24. De Coninck, J., Fernandez-Toledano, J.-C., Dunlop, F., Huillet, T. and Sodji, A., Shape of pendent drops under a tilted surface, Physica D 415 (2021) 132765.
25. Decker, E.L. and Garoff, S., Using Vibrational Noise to Probe Energy Barriers Producing Contact Angle Hysteresis, Langmuir,1996, 12, 8, 2100–2110.
26. Flauraud, V., Mastrangeli, M., Bernasconi, G.D., Butet, J., Alexander, D.T.L., Shahrabi, E., Martin, O.J.F., Brugger, J., Nanoscale topographical control of capillary assembly of nanoparticles, Nature Nanotechnology, 2016, 12, 73–80.
27. de Gennes, P.G., Wetting: statics and dynamics, Rev. Mod. Phys., 1985, 57, 827–863.
28. Dussan E.B., On the spreading of liquids on solid surfaces: static and dynamic contact lines, Ann. Rev. Fluid Mech. 1979, 11, 371–400.
29. Chen, L., Yu, J. and Wang H., Convex Nanobending at a Moving Contact Line: The Missing Mesoscopic Link in Dynamic Wetting, ACS Nano, 2014, 11, 11493–11498.
30. Blake, T.D., The physics of moving wetting lines, J. Colloid Interface Sci. 2006, 299, 1–13.
31. Bonn, D., Eggers, J., Indekeu, J., Meunier, J. and Rolley, E., Wetting and spreading, Rev. Mod. Phys. 2009, 81, 739–805.
32. Blake, T.D. and Haynes, J.M., Kinetics of liquid/liquid displacement, J. Colloid Interface Sci. 1969, 30, 421–423.
33. Blake, T.D., Dynamic contact angles and wetting kinetics, in: Wettability, Marcel Dekker, 1993.
34. Blake, T.D. and De Coninck, J., The influence of solid–liquid interactions on dynamic wetting, Adv. Colloid Interface Sci. 2002, 96, 21–36.
35. Fernandez-Toledano, J.C., Blake, T. and De Coninck, J., Young's equation for a two-liquid system on the nanometer scale, Langmuir, 2017, 33 (11), 2929-2938.
36. Saiz, E. and Tomsia, A., Atomic dynamics and Marangoni films during liquid-metal spreading, Nature Materials, 2004, 3, 903–909.
37. Voinov, O.V., Hydrodynamics of wetting, Fluid Dyn. 1976, 11, 714–721.
38. Cox, R.G., The dynamics of the spreading of liquids on a solid surface. Part 1. Viscous flow, J. Fluid Mech. 1986, 168, 169–194.
39. Rednikov, A. and Colinet, P., Singularity-free description of moving contact lines for volatile liquids, Phys. Rev. E , 2013, 87, 010401(R).
40. Rednikov, A. and Colinet, P., Contact-line singularities resolved exclusively by the Kelvin effect: volatile liquids in air, J. Fluid Mech., 2019,858, 881–916.
41. Dussan , E.B., The moving contact line: the slip boundary condition, J. Fluid Mech. 77 (1976) 665.
42. Kafka, F.Y. and Dussan, E.B., On the interpretation of dynamic contact angles in capillaries, J. Fluid Mech. 1979, 95, 539–565.
43. Petrov, J.G., Ralston, J., Schneemilch, M. ,Hayes, R., Dynamics of partial wetting and dewetting in well-defined systems, J. Phys. Chem. B. 2003, 107, 1634–1645.
44. Seveno, D., Vaillant, A., Riobo, R., Adão, H., Conti, J. and De Coninck, J., Dynamics of wetting revisited, Langmuir, 2009, 25, 13034–13044.
45. Perrin, H., Lhermerout, R., Davitt, K., Rolley, E. and Andreotti, B., Defects at the Nanoscale Impact Contact Line Motion at all Scales, Phys. Rev. Lett., 2016, 116, 184502.
46. Shikhmurzaev, Y.D., The moving contact line on a smooth solid-surface, Int. J. Multiph. Flow. 19 (1993) 589–610.
47. Bertrand, E., Blake, T. and De Coninck, J., Influence of solid-liquid interactions on dynamic wetting: a molecular dynamics study, J. Phys-Condens. Mat., 2009, **21,** 464124.

48. Wang, G.J. and Hadjiconstantinou, N., Universal molecular-kinetic scaling relation for slip of a simple fluid at a solid boundary, Phys. Rev. Fluids, 2019, 4 (6), 064201.
49. Wang, G.J., Damone, A., Benfenati, F. and Hadjiconstantinou, N., Physics of nanoscale immiscible fluid displacement, 2019, Phys. Rev. Fluids, 4(12):124203
50. Fernandez-Toledano, J.C., Blake, T.D. and De Coninck, J., Taking a Closer Look: A Molecular-Dynamics Investigation of Microscopic and Apparent Dynamic Contact Angles, submitted for publication.
51. Petrov, P.G. and Petrov, J.G., A combined molecular-hydrodynamic approach to wetting kinetics, Langmuir, 1992, 8, 1762.
52. Brochard-Wyart, F., and de Gennes, P.G., Dynamics of partial wetting, Adv. Colloid Interface Sci. 1992, 39, 1.
53. de Ruijter, M., De Coninck, J., Oshanin, G., Droplet spreading: partial wetting regime revisited, Langmuir, 1999, 15 (6), 2209–2216.
54. M. de Ruijter, M. Charlot, M. Voué, J. De Coninck, "Experimental evidence of several time scales in drop spreading", Langmuir, 2000, 16 (5), 2363-2368.
55. M.S. Vega, D. Seveno, G. Lemaur, M.H. Adão, J. De Coninck, "Dynamics of the rise around a fibre: experimental evidence of the existence of several time scales", Langmuir, 2005, 21 (21), 9584–9590.
56. Fernandez-Toledano, J.-C., Blake, T.D., De Coninck, J., Contact-line fluctuations and dynamic wetting, J. Colloid Interf. Sci., 2019, 540, 322–329.
57. Fernandez-Toledano, J.-C., Blake, T.D., De Coninck, J., "Moving contact lines and Langevin formalism", J. Colloid Interf. Sci., 2020, 562, 287–292.
58. Fernandez-Toledano, J.-C., Blake, T.D., Kanduc, M., De Coninck, J., The Hidden, Microscopic Life of the Moving Contact Line of a Water-like Liquid, Phys. Rev. F, 2020, 5, 104004.
59. 61. Heslot, F., Fraysse, N. and Cazabat, A. Molecular layering in the spreading of wetting liquid drops. Nature 1989, 338, 640–642.
60. 62. De Coninck, J., D'Ortona, U., Koplik, J., Banavar, J., Terraced spreading of chain molecules via molecular dynamics, Phys. Rev. Lett., 1995, 74 (6), 928 –932.
61. Feernandez-Toledano, J.C., Rednikov, A., Colinet, P, De Coninck, J., work in progress.
62. 64. Backholm, M., Molpeceres, D., Vuckovac, M. et al. Water droplet friction and rolling dynamics on superhydrophobic surfaces. Commun Mater, 2020 **1,** 64
63. Mugele, F. and Baret, J.-C., Electrowetting: from basics to applications, J. Phys.: Cond. Mat., 2005, 17, 28.
64. Baratian, D., Dey, R., Hoek, H., Van Den Ende, D., and Mugele, F., Breath Figures Under Electrowetting: Electrically Controlled Evolution of Drop Condensation Patterns, Phys. Rev. Lett., 2018, 120, 21, 214502.
65. Jarosz, J., Molliex, N., Chenon, G. and Berge B., Adaptive eyeglasses for presbyopia correction: An original variable-focus technology, Optics Express, 2019, 27(8): 10533–10552.
66. Villa, F., Marengo, M. and De Coninck, J., A new model to predict the influence of surface temperature on contact angle, Nature Scientific Reports, 2018, 8, 6549.
67. Gosselin, E., Vanden Eynde, J.J., Petit, A., Conti, J. and De Coninck, J., Designing a high performance, stable spectroscopic biosensor for the binding of large and small molecules, J. Colloid Interf. Sci., 2017, 508, 443–454.
68. Minwei Yao, M., Tijing, L.D., Naidu, G., Kim, S.H., Matsuyama, H., Fane, A.G., Shon,H.K., A review of membrane wettability for the treatment of saline water deploying membrane distillation, Desalination, 2020, 479, 114312.
69. French R.H. (2012) Immersion Lithography Materials. In: Bhushan B. (eds) Encyclopedia of Nanotechnology. Springer, Dordrecht.
70. Barrat, J.L. and Bocquet, L., Influence of wetting properties on hydrodynamic boundary conditions at a fluid/solid interface, Faraday Discuss., 1999, 112, 119–127.
71. Blake, T.D., Fernandez-Toledano, J.-C., Doyen, G. and De Coninck, J., Forced Wetting and Hydrodynamic Assist, Physics of Fluids, 2015, 27, 112101-112125.
72. Bourdon, B., Di Marco, P., Rioboo, R., Marengo, M. and De Coninck, J., Enhancing the onset of pool boiling by wettability modification on nanometrically smooth surfaces, Int. Commun. Heat. Mass., 2013, 45, 11–15.

73. Shen, B., Liu, J., Amberg, G., Do-Quang, M., Shiomi, J., Takahashi, K. and Takata, Y., Contact-line behavior in boiling on a heterogeneous surface: Physical insights from diffuse-interface modeling, Phys. Rev. Fluids, 2020, 5, 033603.
74. Bai, H., Wang, L., Ju, J., Sun, R., Zheng, Y., Jiang, J., Efficient Water Collection on Integrative Bioinspired Surfaces with Star-Shaped Wettability Patterns, Adv. Mater., 2014, 26: 5025–5030.
75. Antonini, C., CHAPTER 11: Fundamentals of Anti-Icing Surfaces, RSC Soft Matter, 2017
76. Alamri, S., Vercillo, V., I Aguilar-Morales, A., Schell, F., Wetterwald, M., Lasagni,A., Bonaccurso, E., Kunze, T., Self-Limited Ice Formation and Efficient De-Icing on Superhydrophobic Micro-Structured Airfoils through Direct Laser Interference Patterning, Advanced Materials Interfaces, 2020, 2001231.
77. Smith, F., and Brutin, D., Wetting and spreading of human blood: Recent advances and applications, 2018, Current Opinion in Colloid & Interface Science, 36, 73.
78. Attinger, D., Moore, C. B., Donaldson, A., Jafari, A. and Stone, H. A., Fluid dynamics topics in bloodstain pattern analysis: Comparative review and research opportunities, 2013, Mechanical Engineering Publications. 135.

Chapter 3
Heat Transfer Enhancement During Dropwise Condensation Over Wettability-Controlled Surfaces

Stefano Bortolin, Marco Tancon, and Davide Del Col

Abstract Dropwise condensation (DWC) is a complex heat transfer process in which vapor phase changes to liquid phase forming discrete droplets on a surface whose temperature is below the dew temperature of the condensing fluid. DWC mode can strongly enhance the heat transfer compared to filmwise condensation (FWC) mode that usually takes place when a vapor condenses over a metallic surface. The wettability of the surface plays a crucial role on the promotion of DWC instead of FWC. This Chapter is focused on heat transfer measurements and modeling during DWC. The first two Sections are dedicated to a short literature review and to the description of the experimental procedures that can be used for the measurement of the heat transfer coefficient. DWC involves millions of droplets per square meter that form the so-called droplet population. Section 3.3 is dedicated to the description of the droplet size distribution. Section 3.4 presents selected models that can be used for the prediction of heat transfer during DWC. Formed droplets can be removed from the condensing surface by gravity or by other external forces. In the literature, most of the DWC experimental data are taken with quiescent vapor and very few works investigate the effect of the vapor drag force on the droplet departing radius and thus on the heat transfer during DWC. Furthermore, the effect of vapor velocity is not accounted for in available DWC models. Therefore, the last Section of this Chapter is focused on heat transfer modeling in the presence of vapor velocity. A recent approach proposed by the present authors to account for the reduction of droplets departing diameter due to vapor velocity is here presented. The model is then used to show the effect of the main parameters on the DWC heat transfer coefficient.

Keywords Dropwise condensation · Vapor velocity · Heat transfer enhancement · Droplet population · Modeling

S. Bortolin · M. Tancon · D. Del Col (✉)
University of Padova, Padova, Italy
e-mail: davide.delcol@unipd.it

M. Marengo and J. De Coninck (eds.), *The Surface Wettability Effect on Phase Change*,
https://doi.org/10.1007/978-3-030-82992-6_3

3.1 Introduction

Condensation is a phase change process encountered in many applications as thermal power plants, desalination of sea water, air conditioning systems, water harvesting, and so on. The surface chemistry and morphology can play a central role to increase the heat transfer coefficient by changing the surface wettability. Wettability can be described by looking at the dynamic contact angles that a liquid (drop) assumes on a surface while moving. Advancing contact angle (θ_a) is defined when the droplet is moving forward to a non-wetted surface and receding contact angle (θ_r) when the contact line is moving backward on a wetted surface [1]. The difference between advancing and receding contact angle is named contact angle hysteresis ($\Delta\theta$). The changing of wettability can determine a different behavior in the interaction between the liquid and solid phases during the condensation process.

In particular, the vapor can condense on a surface in two modes: filmwise condensation (FWC) mode and dropwise condensation (DWC) mode. The DWC mode is promoted on surfaces with controlled wettability (typically on surfaces with low contact angle hysteresis) and it allows an increase of the heat transfer coefficient (HTC) from 5 to 7 times [2] compared to filmwise mode. Starting from the nanoscale up to the macroscale, DWC involves millions of droplets per square meter. DWC is a cyclic process: condensation begins at a molecular scale with drops formation in preferred nucleation sites. Growing by direct condensation at first and later by coalescence, drops reach the critical size at which external forces (e.g., gravity, vapor drag) overcome adhesion forces and they start to move, sweeping the surface and making new nucleation sites available. The process is, then, renewed [3]. The presence of droplets, instead of a continuous liquid film, allows to reduce the thermal resistance compared to the case of FWC thus increasing the condensation heat transfer coefficient.

The heat transfer characteristics of the DWC mode have attracted several researchers for about 100 years. Since the DWC discovery in 1930 by Schmidt et al. [4], the number of experiments related to DWC has grown over the years and a variety of results are reported in the literature as shown by Rose [5]. In fact, several aspects must be considered while performing DWC measurements to obtain reproducible and comparable results, such as the thermal resistance of the coating, the absence of non-condensable gases, an accurate measurement of surface temperature and heat flux. In the following Sections, the main aspects concerning DWC will be presented.

3.1.1 Surface Coatings

Creation of surfaces that can promote dropwise condensation is one of the main issues. Basically, two approaches can be found in the literature. The first one consists of the modification of the surface chemistry for a given substrate by applying a thin coating over the surface where condensation takes place. The second approach instead involves the modification of both the morphology and the chemistry of the surface and it can allow getting the so-called superhydrophobic behavior. In this Section, we will focus on the first approach considering coatings that can be used for the modification of the surface wettability.

Metals are still the family of materials most used in heat transfer applications, from steel [6–9] to copper [10, 11] and aluminum [12]. Clean metallic surfaces are generally wetted by the condensate because of their high surface energy (hydrophilic behavior) [13] and the condensation process occurs in filmwise mode. Coatings can be used to reduce the surface free energy of metals and make them hydrophobic. The main issue, in this case, relies on the robustness of such coatings stressed in harsh environments (high saturation temperature conditions, high heat flux, and high vapor velocity). Coating degradation strongly depends on the coating chemistry, thickness, coating-substrate interfacial adhesion [14, 15], and condensation environment.

Satisfactory results in terms of durability have been obtained with different hydrophobic treatments [2] and copper as substrate. Changing the substrate, the affinity between the treatment and the base material can be very different, thus some materials are more challenging than others in order to get prolonged DWC. DWC on copper has been studied for decades [16–19] and several solutions to promote DWC have been investigated: polymeric coatings [20], self-assembled monolayers (SAMs) [21–23], ion implantation [24], and rare-earth oxide ceramics [25]. An alternative solution to sustain DWC on copper substrates can be graphene coating. Uniform graphene coatings are usually fabricated by bottom-up approaches, such as chemical vapor deposition (CVD) on metals. Among the different top-down approaches, the method that has received the most attention is exfoliation and reduction of graphite oxide (GO), the partially oxidized form of graphene that presents low cost. GO coatings are hydrophilic in nature, but their wettability can be converted to hydrophobicity by chemical or thermal reduction. Rafiee et al. [26] demonstrated that a graphene monolayer on copper, gold, and silicon does not change the wettability of the substrate because of its extreme thickness but, with an increase of graphene layers, it is possible to change the contact angles. Colusso et al. [27] fabricated reduced graphene oxide (rGO) coatings using a solution-based process, by dip-coating copper substrates into a suspension containing graphene oxide flakes obtained by chemical exfoliation of graphite oxide in an aqueous solution. They obtained for water an advancing contact angle of $\sim 84°$ and a receding contact angle of $\sim 35°$. The coating was found to promote DWC of pure steam with heat transfer coefficients of 170 kW m^{-2} K^{-1} measured during condensation at around 100 °C saturation temperature (with an 8-times increase of the HTC compared to FWC). The durability of the rGO coating is reported to be more than 100 h in the tested conditions.

Regarding aluminum substrates, the available studies about DWC promotion are limited compared to other substrates (copper, titanium, and silicon) [28–30]. Steam or liquid at high temperature (>100 °C) etches the aluminum forming boehmite structures [31, 32] enhancing the wettability of the surface and thus efforts must be addressed to protect the surface. Rausch et al. [30] obtained relevant duration results fabricating a hydrophobic aluminum sample by ion implantation: the sample was tested for 8 months but only about 50% of the surface was covered by droplets. At the lowest value of subcooling, the HTC resulted only double as compared to FWC. Paxson et al. [29] obtained interesting results on aluminum processed via initiated chemical vapor deposition (iCVD). The condensation test was performed for 48 h, after which the treatment showed mild signs of degradation; the HTC remained constant around 35 kW m^{-2} K^{-1}. Kim and Jeong [24] implanted chromium ions onto an aluminum substrate. Filmwise condensation occurred on the mirror-polished metal surface with no ion implantation. After irradiation with chromium ions, dropwise condensation was induced. The measured FWC heat transfer coefficient showed good agreement with the theoretical values predicted by Nusselt's film theory whereas, when DWC is promoted, HTCs were two times those expected from Nusselt's theory.

A different approach to achieve DWC on aluminum surfaces is based on hybrid organic–inorganic sol–gel silica coatings containing hydrophobic moieties (e.g., methyl or phenyl group). The sol–gel method has attracted attention due to: simplicity of the production process which does not involve high temperatures or pressures; versatility since the precursors used are in the liquid state; possibility to deposit the coatings on bare substrates with complex shapes; use of compounds that are not harmful for the environment (e.g., fluorinated molecules). Parin et al. [12] developed a surface that can maintain stable DWC, using SiO_2 hybrid organic–inorganic sol–gel coating functionalized with methyl groups over an aluminum substrate. The sol–gel film was deposited on a mirror-polished substrate. The initial aluminum surface displayed hydrophilic characteristic with $\theta_a = 65°$ and $\theta_r = 10°$, whereas, after the coating deposition, the angles were $\theta_a = 89°$ and $\theta_r = 64°$. This surface can be classified as hydrophilic according to the classic definition based on the advancing contact angle being lower than 90° [6]. The strong difference between the treated surface and the untreated one is represented by the receding contact angle and the contact angle hysteresis, which results in a key surface parameter to promote DWC of pure steam. Condensation of steam (at 108 °C saturation temperature and 2.7 m s^{-1} vapor velocity) occurred in dropwise mode with HTCs equal to 150–180 kW m^{-2} K^{-1} (augmentation by 8–10 times compared to FWC) in the heat flux range between 150 and 510 kW m^{-2}. Durability tests performed at constant heat flux equal to 120 kW m^{-2} showed that the coating can sustain DWC for at least 13 h. Recently, Parin et al. [33] investigated three different mixtures of methyl triethoxy silane (MTES) or tetraethyl orthosilicate (TEOS) in combination with phenyl triethoxysilane (PhTES) as reagents for the realization of coatings that are able to promote DWC of saturated steam. The three different coatings displayed similar wettability ($\theta_a \approx 90°$ and $\theta_r \approx 60°$) and coating thickness (between 200 and 420 nm). In particular, the mixture of PhTES and MTES (7/3 molar ratio) baked at 200 °C resulted in the most robust coating sustaining more than 100 h of continuous DWC (saturated steam at 100 °C with heat flux equal to 400 kW m^{-2}) without sign of FWC transition, which is an interesting result obtained on aluminum samples so far. The optimization of the coating's chemistry is important to increase the coating lifetime while maintaining high HTC. Parin et al. [34] used tetraethyl orthosilicate (TEOS) and methyl triethoxy silane (MTES) as silica precursors in order to study the ability of sol–gel MTES/TEOS hybrid coatings for DWC promotion. Six different

MTES/TEOS coatings were considered and all of them promoted pure steam DWC, with a maximum HTC of 300 kW m^{-2} K^{-1} (105 °C saturation temperature, 2.6 m s^{-1} vapor velocity, 400 kW m^{-2} heat flux).

Two important further considerations must be done regarding the contact angle hysteresis and the thermal conductivity of the coating.

Del Col et al. [35] imparted hydrophobic properties over an aluminum substrate by forming onto the metal a low surface energy film (spin coating of a fluorosilane–hexane solution). The substrate was sanded using emery paper. Prior to coating deposition, the advancing contact angle was equal to 32° (hydrophilic) with a contact angle hysteresis of 25°. After coating deposition, the advancing contact angle increased up to 143° (hydrophobic) with 100° of hysteresis. The authors performed saturated steam (at nearly atmospheric pressure) condensation tests at heat flux between 250 and 500 kW m^{-2} and vapor velocity between 2.2 and 6.4 m s^{-1} with both hydrophilic and hydrophobic samples. They found that the condensation mode was purely filmwise, even on the hydrophobic surface, due to the complete flooding of the surface which in turn is caused by the high heat flux and the surface roughness. Although condensation always occurred in filmwise mode, the heat transfer coefficient was higher by 10–45% on the hydrophobic surface as compared to the hydrophilic sample (this increase is however limited compared with the 5–6 times heat transfer coefficient increase expected during DWC). The available literature on liquid flow over hydrophobic surfaces suggests an explanation of the results by assuming some slip of the condensate at the wall. Therefore, it emerges that the DWC is linked to the contact angle hysteresis rather than to the advancing contact angle and thus to the hydrophobic characteristics of the sample.

The use of organic substances as low-surface energy promoters requires strong, long-term adhesion forces between the coating and the metal substrate. Usually, the thicker is the coating, the better its resistance to corrosion/erosion. Each coating has a different thermal resistance depending on its chemistry and thickness, which strongly influences the overall heat transfer coefficient (HTC) as described by Parin et al. [12]. Therefore, from the heat transfer point of view, minimizing the thickness of the layer is essential. In fact, usually, the thermal conductivity of these organic layers (0.2 W m^{-1} K^{-1} [36]) is very low compared to the thermal conductivity of the metallic substrate (200 W m^{-1} K^{-1} for aluminum, 390 W m^{-1} K^{-1} for copper). A wide range of HTCs, up to tens of kW m^{-2} K^{-1} [22, 29, 37, 38], can be found in the literature for DWC of pure steam at atmospheric pressure: the high dispersion of HTC values is also related to the high variety of DWC promoters. Often, the thermal resistance of the hydrophobic layer can be the main resistance in the heat transfer process, greatly influencing the overall HTC during the condensation process. If the thermal conductivity and the thickness of the hydrophobic layer are known, the HTC of the DWC phenomenon itself can be calculated: an evaluation is reported in Rose [2].

On the other hand, in industrial and energy applications, the fundamental parameter is the overall HTC, accounting also for the thermal resistance of the layer. To the best of the authors' knowledge, when comparing results in terms of overall HTC (thermal resistance between steam and metallic substrate), the highest values measured with pure steam are obtained on a copper substrate plated with gold [39] (about 250 kW m^{-2} K^{-1}) or on aluminum samples coated with sol–gel coatings [34] (about 300 kW m^{-2} K^{-1}).

3.1.2 Effects of Saturation Pressure, Heat Flux, and Non-condensable Gases

Several studies have been conducted in order to understand how saturation pressure, heat flux, and non-condensable gas concentration affect the DWC mechanism.

Regarding the effect of saturation pressure on the heat transfer coefficient, for a given saturation-to-wall temperature difference, the heat transfer coefficient increases when the saturation pressure is increased [2]. As reported in Rose [5], this is due to the reduction of the interfacial resistance at the liquid–vapor interface. Furthermore, by increasing the saturation pressure, the surface tension of the fluid decreases and this contributes to a reduction of the droplet adhesion force (see Eq. 3.48) and thus to a reduction of the droplet departing radius [6]. It must be considered that a reduction of the surface tension of the condensing fluid makes the DWC more difficult to be sustained from a given surface because the surface tension of the fluid gets closer to the surface energy of the substrate.

Regarding the effect of the heat flux on the DWC heat transfer coefficient, as reported in Le Fevre and Rose [17], several behaviors can be found in the literature: the heat transfer coefficient increases with the heat flux, decreases with the heat flux, or it is independent on the heat flux. The increase of the heat transfer coefficient with the heat flux can be explained with an increase in the number of active nucleation sites.

During DWC experiments, non-condensable gases can be often present inside the vapor chamber. This is because of two reasons: tests are run at saturation pressure below the atmospheric one and thus air enters the test rig; non-condensable gases can be dissolved into the water. A layer of non-condensable gases accumulates near the condensing surface, introducing a mass transfer resistance to the diffusion of molecules of the vapor phase that have to reach the surface to continue condensation. Due to the low heat transfer resistance expected during DWC, a small quantity of non-condensable gas can dramatically reduce the condensation performance. Therefore, particular care must be taken when performing DWC tests to assure that non-condensable gases are get rid of the apparatus. As reported in Citakoglu and Rose [18], the presence of non-condensable gases can be one of the reasons for discrepancy between data from different laboratories.

3.1.3 Vapor Velocity

The heat transfer during DWC depends on both the heat flux through a single drop and the droplet population density function. Since most of the heat exchanged is associated with small droplets [12], strategies for reducing the average drop size can be interesting for the improvement of the condensation HTC. In particular, the droplets' departing radius (r_{max}) should be decreased. The departing droplet radius r_{max} is the outcome of a force balance between forces that act for moving the droplet (for instance, gravity and drag forces), and the adhesion force, which sticks the droplet to the surface. In ground applications, the gravity force can be varied by changing the orientation of the condensing surface; adhesion force can be reduced decreasing the wettability of the substrate and the drag force can be increased by acting on the vapor velocity. Experimental data show that an increase of the steam mass flux causes a decrease of the droplet departing radius and, at the same time, an increment of the HTC.

Tanner et al. [16] investigated DWC on copper surfaces coated by wax: experimental tests showed an increase of the HTC from 260 to 400 kW m^{-2} K^{-1} (+50%) when the vapor velocity was increased from 3 to 23 m s^{-1}. Similar results were obtained by Tanasawa and Utaka [40] on hydrophobic copper: increasing the steam velocity from 10 to 100 m s^{-1}, the HTC increased by 2.5 times. Moreover, with the intensification of vapor flow, a decrease of the maximum droplet radius was measured and the HTC, once assessed its independency from the heat flux, was found to depend uniquely upon the departing radius. Investigating DWC on plane copper surfaces, Sharma et al. [41] found that an increase in the vapor velocity from 3 to 9 m s^{-1} leads to 20% augmentation of the HTC. Torresin et al. [10] performed DWC experiments for investigating the influence of vapor velocity over nanostructured superhydrophobic copper samples. With an increment in the vapor velocity from 6 to 18 m s^{-1}, the departure droplet size was found to decrease and an increase of the HTC from 35 to 60 kW m^{-2} K^{-1} was measured (+90%). Recently, Tancon et al. [42] measured DWC heat transfer coefficients on aluminum sol–gel coated samples at 107 °C saturation temperature, heat flux of 335 kW m^{-2} and average vapor velocity between 2.7 and 11 m s^{-1}. The sol–gel coated aluminum sample displays advancing contact angle $\theta_a = 87.5°$ and receding contact angle $\theta_r = 63.5°$. In the investigated range of vapor velocity, the authors found that the HTC increases by around 20%, from 104 to 120 kW m^{-2} K^{-1}, with a reduction of the droplet departing radius from 1.44 to 0.89 mm. From high-speed visualization, it emerges as the shape of the droplets is clearly affected by the vapor velocity.

3.1.4 Superhydrophobic Surfaces

Superhydrophobic surfaces have recently been identified as a promising solution to several challenges [43], such as drag reduction, anti-icing [44], and enhancement of two-phase heat transfer performance. A drop placed over a surface may assume different states described by the equations of Wenzel [45] and Cassie-Baxter [46]. Excellent droplet mobility is obtained on superhydrophobic surfaces if the Cassie–Baxter state is reached. Superhydrophobic surfaces present high advancing contact angles, greater than 150°, and low contact angle hysteresis, lower than 10°. Superhydrophobic surfaces can be basically produced by combining two factors: micro-/nanoscale surface roughness and low surface free energy. Proper surface roughness can be obtained through different techniques, as micromachining, chemical etching (e.g., using HCl or NaOH), and deep radiative ion etching. Low surface energy can be obtained by coating the substrate with a thin layer of a material with small surface energy, such as organic substances, polymers, and noble metals (Sect. 3.1.1). These two elements allow water drops to sit over the surface with a quasi-spherical shape and to easily roll off from it (lotus effect). Therefore, superhydrophobic surfaces seem to represent a strategy to promote a more efficient DWC.

In presence of non-condensable gases or with pure saturated vapor at low thermal heat flux, the Cassie-Baxter regime can be obtained [47, 48]. At higher heat flux, with pure vapor, the Wenzel state [49] or even the flooding of the surface may occur. In the Wenzel state, the mobility of the droplets is limited since they are retained within the roughness. Thus, the surfaces with the highest droplet mobility in atmospheric conditions may not guarantee the highest performance during the condensation of steam. In fact, several authors measured higher heat transfer coefficients (HTCs) on smooth hydrophobic surfaces than on structured superhydrophobic surfaces.

Zhong et al. [50] reported heat transfer measurements on a superhydrophobic nanostructured copper sample and compared them to those obtained on a mirror-polished hydrophobic specimen. They found that the nanostructured substrate does not improve the condensation heat transfer performance as expected from the higher contact angle, but better results were achieved with the hydrophobic substrate. Flow condensation tests of saturated vapor on superhydrophobic nanotextured copper surfaces presented by Torresin et al. [10] show that condensing drops form and penetrate into the surface texture, with a reduction of their mobility. Parin et al. [28] fabricated four superhydrophobic aluminum surfaces. The metal substrates were etched using three different strategies to impart nanoscale roughness and, as a second step, a fluorosilane film was deposited over them (by spin coating or immersion) to decrease the surface energy. Experimental tests of pure steam condensation showed that DWC was successfully achieved on the superhydrophobic surfaces, measuring heat transfer coefficients as high as 100 kW m^{-2} K^{-1} (at 105 °C saturation temperature and heat flux around 350 kW m^{-2}). It is interesting to note that such heat transfer coefficients are lower than the ones measured by Parin et al. [12] on a mirror-polished aluminum surface displaying mildly hydrophobic characteristic (with $\theta_a = 89°$ and $\theta_r = 64°$).

In conclusion, as also reported in Miljkovic et al. [51], if the condensation heat flux is relatively low (they referred to a threshold value of 80 kW m^{-2}), a proper designed superhydrophobic surface can exceed or match the heat transfer performance of a smooth dropwise condensing surface having low contact angle hysteresis. Instead, at high heat flux values, a smooth surface (e.g., mirror polished) with low contact angle hysteresis will display better performance.

3.1.5 Low Surface Tension Fluids

In most of the DWC studies, steam is used as the working fluid. The reason is due to the fact that water is a common fluid in many industrial processes and it presents a high value of the surface tension. When considering fluids with relatively low surface tension, the promotion of DWC becomes more challenging because, in this case, the surface tension of the fluid can be comparable with the surface energy of the coating. On the other hand, low surface tension fluids can be found in a variety of industrial applications. For example, HFC (hydrofluorocarbons), HFO (hydrofluoroolefins), and HC (hydrocarbons) are low surface tension fluids largely employed in refrigeration and heat pumps. At 40 °C saturation temperature, the surface tension of HFC-134a is equal to 6.1 mN m^{-1} whereas in the case of water it is equal to 69.6 mN m^{-1}. Furthermore, these low surface tension fluids usually realize in heat exchangers lower heat transfer coefficients compared to water (this is mainly due to their lower thermal conductivity) and thus the possibility to achieve DWC condensation would have important benefits on the efficiency of heat exchangers and systems.

Micro/nanostructured lubricant-infused surfaces (LISs) are studied as a solution to get ultralow contact angle hysteresis and excellent droplet shedding. On a LIS, a lubricant having a low surface energy and vapor pressure is stabilized by capillary forces within a porous or micro/nanostructured surface creating an atomically smooth surface. LIS surfaces present high droplet mobility and thus they are studied as candidates for DWC promotion with low surface tension fluids. However, a LIS surface must satisfy the following two criteria: the lubricant and condensate must be immiscible; the lubricant must be selected to avoid the encapsulation of condensate droplets with the formation of a cloak around them that inhibits droplets growth and shedding. Sett et al. [52] obtained stable DWC of ethanol and hexane on a nanostructured copper oxide (CuO) LIS impregnated with lubricants. For ethanol, the advancing contact angle (θ_a) on the three LISs was $62.4° < \theta_a < 71.1°$ with a maximum contact angle hysteresis, $\Delta\theta = \theta_a - \theta_r \approx 2.7°$. For ethanol on the smooth cooper substrate, the advancing contact angle is equal to 36.5° with contact angle hysteresis of 13.1°. DWC heat transfer coefficient measurements in pure vapor conditions show an increase by 200% when compared to filmwise condensation on smooth copper substrates [52].

Recently, Khalil et al. [53] promoted DWC of low surface tension fluids (ethanol, hexane, and pentane) over titanium and silicon substrates. They used covalently bonded nanometric iCVD films on tubular cylindrical geometries. These iCVD-coated heat exchanger metals are shown to achieve a fourfold to eightfold improvement in vapor-side condenser heat transfer coefficient compared to uncoated surfaces.

Apart from the aforementioned pioneering studies with hydrocarbons, the promotion of DWC with low surface tension fluids is still to be demonstrated.

3.2 Measuring Heat Transfer Coefficients During DWC

Dropwise condensation heat transfer coefficients are usually measured on relatively small heat transfer areas (compared to the heat transfer areas used in industrial heat exchangers) because the objective is often the characterization of the condensation performance of innovative coated samples. The focus is on the heat transfer coefficient measurement and on the visualization of the droplet population. For this reason, almost all the experimental techniques foresee the presence of a glass window that allows the visualization of the condensation process. Analyzing the literature, two main configurations for the condensing surface can be found: in the first case, a vertically oriented surface is maintained below the dew temperature of the fluid, the vapor condenses over it and the liquid is drained by gravity toward the bottom of the vertical surface; in the second case, a horizontal tube is used to promote condensation on its external surface by using a cooling fluid that flows inside the tube. In most of the studies, the condensing surface is placed inside a vapor chamber and the velocity of the vapor is negligible (quiescent vapor). The saturated vapor is supplied to the vapor chamber by an external evaporator [6] or it is produced in the lower part of the chamber [17]. Less frequently the test apparatus can allow the investigation of the effect of vapor velocity on the dropwise condensation. As an example, the experimental technique employed by Bisetto et al. [54] will be explained in detail in Sect. 3.2.2.

3.2.1 Main Measuring Techniques

The quantities that must be measured for the determination of the heat transfer coefficient will be discussed hereafter.

The heat transfer coefficient for a pure fluid is defined as the ratio of heat flux q to saturation-minus-wall temperature difference $(T_{sat} - T_{wall})$:

$$\alpha = \frac{q}{(T_{sat} - T_{wall})} = \frac{Q}{A(T_{sat} - T_{wall})} \tag{3.1}$$

In Eq. 3.1, the heat flux q is the ratio of condensation heat flow rate Q to the condensing surface area A of the sample. This relationship is valid locally or as an average value if the wall temperature can be assumed to be uniform all over the condensing surface.

For pure vapor, the saturation temperature T_{sat} can be obtained with two methods: with a direct measurement of the vapor temperature inserting a temperature sensor (e.g., thermocouple, thermistor) inside the vapor chamber or indirectly from the measurement of the saturation pressure. It must be considered that the presence of some liquid in vapor pressure ports should be avoided since it can introduce pressure measurements errors due to hydrostatic head or to capillary effects related to the presence of meniscus in the tubes. One solution consists of heating pressure ports over the saturation temperature to avoid vapor condensation inside the lines. Using both direct and indirect approaches has the advantage to allow a double-check of the saturation temperature.

A direct contact measurement of the surface temperature T_{wall} of a coated sample (e.g., by soldering thermocouples over it) is not feasible since this would locally modify the surface properties, affecting the condensation process itself. Therefore, the measurement of the wall temperature in Eq. 3.1 is more difficult and it requires the installation of one or more temperature sensors as near as possible to the surface where condensation takes place. Since high values of heat flux are expected during DWC, the measured temperature T_{meas} must be corrected to account for the temperature variation due to thermal conduction inside the sample. Furthermore, considering that often DWC is promoted by applying a proper coating over the surface (Sect. 3.1.1), the thickness and the thermal conductivity of the layer must also be measured for the determination of the actual surface temperature T_{wall}. If this contribution is not included, the measured heat transfer coefficient will consider both the DWC thermal resistance and the resistance due to thermal conduction of the coating (overall heat transfer coefficient). The thickness of modern coatings can be in the order of few nanometers making challenging the measurement of the thermal resistance. For this reason, in the literature, the heat transfer coefficient is often defined using as temperature driving difference the difference between the saturation temperature T_{sat} and the temperature of the substrate material T_{sub} just below the coating (and thus including the resistance due to thermal conduction inside the coating). Test sections are usually designed to obtain a one-dimensional temperature field with a uniform heat flux inside the specimen. In this way, the temperature T_{sub} can be obtained from a linear extrapolation of an array of temperature measurements inside the specimen [10] or from a unique value of temperature T_{meas} (measured at a distance z below the coating) plus the temperature variation due to thermal conduction in the specimen at the distance z:

$$T_{sub} = T_{meas} + \frac{q}{\lambda_{sub}} z \tag{3.2}$$

In Eq. 3.2, λ_{sub} is the thermal conductivity of the substrate material. It must be pointed out that the size of the temperature sensor must be minimized to avoid disturbance in the temperature field and to get , an as much as possible, punctual value of the temperature. The available space for the installation of temperature probes in many cases is limited. In fact, an augmentation of the thickness of the sample has an adverse effect on the overall heat transfer resistance. In analogy with an electrical circuit, considering as potentials the saturation temperature and the temperature of the cooling medium (usually water), there are three main thermal resistances that hinder the heat transfer: the resistance of the DWC process (this, in turn, is made up of several contributions as it will be illustrated in Sect. 3.4), the thermal resistance due to conduction in the sample and the thermal resistance due to forced convection of the cooling fluid. Considering the maximum available temperature difference between saturation and coolant temperature, thickening the sample will result in decreasing the maximum exchangeable heat flux and thus reducing the range of investigable conditions. This is true in particular when using steel or aluminum samples that present a lower thermal conductivity compared to copper. A low value of the thermal resistance on the coolant side is another desirable characteristic. Usually, the cooling fluid is water in forced convection. To reduce the thermal resistance on the secondary fluid side, enhanced/finned surfaces can be used. For example, Torresin et al. [10] realized a cooling system consisting of a copper plate with 39 fins: the ratio between the enhanced cooling surface and the condensation area of the tested samples was around 25. Water flowed in the space between the fins and a staggered arrangement was used for pins in order to continuously alter the water path and break the boundary layer.

To avoid the need for wall temperature measurement (which is more difficult in the case of condensation on a horizontal tube), the Wilson plot method can be used [6]. With this technique, the condensation heat transfer coefficient is obtained starting from the direct measurement of the overall heat transfer coefficient (between the condensing vapor and the coolant) and from the estimation of the coolant side heat transfer coefficient. More details about the application of the Wilson plot method can be found in Del Col et al. [55]. However, since during DWC the main thermal resistance is usually on the coolant side, the Wilson plot method can lead to high experimental uncertainty.

Considering Eqs. 3.1 and 3.2, it is clear that the third quantity needed for the evaluation of the heat transfer coefficient is the heat flux q that can be obtained from the heat flow rate Q. Three main techniques for the measurement of the heat flow rate during condensation can be found in the literature and they can be used together with redundancy. In the first case [54], the heat flow rate is measured from the mass flow rate of the coolant $\dot{m}_{cool}$ and the temperature difference ΔT_{cool} between coolant inlet and outlet:

$$Q = \dot{m}_{cool} c_{cool} \Delta T_{cool} \tag{3.3}$$

In Eq. 3.3, c_{cool} is the specific heat of the cooling medium.

The second method is based on Fourier's law, with the evaluation of heat flow rate from the temperature profile $T(z)$ inside the sample [10]:

$$Q = \lambda_{sub} A \frac{\mathrm{d}T}{\mathrm{d}z} \tag{3.4}$$

where z is the axial coordinate normal to the condensing surface.

When evaluating the heat flow rate Q through Fourier's law, cylindrical metallic blocks (properly insulated to ensure one-dimensional heat conduction) are typically used and condensation takes place over one of the cylinder bases. These blocks are fitted with thermocouples located at different z positions from the condensing surface. The linear interpolating equation $T(z)$ can be determined by the weighted least squares (WLS) regression method [55]. Uncertainties on fitting parameters can arise from both temperature and position uncertainty. The Monte Carlo method can be used for combining the spatial and temperature uncertainties of multiple measurement points in order to obtain the uncertainty of the temperature gradient and thus the uncertainty on the heat flow rate by applying the law of propagation of uncertainty [56]. With this method, to reduce the uncertainty on the heat flux, several thermocouples can be installed in the sample. The drawback is that, increasing the number of wall thermocouples, the height of the block must be increased to allow thermocouples accommodation and, as a consequence, a larger temperature difference between the condensing surface and the coolant side of the cylindrical sample is needed to exchange the same heat flux. Therefore, this method is easier to be implemented with substrate materials having a high thermal conductivity (e.g., copper) but, for materials displaying low thermal conductivity (e.g., stainless steel), it can request temperature of the cooling fluid below 0 °C, excluding the use of pure water as secondary fluid.

The third method for the measurement of the condensation heat flow rate is based on the measurement, at fixed time steps, of the condensate mass by a precision scale. The heat flow rate can be evaluated as:

$$Q = \dot{m}_{cond} h_{lv} \tag{3.5}$$

where h_{lv} is the latent heat of condensation at a given saturation temperature. If the test rig is operating in steady-state conditions, the mass flow of the condensate rate $\dot{m}_{cond}$ can be obtained from a linear interpolation of the measured values of condensate mass plotted versus time. When using this method, particular care must be paid to the fact that all and only the condensate coming from the sample must be collected on the scale. Other liquid that forms at the wall of the vapor chamber must be appropriately separated.

As a last point, it must be remembered that the heat flow rate can affect also the measurement of the surface temperature since the measured temperature must be corrected to account for thermal conduction in the substrate (Eq. 3.2).

3.2.2 Measurements in Presence of Vapor Velocity

In this Section, the experimental technique proposed by Bisetto et al. [54] for the measurement of the DWC heat transfer coefficient during saturated steam condensation over a vertical surface will be illustrated in detail. The reader will find a practical application of some of the concepts previously mentioned.

The technique presents the following main characteristics: possibility to investigate the effect of vapor velocity; reduced thickness of the sample allowing the use of metals with low thermal conductivity (e.g., steel); redundant measurement of the heat flux from Fourier's law (Eq. 3.4) and from the heat balance at the coolant side (Eq. 3.3); possibility to measure local values of heat flux and the wall temperature along the condensing surface and thus to account for heat transfer coefficient variations along the sample; visualization of the DWC process.

The setup consists of a two-phase flow loop operating as a thermosyphon and made up of four main components: a boiling chamber, a test section, a cooling water loop, and a post-condenser. A sketch of the loop is shown in Fig. 3.1a.

Steam is generated in a cylindrical stainless steel boiling chamber by means of four electrical heaters having a maximum power of 4 kW. The electrical power supplied to the heaters is measured using a power analyzer. The pipe connections between the boiler and the test section are insulated and heated by means of an electric resistance installed around the pipe to avoid condensation before the entrance of the test section (the wall temperature is checked through a T-type thermocouple and maintained just above the saturation temperature). The steam enters the test section in saturated conditions. In the test section, the steam is partially condensed over a

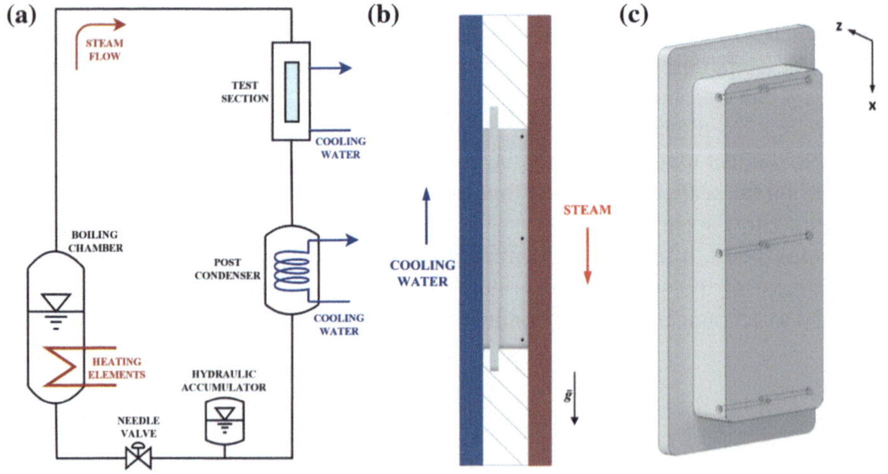

Fig. 3.1 Test apparatus: **a** schematic of the experimental thermosyphon loop for condensation tests; **b** side view of the test section; **c** 3D model of the aluminum sample

metallic sample and the latent heat is removed by cold water coming from a thermostatic bath. The coolant inlet temperature is measured by a T-type thermocouple, while the coolant temperature difference between inlet and outlet is measured by means of a three-junction copper-constantan thermopile. The coolant mass flow rate is measured by a Coriolis-effect mass flow meter. The pressure and temperature of the vapor are measured at the inlet of the test section. Downstream the test section, the two-phase mixture passes through a secondary water condenser where the condensation is completed and the liquid subcooled. The subcooled liquid returns to the boiler driven by the density difference between liquid and vapor and it closes the loop. Since the test apparatus does not have a circulating pump, to guarantee the liquid return to the boiler, a liquid head is necessary and the post-condenser is placed at a higher level with respect to the boiling chamber. The water temperature is measured at the inlet and the outlet of the post-condenser by means of T-type thermocouples, while the water mass flow rate is measured using a magnetic flow meter. The system volume is controlled by means of a hydraulic accumulator installed in the liquid line downstream the post-condenser. A precision needle valve, placed before the boiling chamber, is used to regulate the liquid flow. Before entering the boiling chamber, the temperature of the subcooled liquid is measured by means of a T-type thermocouple. Since even a small concentration of non-condensable gases (NCG) in the vapor could lead to a decrease of the condensation heat transfer coefficient, several actions must be undertaken to avoid the presence of NCG. First of all, the test apparatus works at a pressure higher than the atmospheric one. Furthermore, before each test run, the whole system is vacuumed; then the test rig is charged with deionized (DI) water. When boiling is started, some vapor is released from the top of the loop as well as from a vent valve located in the upper part of the post-condenser in order to get rid of the gases dissolved in the water.

The test section (Fig. 3.1b) allows the measurement of the DWC heat transfer coefficient and the visualization of the droplets. Steam condenses on a vertical rectangular surface (50 mm × 20 mm). The test section consists of two rectangular cross-sectional channels inside which the condensing vapor and the water coolant flow. The vapor channel is 160 mm long (cross-section 30 mm × 5 mm) and it was machined from a PEEK block. One side of the channel is covered by a double glass to allow the visualization of the process whereas the other side of the channel, opposite to the glass, is machined for accommodating the metallic substrate over which vapor condenses. The metallic specimen has 10 mm thickness and it is equipped with an array of six T-type thermocouples placed inside holes obtained by electrical discharge machining. Thermocouples are placed inside the substrate at two different depths z_1 and z_2 from the treated surface (Fig. 3.1c). The thermocouples are used to obtain the local surface temperature at three longitudinal locations along the specimen (in the middle and 2 mm from the up and bottom edge on the sample). The sample is located in between the two channels of the test section: the steam channel is in contact with the frontal face of the specimen beside the water channel is in contact with the back face of the specimen. The length of the coolant channel was determined in order to have a hydrodynamically fully developed flow of the water on the back side of the metallic specimen. In this way, there are no entrance effects and the water heat transfer coefficient can be considered uniform along the whole sample length. The cooling water flows in countercurrent with respect to the steam direction inside the test section.

The heat flux q through the sample can be measured by two different techniques. An average value of the heat flux q_{mean} over the surface can be measured from the coolant side (Eq. 3.3), beside a local value of the heat flux q_{loc} can be evaluated by applying the Fourier's law to the thermocouples placed in correspondence of the three different longitudinal positions shown in Fig. 3.1b, c (Eq. 3.4).

In Eq. 3.4, dT/dz can be replaced with $\Delta T/\Delta z$ where ΔT is the temperature difference between the two thermocouples at the same axial position along the steam direction and Δz is the distance between the two thermocouples. The Fourier law can be applied assuming one-dimensional heat flux.

The steam saturation temperature T_{sat} is obtained from the measurement of the saturation pressure in the test section. The specific enthalpy of the subcooled liquid at the inlet of the boiling chamber $h_{IN,bc}$ is evaluated from the measurements of temperature and pressure. The steam velocity v_v is obtained from

$$v_v = \frac{Q_{bc}}{\left(h_v - h_{IN,bc}\right)} \frac{1}{\rho_v A_c} \tag{3.6}$$

where Q_{bc} is the heat flow rate provided to the boiling chamber (measured using an electrical power meter), h_v is the specific enthalpy of the saturated steam exiting from the boiling chamber, ρ_v is the vapor density, and A_c is the cross-sectional area. By acting on the power supplied to the boiling chamber, it is possible to regulate the mass flow rate of the fluid circulating in the system, thus to perform tests at varying vapor mass velocities.

3.3 Droplet Population

As reported in Sect. 3.1, dropwise condensation is a cyclic process [3]. Condensation begins at the molecular scale with the formation of droplets in favored nucleation sites. The radius of the smallest viable (thermodynamically) drop is called minimum drop radius $r_{\min}$. Drops grow by direct condensation at first and later by coalescence until they reach the departing radius $r_{\max}$ at which body forces overcome adhesion force and they start to move. Once the moving drops clean their path, nucleation of new droplets ensures that the condensation process is cyclic with characteristic timescale, coverage area, and drop size distribution.

3.3.1 Models for Drop Size Distribution

Drop population models are based on the observation that the drop size distribution on a condensing surface is in steady state from the statistical point of view [3]. Le Fevre and Rose [57] introduced for the first time a relationship for the prediction of the droplet population, while Wu and Maa [58], starting from the work by Tanaka [59], proposed the population balance model. Droplets are categorized into small drops and large drops according to the growth mechanism. Small drops grow primarily due to direct condensation of vapor on the drop surface, whereas large drops grow mainly by coalescence with other drops. The distribution of the big and small droplets will be indicated, respectively, with the symbols $N(r)$ and $n(r)$. The radius which separates the two population is called effective radius r_e and its determination is still an open issue due to technological limitations in the visualization techniques. As it will be discussed later, r_e depends on the nucleation sites density distribution, square array or random Poisson distribution (see Sect. 3.4). By integrating the droplet size density function between two radii r_1 and r_2 it is possible to calculate the number of droplets per unit area having radii in between r_1 and r_2.

The drop size distribution of large drops $N(r)$ is obtained using the empirical expression proposed by Le Fevre and Rose [57] and it is given by

$$N(r) = \frac{1}{3\pi r^2 r_{\max}} \left(\frac{r}{r_{\max}} \right)^{-\frac{2}{3}} \tag{3.7}$$

From Eq. 3.7, it can be observed that the large droplet population only depends on the maximum droplet radius $r_{\max}$, which is the outcome of the forces acting on a droplet (see Sects. 3.4 and 3.5).

In order to model the drop size distribution of small drops, the population balance method can be employed. This method assumes that the contact angle of a drop remains the same from the nucleation to the departure from the surface. The drop size distribution of small drops $n(r)$ is evaluated assuming the conservation of the number of drops in a certain size range $r_1 - r_2$. In other words, the number of droplets entering a size range must be equal to the number of drops leaving the same size range. The drop growth rate G is defined as

$$G = \frac{\mathrm{d}r}{\mathrm{d}t} \tag{3.8}$$

Considering a surface area A and an infinitesimal time increment $\mathrm{d}t$, in order to conserve the droplet population in the radius range $r_1 - r_2$, the number of droplets entering this radius range ($An_1G_1\mathrm{d}t$) must be equal to the number of droplets leaving by growth this radius range ($An_2G_2\mathrm{d}t$) plus the number of droplets swept off ($Sn_{1-2}\Delta r\mathrm{d}t$):

$$An_1G_1\mathrm{d}t = An_2G_2\mathrm{d}t + Sn_{1-2}\Delta r\mathrm{d}t \tag{3.9}$$

where n is the number of drops per unit area per unit drop radius, S is the sweeping rate at which the surface is renewed by falling drops, n_{1-2} is the average drop size density in the range $r_1 - r_2$ and Δr is equal to $r_2 - r_1$.

As Δr approaches to zero, n_{1-2} tends to n and Eq. 3.9 can be written as

$$\frac{\mathrm{d}}{\mathrm{d}r}(Gn) + \frac{n}{\tau} = 0 \tag{3.10}$$

where $\tau = A/S$ is the sweeping period. Assuming that all the heat transfer occurs through the drops, the heat transfer rate through a single drop (Q_{dr}) can be equated to the condensation rate of vapor at the drop surface ($\rho_l h_{lv} \mathrm{d}V/\mathrm{d}t$) to obtain the drop growth rate G as

$$Q_{dr} = \rho_l h_{lv} \frac{\mathrm{d}V}{\mathrm{d}t} = \rho_l h_{lv} \pi r^2 (1 - \cos\theta)^2 (2 + \cos\theta) G \tag{3.11}$$

$$G = \frac{Q_{dr}}{\rho_l h_{lv} \pi r^2 (1 - \cos\theta)^2 (2 + \cos\theta)} \tag{3.12}$$

where V is the droplet volume, ρ_l the liquid density, h_{lv} the the vapor–liquid latent heat, and θ the droplet growing contact angle.

Then, the expression of G (Eq. 3.12) can be substituted into Eq. 3.10 and integrated to obtain the drop size density function of small droplets $n(r)$:

$$n(r) = N(r_e) \frac{r(r_e - r_{\min})(A_2 r + A_3)}{r_e(r - r_{\min})(A_2 r_e + A_3)} e^{B_1 + B_2} \tag{3.13}$$

As a first boundary condition, the population of small droplets is imposed to equal the population of large droplets at the effective radius ($n(r_e) = N(r_e)$), providing the following expressions for B_1 and B_2:

$$B_1 = \frac{A_2}{A_1 \tau} \left[\frac{r_e^2 - r^2}{2} + r_{\min}(r_e - r) - r_{\min}^2 \ln\left(\frac{r - r_{\min}}{r_e - r_{\min}} \right) \right] \tag{3.14}$$

$$B_2 = \frac{A_3}{A_1 \tau} \left[(r_e - r) - r_{\min} \ln\left(\frac{r - r_{\min}}{r_e - r_{\min}} \right) \right] \tag{3.15}$$

Imposing $\mathrm{d}(\ln n(r))/\mathrm{d}(\ln r) = \mathrm{d}(\ln N(r))/\mathrm{d}(\ln r) = -8/3$ at $r = r_e$ as a second boundary condition, the sweeping period τ is

$$\tau = \frac{3 r_e^2 (A_2 r + A_3)^2}{A_1 \left[8 A_3 r_e - 14 A_2 r_e r_{\min} + 11 A_2 r_e^2 - 11 A_3 r_{\min} \right]} \tag{3.16}$$

The three coefficients A_1, A_2, and A_3 in Eqs. 3.13–3.16 are obtained analytically from the heat flow exchanged by a single droplet and, being the formulation of Q_{dr}

different for each heat transfer model, the analytical expressions for A_1, A_2, and A_3 are reported in detail in Sect. 3.4.

3.3.2 Measurement of Drop Size Distribution

In Fig. 3.2a, the theoretical drop size density function for a flat vertical hydrophobic surface is reported at varying droplet radius. The dotted curve on the left side of the graph ($r < r_e$) represents the small droplet population $n(r)$, which is calculated by Eq. 3.13–3.16 using the formulation for A_1, A_2, and A_3 provided by Miljkovic et al. [60] (see Sect. 3.4.3). Instead, the continuous curve on the right side of the graph represents the formulation proposed by Le Fevre and Rose [57] (Eq. 3.7) for the drop size density of large drops population $N(r)$. Figure 3.2a shows that DWC involves the simultaneous presence of droplets with variable radius by 6 orders of magnitude, ranging from nanometres (r_{min}) to millimetres (r_{max}). The drop size density function decreases with an increase in drop radius: for a given surface area, there is a huge number of small droplets and a relatively small number of large droplets.

In addition, Fig. 3.2a presents a comparison between theory and experimental visualizations performed by Parin et al. [33] focusing on the large droplet population

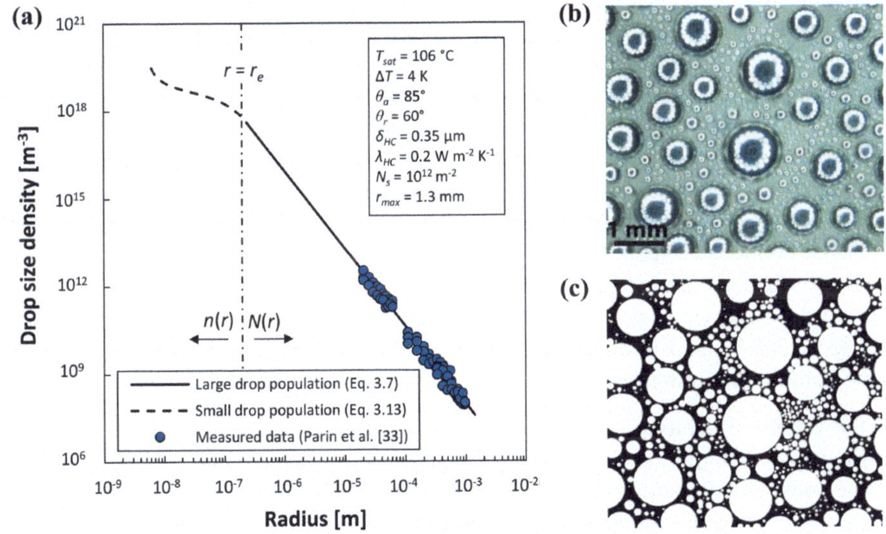

Fig. 3.2 **a** Drop size density function compared with experimental data by Parin et al. [33]. The large droplet population $N(r)$ is obtained by the Le Fevre and Rose [57] equation (Eq. 3.7), whereas the small droplet population $n(r)$ is calculated by Eq. 3.13, using the expressions for A_1, A_2, and A_3 provided by Miljkovic et al. [60] (see Sect. 3.4.3). **b** Enlarge image of DWC taken by a high-speed camera. **c** Resulting black and white image after processing and reconstruction of the observed droplet population

$N(r)$. In their work, Parin et al. [33] mapped the droplet population detecting more than 3 million droplets with radii in the range of 15 μm–1 mm. The measured droplet size distribution resulted to be not affected by the heat flux. The experimental technique consists of a high-speed camera coupled with a torus-shaped illumination system for DWC visualizations and a homemade MATLAB® program for image processing. The torus-shaped light projects its pattern onto each droplet and the external diameter of the torus is proportional to the drop's diameter itself. To detect droplets and to determine their dimensions, the recorded image is then processed by a MATLAB® program, considering the relationship between the external radius of the reflected toroidal light path and the effective drop radius. An example of an image taken by the high-speed camera during DWC and the resulting black and white image after processing and reconstruction by the program are shown, respectively, in Fig. 3.2b, c. As it can be seen from Fig. 3.2a, a satisfactory agreement between the equation proposed by Le Fevre and Rose [57] and the experimental data taken by Parin et al. [33] during DWC of steam was obtained down to tens of microns; the mean deviation between measurements and predicted values is below 20%.

3.4 Heat Transfer Models for DWC with Quiescent Vapor

Different models in the literature are aimed at describing the complex phenomena that take place during DWC: the nucleation of a droplet until its departure, the heat exchanged by the drop during its lifetime, and the droplet population on the surface.

Because of the unsteady behavior of DWC, researchers have usually adopted a statistical approach to model the dropwise condensation heat transfer [2, 36, 60]. This method is based on the experimental observation that the overall drop size distribution is constant with time even though the individual drop growth is an unsteady phenomenon. In the literature, several heat transfer models have been proposed for dropwise condensation on flat or structured surfaces and, among them, four different studies have been selected in the present work: Le Fevre and Rose [57], Kim and Kim [36], Miljkovic et al. [60], and Chavan et al. [32]. The models are presented in the Sects. 3.4.1–3.4.4 in chronological order. The models were developed with the same common assumptions:

- the vapor temperature is uniform and equal to the saturation temperature;
- the vapor is in quiescent conditions (negligible vapor velocity);
- the substrate is assumed as a semi-infinite body at uniform temperature;
- the presence of non-condensable gases is neglected.

In the statistical approach, the heat transfer through a single drop of a given radius r is multiplied by the respective drop size density function ($n(r)$ or $N(r)$) and the product is then integrated between r_{min} and r_{max} to obtain the overall condensation heat flux q transferred during steady-state DWC:

$$q = \int_{r_{\min}}^{r_e} Q_{dr}(r)n(r)dr + \int_{r_e}^{r_{\max}} Q_{dr}(r)N(r)dr \tag{3.17}$$

As already reported in Sect. 3.3, $n(r)$ and $N(r)$ are, respectively, the small droplet population ($r_{\min} \le r \le r_e$) and the population of large droplets ($r_e \le r \le r_{\max}$); r_e is the effective radius and it denotes the drop radius at the boundary between small and large drops, whereas Q_{dr} is the heat transfer through a drop of radius r.

The lower limit of the first integral ($r_{\min}$) is the minimum radius of a stable droplet, which can be calculated as [61]:

$$r_{\min} = \frac{2\sigma T_{sat} v_l}{h_{lv} \Delta T} \tag{3.18}$$

where ΔT is the degree of subcooling (that is the temperature difference between the saturated steam and the surface), σ is the liquid-vapor surface tension, v_l is the specific volume of liquid, T_{sat} is the saturation temperature, and h_{lv} is the vapor–liquid latent heat. Equation 3.18 is used in all the selected models. In the second integral, $r_{\max}$ is the departing radius which is the maximum dimension assumed by drops before sliding on the condensation surface. In quiescent vapor, $r_{\max}$ is obtained by applying a balance between adhesion force (which retains the drop) and gravity force (which works for moving the drop) [36, 42, 62]. In the detailed discussion of the models (Sects. 3.4.1–3.4.4), the specific formulations developed for the evaluation of the departing radius and the effective radius will be presented.

The heat transfer coefficient (HTC) is obtained by dividing the condensation heat flux by the saturation-to-wall temperature difference (degree of subcooling):

$$\text{HTC} = \frac{q}{\Delta T} \tag{3.19}$$

To estimate the heat transfer through a single drop Q_{dr}, a network of thermal resistances is employed. The different assumptions done by the authors of the models can provide different results in terms of overall heat flux q (Eq. 3.17). As already reported in Sect. 3.3.1, the formulation of Q_{dr} modifies the growth rate G (Eq. 3.12) and thus the expression of the small droplet population (Eq. 3.13).

3.4.1 Le Fevre and Rose (1966) Model

In the model by Le Fevre and Rose [57], the temperature drop due to the droplet curvature is considered and the following thermal resistances are accounted for: the liquid–vapor interfacial resistance, the conduction resistance through the drop, and the resistance of the coating. Their model assumed that within the drop, convection is negligible and conduction is the dominant heat transfer mechanism. However, in the model, all the drops are considered hemispherical with a contact angle of 90°.

The heat flow rate through a single drop is calculated as

$$Q_{dr}(r) = \frac{\Delta T - \frac{2\sigma T_{sat}}{r\rho_l h_{lv}}}{K_1 \frac{r}{\lambda_l} + K_2\left(\frac{0.627}{0.664}\right)\frac{T_{SAT}}{h_{lv}^2\rho_v}\frac{\gamma+1}{\gamma-1}\left[\frac{RT_{sat}}{2\pi}\right]^{0.5}} \tag{3.20}$$

where K_1 and K_2 are constants equal to 2/3 and 1/2, respectively, λ_l is the liquid conductivity, ρ_l is the liquid density, γ is the ratio of the specific heat capacities and R is the specific ideal-gas constant. At the denominator, the first term is the conduction resistance and the second term accounts for the liquid–vapor interfacial resistance plus the conduction resistance through the coating (included in K_2).

In this model, the heat transfer through a single drop Q_{dr} (Eq. 3.20) is combined with the drop size distribution of large droplets $N(r)$ (Eq. 3.7) to obtain the average heat flux as

$$q = \int_{r_{\min}}^{r_{\max}} Q_{dr}(r)N(r)dr \tag{3.21}$$

where $r_{\min}$ is the minimum droplet radius (Eq. 3.18) and $r_{\max}$ is the maximum droplet radius evaluated as

$$r_{\max} = K_3\left[\frac{\sigma}{\rho_l g}\right]^{0.5} \tag{3.22}$$

where K_3 is a constant equal to 0.4 that was determined experimentally.

3.4.2 *Kim and Kim (2011) Model*

The model by Kim and Kim [36] computes the heat transfer through a single drop incorporating the various thermal resistances from the vapor to the surface and considers both the populations of small and large droplets. Kim and Kim [36] improved the model by Le Fevre and Rose [57] accounting for the effect of the contact angle on the heat transfer performance. In particular, they modeled the conduction resistance through droplets exhibiting larger growing contact angles ($\theta > 90°$). Although superhydrophobic surfaces are considered, the surface morphology is neglected. In this model, the thickness of the coating layer and the number of nucleation sites are also accounted for.

In terms of temperature drop, the total temperature difference between the vapor and the surface is expressed as

$$\Delta T = \Delta T_i + \Delta T_c + \Delta T_d + \Delta T_{HC} \tag{3.23}$$

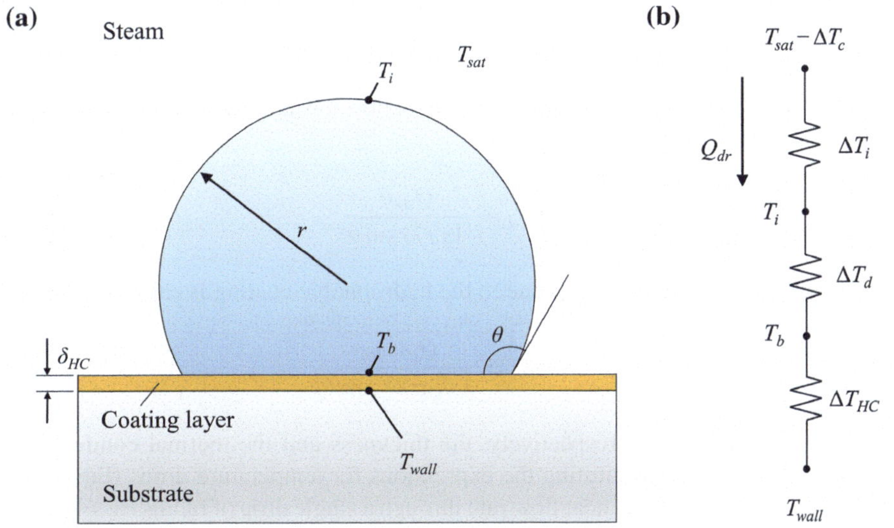

Fig. 3.3 **a** Schematic representation of a droplet sitting on the condensing surface coated with a hydrophobic layer. **b** Resistance network and temperature drop contributions due to the liquid–vapor interface (ΔT_i), the droplet curvature (ΔT_c), the conduction through the droplet (ΔT_d), and the coating layer (ΔT_{HC}); T_i is the temperature of the liquid–vapor interface and T_b is the temperature at the droplet base

where ΔT_i, ΔT_c, $\Delta T_{d,}$ and ΔT_{HC} are respectively the temperature drops due to the liquid–vapor interface, the droplet curvature, the thermal conduction through the droplet and the coating layer (Fig. 3.3). In particular, the temperature drop due to the interfacial resistance is given by

$$\Delta T_i = \frac{Q_{dr}}{h_i 2\pi r^2 (1 - \cos\theta)} \tag{3.24}$$

where h_i is the interfacial heat transfer coefficient given by [63]:

$$h_i = \frac{2\alpha}{2-\alpha} \frac{1}{\sqrt{2\pi R_g T_{sat}}} \frac{h_{lv}^2}{v_v T_{sat}} \tag{3.25}$$

In Eq. 3.25, R_g is the specific ideal gas constant, ν_v is the vapor specific volume, and α is the accommodation coefficient. The accommodation coefficient α is the ratio of vapor molecules captured by the liquid phase to the total number of vapor molecules approaching the surface ($0 < \alpha < 1$). A α close to 0 indicates a high concentration of NCG, while α close to 1 means the absence of NCG [63].

The temperature drop caused by the droplet curvature is evaluated as

$$\Delta T_c = \frac{2T_{sat}\sigma}{h_{lv} r \rho_l} \tag{3.26}$$

The thermal resistance due to heat conduction through the droplet causes the following temperature drop:

$$\Delta T_d = \frac{Q_{dr}\theta}{4\pi r \lambda_l \sin\theta} \tag{3.27}$$

Finally, the temperature drop due to the hydrophobic coating is calculated as

$$\Delta T_{HC} = \frac{Q_{dr}\delta_{HC}}{\lambda_{HC}\pi r^2 \sin^2\theta} \tag{3.28}$$

where δ_{HC} and λ_{HC} are, respectively, the thickness and the thermal conductivity of the coating layer. Substituting the expressions for temperature drops (Eqs. 3.24, 3.26–3.28) in Eq. 3.23, the heat flow rate through a single drop of radius r is obtained as

$$Q_{dr}(r) = \frac{\pi r^2\left(\Delta T - \frac{2T_{sat}\sigma}{r h_{lv}\rho_l}\right)}{\left(\frac{\delta_{HC}}{\lambda_{HC}\sin^2\theta} + \frac{r\theta}{4\lambda_l \sin\theta} + \frac{1}{2h_i(1-\cos\theta)}\right)} \tag{3.29}$$

In the model by Kim and Kim [36], the heat transfer through a single drop $Q_{dr}(r)$ (Eq. 3.29) is combined with the drop size density functions of large droplets $N(r)$ and small droplets $n(r)$ to obtain the overall heat flux during DWC (Eq. 3.17). With regard to the droplet population, $N(r)$ is obtained using the empirical relation given by Le Fevre and Rose [57] (Eq. 3.7), while $n(r)$ is calculated using Eq. 3.13 (see Sect. 3.3).

In Eq. 3.17, the limits of integration are calculated using Eq. 3.18 for $r_{\min}$ and Eqs. 3.30 and 3.31 respectively for r_e and $r_{\max}$:

$$r_e = \frac{1}{\sqrt{4N_s}} \tag{3.30}$$

$$r_{\max} = \sqrt{\frac{6k_c(\cos\theta_r - \cos\theta_a)\sin\theta}{\pi\left(2 - 3\cos\theta + \cos^3\theta\right)}\frac{\sigma}{\rho_l g}} \tag{3.31}$$

In Eq. 3.30, N_s is the nucleation site density; in Eq. 3.31, θ_a is the advancing contact angle, θ_r is the receding contact angle, and k_c is the retention factor, a constant depending on the droplet geometry. The present formulation for the effective radius r_e assumes that the nucleation sites have a uniform distribution and form a square array over the surface, while the departing radius is deduced by equating the gravity force and the adhesion force.

The coefficients for the population of small droplets (Eqs. 3.13–3.16) were obtained analytically as

$$A_1 = \frac{\Delta T}{2\rho_l h_{lv}} \tag{3.32}$$

$$A_2 = \frac{\theta(1-\cos\theta)}{4\lambda_l \sin\theta} \tag{3.33}$$

$$A_3 = \frac{1}{2h_i} + \frac{\delta_{HC}(1-\cos\theta)}{\lambda_{HC}\sin^2\theta} \tag{3.34}$$

3.4.3 Miljkovic et al. (2013) Model

Miljkovic et al. [60] modified the model by Kim and Kim [36] by taking into account the surface morphology with the aim of estimating the condensation heat transfer on micro/nanostructured surfaces. This model extends the previously developed droplet size distribution theory to both constant and non-constant contact angle droplets growing processes. It is worth noting that assuming a constant droplet contact angle during growth is suitable for dropwise condensation on flat hydrophobic surfaces; however, this assumption is not valid for structured superhydrophobic surfaces, since the droplet contact angles have been observed to vary during droplet growth [60]. As for the models described in Sects. 3.4.1 and 3.4.2, Miljkovic et al. [60] combines the single drop heat transfer with the droplet population for the estimation of the average heat flux during DWC (Eq. 3.17). In particular, the model accounts for the temperature drop contributions due to the liquid–vapor interfacial resistance, the droplet curvature, the conduction resistance of the drop, the thermal resistance of the micro/nanostructure (if present), and the thermal resistance of the coating. The heat transfer rate exchanged by a single droplet Q_{dr} is calculated as follows:

$$Q_{dr}(r,\theta) = \frac{\pi r^2\left(\Delta T - \frac{2T_{sat}\sigma}{r h_{lv}\rho_l}\right)}{\frac{1}{2h_i(1-\cos\theta)} + \frac{r\theta}{4\lambda_l\sin\theta} + \frac{1}{\lambda_{HC}\sin^2\theta}\left[\frac{\lambda_p\varphi}{\delta_{HC}\lambda_p + l_p\lambda_{HC}} + \frac{\lambda_l(1-\varphi)}{\delta_{HC}\lambda_l + l_p\lambda_{HC}}\right]^{-1}} \tag{3.35}$$

where φ and l_p are geometrical parameters that, in the case of flat surfaces, are equal to 0. For details about the calculation of φ and l_p in the case of structured surfaces, the reader can refer to the work by Miljkovic et al. [60].

For large drops growing mainly by coalescence, the droplet size distribution $N(r)$ is evaluated by the expression proposed by Le Fevre and Rose [57] (Eq. 3.7), where the departing radius is calculated as follows:

$$r_{\max} = \sqrt{\frac{6(\cos\theta_r - \cos\theta_a)\sin\theta_e}{\pi\left(2 - 3\cos\theta_e + \cos^3\theta_e\right)}\frac{\sigma}{\rho_l g\cos\beta}} \tag{3.36}$$

where β is the inclination of the condensing surface (90° corresponds to a horizontal surface and 0° corresponds to a vertical surface) and θ_e is the equilibrium contact angle defined as $\theta_e = \cos^{-1}(0.5\cos\theta_a + 0.5\cos\theta_r)$.

To solve the first integral in Eq. 3.17 (which includes the small droplet population), $r_{\min}$ is calculated by Eq. 3.18 and, assuming the nucleation sites randomly distributed on the condensation surface (Poisson distribution), r_e is given by

$$r_e = \frac{1}{4\sqrt{N_s}} \tag{3.37}$$

In accordance with the expression for the heat transfer through a single droplet developed by Miljkovic et al. [60] (Eq. 3.35), the small droplet population (Eqs. 3.13–3.16) is calculated with the following coefficients:

$$A_1 = \frac{\Delta T}{\rho_l h_{lv}(1 - \cos\theta)^2(2 + \cos\theta)} \tag{3.38}$$

$$A_2 = \frac{\theta}{4\lambda_l \sin(\theta)} \tag{3.39}$$

$$A_3 = \frac{1}{2h_i(1 - \cos\theta)} + \frac{1}{\lambda_{HC}\sin^2\theta}\left[\frac{\lambda_p\varphi}{\delta_{HC}\lambda_P + l_p\lambda_{HC}} + \frac{\lambda_l(1 - \varphi)}{\delta_{HC}\lambda_l + l_p\lambda_{HC}}\right]^{-1} \tag{3.40}$$

3.4.4 *Chavan et al. (2016) Model*

Recently, Chavan et al. [32] proposed a procedure for the calculation of the heat transfer through a droplet that is based on the results of steady-state 2D axisymmetric numerical simulations of the droplet growth. The models reported in Sects. 3.4.1–3.4.3 make the assumption of constant temperature at the liquid–vapor interface (droplet surface) and at the solid–liquid interface (droplet base) for the calculation of the conduction thermal resistance of the single droplet [36, 57, 60]. Instead, Chavan et al. [32] solved the heat equation through a single droplet by means of a numerical model based on the finite element method and replacing the constant temperature boundary condition at the liquid–vapor interface with a convective boundary condition (fixing a constant value of the heat transfer coefficient h_i). The simulations showed that the local heat flux at the three-phase contact line resulted to be four orders of magnitude higher than at the droplet top and this phenomenon was not taken into account by the previous formulations (Eqs. 3.29 and 3.35).

The Chavan et al. [32] model is based on three dimensionless quantities governing the heat transfer through a single drop: the Biot number (Bi), the apparent advancing contact angle (θ_a), and the droplet Nusselt number (Nu). The Nusselt number can be evaluated as a function of the other two dimensionless groups as $\mathrm{Nu} = \mathrm{f}(\mathrm{Bi}, \theta_a)$. The Biot and Nusselt numbers can be expressed in terms of the droplet base radius (r_b) as

$$\mathrm{Bi} = \frac{h_i r_b}{\lambda_l} \tag{3.41}$$

$$\mathrm{Nu} = \frac{Q_{dr}}{\lambda_l r_b (T_{sat} - T_{wall})} \tag{3.42}$$

where the interfacial heat transfer coefficient h_i is calculated from Eq. 3.25. For an easy implementation of their method in DWC heat transfer models, the authors provided the following expressions for the estimation of Nu (θ_a is in radians):

$$\mathrm{Nu} = 3\theta_a^{0.65}\mathrm{Bi}^{0.83} + 0.007\theta_a^{5.1}\,\mathrm{Bi}^{-0.23} \quad \mathrm{Bi} \leq 0.5 \tag{3.43}$$

$$\mathrm{Nu} = 0.29\theta_a^{2.24}\mathrm{Bi}^{-0.17} + 3.33\theta_a^{-0.3}\mathrm{Bi}^{0.72} \quad 0.5 < \mathrm{Bi} \leq 2 \tag{3.44}$$

$$\mathrm{Nu} = 5.76e^{-0.28\theta_a^{0.68}}\ln\left(1 + 5\mathrm{Bi}^{0.82} - 2.79\mathrm{Bi}^{0.83}\right) \quad 2 < \mathrm{Bi} \leq 10^5 \tag{3.45}$$

Replacing the liquid–vapor interfacial thermal resistance and the conduction resistance through the droplet with the thermal resistance obtained by numerical simulations (Eq. 3.42), the heat flow rate through a single droplet can be calculated as

$$Q_{dr}(r, \theta) = \frac{\Delta T - \frac{2T_{sat}\sigma}{r h_{lv}\rho_l}}{\frac{1}{\mathrm{Nu}\lambda_l r \sin\theta_a} + \frac{\delta_{HC}}{\lambda_{HC}\pi r^2 \sin^2\theta_a}} \tag{3.46}$$

where Nu is the results of Eqs. 3.43–3.45 whereas the thermal resistance of the coating (the second term at denominator) was included in Eq. 3.35 by Birbarah et al. [64] as an additional thermal resistance in series with the one developed by Chavan et al. [32].

In the model by Chavan et al. [32], to obtain the overall condensation heat flux during DWC (Eq. 3.17), the heat transfer through a single droplet (Eq. 3.46) is combined with the drop size distribution. In particular, the authors calculate the drop size density function of large drops $N(r)$ and small drops $n(r)$ accordingly to the model by Miljkovic et al. [60]. For the large droplet population, $N(r)$ is obtained by Eq. 3.7 together with the expression for the departing radius provided by Eq. 3.36.

Instead, for the small droplet population, $n(r)$ is evaluated by Eqs. 3.13–3.16 using the coefficients reported in Eqs. 3.38–3.40. With regard to the minimum drop radius and the effective radius, Eqs. 3.18 and 3.37 are, respectively, used.

When the overall condensation heat flux during DWC is calculated using the equation by Chavan et al. [32] for the heat transfer through a single drop, it results that the previous models (Kim and Kim [36] and Miljkovic et al. [60]) underpredict the overall heat transfer.

3.5 Effect of Vapor Velocity on DWC Heat Transfer Coefficient

Vapor velocity is expected to affect the drop size distribution on the condensing surface during DWC. In particular, an increase in vapor velocity causes a decrease of the droplet departing radius and, at the same time, it leads to higher HTC [16, 40, 42]. As reported in Sect. 3.4, droplets grow from the nucleation radius to the departing radius, which is the result of a balance between retentive forces (droplet adhesion) and external forces which promote droplet movement (gravity and drag). The classical formulation of the droplet departing radius (Eqs. 3.31 and 3.36) derives from a force balance between droplet adhesion and gravity. Recently, Tancon et al. [42] proposed a strategy for modeling the effect of vapor velocity during DWC based on accounting for the drag force into the expression for the maximum droplet radius. After the presentation of the Tancon et al. [42] model, the equation accounting for vapor velocity will be included in the Chavan et al. [32] and in the Miljkovic et al. [60] models; the predicted HTC will be compared with datasets from independent laboratories [16, 41, 42].

3.5.1 Description of the Model by Tancon et al. (2021)

When the drop reaches the maximum dimension before sliding, the sum of drag force plus gravity force must equal the adhesion force:

$$F_{ad}(r_{\max}) = F_d(r_{\max}) + F_g(r_{\max}) \tag{3.47}$$

Assuming a circular drop, the adhesion force F_{ad} is evaluated as

$$F_{ad}(r) = 2k_c\sigma \sin\theta_e(\cos\theta_r - \cos\theta_a)r \tag{3.48}$$

where θ_a and θ_r are the advancing and the receding contact angles, θ_e is the equilibrium contact angle calculated as $\theta_e = \cos^{-1}(0.5\cos\theta_a + 0.5\cos\theta_r)$, σ is the surface tension of the condensing fluid and k_c is the retention factor, which can be analytically calculated and, for a circular shaped droplet, it is equal to 2/π [65].

The gravity force F_g acting on a droplet is calculated from the droplet volume as

$$F_g(r) = \frac{2 - 3\cos\theta_e + \cos^3\theta_e}{3}\pi\rho_l g r^3 \tag{3.49}$$

where g is the gravity acceleration and a vertical orientation of the condensing surface is assumed.

The drag force F_d on a droplet due to the action of vapor flow is expressed as [66]:

$$F_d(r) = \frac{1}{2}\rho_v v_v^2 C_d(\theta_e - \sin\theta_e\cos\theta_e)r^2 \tag{3.50}$$

where ρ_v is the density of the vapor, v_v is the vapor mean velocity in the channel where vapor flows and C_d is the drag coefficient. In Eq. 3.50, θ_e is expressed in radians. To estimate the drag coefficient of a droplet, Tancon et al. [42] performed CFD numerical simulations. They found that, for the specific case of a droplet placed on the wall of a rectangular cross-sectional channel (characterized by a large width-to-height ratio, equal to six in their case), the drag coefficient C_d can be expressed as a product of only two dimensionless groups: the ratio of channel height to droplet height L_c/l_{dr} and the droplet Reynolds number defined as $\mathrm{Re}_{dr} = l_{dr}v_v\rho_v/\mu_v$.

$$C_d = 5.6053\left[(L_c/l_{dr})^{-4/3}\mathrm{Re}_{dr}^{-1/6}\right] + 0.1754 \tag{3.51}$$

Substituting the expressions of F_{ad}, $F_{g,}$ and F_d (Eqs. 3.48–3.50) into the force balance equation (Eq. 3.47), the droplet departing radius $r_{\max}$ in the case of non-negligible vapor velocity can be calculated as

$$r_{\max} = \frac{-C + \sqrt{C^2 + 4AB}}{2B} \tag{3.52}$$

where the coefficients *A*, *B,* and *C* are equal to

$$A = 2k_c\sigma\sin\theta_e(\cos\theta_r - \cos\theta_a) \tag{3.53}$$

$$B = \frac{2 - 3\cos\theta_e + \cos^3\theta_e}{3}\pi\rho_l g \tag{3.54}$$

$$C = \frac{1}{2}\rho_v v_v^2 C_d(\theta_e - \sin\theta_e\cos\theta_e) \tag{3.55}$$

The present method for the determination of r_{max} requires an iterative procedure: a guess value of the droplet departing radius r_{max} (Eq. 3.52) is needed to estimate the drag force (Eq. 3.50). A first attempt value for r_{max} is calculated from Eq. 3.47 assuming the vapor shear stress component equal to zero. With this initial value of r_{max}, C_d is obtained by Eq. 3.51 to estimate the drag force on the droplet, thus the force balance equation is solved a second time and a new value of r_{max} is calculated (Eq. 3.52). The convergence is achieved when the difference between two consecutive r_{max} is lower than an established value (e.g., 1 μm).

When the formulation for the departing radius proposed by Tancon et al. [42] is included in the models reported in Sect. 3.4, the effect of the vapor velocity on the overall heat transfer can be accounted for.

3.5.2 *Comparison Against Experimental Data*

The Miljkovic et al. [60] and the Chavan et al. [32] models (described in Sect. 3.4), in their original formulation and modified with the expression for the departing radius proposed by Tancon et al. [42] (Eq. 3.52), have been used to predict the HTC during DWC in presence of vapor velocity. The models are compared against a database composed of the heat transfer data by Tancon et al. [42], Sharma et al. [41] and Tanner et al. [16]. Datasets refer to vapor velocity conditions between 3 and 24 m s^{-1}. The results of the comparison, using the input parameters listed in Table 3.1, are reported in Fig. 3.4, where the ratio of the calculated to experimental HTC is plotted against vapor velocity.

As expected, since the original models (Miljkovic et al. [60] and Chavan et al. [32]) predict a constant value of heat transfer coefficient with vapor velocity, the experimental HTC is underpredicted at high vapor velocity (Fig. 3.4a). Instead, the

Table 3.1 List of input parameters, for each set of experimental data, used in the models by Miljkovic et al. [60] and Chavan et al. [32] for the evaluation of the DWC heat transfer coefficient

Parameter	Input values		
	Tancon et al. [42]	Sharma et al. [41]	Tanner et al. [16]
T_{sat} [°C]	107	111	108
ΔT [K]	3.2–2.7	4.1	2.2–1.7
θ_a [°]	87.5	145	108
θ_r [°]	63.5	51	100
v_v [m s^{-1}]	3–11	3–9	3–23
δ_{HC} [nm]	190	125	550
λ_{HC} [W m^{-1} K^{-1}]	0.2	0.1	0.15
N_s [m^{-2}]	8×10^{14} (Miljkovic et al.) 10^{11} (Chavan et al.)	8×10^{15} (Miljkovic et al.) 2×10^{11} (Chavan et al.)	2×10^{15} (Miljkovic et al.) 10^{12} (Chavan et al.)

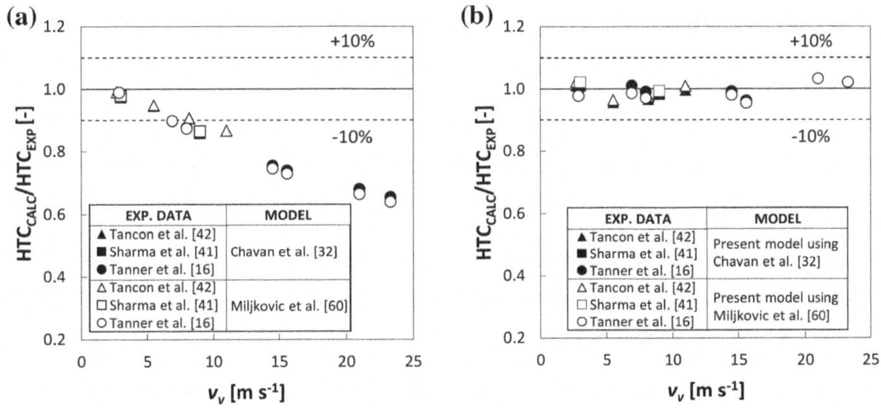

Fig. 3.4 Experimental data by Tancon et al. [42], Tanner et al. [16], and Sharma et al. [41] compared with models predictions at varying vapor velocity: **a** predictions by the original Chavan et al. [32] and Miljkovic et al. [60] models; **b** predictions by the present model, coupled with Chavan et al. [32] and Miljkovic et al. [60] models, to account for the effect of vapor velocity. Models input used for the comparison are reported in Table 3.1

present model, obtained coupling the r_{max} formulation proposed by Tancon et al. [42] with Miljkovic et al. [60] and Chavan et al. [32] models, is able to predict the heat transfer coefficient increase due to vapor velocity (Fig. 3.4b). For each calculated value, the deviation with respect to measured data is below 10% and the whole experimental dataset is predicted with a mean relative deviation below 4%.

3.6 Effect of Main Parameters on the Heat Transfer Coefficient

After validation, the modified Miljkovic et al. [60] model with the formulation by Tancon et al. [42] for the droplet departing radius is here used to study the effect of the main parameters affecting the heat transfer coefficient during DWC. The following input model parameters have been chosen as reference values: saturation temperature 107 °C, heat flux 335 kW m^{-2}, coating thermal resistance 1 m^2 K MW^{-1}, advancing contact angle 90°, contact angle hysteresis 20°, vapor velocity $v_v = 3$ m s^{-1}. In addition, the condensing surface is assumed to be in a vertical position and it does not present a micro/nanostructure (which implies the input parameters φ and l_p in the Miljkovic et al. [60] model are equal to 0).

3.6.1 Temperature Drops and Cumulative Normalized Heat Flux Distribution

The temperature drops that arises from the thermal resistances affecting DWC together with the cumulative heat flux distribution function are depicted in Fig. 3.5 for two values of the coating thermal resistance ($\delta/\lambda = 1$ m^2 K MW^{-1} and $\delta/\lambda =$ 0.2 m^2 K MW^{-1}). Considering a total temperature drop $\Delta T = 3$ K, Fig. 3.5a shows the temperature drops due to the four thermal resistances as a function of the droplet radius. As already reported in Sect. 3.4.3, the Miljkovic et al. [60] model considers the following thermal resistances: liquid–vapor interfacial resistance, droplet curvature resistance, conduction resistance of the drop, and conduction resistance of the coating. The higher the coating resistance, the higher its contribution to the total temperature drop and, consequently, the larger the radius interval at which the conduction through the coating is the dominant thermal resistance (up to 0.5 μm in case of $\delta/\lambda = 0.2$ m^2 K MW^{-1} and up to 2 μm in case of $\delta/\lambda = 1$ m^2 K MW^{-1}). The coating thermal resistance is the most important resistance up to 0.5–2 μm of drop radius, whereas the conduction resistance through the droplet is dominant for higher values of droplet radius. It should be noted that, in the early stage of drop growth, when the radius is lower than some tens of nanometers, the thermal resistance due to droplet curvature gives a considerable contribution to the total temperature drop.

The percentage of the heat flux exchanged by droplets smaller than a certain value r is shown in Fig. 3.5b where the cumulative normalized heat flux distribution is plotted versus droplet radius. The cumulative normalized heat flux $F_{heat\,flux}$ is defined as

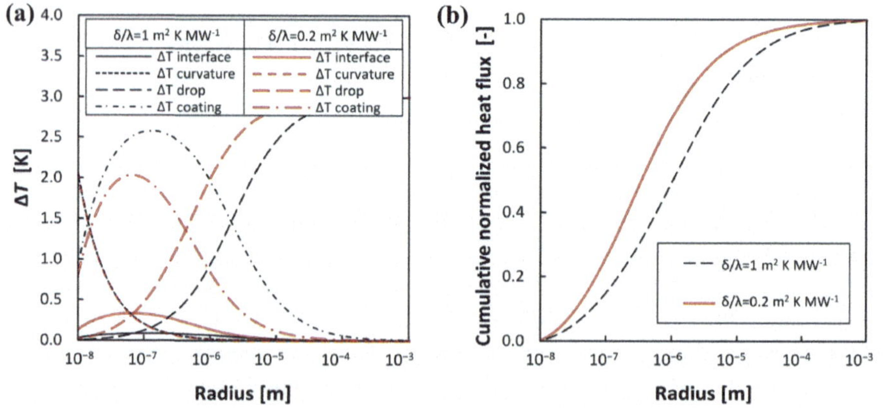

Fig. 3.5 **a** Temperature drops due to the liquid–vapor interface, droplet curvature, conduction through the drop, and through the coating plotted against droplet radius. **b** Cumulative normalized heat flux distribution versus droplet radius. Input parameters for the modified Miljkovic et al. [60] model using the Tancon et al. [42] formulation for the departing radius: $\Delta T = 3$ K, $T_{sat} = 107$ °C, $\theta_a = 90°$, $\Delta\theta = 20°$, $\delta/\lambda = 1$ m^2 K MW^{-1} and $\delta/\lambda = 0.2$ m^2 K MW^{-1}, $v_v = 3$ m s^{-1}

$$F_{heatflux}(r) = \frac{\int_{r_{\min}}^{r} Q_{dr}(r)n(r)dr}{q} \quad r_{\min} \le r \le r_e \tag{3.56}$$

$$F_{heatflux}(r) = \frac{\int_{r_{\min}}^{r_e} Q_{dr}(r)n(r)dr + \int_{r_e}^{r} Q_{dr}(r)N(r)dr}{q} \quad r_e < r \le r_{\max} \tag{3.57}$$

where q is the total condensation heat flux, $Q_{dr}(r)$ is the heat exchanged by a single droplet (Eq. 3.35), $n(r)$ is the drop size density for the small droplet population (Eq. 3.13) and $N(r)$ is the drop size density for the population of large droplets (Eq. 3.7). In both the considered cases ($\delta/\lambda = 1$ m^2 K MW^{-1} and $\delta/\lambda = 0.2$ m^2 K MW^{-1}), around 60% of the total heat flux is exchanged in the radius range where conduction through the coating is the dominant thermal resistance (Fig. 3.5a). This result shows the strong effect of the coating thermal resistance on the total heat transfer. Reducing the promoter thermal resistance by lowering the coating thickness can be a strategy to improve the HTC during DWC. However, the coating thickness affects the lifetime of the treatments. The thickness should be increased in order to increase the coating lifetime, as it is reported in the literature [2, 14], but the advantage in terms of HTC would be adversely affected [12].

3.6.2 Predicted Effect of Contact Angle Hysteresis, Coating Thermal Resistance, Heat Flux and Vapor Velocity on the Heat Transfer Coefficient

In Fig. 3.6, the HTC and the departing radius, calculated by the modified Miljkovic et al. [60] model using Eq. 3.52, are plotted against vapor velocity v_v at different values of coating thermal resistance, contact angle hysteresis, and heat flux. As already shown in Sect. 3.6.1, the coating thermal resistance is an important aspect to be considered during DWC since it can affect the overall HTC. In Fig. 3.6a, the HTC is plotted versus vapor velocity v_v considering three different values of the coating thermal resistance, ranging between 0.2 and 5 m^2 K MW^{-1}. For a given vapor velocity, the HTC is strongly affected by changing the coating thermal resistance, whereas the droplet departing radius remains the same. When the thermal resistance passes from 5 to 1 m^2 K MW^{-1}, the HTC is doubled. In Fig. 3.6b, the HTC is plotted for different values of the contact angle hysteresis ($\Delta\theta$); the advancing contact angle θ_a is kept fixed whereas the receding contact angle is varied. For a fixed value of vapor velocity, the droplet departing diameter $r_{\max}$ increases when increasing the contact angle hysteresis $\Delta\theta$ and the HTC decreases. Improving the vapor velocity leads to an HTC increase and this effect is more prominent at low $\Delta\theta$ values. The effect of the heat flux q is reported in Fig. 3.6c. The HTC, according to the literature [2, 67], shows a weak increase with the heat flux for a fixed vapor velocity v_v, while the departing radius $r_{\max}$ is not affected by the heat flux. Increasing the heat flux from 200 to 1000 kW m^{-2} leads to an augmentation of the HTC by 10%.

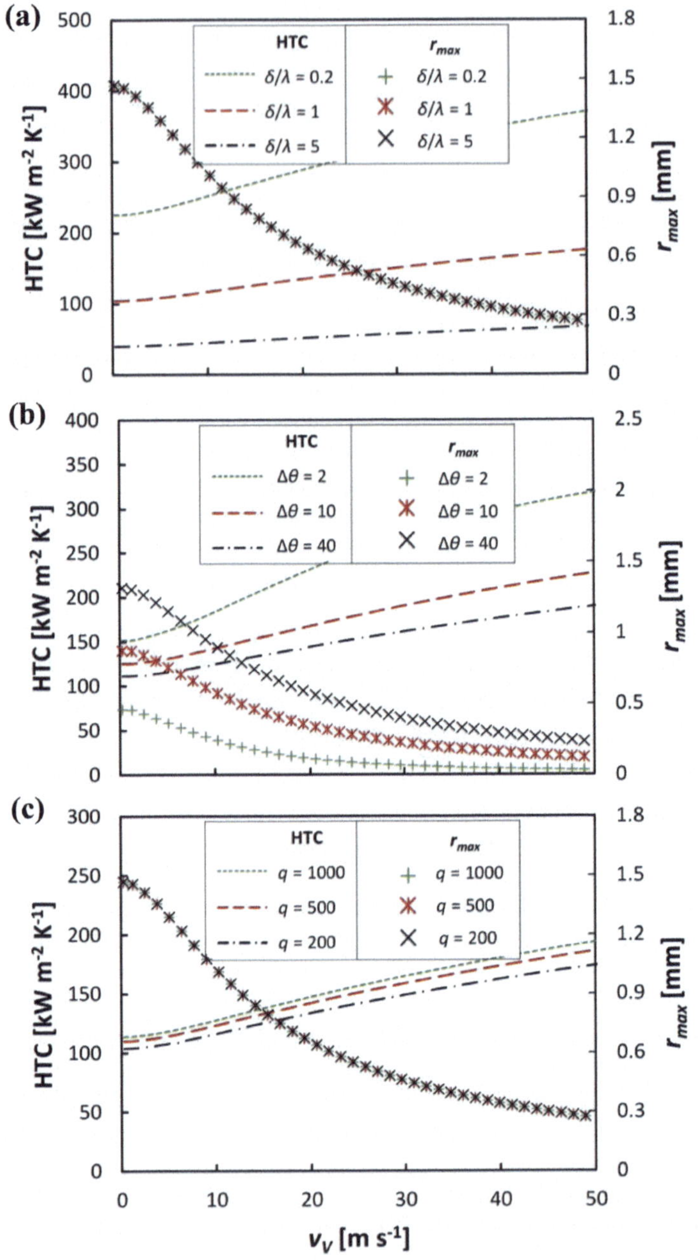

Fig. 3.6 Calculated HTC and droplet departing radius using the modified Miljkovic et al. [60] model with the expression by Tancon et al. [42] for vapor velocity: **a** effect of coating thermal resistance δ/λ [m^2 K MW^{-1}] ($\Delta\theta = 24°$ and $q = 335$ kW m^{-2}); **b** effect of contact angle hysteresis $\Delta\theta$ [°] ($q = 335$ kW m^{-2} and $\delta/\lambda = 1$ m^2 K MW^{-1}); **c** effect of heat flux q [kW m^{-2}] ($\Delta\theta = 24°$ and $\delta/\lambda = 1$ m^2 K MW^{-1})

In particular, it has been calculated from the modified Miljkovic et al. model that, for a given value of contact angle hysteresis ($\Delta\theta = 24°$), coating thermal resistance ($\delta/\lambda = 1$ m^2 K MW^{-1}) and heat flux ($q = 335$ kW m^{-2}), the vapor velocity must be increased from 1 m s^{-1} to 25 m s^{-1} to improve the HTC by 50%.

Figure 3.6 shows that, depending on the application, different strategies can be adopted to enhance the heat transfer during DWC. Increasing the vapor velocity positively affects the HTC, but the pressure drop will be higher. Other actions may address the promoter thermal resistance, by using advanced coating methods with very low thermal resistance to achieve HTC increase. Another approach regards the modification of surface wettability, developing coatings with very low contact angle hysteresis [68, 69].

3.7 Conclusions

The present Chapter is focused on measurements and modeling of heat transfer during dropwise condensation (DWC). Despite DWC was initially studied back in 1930, it is still capturing the interest of many researchers. This is mainly related to the recent advances in material sciences and the availability of experimental techniques that allow the investigation of DWC process at micro/nanoscale.

However, lessons learned from the past must be taken into consideration. Accurate heat transfer coefficient measurements that ensure reproducible results among different laboratories are very important. Particular care must be paid to the measurement of heat flux and surface temperature. Furthermore, the presence of non-condensable gases must be avoided when condensing pure vapor since it causes a strong reduction of the heat transfer coefficient.

Considering smooth surfaces, in the literature are available thin coatings that promote DWC with overall heat transfer coefficients (including the thermal resistance of the coating) up to 300 kW m^{-2} K^{-1} (and heat flux of 400 kW m^{-2}) with aluminum substrates. The duration of the coatings (especially in harsh environments as it is the case of steam condensation at atmospheric pressure) needs to be improved. In particular, it is important to develop more robust coatings that, at the same time, present a low thermal resistance.

Superhydrophobic surfaces can provide low contact angle hysteresis in air ambient conditions but, when employed during condensation tests with saturated steam, their heat transfer performance can be strongly penalized by flooding.

Different models have been proposed over the years to model the heat transfer during DWC. The small droplet population has been introduced and more complex models have been developed that can be used also with structured surfaces. These models consider DWC in absence of vapor velocity. The shape of the droplets and the droplet departing diameter are clearly affected by the vapor velocity. A modification proposed by the present authors to consider the effect of vapor velocity on droplet departing diameter and thus on the resulting heat transfer coefficient has been illustrated in the present Chapter. The model is then used to study the effect of the main parameters on DWC heat transfer. It results that around 60% of the total heat flux occurs in the droplets radius interval where conduction through the coating represents the dominant thermal resistance. This again confirms the importance of the thermal resistance of the coating. Furthermore, the nucleation site density remains one important parameter that must be set as a boundary condition for the models and its value can strongly affect the predicted values. Strategies devoted to increase the heat transfer coefficient during DWC on flat surfaces should combine low contact angle hysteresis, low thermal resistance of the coating and increased vapor velocity.

Acknowledgements This work was supported by the European Space Agency through the MAP Condensation program ENCOM (AO-2004-096).

References

1. P.G. de Gennes, F. Brochard-Wyart, D. Quere, Capillarity and Wetting Phenomena: Drops, Bubbles, Pearls, Waves, Springer New York, 2003.
2. J.W. Rose, Dropwise condensation theory and experiment: A review, Proc. Inst. Mech. Eng. Part A J. Power Energy. 216 (2002) 115–128. https://doi.org/10.1243/09576500260049034.
3. S. Khandekar, K. Muralidhar, K. Sameer, M. Krishnamurthy, Dropwise Condensation on Inclined Textured Surfaces, Springer, 2014. https://doi.org/10.1007/978-1-4614-8447-9.
4. E. Schmidt, W. Schurig, W. Sellschopp, Versuche über die Kondensation von Wasserdampf in Film- und Tropfenform, Tech. Mech. Und Thermodyn. 1 (1930) 53–63. https://doi.org/10.1007/BF02641051.
5. J.W. Rose, Personal reflections on fifty years of condensation heat transfer research, J. Enhanc. Heat Transf. 22 (2015) 89–120. https://doi.org/10.1615/JEnhHeatTransf.2015012451.
6. A. Bani Kananeh, M.H. Rausch, A.P. Fröba, A. Leipertz, Experimental study of dropwise condensation on plasma-ion implanted stainless steel tubes, Int. J. Heat Mass Transf. 49 (2006) 5018–5026. https://doi.org/10.1016/j.ijheatmasstransfer.2006.04.039.
7. A.B. Kananeh, M.H. Rausch, A. Leipertz, A.P. Fröba, Dropwise condensation heat transfer on plasma-ion-implanted small horizontal tube bundles, Heat Transf. Eng. 31 (2010) 821–828. https://doi.org/10.1080/01457630903547545.
8. W.T. Choi, K. Oh, P.M. Singh, V. Breedveld, D.W. Hess, Wettability control of stainless steel surfaces via evolution of intrinsic grain structures, J. Mater. Sci. 51 (2016) 5196–5206. https://doi.org/10.1007/s10853-016-9821-y.
9. N. Watanabe, M. Aritomi, Correlative relationship between geometric arrangement of drops in dropwise condensation and heat transfer coefficient, Int. J. Heat Mass Transf. 105 (2017) 597–609. https://doi.org/10.1016/j.ijheatmasstransfer.2016.09.104.
10. D. Torresin, M.K. Tiwari, D. Del Col, D. Poulikakos, Flow condensation on copper-based nanotextured superhydrophobic surfaces, Langmuir. 29 (2013) 840–848. https://doi.org/10.1021/la304389s.
11. M. Roudgar, J. De Coninck, Condensation heat transfer coefficient versus wettability, Appl. Surf. Sci. 338 (2015) 15–21. https://doi.org/10.1016/j.apsusc.2015.02.087.
12. R. Parin, M. Sturaro, S. Bortolin, A. Martucci, D. Del Col, Heat transfer during dropwise condensation of steam over a mirror polished sol-gel coated aluminum substrate, Int. J. Therm. Sci. 144 (2019) 93–106. https://doi.org/10.1016/j.ijthermalsci.2019.05.017.
13. M.E. Schrader, Wettability of clean metal surfaces, J. Colloid Interface Sci. 100 (1984) 372–380. https://doi.org/10.1016/0021-9797(84)90442-9.

14. J. Ma, H. Cha, M.K. Kim, D.G. Cahill, N. Miljkovic, Condensation Induced Delamination of Nanoscale Hydrophobic Films, Adv. Funct. Mater. 29 (2019) 1–10. https://doi.org/10.1002/adfm.201905222.
15. K.M. Holden, A.S. Wanniarachchi, P.J. Marto, D.H. Boone, J.W. Rose, The use of organic coatings to promote dropwise condensation of steam, J. Heat Transfer. 109 (1987) 768–774. https://doi.org/10.1115/1.3248156.
16. D.W. Tanner, C.J. Potter, D. Pope, D. West, Heat transfer in dropwise condensation-Part I The effects of heat flux, steam velocity and non-condensable gas concentration, Int. J. Heat Mass Transf. 8 (1965) 419–426. https://doi.org/10.1016/0017-9310(65)90005-0.
17. E.J. Le Fevre, J.W. Rose, An experimental study of heat transfer by dropwise condensation, Int. J. Heat Mass Transf. 8 (1965) 1117–1133. https://doi.org/10.1016/0017-9310(65)90139-0.
18. E. Citakoglu, J.W. Rose, Dropwise condensation-some factors influencing the validity of heat-transfer measurements, Int. J. Heat Mass Transf. 11 (1968) 523–537. https://doi.org/10.1016/0017-9310(68)90094-X.
19. A.W. Neumann, A.H. Abdelmessih, A. Hameed, The role of contact angles and contact angle hysteresis in dropwise condensation heat transfer, Int. J. Heat Mass Transf. 21 (1978) 947–953.https://doi.org/10.1016/0017-9310(78)90186-2.
20. D.E. Kim, H.S. Ahn, T.S. Kwon, Experimental investigation of filmwise and dropwise condensation inside transparent circular tubes, Appl. Therm. Eng. 110 (2017) 412–423. https://doi.org/10.1016/j.applthermaleng.2016.08.175.
21. R.W. Bonner, Dropwise condensation life testing of self assembled monolayers, 2010 14th Int. Heat Transf. Conf. IHTC 14. 2 (2010) 221–226. https://doi.org/10.1115/IHTC14-22936.
22. S. Vemuri, K.J. Kim, B.D. Wood, S. Govindaraju, T.W. Bell, Long term testing for dropwise condensation using self-assembled monolayer coatings of n-octadecyl mercaptan, Appl. Therm. Eng. 26 (2006) 421–429. https://doi.org/10.1016/j.applthermaleng.2005.05.022.
23. B.J. Zhang, C. Kuok, K.J. Kim, T. Hwang, H. Yoon, Dropwise steam condensation on various hydrophobic surfaces: Polyphenylene sulfide (PPS), polytetrafluoroethylene (PTFE), and self-assembled micro/nano silver (SAMS), Int. J. Heat Mass Transf. 89 (2015) 353–358. https://doi.org/10.1016/j.ijheatmasstransfer.2015.05.060.
24. K. Kim, J.H. Jeong, Steam condensate behavior and heat transfer performance on chromium-ion-implanted metal surfaces, Int. J. Heat Mass Transf. 136 (2019) 681–691. https://doi.org/10.1016/j.ijheatmasstransfer.2019.03.019.
25. G. Azimi, R. Dhiman, H.M. Kwon, A.T. Paxson, K.K. Varanasi, Hydrophobicity of rare-earth oxide ceramics, Nat. Mater. 12 (2013) 315–320. https://doi.org/10.1038/nmat3545.
26. J. Rafiee, X. Mi, H. Gullapalli, A. V. Thomas, F. Yavari, Y. Shi, P.M. Ajayan, N.A. Koratkar, Wetting transparency of graphene, Nat. Mater. 11 (2012) 217–222. https://doi.org/10.1038/nmat3228.
27. E. Colusso, M. Tancon, L. Cazzola, R. Parin, S. Agnoli, F. De Boni, M.G. Pelizzo, E. Della Gaspera, D. Del Col, A. Martucci, Solution-processed graphene oxide coatings for enhanced heat transfer during dropwise condensation of steam, Nano Sel. 2 (2021) 61–71. https://doi.org/10.1002/nano.202000105.
28. R. Parin, A. Martucci, M. Sturaro, S. Bortolin, M. Bersani, F. Carraro, D. Del Col, Nano-structured aluminum surfaces for dropwise condensation, Surf. Coatings Technol. 348 (2018) 1–12. https://doi.org/10.1016/j.surfcoat.2018.05.018.
29. A.T. Paxson, J.L. Yagüe, K.K. Gleason, K.K. Varanasi, Stable dropwise condensation for enhancing heat transfer via the initiated chemical vapor deposition (iCVD) of grafted polymer films, Adv. Mater. 26 (2014) 418–423. https://doi.org/10.1002/adma.201303065.
30. M.H. Rausch, A.P. Fröba, A. Leipertz, Dropwise condensation heat transfer on ion implanted aluminum surfaces, Int. J. Heat Mass Transf. 51 (2008) 1061–1070. https://doi.org/10.1016/j.ijheatmasstransfer.2006.05.047.
31. L. Liu, J. Zhao, Y. Zhang, F. Zhao, Y. Zhang, Fabrication of superhydrophobic surface by hierarchical growth of lotus-leaf-like boehmite on aluminum foil, J. Colloid Interface Sci. 358 (2011) 277–283. https://doi.org/10.1016/j.jcis.2011.02.036.

32. S. Chavan, H. Cha, D. Orejon, K. Nawaz, N. Singla, Y.F. Yeung, D. Park, D.H. Kang, Y. Chang, Y. Takata, N. Miljkovic, Heat Transfer through a Condensate Droplet on Hydrophobic and Nanostructured Superhydrophobic Surfaces, Langmuir. 32 (2016) 7774–7787. https://doi.org/10.1021/acs.langmuir.6b01903.
33. R. Parin, M. Tancon, M. Mirafiori, S. Bortolin, L. Moro, L. Zago, F. Carraro, A. Martucci, D. Del Col, Heat Transfer and Droplet Population During Dropwise Condensation on High Durability Coatings, Appl. Therm. Eng. 179 (2020) 115718. https://doi.org/10.1016/j.applthermaleng.2020.115718.
34. R. Parin, M. Rigon, S. Bortolin, A. Martucci, D. Del Col, Optimization of hybrid sol-gel coating for dropwise condensation of pure steam, Materials (Basel). 13 (2020). https://doi.org/10.3390/ma13040878.
35. D. Del Col, R. Parin, A. Bisetto, S. Bortolin, A. Martucci, Film condensation of steam flowing on a hydrophobic surface, Int. J. Heat Mass Transf. 107 (2017) 307–318. https://doi.org/10.1016/j.ijheatmasstransfer.2016.10.092.
36. S. Kim, K.J. Kim, Dropwise condensation modeling suitable for superhydrophobic surfaces, J. Heat Transfer. 133 (2011) 1–8. https://doi.org/10.1115/1.4003742.
37. R. Parin, D. Del Col, S. Bortolin, A. Martucci, Dropwise condensation over superhydrophobic aluminium surfaces, J. Phys. Conf. Ser. 745 (2016). https://doi.org/10.1088/1742-6596/745/3/032134.
38. D.J. Preston, D.L. Mafra, N. Miljkovic, J. Kong, E.N. Wang, Scalable graphene coatings for enhanced condensation heat transfer, Nano Lett. 15 (2015) 2902–2909. https://doi.org/10.1021/nl504628s.
39. H. Shigeo, T. Hiroaki, Dropwise condensation of steam at low pressures, Int. J. Heat Mass Transf. (1987). https://doi.org/10.1016/0017-9310(87)90264-X.
40. I. Tanasawa, Y. Utaka, Measurement of Condensation Curves for Dropwise Condensation Heat Transfer., Am. Soc. Mech. Eng. Appl. Mech. Div. AMD. 30 (1979) 63–68.
41. C.S. Sharma, C. Stamatopoulos, R. Suter, P.R. Von Rohr, D. Poulikakos, Rationally 3D-Textured Copper Surfaces for Laplace Pressure Imbalance-Induced Enhancement in Dropwise Condensation, ACS Appl. Mater. Interfaces. 10 (2018) 29127–29135. https://doi.org/10.1021/acsami.8b09067.
42. M. Tancon, R. Parin, S. Bortolin, A. Martucci, D. Del Col, Effect of steam velocity during dropwise condensation, Int. J. Heat Mass Transf. 165 (2021) 120624. https://doi.org/10.1016/j.ijheatmasstransfer.2020.120624.
43. R. Rioboo, I. Demnati, M. Amin Ali, R. Sevkan, J. De Coninck, Superhydrophobicity of composite surfaces created from polymer blends, J. Colloid Interface Sci. 560 (2020) 596–605. https://doi.org/10.1016/j.jcis.2019.10.043.
44. D. Mangini, C. Antonini, M. Marengo, A. Amirfazli, Runback ice formation mechanism on hydrophilic and superhydrophobic surfaces, Cold Reg. Sci. Technol. 109 (2015) 53–60. https://doi.org/10.1016/j.coldregions.2014.09.012.
45. R.N. Wenzel, Resistance of solid surfaces to wetting by water, Ind. Eng. Chem. (1936). https://doi.org/10.1021/ie50320a024.
46. A.B.D. Cassie, S. Baxter, Wettability of porous surfaces, Trans. Faraday Soc. (1944). https://doi.org/10.1039/tf9444000546.
47. R. Enright, N. Miljkovic, J.L. Alvarado, K. Kim, J.W. Rose, Dropwise condensation on micro- and nanostructured surfaces, Nanoscale Microscale Thermophys. Eng. 18 (2014) 223–250. https://doi.org/10.1080/15567265.2013.862889.
48. N. Miljkovic, R. Enright, Y. Nam, K. Lopez, N. Dou, J. Sack, E.N. Wang, Jumping-droplet-enhanced condensation on scalable superhydrophobic nanostructured surfaces, Nano Lett. 13 (2013) 179–187. https://doi.org/10.1021/nl303835d.
49. R. Enright, N. Miljkovic, A. Al-Obeidi, C. V. Thompson, E.N. Wang, Condensation on superhydrophobic surfaces: The role of local energy barriers and structure length scale, Langmuir. 28 (2012) 14424–14432. https://doi.org/10.1021/la302599n.
50. L. Zhong, M. Xuehu, W. Sifang, W. Mingzhe, L. Xiaonan, Effects of surface free energy and nanostructures on dropwise condensation, Chem. Eng. J. 156 (2010) 546–552. https://doi.org/10.1016/j.cej.2009.04.007.

51. N. Miljkovic, D.J. Preston, E.N. Wang, Recent Developments in Altered Wettability for Enhancing Condensation, in: J.R. Thome (Ed.), Encycl. Two-Phase Heat Transf. Flow II, First, World Scientific, 2015: pp. 85–131. https://doi.org/10.1142/9311-vol3.
52. S. Sett, P. Sokalski, K. Boyina, L. Li, K.F. Rabbi, H. Auby, T. Foulkes, A. Mahvi, G. Barac, L.W. Bolton, N. Miljkovic, Stable Dropwise Condensation of Ethanol and Hexane on Rationally Designed Ultrascalable Nanostructured Lubricant-Infused Surfaces, Nano Lett. 19 (2019) 5287–5296. https://doi.org/10.1021/acs.nanolett.9b01754.
53. K. Khalil, D. Soto, T. Farnham, A. Paxson, A.U. Katmis, K. Gleason, K.K. Varanasi, Grafted Nanofilms Promote Dropwise Condensation of Low-Surface-Tension Fluids for High-Performance Heat Exchangers, Joule. 3 (2019) 1377–1388. https://doi.org/10.1016/j.joule.2019.04.009.
54. A. Bisetto, S. Bortolin, D. Del Col, Experimental analysis of steam condensation over conventional and superhydrophilic vertical surfaces, Exp. Therm. Fluid Sci. 68 (2015) 216–227. https://doi.org/10.1016/j.expthermflusci.2015.04.019.
55. D. Del Col, S. Bortolin, M. Azzolin, Measuring Heat Transfer Coefficient during Condensation Inside Channels, in: J.P. Meyer, M. De Paepe (Eds.), Art Meas. Therm. Sci., First, CRC Press, Taylor & Francis Group, LLC, Boca Raton, FL and Abingdon, UK, 2021: pp. 385–407.
56. R. Kempers, A. Robinson, Heated Meter Bar Techniques: What You Should Know and Why, in: J. Meyer, M. De Paepe (Eds.), Art Meas. Therm. Sci., First, CRC Press, Taylor & Francis Group, LLC, Boca Raton, FL and Abingdon, UK, 2021: pp. 313–335.
57. E.J. Le Fevre, J.W. Rose, A theory of heat transfer by dropwise condensation, in: Proc. 3rd Int. Heat Transf. Conf. Vol. 2, 1966.
58. W.H. Wu, J.R. Maa, On the Heat Transfer in Dropwise Condensation, Chem. Eng. J. 12 (1976) 225–231.
59. H. Tanaka, A theoretical study of dropwise condensation, J. Heat Transfer. 97 (1975) 72–78. https://doi.org/10.1115/1.3450291.
60. N. Miljkovic, R. Enright, E.N. Wang, Modeling and optimization of superhydrophobic condensation, J. Heat Transfer. 135 (2013) 111004. https://doi.org/10.1115/1.4024597.
61. Van P. Carey, Liquid Vapor Phase Change Phenomena, An Introduction to the Thermophysics of Vaporization and Condensation Processes in Heat Transfer Equipment, Hemisphere Pub. Corp., Washington, DC, 1992: pp. 399–452.
62. X. Chen, M.M. Derby, Droplet departure modeling and a heat transfer correlation for dropwise flow condensation in hydrophobic mini-channels, Int. J. Heat Mass Transf. 125 (2018) 1096–1104. https://doi.org/10.1016/j.ijheatmasstransfer.2018.04.140.
63. A. Umur, P. Griffith, Mechanism of Dropwise Condensation, J. Heat Transfer. 87 (1965) 275–282.
64. P. Birbarah, S. Chavan, N. Miljkovic, Numerical Simulation of Jumping Droplet Condensation, Langmuir. 35 (2019) 10309–10321. https://doi.org/10.1021/acs.langmuir.9b01253.
65. A.I. ElSherbini, A.M. Jacobi, Retention forces and contact angles for critical liquid drops on non-horizontal surfaces, J. Colloid Interface Sci. 299 (2006) 841–849. https://doi.org/10.1016/j.jcis.2006.02.018.
66. G.K. Batchelor, An Introduction to Fluid Dynamics, Cambridge University Press, 2000. https://doi.org/10.1017/cbo9780511800955.
67. B.S. Sikarwar, S. Khandekar, S. Agrawal, S. Kumar, K. Muralidhar, Dropwise condensation studies on multiple scales, Heat Transf. Eng. 33 (2012) 301–341. https://doi.org/10.1080/01457632.2012.611463.
68. S. Sett, X. Yan, G. Barac, L. Bolton, N. Miljkovic, Lubricant-Infused Surfaces for Low Surface Tension Fluids: Promise vs Reality, ACS Appl. Mater. Interfaces. (2017) acsami.7b10756. https://doi.org/10.1021/acsami.7b10756.
69. D.J. Preston, Z. Lu, Y. Song, Y. Zhao, K.L. Wilke, D.S. Antao, M. Louis, E.N. Wang, Heat Transfer Enhancement during Water and Hydrocarbon Condensation on Lubricant Infused Surfaces, Sci. Rep. 8 (2018) 1–9.https://doi.org/10.1038/s41598-017-18955-x.

Chapter 4
About Phenomenology and Modeling of Dropwise Condensation

J. Lethuillier, P. Lavieille, F. Topin, and M. Miscevic

Abstract The modeling of dropwise condensation remains, at the present time, difficult. Indeed, in order to predict the heat transfers in this regime, it is necessary to know the heat flux which crosses each drop according to its size as well as the drop-size distribution on the surface considering radii that can vary from a few nanometers to several centimeters. As this distribution is a function of the life cycle of each drop, an overview of a drop's lifecycle has been first done in order to better understand the phenomena underlying dropwise condensation. Particular attention has been paid on the drop growth rate modeling. Secondly, a description of the drop-size distribution models has been done. Due to the very large number of drops, very fast dynamics and the difference in drop-sizes, only two types of modeling are available. The first approach is based on a semi-empirical law to model the distribution of the largest drops (i.e., those with a radius greater than a few microns) together with a population balance for the size distribution of the smallest drops. The second approach consists in the following of all the drops along time in order to determine the stationary drop-size distribution. This approach already succeed to predict the size distribution of the big drops many times. Based of these overviews, an individual-based modeling has been developed and computed, focusing on the behavior of the smallest droplets. A comparison of the results obtained with this model with the ones obtained with classical population-balanced approach has then be realized. Discrepancies of several orders of magnitude have been found on drop-size distribution. This important difference is attributed to one of the hypothesis of population balance modeling, i.e., the hypothesis of constant renewal rate whatever the drop radius. The impact of such deviations in the drop-size distribution on global heat transfer has then be quantified. In most of the configurations studied, the population balance approach predicts global heat fluxes about 30% higher compared to the individual-based model's ones. Finally, a parametric study has been done considering three parameters that can be potentially controlled in experimental works: advancing contact angle, nucleation sites density, and departure radius.

J. Lethuillier · P. Lavieille · F. Topin · M. Miscevic (✉)
LAPLACE, UMR CNRS-INP-UPS n°5213, University of Toulouse, Toulouse, France
e-mail: marc.miscevic@laplace.univ-tlse.fr

F. Topin
IUSTI, UMR CNRS-AMU n°7343, Aix-Marseille Université, Marseille, France

M. Marengo and J. De Coninck (eds.), *The Surface Wettability Effect on Phase Change*,
https://doi.org/10.1007/978-3-030-82992-6_4

4.1 Dropwise Condensation: An Effective Way to Transfer Heat

Two regimes of condensation can exist according to the experimental configuration. The first one, commonly encountered in heat exchangers, is filmwise condensation. In that regime, the condensates form a continuous film that cover the cold substrate. In the second regime, i.e., dropwise regime, the condensates form small droplets separated from each other. The latter regime can appear in some portions of space and/or time, but is often reported in literature as difficult to maintain on an extended surface during long time.

About 90 years ago, Schmidt et al. [41] were the first ones to demonstrate experimentally that dropwise condensation leads to very high heat transfer coefficient values, up to one order of magnitude higher than the heat transfer coefficient in filmwise condensation.

This observation has then been experimentally confirmed several times during the last half century [34, 39]. For instance, very recently, heat transfer coefficient up to 250 kW m^{-2} K^{-1} have been measured by Parin [35] on hydrophobic aluminum substrate with 0.2 μm silica layer thickness.

The main reasons generally given by authors in literature to explain such a high value of the heat transfer coefficient is the very small dimension of the drops that appear on the cooled substrate, as well as their big number.

Thus, the transport mechanisms (heat transfer, fluid flow, phase change) should be identified and understood at both the drop scale and the macroscale (i.e., the scale of whole the population of drops). So, individual drop's lifecycle as well as resulting population distribution must be analyzed and modeled.

4.1.1 The Drop's Lifecycle

The lifecycle of a single drop can be divided into four main stages: the drop is born, then grows due to condensation, interacts with other drops and disappears. A phenomenology overview of each of these four stages is proposed in the following paragraphs along with summary of some geometrical characteristics of drops and a literature brief about associated heat transfer models.

4.1.1.1 The Drop's Birth

Following the work of Schmidt et al. in 1930 [41], a first mechanism was proposed for the drops birth: it form from a very thin liquid film (usually not visible) adsorbed on the surface. Jacob in 1936 [20] postulated the rupture of such a liquid film due to hydrodynamic instabilities, leading to the appearance of droplets with a spherical cap shape to minimize the surface energy.

A second mechanism was also proposed nearly at the same time: Von Euken [16], using considerations about the heat transfer rate and the thermal resistance due to an adsorbed film, attributed the formation of drops to a mechanism of nucleation, i.e., the formation of droplets at atomic scale followed by their growth at continuum level. Almost 30 years later, Umur and Griffith [50] showed experimentally that no film exists between the drops and confirmed the nucleation mechanism model.

Classical theory of nucleation considers the Gibbs free energy variation to determine the equilibrium radius r_{eq} of a drop that can appear in an isothermal subcooled vapor (see for example [9]):

$$r_{\min} \approx r_{eq} = \frac{2\sigma_{lv}}{\Delta T} \frac{T_{sat}}{\rho_v \mathrm{L}_{lv}} \tag{4.1}$$

This equilibrium is unstable: a drop forming with a radius less than r_{eq} will quasi-instantaneously disappear while a drop forming with a radius greater than r_{eq} will grow by vapor condensation process. So, the minimum radius $r_{\min}$ of liquid nuclei forming in the subcooled vapor is expected to be close to r_{eq}.
In the presence of a solid substrate, the probability to form a cluster of molecules having a liquid–vapor interface with a curvature radius greater than the equilibrium radius r_{eq} is higher in the case of a spherical cap rather than in the case of a sphere, due to the reduction of the number of molecules involved. In the case of a "real" surface, i.e., a surface with roughnesses, small amount of liquid can be trapped in the pits, forming pre-existing embryos. The minimum radius of nuclei is then governed by the characteristic size of those pits. The correlation with the subcooling of the solid wall can thus be evaluated using Eq. 4.1, in which ΔT is the temperature difference between the saturation temperature and the wall temperature ("wall subcooling"). In case of heterogeneous nucleation, the initial nuclei of radius $r > r_{eq}$ that formed on specific locations grow and form spherical caps whose volume is a function of the wettability of the liquid on the solid substrate:

$$V = \frac{4}{3}\pi r_{eq}^3 \times \left(\frac{2 - 3\cos\theta + \cos^3\theta}{4}\right) \tag{4.2}$$

It can be noticed that the expression of the radius r_{eq} in Eq. 4.1 is derived assuming an isothermal vapor phase. Several studies have been conducted to take into account temperature gradients in the continuous vapor phase close to the solid wall [19] or within the liquid droplet [24]. As the value of $r_{\min}$ is generally very small (in the order of few nanometers to few tens of nanometers) the effect of temperature gradients on the thermodynamic equilibrium's change is often negligible, at least for configurations other than superhydrophobics. For instance, Liu and Cheng [24] found that the minimum radius $r_{\min}$ is affected by temperature gradients only for case with simultaneously high values of the contact angle ($\theta = 150°$) and low wall subcooling (less than 1.5 °C).

Mc Cormick and Westwater [29] have shown experimentally that the nuclei appear on specific locations on a substrate. These authors repeated condensation experiments on two copper disks covered by an adsorbed monolayer of benzyl mercaptan, leading to advancing contact angle of 92°. They observed that 21 previously identified nucleation sites were systematically activated again during each of the 8 following condensation experiments. So, nucleation occurs on very specific places on the surface, indicating that nucleation sites are located on surface accidents, such as cavities or pits. This conclusion was confirmed by doing specific condensation experiments on a substrate on which artificial nucleation sites were made by a spark erosion technique. Observations were then focused on the artificially produced pits during five successive condensations, clearly showing that each of these artificially produced pits were active sites for the formation of drops.

As droplets appear on specific nucleation sites, the number of nuclei can vary greatly depending on the substrate nature or microstructures (inclusions, pits, cracks, etc.). A first evaluation of the nucleation sites density was proposed by Rose [37] in 1976. Considering the drop-size distribution proposed several years earlier [22] and making some assumptions on the mean radius of droplets at the first coalescence event, he found

$$N_s = \frac{0.037}{r_{\min}^2} \tag{4.3}$$

where $r_{\min}$ is the radius of the nuclei estimated using Eq. 4.1.

It can be highlighted that this law was determined considering drop-size distribution's law established for droplet radii typically greater than few microns. The minimum radius of liquid embryos according to Eq. 4.1 is about two orders of magnitude less. So, evaluating N_s with Eq. 4.3 may lead to significant discrepancies with real values, as pointed out for instance by Liu and Cheng [24].

From a study to another, the reported nucleation sites density values are spread on 6 orders of magnitude, i.e., from 10^9 to 10^{15} m^{-2} [2]. This huge difference in the reported nucleation sites density highlights the difficulty to predict its value accurately, as well as the strong influence of the surface microscale geometry. Moreover, several other parameters may also affect the density of active sites N_s such as surface subcooling, substrate history, etc., or the detection threshold linked to the magnification of the diagnostic tool used to detect the smallest droplets. Consequently, the accurate determination of the nucleation sites density remains, nowadays, a very challenging problem and this parameter will receive a special attention in the following.

4.1.1.2 The Growth of Drop by Condensation Process

Following the nuclei appearance and, due to the temperature difference between the vapor phase and the subcooled substrate, the vapor condenses at the liquid–vapor interfaces inducing the drops growing. Fatica and Katz [18] were the first to propose a heat transfer model through a single droplet. They proposed that drop growth occurs by condensation on drop interface with the latent heat conducted through the drop to the solid surface. They developed a stationary heat transfer model based on axisymmetric heat flux configuration and derived a semi-analytic expression of drop thermal resistance. They already remarked that, due to the low thermal conductivity of the liquid and spherical cap shape, the heat flux is low except close to the triple line. Based on this work, several theoretical studies have been conducted in which the following assumptions are considered:

- The drop-sizes considered being much smaller than the capillary length $l_c = \sqrt{\frac{\sigma_{lv}}{\rho_l g}}$, the drops are supposed to have a spherical cap shape;
- The substrate's wall is isothermal;
- The gas phase is constituted by pure vapor (without non-condensable gas);
- The only heat transfer mechanism into the drop is thermal conduction (without taking into account thermal inertia);
- The liquid phase is immobile inside the drop (no convection nor Marangoni effect).

In 1966, Le Fevre Rose [22] extended this 1D quasi-static heat transfer model through a single drop by adding several terms. Their model is then constituted by four thermal resistances placed in series between the vapor phase and the substrate's wall, which involve four temperature jumps:

$$\Delta T = T_{sat} - T_s = \Delta T^i + \Delta T_{curv} + \Delta T_l + \Delta T_{coat} \tag{4.4}$$

These temperature jumps are respectively linked to:

- For ΔT^i: the vapor–liquid interfacial thermal resistance,
- For ΔT_{curv}: the modification of the saturation conditions induced by the curvature of the interface,
- For ΔT_l: the thermal conduction in the liquid within the droplet,
- For ΔT_{coat}: the thermal conduction in the coating of the substrate.

It can be noticed that the model of Le Fevre and Rose considered only hemispherical drops (and thus a contact angle of 90°), and was expressed using 2 constants that gather both shape and thermophysical parameters.

This model was used by Wen and Jer [52] in theoretical developments aimed to determine the drop-size distribution of the small droplets, and the macroscopic heat transfer coefficient. In their model, the effect of the coating thermal resistance was not taken into account. This latter was then reintroduced by Abu-Orabi [1]. About 10 years ago, Kim and Kim [21] modified the expression of the drop conductive thermal resistance in order to calculate it for an arbitrary contact angle on a hydrophobic surface. Starting from a constant interface temperature hypothesis they imposed that isotherms are spherical caps passing by triple line and whose curvature radii increase from interface to solid surface. They considered that the average distance between two consecutive isotherms is half of their max distance and calculated the drop temperature difference considering that the mean heat flux through the drop is proportional to this average distance. This liquid thermal resistance expression combined with Eq. 4.4 is currently the most widely used model in the literature, even in the case of hydrophilic surface. The most commonly used expressions of the four temperature jumps are nowadays expressed as

$$\Delta T_l = \frac{Q_d \theta_{adv}}{4\pi r \mathrm{k}_l \sin\theta_{adv}} \tag{4.5}$$

$$\Delta T^i = \frac{Q_d}{2\pi r^2 h^i (1 - \cos\theta_{adv})} \tag{4.6}$$

with h^i the heat transfer coefficient at the liquid–vapor interface, calculated from the kinetic model of Schrage [42]:

$$h^i = \frac{2f}{2-f} \frac{1}{\sqrt{2\pi R_g T_{sat}}} \frac{\rho_v \mathrm{L}_{lv}^2}{T_{sat}} \tag{4.7}$$

where f is the condensation coefficient corresponding to the ratio between the rate of molecules that cross the liquid–vapor interface by the total rate of molecules that hit the interface. Its value is generally set to 1 for pure water vapor close to the atmospheric pressure but could vary over several decades depending on conditions or fluid nature. For water, Tanasawa [46] reported values for h^i of 0.383, 2.57, and 15.7 MW m^{-2} K^{-1} for vapor pressure of 0.01, 0.1, and 1 bar, respectively.

From thermodynamic equilibrium considerations, the modification of the saturation conditions due to the curvature of the interface can be expressed as

$$\Delta T_{curv} = \frac{2T_{sat}\sigma_{lv}}{\mathrm{L}_{lv}\rho_v \Delta T} = \frac{r_{\min}}{r}\Delta T \tag{4.8}$$

Finally, the temperature jump due to the thermal conduction through the coating is simply

$$\Delta T_{coat} = \frac{Q_d \delta_{coat}}{\mathrm{k}_{coat}\pi r^2 \sin^2\theta_{adv}} \tag{4.9}$$

For illustration, a comparison of the relative contribution of these four terms to the global temperature jump is shown on Fig. 4.1 in the case of pure water at atmospheric pressure (f = 1), considering an advancing contact angle of 85°, a wall subcooling of 1 °C and with typical values of the thickness and the thermal conductivity of a coating. As expected, the curvature effect is significant for the very small drops whereas the thermal conduction effect is predominant for the biggest drops. The temperature jump due to the interfacial resistance is very low for the selected conditions, as f has been set to one. Obviously, for smaller values of f this resistance may becomes dominant as discussed in the following. Due to a simple geometrical effect (drop is a spherical cap while coating is considered as a wall limited by 2 isothermal surfaces), the temperature jump created by the coating resistance presents an asymmetric bell shape with respect to drop radius. Here, the coating thermal resistance is rather important due to the low conductivity of this layer. For the selected conditions, it has a marked effect for drops radius around 100 nanometers.

In the case of a small value of f (i.e., for low-pressure conditions) the distribution of the temperature jumps is very different (Fig. 4.2). For f = 0.01, the temperature

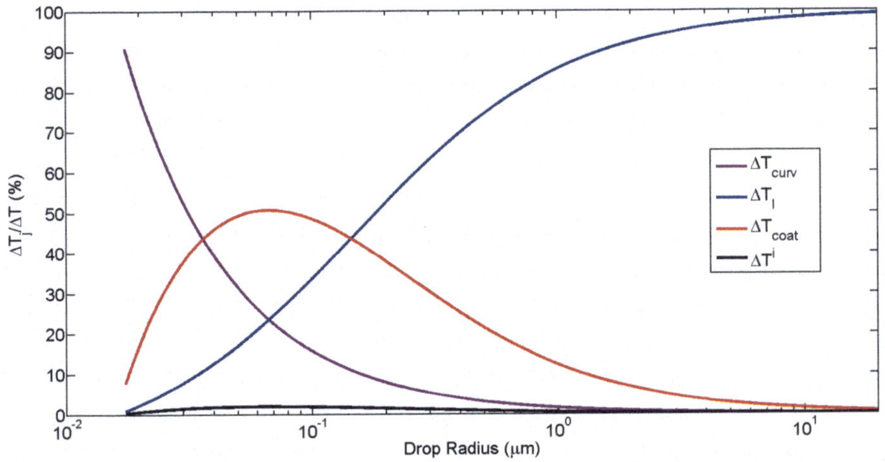

Fig. 4.1 Variation of the dimensionless temperature jumps according to the curvature drop radius in the case of pure water for f = 1 ($\theta_{adv} = 85°$, $\Delta T = 1$ K, $T_{sat} = 373$ K, $\delta_{coat} = 100$ nm, $k_{coat} = 2$ W m^{-1} K^{-1}, and $h^i = 15.7$ MW m^{-2} K^{-1} (f = 1))

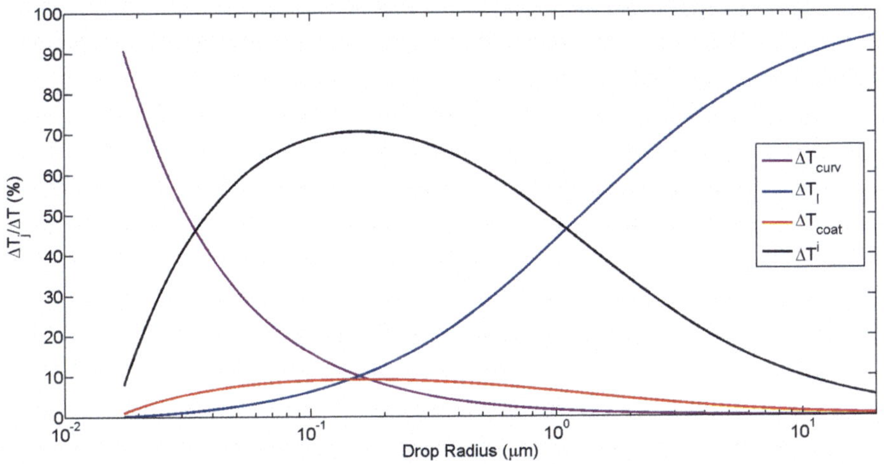

Fig. 4.2 Variation of the dimensionless temperature jumps according to the curvature drop radius in the case of pure water for f = 0.01 ($\theta_{adv} = 85°$, $\Delta T = 1$ K, $T_{sat} = 373$ K, $\delta_{coat} = 100$ nm, $k_{coat} = 2$ W m^{-1} K^{-1}, and $h^i = 78.9$ kW m^{-2} K^{-1} (f = 0.01))

jump at the interface represents up to 70% of the total temperature jump for $r \approx 0.2$ μm and remains non-negligible for radii up to about 10 μm.

Using Eqs. 4.4–4.9 and rearranging the expressions of the temperature jumps, the heat transfer rate Q_d through a single drop is commonly expressed as

$$Q_d = \frac{\Delta T \pi r^2 (1 - \frac{r_{min}}{r})}{\frac{\delta_{coat}}{\mathrm{k}_{coat} \sin^2 \theta_{adv}} + \frac{r\theta_{adv}}{4\mathrm{k}_l \sin\theta_{adv}} + \frac{1}{2h^i(1-cos\theta_{adv})}} \tag{4.10}$$

The growth rate $G = \frac{dr}{dt}$ of any given drop is usually deduced from a simplified quasi-static energy balance using Eq. 4.2 for the expression of the volume of the drop:

$$Q_d = \frac{d}{dt}\left(\rho_l \frac{4}{3}\pi r^3 \left(\frac{2 - 3\cos\theta_{adv} + \cos^3\theta_{adv}}{4}\right)\right)\mathrm{L}_{lv} = \rho_l \mathrm{L}_{lv} \pi r^2 (2 - 3\cos\theta_{adv} + \cos^3\theta_{adv})G \tag{4.11}$$

As already pointed out by Miljkovic et al. [31], this model is valid for both hydrophobic and hydrophilic cases as the volume of the drop is calculated exactly for a spherical cap of any contact angle. On the other hand, the expression used by Kim and Kim [21], based on an approximate rate of volume variation, is valid only for hydrophobic configurations.

The growth rate of a drop can then be deduced from Eqs. 4.10 and 4.11:

$$G = \frac{dr}{dt} = \frac{A_1(1 - \frac{r_{min}}{r})}{A_2 r + A_3} \tag{4.12}$$

with:

$$A_1 = \frac{\Delta T}{\rho \mathrm{L}_{lv}(2 - 3\cos\theta_{adv} + \cos^3\theta_{adv})} \tag{4.13}$$

$$A_2 = \frac{\theta}{4\mathrm{k}_l \sin\theta_{adv}} \tag{4.14}$$

$$A_3 = \frac{\delta_{coat}}{\mathrm{k}_{coat} \sin^2\theta_{adv}} + \frac{1}{2h^i(1 - \cos\theta_{adv})} \tag{4.15}$$

An example of the growth dynamic of a single drop is reported on Fig. 4.3 for different advancing contact angles and the same other parameters than the ones considered in Fig. 4.1. For all cases, after a small delay, the drop starts to grow in a slightly S-shaped manner that tends to a linear asymptote (in log–log representation). The asymptotes are roughly parallel (for high drop radii the conduction effect within the drops is prevalent) and the curves are translated to the right according to the contact angle. The growth delay at small radius increases with the contact angle as the curvature effect (in particular) reduces the growth rate at low radii. More the surface is hydrophilic, smaller is the time necessary to reach a given radius. For

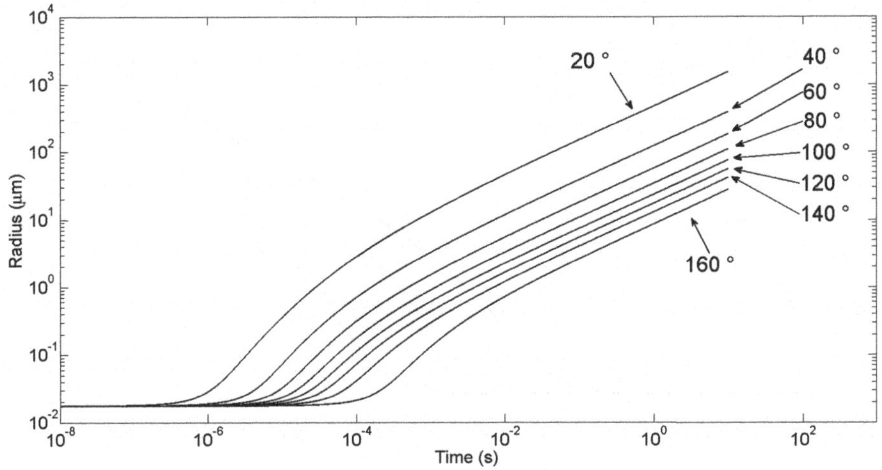

Fig. 4.3 Evolution of drop curvature radius according to time for different advancing contact angles ($\Delta T = 1$ K, $T_{sat} = 373$ K, $\delta_{coat} = 100$ nm, $k_{coat} = 2$ W m^{-1} K^{-1}, and $h^i = 15.7$ MW m^{-2} K^{-1} (f = 1))

example, it will take about 10^{-3} s for a nucleus to grow up to 10 µm for an advancing contact angle of 20°, while this time rises to more than 10 s for an advancing contact angle of 160°. This difference is mainly due to the thermal conduction resistance through the drop, that is more important for high contact angle. For a given curvature radius, drops on a hydrophobic substrate present bigger liquid volumes than on a hydrophilic one. As the total volume needed to increase the drop radius is higher in hydrophobic configuration, a more long time is necessary to condense.

So, in order to maximize the heat transfer, it appears at first that it is more interesting to use hydrophilic surface. However, in that case, a major issue will be the evacuation of the biggest drops. This evacuation is mandatory in order to renew the nucleation sites (see Sect. 4.1.1.4) and to avoid the formation of a continuous liquid film.

The second main parameter which may have a strong effect on the growth dynamic is the condensation coefficient f, especially in the case it has a small value (Fig. 4.4). Indeed, in the case of pure water (i.e., without non-condensable gas) it is usually set to one at atmospheric pressure but it can be less at lower pressure [4, 28] (it can be noticed that even a small amount of non-condensable gas can change significantly the value of the interfacial heat transfer coefficient h^i). So, the precise determination of f remains mandatory to well predict the heat transfer rate through small droplets. For instance, according to Fig. 4.4, the time necessary to reach a radius of 1 µm is more than one order of magnitude higher for f = 0.01 than for f = 1.

Experimentally, this individual growth rate model is difficult to validate, because when a drop becomes observable (i.e., typically when it reaches a radius of about a few microns) it has already coalesced a great number of times with other non-observable drops; the growth rate is no more governed only by condensation process. So, in

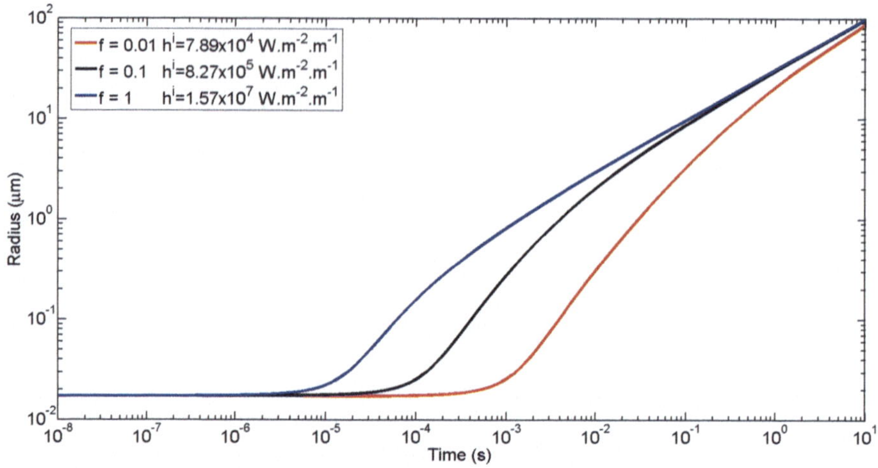

Fig. 4.4 Evolution of drop radius according to time for different condensation coefficients ($\theta_{adv} = 85°$, $\Delta T = 1$ K, $T_{sat} = 373$ K, $\delta_{coat} = 100$ nm, $k_{coat} = 2$ W m^{-1} K^{-1} and $h^i = 15.7$ MW m^{-2} K^{-1} (f = 1))

order to obtain reference data of the growth rate of a single droplet, some authors have conducted numerical works.

By order of complexity increases, the available approaches include:

- Neumann boundary condition on drop liquid–vapor interface (in replacement of classic imposed temperature one), along with simple quasi-static conduction and Dirichlet condition on the liquid–solid interface,
- Thermocapillary convection considering the flow inside the drop in quasi-static regime,
- The dynamic of the growth (i.e., the deformation of the liquid drop with time) with transient convection and diffusion phenomena.

For example, Phadnis and Rykaczewski [36] analyzed more particularly the effect of stationary Marangoni convection on heat transfer in a single droplet using finite elements method. They compared 2 numerical situations for static drops (the condensation dynamic process was not considered). In the first one, they computed a pure conduction case with a convective boundary condition at the liquid–vapor interface and in the second one, they took into consideration the Marangoni effect. They obtained in this latter case an increase in the heat transfer rate by a factor up to 6 (compared to the case without Marangoni effect) in the case of a single drop of radius 1 mm. Nevertheless, as the increase is far more limited for the smaller drops and—see below—the big drops are rare, the global heat transfer rate increase is found to be limited: they reported an increase of less than 30% in this global heat transfer coefficient compared to the pure conduction case.

The transient resolution of the drop growth dynamic was only recently achieved due mostly to the important computational resources involved in such shape evolving coupled non linear problem. Xu and et al. [55] solved the heat transfer problem in the liquid phase coupled to the flow problem in both vapor and liquid phases, taking into account interfacial mass transfer and drop deformation (prediction of interface motion along with liquid domain deformation). This comprehensive model was used to study the impact of fluid motion inside the drop on heat transfer. The authors shown that flow pattern in growing drop differs strongly from the quasi-static ones for pure vapor condensation. An increase up to a factor 4 in individual drop heat transfer compared to pure conduction case was obtained for large drops. They also showed that the mass flow through the liquid–vapor interface is the dominant factor responsible of the strong convection. The critical radius where convection starts to have a significant influence on droplet growth was then determined for different subcooling temperatures and contact angles. The criterion chosen to define this critical radius was a difference in the heat transfer rate greater than 5% compared to the pure conduction hypothesis. For contact angles from $\theta_{adv} = 90°$ to $\theta_{adv} = 140°$ and sub-cooling between 1 and 7° C, the critical radius ranges from 0.5 to 20 μm. Droplets have smaller critical radii under larger subcooling temperature or larger contact angle.

Moreover, no model are yet available taking into account the thermal conduction into the substrate and the direct heat transfers between the vapor and the substrate. However, taking them into account could lead to non-negligible modifications of both the temperature field near the droplet and the heat flux at/near the triple line. This point remains to be analyzed more deeply in future works.

It is now clear that simple models that lead to explicit and tractable expressions of drop growth rates are probably oversimplified and may present bias or give unrealistic growth rate values. On the other hand, comprehensive models, even if they are convenient for the detailed study of a single drop, have not yet allow to generate simple expressions of heat and mass transfers at drop scale.

4.1.1.3 The Interactions with Other Drops

As they grow or if they move, the drops interact with many other drops and merge with them. The coalescence of two drops begins when they come into contact, the point of contact being located on the substrate, in the triple line region, for hydrophilic cases, while it is at a given height for hydrophobic configurations. Calling d_{ij} the minimum distance between liquid–vapor interfaces of drop i and drop j, the contact criterion can be written as follows for hydrophilic and hydrophobic surfaces, respectively:

$$d_{ij} - (r_i + r_j) \sin \theta_{adv} \leq 0 \tag{4.16}$$

$$d_{ij} - \sqrt{(r_i + r_j)^2 - (r_i - r_j)^2 \cos^2 \theta_{adv}} \leq 0 \tag{4.17}$$

Generally, the drop-size distribution covers about 7 orders of magnitude, namely from a few nanometers to several centimeters. So, coalescence events imply mostly drops of very different radii and thus, huge differences in volume (volume distribution can be spread on more than 20 orders of magnitude). For sake of simplicity, let us consider first the coalescence of two identical drops. The two drop surfaces touch each other at the contact point and a liquid bridge forms. The surface enclosing this bridge is curved differently than the remaining of the drops surface: in the liquid bridge zone, the surface tension force differs and the liquid starts to move toward the contact zone resulting in the growth of the liquid bridge and leading eventually to an elongated ellipsoidal shape of the resulting drop. This latter will then progressively return to a spherical cap shape in absence of external forces or for small enough drops. Three main forces drive the liquid behavior and thus the coalescence duration:

- The interfacial tension force that induces liquid pressure gradients and thus liquid movement such as minimizing the interfacial area, regularizing the curvature and accommodating static or dynamic contact angle constraints;
- The viscous force that impedes fluid flow;
- Inertia force that limits fluid accelerations.

The two latter effects act as damping factor and thus limit the fluid displacement. As the main driving force acts on drop interface, it is thus expected, as it has been experimentally observed [5, 10], that the center of mass of the resulting drop will be located at the center of mass of the parent droplets.

In the case of an asymmetric coalescence event between two droplets of different radii, an additional effect takes place. Indeed, the internal pressure in the two drops are different. As the liquid pressure varies like the inverse of the radius, it is higher when the drop is small. The more the drop-size ratio is significant, the more the movement of the fluid could be qualitatively described as the liquid contained in the smallest drop is "injected" into the largest one. In other words, momentum conservation associated with the surface tension force leads the smallest drop to move toward the mass center of the big one.

Most of the existing dropwise condensation models consider instantaneous coalescence events although coalescence is a temporal phenomenon whose duration depends on the size of both the parent drops, the coalescence time being longer when the drops are large and close in size. Recently, numerical simulations of coalescence events by a VOF method were carried out by Adhikari [3] in order to analyze the validity of the quasi-instantaneous coalescence hypothesis, with particular attention on the impact of this hypothesis on heat transfers. The author identified two mechanisms that impact heat transfers: a direct mechanism linked to the stabilization time of the drop following coalescence and an indirect mechanism linked to the oscillation of the foot of the resulting drop on the surface, causing cleaning of the area close to the drop (and thus modifying the heat transfers). Taking these two mechanisms into account leads to a 15–20% increase in heat transfer compared to the instantaneous coalescence model in the most unfavorable case tested (i.e., coalescence between two large drops). Thus, it could be necessary to take into account the effect of the transient phase of the coalescence on the heat transfers if the number of coalescence events between large drops is high.

4.1.1.4 The Drop's Departure

The critical radius $r_{\max}$ beyond which drops start to move on the surface is an essential parameter, which drives the global heat transfers during condensation in dropwise regime (see next section), as well as the critical heat flux leading to the transition from dropwise to filmwise regime. The drops are pinned on the surface by the force at the triple line. The critical radius $r_{\max}$ is the radius at which another force unbalance this pinning force.

Several forces can displace the drops, the most common ones being the gravity and the shear stress. Rose [38] in 1988 proposed a simple correlation based on capillary length to evaluate $r_{\max}$ for droplet in a gravity field:

$$r_{\max} = K\sqrt{\frac{\sigma_{lv}}{\rho_l g}} \tag{4.18}$$

where K is equal to 0.4 for steam condensation on a vertical flat plate.
Taking into account the contact angle hysteresis and considering that the triple line is an ellipse, the formulation of the equivalent critical radius for a single drop on a vertical plane is expressed as follows [21]:

$$r_{\max} = \sqrt{\frac{6c\,(\cos\theta_{rec} - \cos\theta_{adv})\sin\theta}{\pi(2 - 3\cos\theta + \cos^3\theta)}\frac{\sigma_{lv}}{\rho_l g}} \tag{4.19}$$

where c is a constant that depends on the shape of the drop and the inclination of the substrate surface. Different values of c were then proposed by several authors, as reported in Table 4.1. β in the work of Extrand and Kumagai [17] represents the length to width ratio of the substrate area wetted by the drop, derived from the assumption that the contact line is an ellipse and experimentally validated on several fluid–substrate couples.

Thus, in order to reduce this radius of departure $(\cos\theta_r - \cos\theta_{adv})$ must be as small as possible. So far, in literature, the smallest values of $r_{\max}$ have been obtained on superhydrophobic surfaces.

When gravity force is too small or does not exist as it is the case for instance in horizontal configuration or in space applications, a wettability gradient on the surface can be used to create an additional driving force. In the case of a 1D wettability gradient in the **x** direction, Mancio Reis et al. [27] established the following relation allowing to correlate the driving force to the contact angle hysteresis:

Table 4.1 Values of the constant c in Eq. 4.19 according to different authors

Ref.	c
Extrand and Kumagai [17]	$0.23 + 1.04\beta$
Brown et al. [7]	$\pi/2$
Dussan and Chow [14]	1.0
Wolfram and Faust [53]	π
El Sherbini and Jacobi [15]	$48/\pi^3$

$$F_\theta(x_G, t) = \frac{\sigma_{lv} R(x_G, t)\pi}{2}[\cos\theta_{adv}(x_G + R(x_G, t)) - \cos\theta_{rec}(x_G - R(x_G, t))] \tag{4.20}$$

where x_G is the position of the center of mass of the drop and R is the footprint radius. The drop is then set in motion when the footprint radius satisfies $F_\theta > 0$.

The use of a wettability gradient to displace a drop has been successfully realized in several studies [25, 26, 32] for drops radius down to 1 mm. Nevertheless several limitations can be highlighted:

- The length of displacement is limited to a few times the footprint diameter of the drops when the contact angle hysteresis is not weak;
- The coating used to obtain the wettability gradient usually exhibits a limited lifetime.

Shear stress on the drop liquid–vapor interface induced by an imposed gas flow is another way to decrease the critical departure radius. Among the pioneering work, one can cite the ones of Tanner et al. [48] and O'Bara et al. [33] who observed significant enhancement of the heat transfer along with an early removal of the drops according to the increase of the vapor velocity. However, O'Bara et al. [33] also shown that high velocities can also change the drop shape and flatten them leading to the reduction of the heat transfer coefficient due to the increase of the surface covered by these big drops.

Over time, several studies on the effect of vapor velocity [11–13, 49] have been carried out using various substrates and producing different conclusions. Although it is well accepted that increasing the vapor velocity reduces the departure radius (inducing, most of the time, an improvement in the heat transfer), the extent of the improvement and the dependence of this improvement to the steam velocity remain unclear. It is very difficult to rigorously compare these studies as the drag force depends on the velocity profile in the vicinity of the drops which is usually not measured directly, the usual measured data being flow rate or punctual readings in a cross section. Tancon et al. [47], recently proposed a neat evaluation of the effect of the vapor velocity on the departure radius of the droplets. The experiment were conducted with pure steam, on a vertical sol-gel-coated aluminum substrate leading to an advancing contact angle of around 87°. The vapor average velocity ranged from 2.7 to 11 m s^{-1} along with the centimetric dimension of the test channels led to turbulent regime. The drag force acting on the drop was classically expressed as the product between the fluid kinetic energy, the drop frontal area and the drag coefficient C_d:

$$F_d = \frac{1}{2}\rho_v v_v^2 C_d \left(\theta_e - \sin\theta_e \cos\theta_e\right) r^2 \quad (4.21)$$

where $\theta_e = \cos^{-1}\left(0.5\cos\theta_{adv} + 0.5\cos\theta_{rec}\right)$ (in radian) and v_v the mean vapor velocity in the cross section.

The drag coefficient C_d of the droplet was obtained from a fit of the data generated using CFD calculations of the flow in the channel for low Reynolds numbers (i.e., in the range 100–1000) in turbulent regime. The expression of C_d was then:

$$C_d = 5.6053\left[\left(L_c/l_{dr}\right)^{-4/3} Re_{dr}^{-1/6}\right] + 0.1754 \quad (4.22)$$

where L_c/l_{dr} is the ratio between channel height and droplet height and $Re_{dr} = l_{dr} v_v \rho_v/\mu_v$. Eq. 4.22 is in agreement with Brown–Lawler equation [6].

Then the balance between gravity force, adhesion force, and drag force allow to express the drop departing radius:

$$r_{\max} = \frac{-C + \sqrt{C^2 + 4AB}}{2B} \quad (4.23)$$

with

$$A = 2c\sigma_{lv}\sin\theta_e\left(\cos\theta_{rec} - \cos\theta_{adv}\right) \quad (4.24)$$

$$B = \frac{2 - 3\cos\theta_e + \cos^3\theta_e}{3}\pi\rho_l g \quad (4.25)$$

$$C = \frac{1}{2}\rho_v v_v^2 C_d\left(\theta_e - \sin\theta_e\cos\theta_e\right) \quad (4.26)$$

This mechanical force balance has been validated with direct visualization results. The authors show that for a velocity of 11 m s^{-1} the drag force has the same magnitude as gravity and allows a reduction of the drop departure radius of more than 33%

4.1.2 Drops Population Models

The drop population on a substrate is usually described in term of drop-size (mostly the curvature radius) distribution. Several difficulties can then be highlighted to model this drop-size distribution.

- The first one is the huge number of droplets that have to be considered. Indeed, the nucleation sites density is in the order of 10^9–10^{15} m^{-2}. Even if these nucleation

sites are not all active simultaneously, the total number of drops per unit of substrate area is expected to be very high.
- The second main difficulty is that the radii of the drops vary on several orders of magnitude, typically from 10^{-8} to 10^{-2} m.
- The third difficulty is the dynamic nature of the drops life and the very small timescales associated (e.g., interface displacement, sweeping, heat diffusion…) in the range 10^{-9}–10^{0} s.

Indeed, a given nascent drop can collide with a big one during the first coalescence event and thus reach very quickly the critical drop radius leading to its departure (and maybe other coalescence events). By contrast, another nascent drop can coalesce many times with other small drops before the resulting one reaches $r_{\max}$. The resulting drop-size distribution on the substrate is thus a function of each of the drops' lifecycle. One should note that "drop population" design in fact a "stationary" or "time averaged" distribution. There are 2 possible ways to tackle these classes of problem: (a) use hypothesis to write a macroscale model (e.g., Population Balance Models) then solve it, (b) use a "tracking of events" strategy over an extended period of time then extract macroscale data; the models of such distribution should take into account a large number of events. Furthermore, the modeling is constrained by the multiple time and space scales that should be respected to calculate the considered phenomena. Consequently, direct CFD-like approaches of dropwise condensation are not feasible. On the other hand, a way to model the drop-size distribution on the substrate by considering drops at individual level remains still possible if strong assumptions can be made. Many papers have been published developing such Individual-Based Models (IBM) [8, 23, 30, 43, 45, 54, 56] (sometimes named lagrangian models). In all these works the following main assumptions are made:

- The nucleation sites density is known a priori and nucleation sites are then randomly distributed on the surface;
- The substrate temperature is supposed to be constant and uniform;
- The contact angle is imposed to the advancing contact angle value during all the drop's growth process;
- If a nucleation site is not covered by a drop, a nucleus instantaneously appears on it;
- The nuclei are formed with a radius $r_{\min}$ corresponding to the unstable equilibrium radius of a drop forming on a cold substrate with an imposed contact angle θ.
- The individual drop grows according Eq. 4.12 between two successive interactions.
- Only binary drop interactions are considered (either due to growth or to sweeping). Nevertheless, cascading coalescences can take place, i.e., several successive coalescence events can be considered during a time step.
- When two drops collide, they coalesce instantaneously to form a new drop located at the center of mass of the two parent drops.
- When the radius of a drop reaches an imposed value $r_{\max}$, this drop is set in motion in a given direction, and then sweeps the surface of the substrate.

Despite these simplifying assumptions, as well as parallel and adapted numerical implementation, the calculation time is usually very important or even prohibitive.

Additionally, as the nucleation sites density is rather high and the number of individuals to consider grows like the square of the domain size, IBM approaches are limited in term of domain size and thus in term of maximal drop radius. Moreover, very small drops imply using an extremely small time step (see Fig. 4.3). As the observation time should be long enough to obtain the "stationary" distribution, a compromise should be made between the number of nucleation sites and the determination of the size distribution of the smallest drops. For these reasons, very few data are available in literature on the distribution including drop-size less than one micrometer; moreover no experimental data are available for validation purpose.
So, most of the literature reported data are limited to the distribution of drop of radius greater than a few micrometers. In that case, the calculated distribution can be compared with the one obtained with the reference law developed by Rose and coworkers. Noticing that experimental observations of drop population remains roughly identical whatever the magnification of the camera, and using geometrical argument on how to pack circles on a plane, Le Fevre and Rose [22] correlated the fraction of the substrate area covered by drops to $(r/r_{\max})^{1/\gamma}$, where γ is an empirical constant to determine. Few years later, Rose and Glicksman [40] derived an expression of the drop-size distribution for "visible" drops (i.e., drops greater than few micrometers):

$$N(r) = \frac{1}{3\pi r_{\max} r^2} \left(\frac{r}{r_{\max}} \right)^{-2/3} \tag{4.27}$$

All papers reporting results obtained from individual-based modeling highlight a good adequacy with this law for drops greater than 10 μm, whatever are the conditions considered in the simulations or the experiments (wettability, subcooling, ...).

As the objective of the modeling is to access to the drop-size distribution, some authors proposed to not follow each individual drop but to write a balance equation directly in terms of drop population density. These statistical type approaches were first used to determine the drop-size distribution of the smallest drops, i.e., smaller than the ones considered in the Rose and Glicksman correlation. This population balance model (PBM) has been proposed by Wen and Jer [52] in 1976 and is widely used in the literature [1, 21, 31, 44, 51]. The common assumptions associated to PBM are:

- The nucleation sites are regularly placed on the substrate and are separated by a distance equal to the average distance derived from the nucleation sites density: $r_e = \frac{1}{\sqrt{4N_s}}$;
- Only the distribution of drops whose size is in the range $r_{\min} < r < r_e$ are modeled. This implies that modeled drops do not move from their nucleation sites and do not coalesce with other pinned drops (i.e., they can coalesce with droplets that sweep the substrate).
- The growth of the modeled drops is only due to the condensation of pure vapor.
- The drops may be swept away by the moving drops of radius equal to $r_{\max}$ with a constant renewal characteristic time.

– The distribution function N(r) of Rose and Glicksman (Eq. 4.27) is used for the large drops ($r > r_e$).

Using these hypotheses, a simple balance equation was established to determine the drop-size distribution $n(r)$ according to the drop radius from the following reasoning. Considering a class of drop-size comprised between r_1 and r_2, at steady state, the balance of the number of individuals belonging this size class is given by Eq. 4.28:

$$A\,n(r_1)\,G(r_1) = A\,n(r_2)\,G(r_2) + S\,\bar{n}\,\delta r \tag{4.28}$$

The rate of drops entering this class due to the growth by condensation (left-hand side) is equal to the sum of the rate of drops leaving this class because of the condensation (first right-hand side term) plus the rate of drops' disappearance (second right-hand side term). This latter term is due to the sweeping of the surface by the moving drops of radius above or equal to r_{max} (see Sect. 4.1.1.4). In Eq. 4.28, A is the condensing surface area, S is the surface renewal rate by moving drops ($m^2\ s^{-1}$) and $\bar{n}$ is the average of n over the interval $[r_1, r_2]$.

By reducing the width δr between r_1 and r_2 toward zero, the following differential equation is obtained:

$$\frac{d(G(r)n(r))}{dr} + \frac{n(r)}{\tau} = 0 \tag{4.29}$$

where $n(r)$ is the drop-size distribution (defined only for $r_{min} < r < r_e$) and $\tau = \frac{A}{S}$ is the renewal characteristic time. This latter is defined as the time required by moving drops to sweep the entire surface area. As the sweeping of the surface affects all the drops whatever are their sizes, this renewal characteristic time is supposed to be independent of the drop radius, and thus is constant.

To solve this differential equation, as τ is unknown, two additional conditions are needed. For that purpose, the continuity of the drop-size distributions $N(r)$ and $n(r)$, corresponding, respectively, to large and small drops, as well as the continuity of their derivatives are imposed at $r = r_e$, the drop-size distribution of Rose and Glicksman (Eq. 4.27) being used for the large drops:

$$n(r_e) = N(r_e) = \frac{1}{3\pi r_{max} r_e^2}\left(\frac{r_e}{r_{max}}\right)^{-2/3} \tag{4.30}$$

$$\left(\frac{d\log n(r)}{d\log r}\right)_{r=r_e} = \left(\frac{d\log N(r)}{d\log r}\right)_{r=r_e} = -\frac{8}{3} \tag{4.31}$$

Using Eq. 4.12 for the growth rate $G(r)$, along with the associated A_1, A_2 and A_3 constant expressions (Eqs. 4.13–4.15), the resolution of Eq. 4.29 leads to the following expression of the drop-size distribution $n(r)$:

$$n(r) = N(r_e)\frac{G(r_e)}{G(r)}\,\mathrm{e}^{B(r)+C(r)} \tag{4.32}$$

with:

$$B(r) = \frac{A_2}{\tau A_1}\left[\frac{r_e^2 - r^2}{2} + r_{\min}(r_e - r) - r_{\min}^2 \ln\left(\frac{r_e - r_{\min}}{r - r_{\min}}\right)\right] \tag{4.33}$$

$$C(r) = \frac{A_3}{\tau A_1}\left[(r_e - r) - r_{\min} \ln\left(\frac{r_e - r_{\min}}{r - r_{\min}}\right)\right] \tag{4.34}$$

and with:

$$\tau = \frac{3r_e^2(A_2\, r_e + A_3)^2}{A_1(11\, A_2\, r_e^2 - 14\, A_2\, r_e\, r_{\min} + 8\, A_3\, r_e - 11\, A_3\, r_{\min})} \tag{4.35}$$

It is noteworthy to compare the IBM and PBM. IBM relies on tracking individual drops over space and time until a global stationary state is reached; then the drop-size distribution can be derived by post-processing the results. On the other hand, PBM is based on the a priori construction of a model of drops population and its resolution. However, both models consider the same modeling of the heat transfers through a single droplet (see Sect. 4.1.1.2) and use several other common hypotheses. Indeed, in both approaches:

- The substrate temperature is supposed to be constant and uniform;
- The drops nucleate with a radius $r_{\min}$ and grow with an imposed contact angle (advancing contact angle) during condensation process;
- The drops grow at a rate given by a simplified drop energy balance (e.g., Eq. 4.12) between two successive coalescences;
- When the radius of a drop reaches a given value $r_{\max}$, this drop is set in motion in a given direction, and then sweeps the surface of the substrate.

It is noteworthy to review also the main differences between IBM and PBM:

- IBM use the same mechanisms for all drops, while PBM is constructed by juxtaposing $n(r)$ up to r_e and $N(r)$ above and up to $r_{\max}$, these two latter distributions being obtained from completely different frameworks;
- While drops of any sizes may coalesce in IBM either with a neighboring drop pinned on the surface or with a moving one, the PBM approach prohibits coalescence for the drops of radius smaller than r_e (these ones can interact only with moving drops of radius $r_{\max}$).

The results obtained with both approaches are compared in the following section. Analyses of the effect of some of the assumptions made on the drops life cycle are also proposed for both the drop-size distribution and the heat transfers.

4.2 Drop-Size Distribution According to Individual-Based and Population-Based Models

As already pointed out, IBM and PBM are used to calculate the drop-size distribution. As experimental drop-size distribution for the smallest droplets are not yet available, only cross comparisons are possible. To compare the results obtained using both IBM and PBM, it is mandatory to use configurations as close as possible. Thus, a PBM and a IBM [23] were implemented using the classical assumptions described in previous Sect. 4.1.2 with a specific care taken to limit as much as possible the differences between the two simulated configurations. Thus, for both IBM and PBM, the following assumptions have been used:

- The wall substrate is isothermal;
- The gas phase is constituted by pure vapor at atmospheric pressure;
- The drops form with a radius $r_{\min}$ calculated from Eq. 4.1;
- The nucleation sites density value N_s is set (we emphasize that fixing N_s is equivalent to impose the average distance $2r_e$ between 2 neighboring nucleation sites);
- For IBM calculations the nucleation sites are randomly distributed with respect to this average value;
- The drop growth model used is the one described in Sect. 4.1.1.2.

Note that one of the mains drawback of IBM is the computation time. Indeed, the necessity of simulating large substrate area to get a sufficient sampling of the biggest drops together with the high value of nucleation sites density lead to a very high number of drops to simulate. Moreover, these drops interact with each other at a very high frequency. So, the time step used in the simulation has to be as small as possible to take into account all coalescence events. Finally, the simulation should be continued until a global stationary regime is obtained.

Most of the studies available in the literature consider relatively high maximum radii of the drops (and so large surfaces) in order to compare numerical results with experimental ones. In consequence, the time step δt is generally set to a relatively high value and the distribution of the smallest drop-sizes cannot be determined. As the aim is to compare our IBM approach to the PBM one, it is mandatory to use a time step small enough to capture all coalescence events for drops of size as small as $r_{\min}$. At the end of each time step, the next coalescence event is located and the time step is adjusted accordingly; thus all coalescence events are taken into account. However, to keep the computation time reasonable, a minimum value of δt is imposed, typically $\delta t_{\min} = 10^{-5}$ s (see [23] for details).

Some differences cannot be bridged between the used IBM and PBM. For instance, the nucleation sites density is the same for both models but for IBM the site are randomly placed to obtain an homogeneous distribution of the distance between two neighboring sites while they are equally spaced for PBM. It can be also noticed that in the IBM, when a drop reaches the radius $r_{\max}$, it is set in motion on the surface at an imposed velocity in a given direction and interacts with other drops until it leaves the modeled domain.

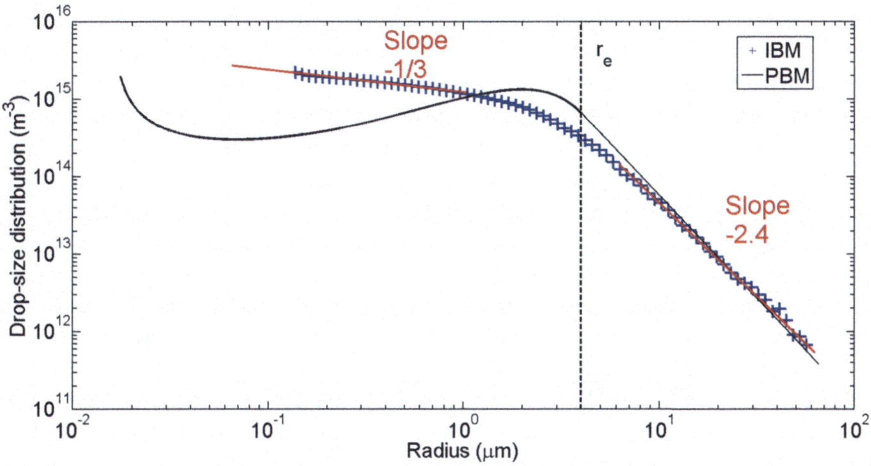

Fig. 4.5 Comparison of drop-size distributions obtained from both models ($\theta_{adv} = 85°$, $\Delta T = 1$ K, $T_{sat} = 373$ K, $\delta_{coat} = 100$ nm, $k_{coat} = 2$ Wm^{-1}K^{-1} and $h^i = 15.7$ MW m^{-2} K^{-1})

We defined and set a reference case that is suitable for quantitative comparisons between IBM and PBM while maintaining an acceptable computation time (i.e., few days on a simple workstation using a standard Matlab® code). The computational domain is a square of edge 340 μm with 2025 nucleation sites (leading to a nucleation sites density $N_s = 1.56 \times 10^{10}$ m^{-2}). The advancing contact angle is $\theta_{adv} = 85°$ and the subcooling is $\Delta T = 1$ K. As the fluid considered is pure water, the condensation coefficient is set to $f = 1$. Typical value of coating parameters are arbitrarily imposed, i.e., a thermal conductivity $k_{coat} = 2$ W m^{-1} K^{-1} and a thickness $\delta_{coat} = 100$ nm. The maximum drop radius is set to $r_{\max} = 65$ μm and the velocity of the drops that sweep the surface is fixed at 0.01 ms^{-1}.

First, the reproducibility of the IBM results has been checked, as well as their independence to the numerical parameter $\delta t_{\min}$ and the total number of initial nucleation sites N_{ns} (at constant nucleation sites density) [23]. A video showing the behavior of the calculated drops on the surface can be seen on Youtube.[1] It can be noticed that the number of coalescence events is very large, in the order of one million per second, although the modeled domain is limited.

The drop-size distribution obtained from both approaches are reported on Fig. 4.5. It can be pointed out that for the biggest drops (radius greater than 10 μm), the IBM results are consistent with the Rose and Glicksman law (Eq. 4.27) that has been experimentally validated many times. This confirms the relevance of the IBM approach, even if it does not ensure that the calculated distribution of the small drops is right.

Below 10 μm, the 2 models start to disagree. Overall, the differences between the 2 distributions increase by going towards the small radii and reach more than one

[1] https://www.youtube.com/watch?v=W4sgNEoCwuo.

order of magnitude for r around hundred nanometers. The IBM exhibits 3 different behaviors. For the smallest drops (below one micron), the drops-size distribution follows a power law with a weak absolute value of the exponent ($-\frac{1}{3}$). Indeed, such very small drops may coalesce only with rather big or moving ones and those events are not very frequent. So, they grow mainly by vapor condensation . For the drops greater than 10 µm, the drop-size distribution is almost identical to the one predicted by the Rose and Glicksman law which involves both mechanisms of growing (i.e., coalescence and condensation). However a weak difference can be highlighted in the slopes: −2.4 for IBM while Rose and Glicksman gives $-\frac{8}{3}$ (see Eq. 4.31). Finally, between these sizes, the contribution of the coalescence in the growth dynamic increases progressively and the transition from one driving process of growing to the other is progressive.

Between $r_{\min}$ and r_e, PBM results exhibit a N-shaped curve with one maximum at $r_{\min}$ and the other slightly below r_e. The IBM gives clearly different results than PBM for drop radius less than r_e. This latter approach (Eq. 4.29) assumes a constant renewal characteristic time whatever the drop-size, that only depends of the sweeping rate by the moving drops. This hypothesis seems to be the main reason which can explain the discrepancies between the 2 models. Indeed, this constant value of τ implies that the small drops ($r < r_e$) do only coalesce with the ones that sweep the surface. But one can reasonably expects that even 2 small drops can coalesce if the distance between their respective nucleation sites is less than $2r_e$. Also this drop may also coalesce with a second one (small or not) that has already coalesced and is thus no more centered on its nucleation site. So, the second drop could be very close to the first (small) drop.

Using Eq. 4.29, the renewal characteristic time τ was calculated according to the drop radius r from the IBM results (Fig. 4.6). The value of τ obtained from Eq. 4.35 is also reported in its definition domain (i.e., from $r_{\min}$ to r_e). As it can be seen on the figure, the renewal characteristic time calculated from the IBM results increases from $10^{-4}s$ for the smallest droplets to about 10^{-2} s at $r = r_e$, and to more than $1s$ for the biggest drops. According to Eq. 4.35, as illustrated, τ should be about 10^{-2} s for all drops of radius less than r_e. So, taking into account the coalescences of small drops (lower than r_e) leads to a smaller renewal characteristic time τ. The difference is important and goes up to 2 orders of magnitude for the smallest drop-sizes.

To continue this analysis and better understand the role of coalescence on the small drops renewal characteristic time, the disappearing term $\frac{n}{\tau}$ in Eq. 4.29 is broken down into two terms according to the involved mechanism. The differential equation which governs the drop-size distribution is rewriten as:

$$\frac{d(G(r)n(r))}{dr} + n(r)\left[\frac{1}{\tau_c(r)} + \frac{1}{\tau_{sw}(r)}\right] = 0 \tag{4.36}$$

where the renewal characteristic times τ_c and τ_{sw} related to drop radius r are associated to coalescence events between pinned drops (called “coalescence” in the next §) and to the sweeping loss, i.e., coalescence events between a pinned drop and a moving drop (called “sweeping” in the next §), respectively. These 2 characteristic

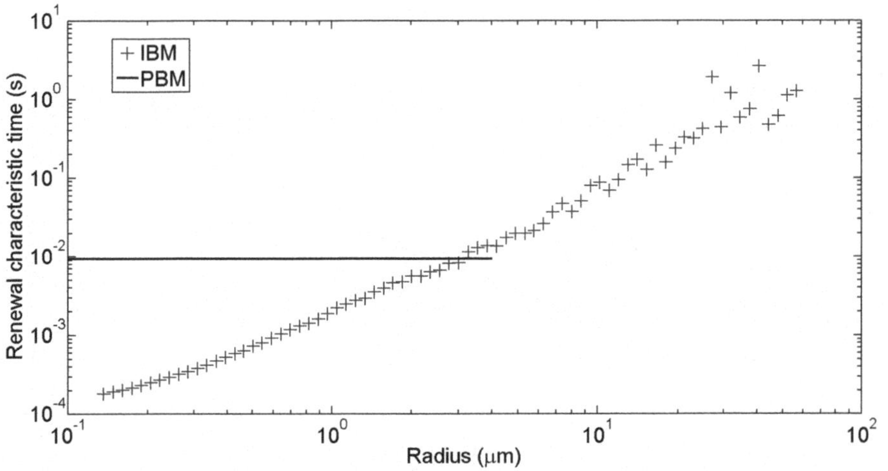

Fig. 4.6 Global renewal characteristic time obtained from both models ($\theta_{adv} = 85°$, $\Delta T = 1$ K, $T_{sat} = 373$ K, $\delta_{coat} = 100$ nm, $k_{coat} = 2$ W m^{-1} K^{-1} and $h^i = 15.7$ MW m^{-2} K^{-1})

times were derived from the IBM simulation as we have counted separately the coalescence and sweeping events for each drop-size. When a global stationary regime is reached, the rate of disappearance (or appearance) due to coalescence or sweeping are extracted. Each characteristic time is then simply the ratio between the number of individuals in the considered size-class and this rate of disappearance. The variations of these different characteristic times with respect to the drops radius are reported on Fig. 4.7.

Except for the largest drops, the coalescence characteristic time τ_c is largely smaller than the sweeping one τ_{sw}. This indicates that the coalescence events are much more frequent than the sweeping ones. This is particularly true for small drops. Moreover, the global renewal characteristic time τ (Fig. 4.6) calculated using Eq. 4.29 is almost identical to the coalescence characteristic time τ_c (Fig. 4.7). Such a behavior shows that the sweeping is very often negligible in the drops renewal rate, except for the biggest drops (i.e., $r > 20$ μm). For example, the drop population in the class around $1 \mu m$ is renewed with a characteristic time τ_c of 10^{-3} s while it is about 0.1 s for τ_{sw}. Thus the drops are renewed only one times due to the sweeping when they are renewed one hundred of times due to the coalescence events. The assumption that small droplets do not coalesce in the PBM appears therefore not acceptable in that case.

The influence of the surface wettability on the renewal characteristic time τ is reported on Fig. 4.8 according to the drop radius. Whatever the advancing contact angle in the range [45°; 140°], the conclusion made previously about the great importance of coalescence events in the small drops-size distribution remains valid. Indeed, the renewal characteristic time grows by several orders of magnitude when the drops radius is increased from $r_{\min}$ to $r_{\max}$. A vertical shift of the curves is observed with

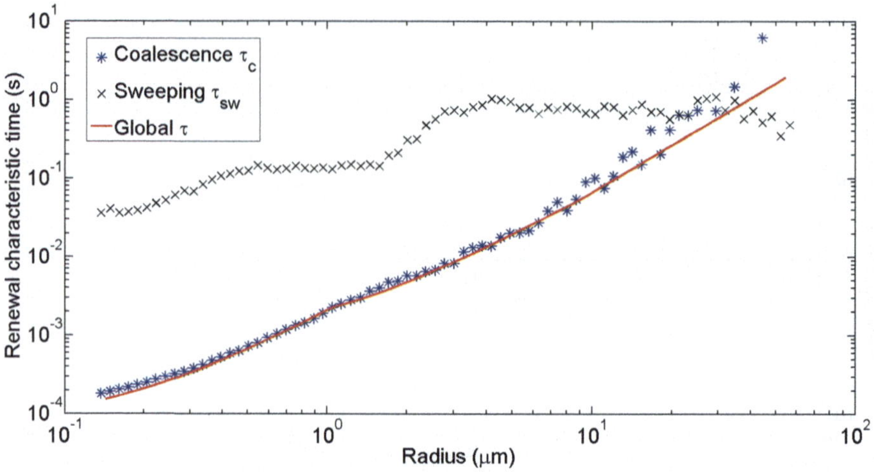

Fig. 4.7 Characteristic times of coalescence and sweeping obtained thanks to the IBM numerical results as a function of the drop radius ($\theta_{adv} = 85°$, $\Delta T = 1$ K, $T_{sat} = 373$ K, $\delta_{coat} = 100$ nm, $k_{coat} = 2$ W m^{-1} K^{-1} and $h^i = 15.7$ MW m^{-2} K^{-1})

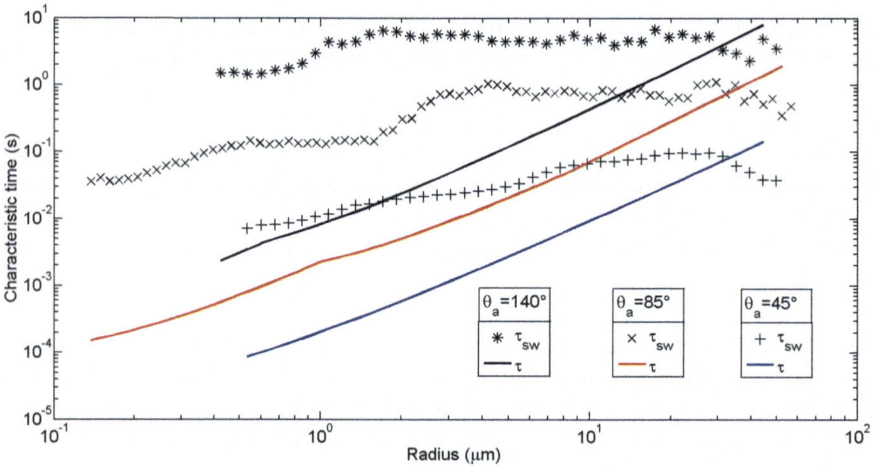

Fig. 4.8 Global and sweeping renewal characteristic times obtained for different contact angles as a function of the drop radius ($\Delta T = 1$ K, $T_{sat} = 373$ K, $\delta_{coat} = 100$ nm, $k_{coat} = 2$ W m^{-1} K^{-1} and $h^i = 15.7$ MW m^{-2} K^{-1})

respect to the wettability: more the surface is hydrophilic, more the renewal characteristic time is weak. This is due to the growth dynamics of each single droplet which is faster when the contact angle is small (Fig. 4.3). As a consequence, the frequency of interaction between the drops is greater. In any case, the effect of the sweeping is always negligible, except for the drops of radius close to $r_{\max}$.

So, the 2 approaches allowing to determine the drop-size distribution conduct to differences in the results up to several orders of magnitude, especially for the distribution of the smallest drop radii. These differences can be explained by the use of a constant renewal characteristic time whatever is the drop radius in the PBM. In the following, the influence of the differences between the drop-size distributions obtained with IBM and PBM on the global heat transfer prediction is quantified and analyzed.

4.3 Heat Transfer

4.3.1 Heat Flux Distribution According to Drop-size

From any given drop-size distribution $\bullet(r)$ together with the model of heat transfer through a single droplet described in Sect. 4.1.1.2, it is possible to calculate the heat flux ($\frac{W}{m^2}$) associated to any radius range $[r_1, r_2]$:

$$q(r_1, r_2) = \int_{r_1}^{r_2} \bullet(r) Q_d(r)\, \mathrm{d}r \tag{4.37}$$

When related to IBM, $\bullet(r)$ is simply $n(r)$ while with PBM $\bullet(r)$ is a piecewise function: $n(r)$ (Eq. 4.32) below r_e and $N(r)$ (Eq. 4.27) above. For instance, the global heat flux can then be determined by:

$$q = \int_{r_{\min}}^{r_{\max}} n(r) Q_d(r)\, dr \qquad \text{for IBM} \tag{4.38}$$

$$q = \int_{r_{\min}}^{r_e} n(r) Q_d(r)\, dr + \int_{r_e}^{r_{\max}} N(r) Q_d(r)\, dr \qquad \text{for PBM} \tag{4.39}$$

The impact of drop-size distribution on heat flux variation according to drop radius can then be analyzed. As the drop-size distributions obtained using IBM and PBM clearly differ, particularly for the small radii (see Sect. 4.2), significant differences in terms of heat flux are expected. The heat flux associated to each drop radii were computed and results are reported on Fig. 4.9. The frontier between "small" and "large" drops (according to PBM) is represented by the dashed line. The curves behavior is fairly similar, but, as it was expected, an important discrepancy between the 2 approaches for the "small" drops and a relatively good agreement for the "large" drops can be observed.

The contribution of each of the drops' family (i.e., small drops and large drops) to the global heat flux can be determined by integrating from $r_{\min}$ to r (with r varying from $r_{\min}$ to $r_{\max}$) the curves represented on Fig. 4.9 using Eq. 4.37. We obtain the cumulatives of the heat flux reported on Fig. 4.10. It can be highlighted that in the

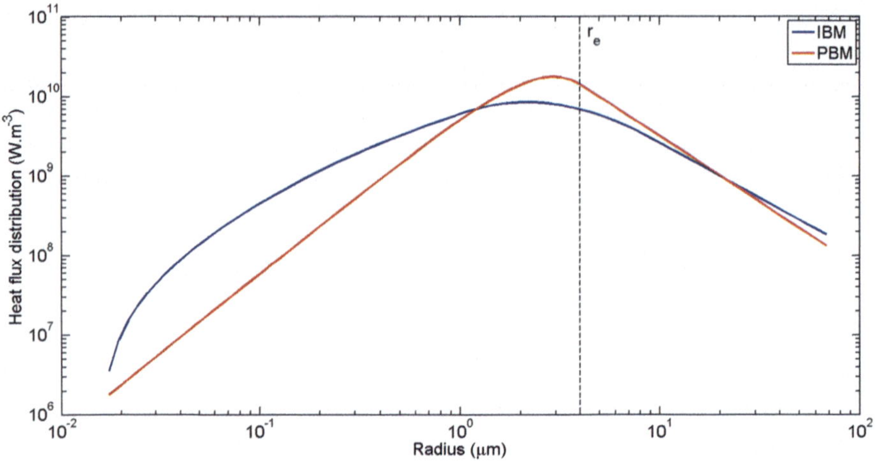

Fig. 4.9 Heat flux distribution according to r obtained thanks to the 2 different approaches ($\theta_{adv} = 85°$, $r_e = 4$ μm, $r_{\max} = 65$ μm, $\Delta T = 1$ K, $T_{sat} = 373$ K, $\delta_{coat} = 100$ nm, $k_{coat} = 2$ W m^{-1} K^{-1} and $h^i = 15.7$ MW m^{-2} K^{-1})

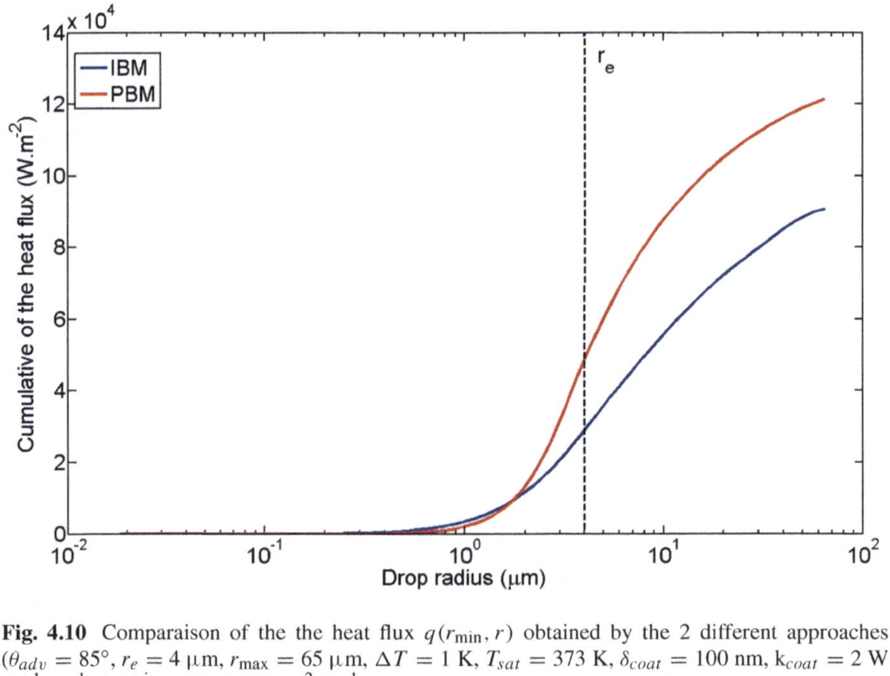

Fig. 4.10 Comparaison of the the heat flux $q(r_{\min}, r)$ obtained by the 2 different approaches ($\theta_{adv} = 85°$, $r_e = 4$ μm, $r_{\max} = 65$ μm, $\Delta T = 1$ K, $T_{sat} = 373$ K, $\delta_{coat} = 100$ nm, $k_{coat} = 2$ W m^{-1} K^{-1} and $h^i = 15.7$ MW m^{-2} K^{-1})

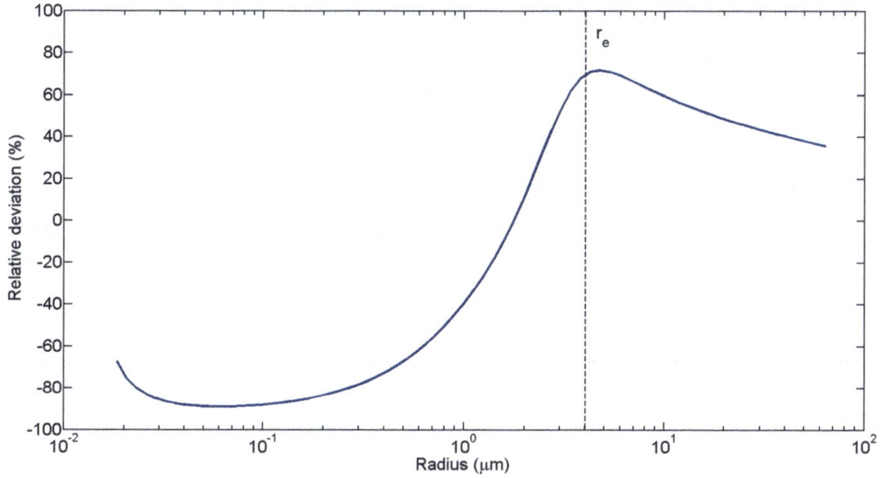

Fig. 4.11 Relative deviation (IBM basis) between the cumulative heat fluxes (PBM approach minus IBM one) as a function of drop radius ($\theta_{adv} = 85°$, $r_e = 4$ μm, $r_{\max} = 65$ μm, $\Delta T = 1$ K, $T_{sat} = 373$ K, $\delta_{coat} = 100$ nm, $k_{coat} = 2$ W m^{-1} K^{-1} and $h^i = 15.7$ MW m^{-2} K^{-1})

considered configuration both models predict that the heat flux is mainly evacuated by drops between 1 and 10 μm (70% for the PBM and 60% for the IBM). However, in this drop radius range, the 2 curves plotted on Fig. 4.9, although they both present a maximum in this range, differ significantly: the PBM curve presents a more sharp shape than the IBM one which is more smooth and below the PBM curve. Thus, as these differences lead to important changes on the cumulative heat fluxes (Fig. 4.10), it appears particularly important to use accurate model of the drop population in this drop-size range in order to predict the global heat flux correctly.

To go further in the analysis of the effect of the difference between the IBM and PBM drop-size distributions, we calculated the relative deviation between the cumulative heat flux obtained by both approaches. The results are reported on Fig. 4.11. For the small drops, the relative deviation reaches more than -85% for $r \approx 0.1$ μm. This means that for drops below this size, IBM predicts heat flux much higher than PBM. Fortunately, in the considered configuration, these drop-sizes have a relatively low contribution to the total heat flux.

Between approximately $r = 1$ μm and $r = r_e$, the relative deviation varies sharply because (i) the PBM predicts higher population in the drop classes close to r_e (see Fig. 4.5) and (ii) because these drop classes are those that mainly contribute to the global heat flux (see Fig. 4.9 and/or Fig. 4.10).

Above r_e, the relative deviation decreases as the slopes of drop population distributions differ (i.e., $-\frac{8}{3}$ for PBM (Eq. 4.31) and -2.4 for IBM). It reaches a value of about 34% when all drop-sizes have been considered (i.e., $q = 121.4$ kW m^{-2} for PBM and $q = 90.6$ kW m^{-2} for IBM). This value remains relatively low because of compensations between the different drop-size zones. Indeed, it can be observed that the relative difference varies from −85% up to 70% to finally end at 34%. Compared to IBM, PBM strongly overestimates the contribution to the global heat flux of the drops included in the interval [0.1 μm, r_e] and clearly underestimates this contribution when considering the drops below or above this radius interval.

From these results, several conclusions can be made:

- The 2 approaches are overall disagreeing to each other, especially for the "small" drops. The overall relative deviation in the global heat flux is non-negligible, even if it remains relatively moderate (i.e., 34%) for the considered configuration;
- This moderate global deviation is due to compensations between the different drop-size zones, the "local" deviation being much more important;
- Both approaches predict that the majority of the heat flux is evacuated by drops having a radius close to r_e.

4.3.2 Parametric Analysis

In order to go further in the analysis, the influence of 3 main parameters (i.e., the ones that could a priori be adjusted experimentally) has been determined: advancing contact angle θ_{adv}, nucleation sites density N_s and maximum drop radius $r_{\max}$.

The variations of the heat flux obtained thanks to both approaches as a function of the advancing contact angle are reported on Fig. 4.12. Both IBM and PBM predict that decreasing θ_{adv} leads to a strong increase of the heat flux. For instance, the heat flux value is approximately 3 times higher for $\theta_{adv} = 45°$ than for $\theta_{adv} = 140°$.

For a given drop radius, the drop volume (and thus the thickness of the liquid layer) is much lower in hydrophilic configuration. As a consequence, the thermal resistance by conduction within the liquid is lower, and the heat flux is thus higher. Note that here, θ_{adv} and $r_{\max}$ are decoupled and $r_{\max}$ is held constant, which is generally not the case in experiments (see Sect. 4.1.1.4). The differences between PBM and IBM results are found to be significants for θ_{adv} varying from 60° to 110°, with a maximum deviation of about 35%.

As pointed out previously, the 2 approaches are mainly in disagreement for the size distribution of the small drops. So, an analysis of the contribution of these small drops to the total heat flux has been conducted for the 2 approaches. The results are reported on Fig. 4.13. PBM predicts that, despite a global heat flux that varies by a factor 3 (see Fig. 4.12), the contribution of the small drops remains constant at about 37% of the total heat flux for any advancing contact angle θ_{adv}. On the other hand, IBM approach shows an increase of the contribution of the small drops with the contact angle (from 20% for $\theta_{adv} = 45°$ to almost 45% for $\theta_{adv} = 140°$). So, in hydrophilic situation the global heat flux is high but the contribution of small droplets is low. In hydrophobic situation it is the contrary: the global heat flux is low and the contribution of the small drops is much higher. As a result, the heat flux evacuated

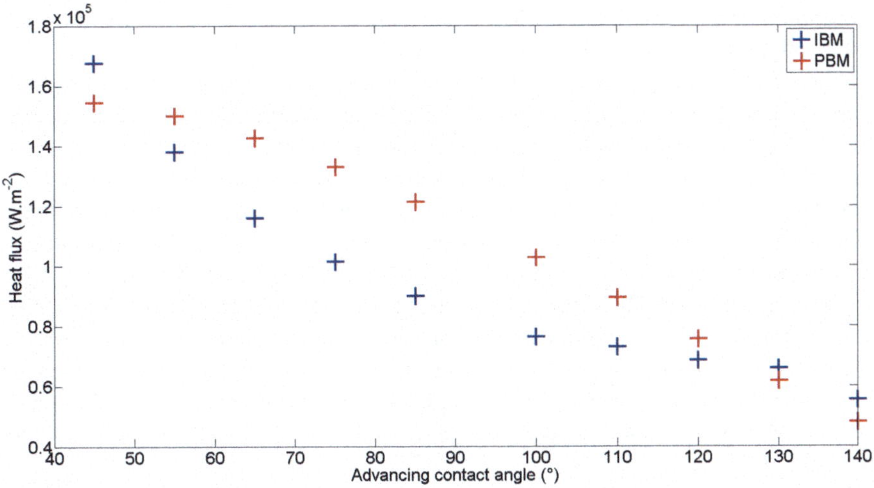

Fig. 4.12 Variations of heat fluxes versus advancing contact angle—IBM/PBM comparison. (r_e = 4 μm, $r_{\max}$ = 65 μm, ΔT = 1 K, T_{sat} = 373 K, δ_{coat} = 100 nm, k_{coat} = 2 W m^{-1} K^{-1} and h^i = 15.7 MW m^{-2} K^{-1})

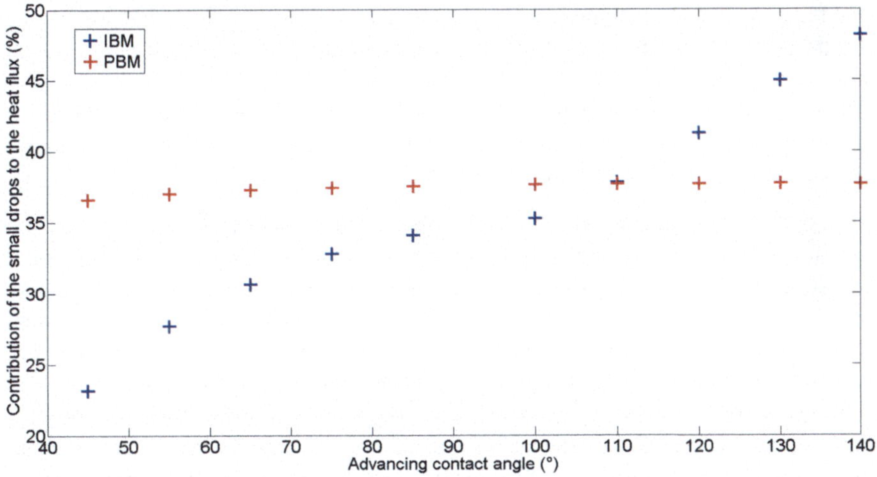

Fig. 4.13 Contribution of small droplets to the global heat flux in function of the advancing contact angle—IBM/PBM comparison. (r_e = 4 μm, $r_{\max}$ = 65 μm, ΔT = 1 K, T_{sat} = 373 K, δ_{coat} = 100 nm, k_{coat} = 2 W m^{-1} K^{-1} and h^i = 15.7 MW m^{-2} K^{-1})

by the small droplets is higher for low contact angles (i.e., $\theta_{adv} < 90°$) than for high contact angles (i.e., $\theta_{adv} > 90°$).

In order to better quantify the deviation between the contributions of the small drops to the total heat flux obtained with the two approaches, the relative deviations of PBM results to IBM results have been determined and are reported on Fig. 4.14. The deviations between the total heat fluxes are also reported on the same figure. As already mentioned, the most important deviation between the global heat fluxes are found at intermediate contact angles (i.e., around $\theta_{adv} = 90°$). For small drops, the deviation is large for hydrophilic surfaces (about 70%) and then decreases for hydrophobic surfaces. This remains consistent with the previous results (Fig. 4.13): for hydrophilic surface the contribution of small droplets is weak, so even a great difference between the two approaches leads to a moderate discrepancy between the global heat fluxes. By contrary, for hydrophobic configurations, the contribution of small drops is higher, and thus the two curves on Fig. 4.13 have the same behaviors for $\theta_{adv} > 120°$.

To summarize the results of this analysis on the effect of θ_{adv} on the global heat transfer:

- The highest heat fluxes are obtained for hydrophilic surfaces (about 3 times greater for an advancing contact angle of 45° than for a advancing contact angle of 140°);
- For the considered configuration ($r_{max} = 65$ μm and $N_s = 1.56 \times 10^{10}$ m^{-2}), the contribution of the small droplets remains moderate (i.e., less than 50%) for all contact angles;
- The global deviation between the 2 approaches is the most important for the intermediate contact angles (around 90°). For small drops, the deviation is particularly important when the contact angle is weak.

The other parameters on which it is possible a priori to act experimentally are the maximum drop radius (by imposing for instance an external force) and the density of nucleation sites (by realizing micro or nano structures on the surface). The respective deviations between the heat fluxes, as well as between the contributions of small droplets to these heat fluxes, obtained with population balance approach to the ones obtained with individual-based approach are reported on Figs. 4.15 and 4.16. The deviations increase when r_{max} and N_s increase, with a slight nuance for the r_{max} parameter where the overall deviation reaches an asymptotic-like value of about 45% for the global heat flux and 70% for the contribution of small droplets.

Both approaches show that (i) increasing N_s and (ii) decreasing r_{max} lead to an important increase of the heat flux (Figs. 4.17 and 4.18). In order to maximize the heat transfers, it is therefore interesting to increase N_s and decrease r_{max} as much as possible. The heat flux enhancement is more pronounced for the smallest values of r_{max} which may be difficult to obtain experimentally. Finally, the heat flux improvement is more or less constant with the increase of the nucleation sites density: an increase of one order of magnitude of N_s leads to a gain of more than 100% for the heat flux. However, if the decrease of r_{max} reduces the deviation between the 2 approaches, conversely, the increase of N_s causes an increase in this deviation. Regarding this latter remark, it must be remembered that variations of N_s lead to

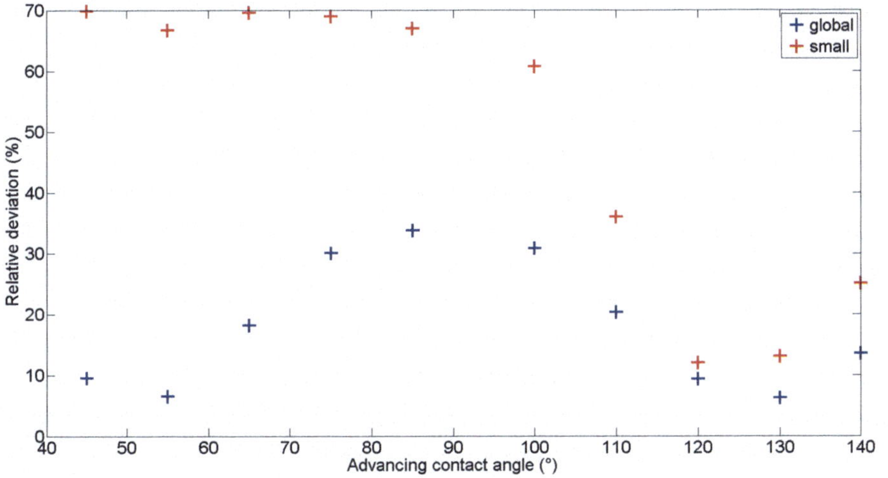

Fig. 4.14 Relative deviation (IBM basis) between the global heat fluxes as well as the contribution of small drops in function of the advancing contact angle ($r_e = 4$ μm, $r_{max} = 65$ μm, $\Delta T = 1$ K, $T_{sat} = 373$ K, $\delta_{coat} = 100$ nm, $k_{coat} = 2$ W m^{-1} K^{-1} and $h^i = 15.7$ MW m^{-2} K^{-1})

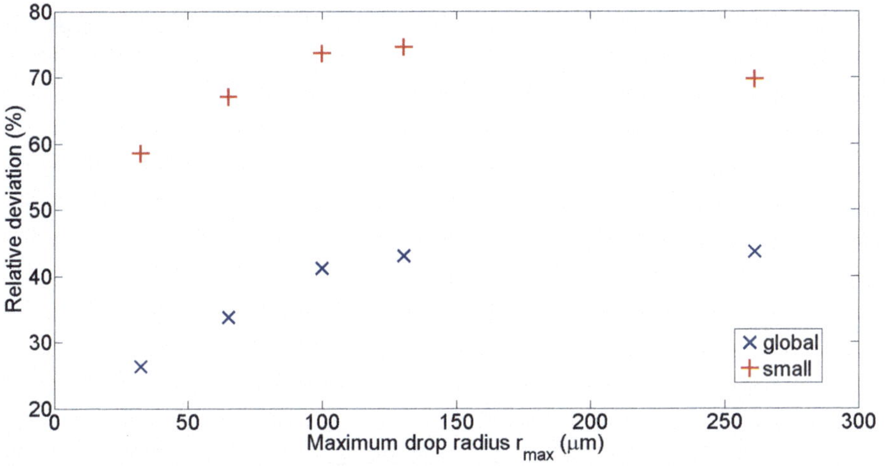

Fig. 4.15 Relative deviation (IBM basis) between the global heat fluxes and between the contributions of small drops as a function of the maximum drop radius r_{max} ($\theta_{adv} = 85°$, $r_e = 4$ μm, $\Delta T = 1$ K, $T_{sat} = 373$ K, $\delta_{coat} = 100$ nm, $k_{coat} = 2$ W m^{-1} K^{-1} and $h^i = 15.7$ MW m^{-2} K^{-1})

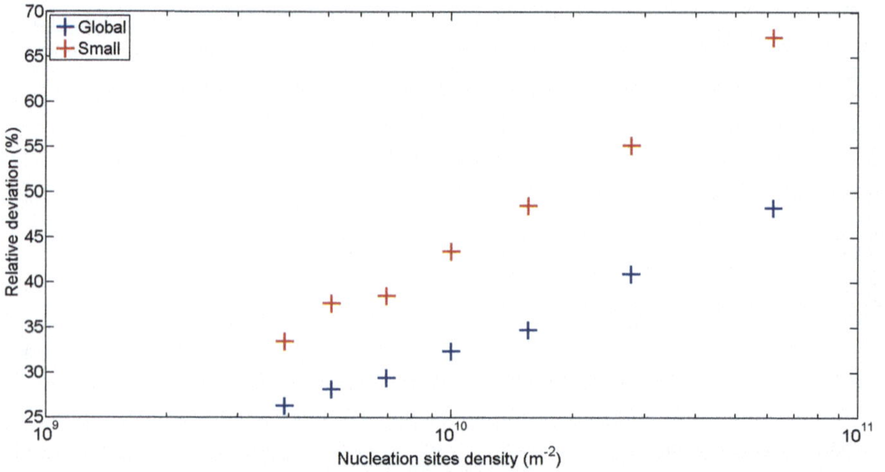

Fig. 4.16 Relative deviation (IBM basis) between the global heat fluxes and between the contributions of small drops as a function of the nucleation sites density N_S ($\theta_{adv} = 85°$, $r_{\max} = 65\ \mu$m, $\Delta T = 1$ K, $T_{sat} = 373$ K, $\delta_{coat} = 100$ nm, k$_{coat} = 2$ W m^{-1} K^{-1} and $h^i = 15.7$ MW m^{-2} K^{-1})

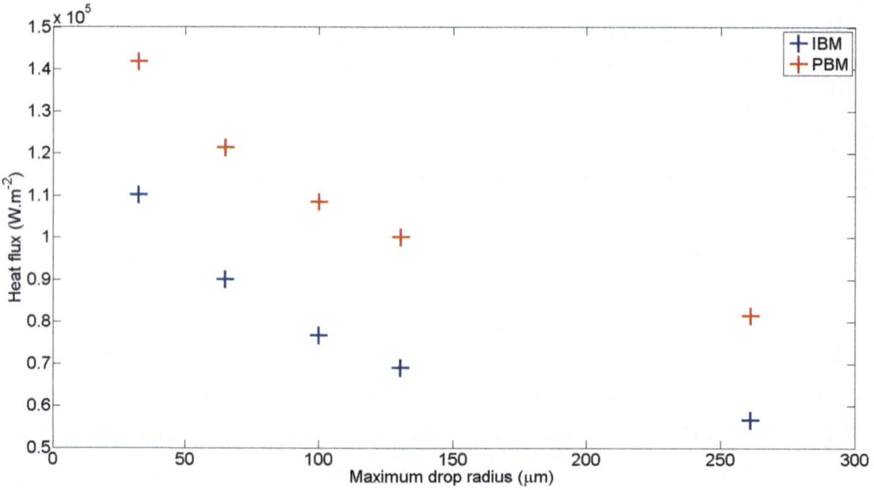

Fig. 4.17 Variations of heat fluxes versus maximum drop radius $r_{\max}$—IBM/PBM comparison. ($\theta_{adv} = 85°$, $r_e = 4\ \mu$m, $\Delta T = 1$ K, $T_{sat} = 373$ K, $\delta_{coat} = 100$ nm, k$_{coat} = 2$ W m^{-1} K^{-1} and $h^i = 15.7$ MW m^{-2} K^{-1})

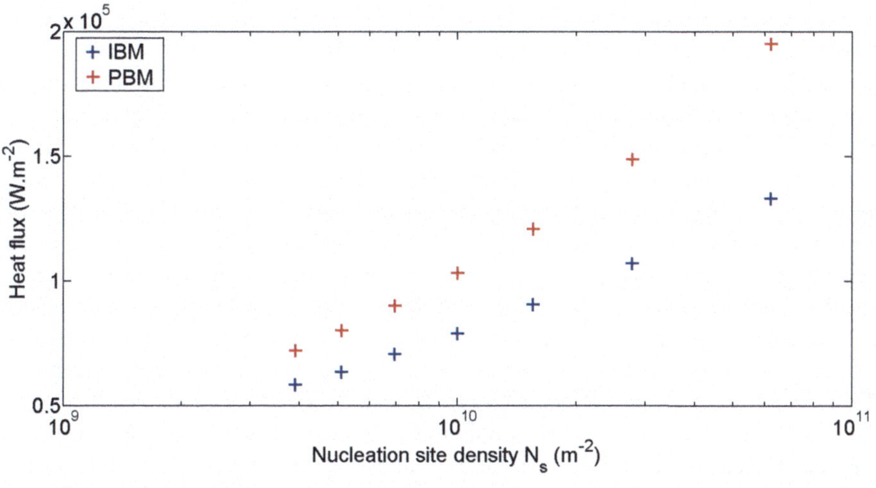

Fig. 4.18 Variations of heat fluxes versus nucleation sites density N_s—IBM/PBM comparison ($\theta_{adv} = 85°$, $r_{\max} = 65$ µm, $\Delta T = 1$ K, $T_{sat} = 373$ K, $\delta_{coat} = 100$ nm, $k_{coat} = 2$ W m^{-1} K^{-1} and $h^i = 15.7$ MW m^{-2} K^{-1})

a modification of the frontier between small and large drops (the frontier being at $r_e = \frac{1}{\sqrt{4N_s}}$). The increase of N_s leads to a smaller contribution of the small drops to the global heat flux, because the drop-sizes that mainly contribute to the heat flux remain around few µm (and therefore gradually belong to the class of large drops when N_s is increased).

4.4 Conclusion

Dropwise condensation allows to reach very high heat transfer coefficients, up to several hundred thousand of W m^{-2} K^{-1}. Modeling the heat transfer in such a regime implies to predict the drop-size distribution on the surface, with drop radii spread over 6 or 7 orders of magnitude. Moreover, a heat transfer law through each single drop is needed whatever the size of this single drop. For all these reasons, CFD numerical simulations at drop scale of dropwise condensation are prohibited because of calculation time.

From experimental point of view, the drop-size distribution is very difficult to access because of both small characteristic times and huge difference between the radii that must be measured simultaneously. Only the distribution of drops greater than few microns have been measured up to now, well predicted by the correlation of Rose and Glicksman [40].

So, drastically simplified theoretical approaches have been developed to calculate both the heat flux through a single drop and the size distribution of the smallest

drops (population balance model). Unfortunately, results given by these theoretical approaches have not yet been validated since no reference data exist.
So, during last decade, with the increase in computing capabilities, individual-based models were developed as an alternative to the population balance model to compute drop-size and heat transfer. The individual-based model consists in following each drop in its life cycle (born, growth, interaction with other drops and departure), making strong simplifying assumptions in order to limit the calculation time. Comparison of individual-based model and population balance model was thus carried out in the present paper, using a common set of assumptions. Important discrepancies have been highlighted in the drop-size distributions that can be critical from heat transfer point of view especially for radii close to half of the mean distance between two nucleation sites. Without prejudging the accuracy of the individual-based approach. The assumption of a renewal characteristic time due only to sweeping and thus independent of drop radius used in the population balance models has been demonstrated to be main cause of discrepancy and thus is questionable to dropwise condensation modelling. Although the important differences in the drop-size distributions, the prediction of the global heat transfer by both approaches can be fairly close, depending of the input parameter set.
Thus, further works are needed to make the dropwise condensation models more reliable and their predictions more accurate in the future, particularly for situations where the improvement of the heat transfer (by increasing the nucleation sites density and/or reducing the departure radius) will make more crucial the role of the smallest drops in the global heat transfer coefficient.

Acknowledgements This work was funded by the European Space Agency, MAP ENCOM contract n°4200020276.

References

1. Abu-Orabi, M.: Modeling of heat transfer in dropwise condensation. International Journal of Heat and Mass Transfer. **41(1)**, 81–87 (1998)
2. Adhikari, S.: A Study on Heat Transfer in Dropwise Condensation. PhD thesis, Pennsylvania State University, (2020)
3. Adhikari, S.: Heat transfer during condensing droplet coalescence. International Journal of Heat and Mass Transfer. **127**, 1159–1169 (2018)
4. Anand, S., Young Son, S.: Sub-micrometer dropwise condensation under superheated and rarefied vapor condition. Langmuir **26(22)**, 17100–17110 (2010)
5. Andrieu, C., Beyens, D. A. Nikolayev, V. S., Pomeau, Y.: Coalescence of sessile drops. Journal of Fluid Mechanics **453**, 427–438 (2002)
6. Brown, P., Lawler, D. F.: Sphere drag and settling velocity revisited. Journal of Environmental Engineering **129(3)**, 222–231 (2003)
7. Brown, R. A., Orr, F. M. Scriven, L. E.: Static drop on an inclined plate: Analysis by the finite element method. Journal of Colloid and Interface Science **73(1)**, 76–87 (1980)
8. Burnside, B. M., Hadi, H. A.: Digital computer simulation of dropwise condensation from equilibrium droplet to detectable size. International Journal of Heat and Mass Transfer **42(16)**, 3137–3146 (1999)

9. Carey, V. P.: Liquid-vapor phase-change phenomena: an introduction to the thermophysics of vaporization and condensation processes in heat transfer equipment. CRC Press (2020)
10. Cha, H., Min Chun, J, Sotelo, H., Miljkovic, N.: Focal plane shift imaging for the analysis of dynamic wetting processes. ACS Nano **10(9)**, 8223–8232 (2016)
11. Chatterjee, A., Derby, M. M., Peles, Y., Jensen, M. K.: Condensation heat transfer on patterned surfaces. International Journal of Heat and Mass Transfer **66**, 889–897 (2013)
12. Chatterjee, A., Derby, M. M., Peles, Y., Jensen, M. K.: Enhancement of condensation heat transfer with patterned surfaces. International Journal of Heat and Mass Transfer **71**, 675–681 (2014)
13. Cuthbertson, G., McNeil, D. A., Burnside, B. M.: Dropwise condensation of steam on a small tube bundle at turbine condenser conditions. Experimental Heat Transfer **13(2)**, 89–105 (2000)
14. Dussan V, R., Tao-Ping Chow, E. B.: On the ability of drops or bubbles to stick to non-horizontal surfaces of solids. Journal of Fluid Mechanics (1983)
15. ElSherbini, A. I., Jacobi, A. M.: Retention forces and contact angles for critical liquid drops on non-horizontal surfaces. Journal of colloid and interface science **299(2)**, 841–849 (2006)
16. Eucken, A. V.: Energie-und stoffaustausch an grenzflächen. Naturwissenschaften **25(14)**, 209–218 (1937)
17. Extrand, C. W., Kumagai, Y.: Liquid drops on an inclined plane: The relation between contact angles, drop shape, and retentive force. Journal of colloid and interface science **170(2)**, 515–521 (1995)
18. Fatica, N., Katz, D. L.: Dropwise condensation. Chemical Engineering Progress **45**, 661–674 (1949)
19. Hsu, Y. Y.: On the size range of active nucleation cavities on a heating surface. Journal of Heat Transfer, trans ASME **84(3)**, 207–213 (1962)
20. Jacob, M.: Heat transfer in evaporation and condensation ii. Mech. Eng **58**, 729–740 (1936)
21. Kim, S., Kim, K. J.: Dropwise condensation modeling suitable for superhydrophobic surfaces. Journal of Heat Transfer, trans ASME **133(8)**, 081502 (2011)
22. Le Fevre, E. J, Rose, J. W.: A theory of heat transfer by dropwise condensation. Proceeding of International Heat Transfer Conference 3 (1966)
23. Lethuillier, J., Lavieille, P., Miscevic, M.: About the role of falling droplets' sweeping in surface renewal during dropwise condensation. Langmuir **36(43)**, 12877–12886 (2020)
24. Liu, X., Cheng, P.: Dropwise condensation theory revisited part II. droplet nucleation density and condensation heat flux. International Journal of Heat and Mass Transfer **83**, 842–849 (2015)
25. Macner, A. M., Daniel, S., Steen, P. H.: Condensation on surface energy gradient shifts drop size distribution toward small drops. Langmuir **30(7)**, 1788–1798 (2014)
26. Mancio Reis, F. M.: Dynamique d'une goutte sur une surface á mouillabilitéhétérogéne : application á l'intensification des transferts de chaleur avec changement d'état. PhD thesis, Toulouse University (2015)
27. Mancio Reis, F. M., Lavieille, P., Miscevic, M.: Toward enhancement of water vapour condensation using wettability gradient surface. Experimental Thermal and Fluid Science **67**, 70–74 (2015)
28. Marek, R., Straub, J.: Analysis of the evaporation coefficient and the condensation coefficient of water. International Journal of Heat and Mass Transfer **44(1)**, 39–53 (2001)
29. McCormick, J. L., Westwater, J. W.: Nucleation sites for dropwise condensation. Chemical Engineering Science **20(12)**, 1021–1036 (1965)
30. Mei, M., Hu, F., Han, C., Cheng, Y.: Time-averaged droplet size distribution in steady-state dropwise condensation. International Journal of Heat and Mass Transfer **88**, 338–345 (2015)
31. Miljkovic, N., Enright, E., Wang, E. N.: Modeling and optimization of superhydrophobic condensation. Journal of Heat Transfer, trans ASME **135(11)**, (2013)
32. Moumen, N., Subramanian, R. S., McLaughlin, J. B.: Experiments on the motion of drops on a horizontal solid surface due to a wettability gradient. Langmuir **22(6)**, 2682–2690 (2006)
33. O'Bara, J. T., Killian, E. S., Roblee, L. H. S.: Dropwise condensation of steam at atmospheric and above atmospheric pressures. Chemical Engineering Science **22(10)**, 1305–1314 (1967)

34. Parin, R., Del Col, D., Bortolin, S., Martucci, A.: Dropwise condensation over superhydrophobic aluminium surfaces. Journal of Physics: Conference Series **745**, 032134 (2016)
35. Parin, R., Tancon, M., Mirafiori, M., Bortolin, S., Moro, M., Zago, L., Carraro, F., Martucci, A., Del Col, D.: Heat transfer and droplet population during dropwise condensation on durable coatings. Applied Thermal Engineering **179**, 115718 (2020)
36. Phadnis, A., Rykaczewski, K.: The effect of marangoni convection on heat transfer during dropwise condensation on hydrophobic and omniphobic surfaces. International Journal of Heat and Mass Transfer **115**, 148–158 (2017)
37. Rose, J. W.: Further aspects of dropwise condensation theory. International Journal of Heat and Mass Transfer **19(12)**, 1363–1370 (1976)
38. Rose, J. W.: Some aspects of condensation heat transfer theory. International Communications in Heat and Mass Transfer **15(4)**, 449–473 (1988)
39. Rose, J. W.: Condensation theory and experiment: A review. Proceedings of the Institution of Mechanical Engineers, Part A: Journal of Power and Energy **216(2)**, 115–128 (2002)
40. Rose, J. W., Glicksman, L. R.: Dropwise condensation, the distribution of drop sizes. International Journal of Heat and Mass Transfer **16(2)**, 411–425 (1973)
41. Schmidt, E., Schurig, W., Sellschopp, W.: Versuche über die kondensation von wasserdampf in film- und tropfenform. Technische Mechanik und Thermodynamik **1(2)**, 53–63 (1930)
42. Schrage, R. W.: A Theoretical Study of Interphase Mass Transfer. Columbia University Press (1953)
43. Sikarwar, B. S., Khandekar, S., Muralidhar, K.: Mathematical modelling of dropwise condensation on textured surfaces. Sadhana **38(6)**, 1135–1171 (2013)
44. Singh, M., Kondaraju, S., Bahga, S. S.: Mathematical model for dropwise condensation on a surface with wettability gradient. Journal of Heat Transfer, trans ASME **140(7)**, 071502 (2018)
45. Stevens, K. A., Crockett, J., Maynes, D., Iverson, B. D.: Simulation of drop-size distribution during dropwise and jumping drop condensation on a vertical surface: Implications for heat transfer modeling. Langmuir **35(39)**, 12858–12875 (2019)
46. Tanasawa, I.: Advances in condensation heat transfer. Advances in Heat Transfer **21**, 55–139 (1991)
47. Tancon, M., Parin, R., Bortolin, S., Martucci, A., Del Col, D.: Effect of steam velocity during dropwise condensation. International Journal of Heat and Mass Transfer **165**, 120624 (2020)
48. Tanner, D. W., Potter, C. J., Pope, D., West, D.: Heat transfer in dropwise condensation—part i the effects of heat flux, steam velocity and non-condensable gas concentration. International Journal of Heat and Mass Transfer **8(3)**, 419–426 (1965)
49. Torresin, D., Tiwari M. K., Del Col, D., Poulikakos, D.: Flow condensation on copper-based nanotextured superhydrophobic surfaces. Langmuir **29(2)**, 840–848 (2013)
50. Umur, A., Griffith, P.: Mechanism of dropwise condensation. Journal of Heat Transfer, trans ASME **87(2)**, 275–282 (1965)
51. Vemuri, S., Kim, K. J.: An experimental and theoretical study on the concept of dropwise condensation. International Journal of Heat and Mass Transfer **49(3-4)**, 649–657 (2006)
52. Wen, H. W., Jer, R. M.: On the heat transfer in dropwise condensation. The Chemical Engineering Journal **12(3)**, 225–231 (1976)
53. Wolfram, E., Faust, R.: Liquid drops on a tilted plate, contact angle hysteresis and the young contact angle. In Wetting, spreading, and adhesion, 213–222. Editor JF Padday (1978)
54. Xu, W., Lang, Z., Liu, Q., Du, B., Ma, X.: Droplet size distributions in dropwise condensation heat transfer: Consideration of droplet overlapping and multiple re-nucleation. International Journal of Heat and Mass Transfer **127**, 44–54 (2018)
55. Xu, Z., Zhang, L., Wilke, K., Wang, E. N.: Multiscale dynamic growth and energy transport of droplets during condensation. Langmuir **34(30)**, 9085–9095 (2018)
56. Zheng, S., Eimann, F., Philipp, C., Fieback, T., Gross, U.: Modeling of heat and mass transfer for dropwise condensation of moist air and the experimental validation. International Journal of Heat and Mass Transfer **120**, 879–894 (2018)

Chapter 5
Spreading, Wetting and Drying of Human Blood

Houssine Benabdelhalim and David Brutin

Abstract We emphasize the recent works on human blood drops and pools spreading then drying. Several studies have been carried out to investigate these coupled phenomena on crime scenes and for biomedical obvious interest. The aim of the forensic studies is to improve Bloodstain Pattern Analysis (BPA) methodologies in order to improve the elucidation rates, while the aim of biomedical studies is to detect disease using a drop of whole human blood. After presenting the blood physical and biological properties, we present the different mechanisms involved in the spreading and the evaporation of this complex fluid composed of bio-colloidal particles. Later the drying dynamics of drops and pools is be detailed. The effect of the surrounding relative humidity and temperature on the spreading and drying dynamics is explained. Also, the influence of substrate nature on the spreading dynamic and the final pattern is compared, since we are dealing with blood wetting surfaces of totally different nature. While substrates are mainly clean and smooth with microscope glass plates with or without coatings for biomedical studies, however in forensic science, the substrates are often rough with wood floors, tiles.

Keywords Pool · Drop · Drying · Spreading · Bloodstain pattern analysis

5.1 Human Blood Properties

Blood is a fluid tissue, permanently conveyed in the circulatory system in a unidirectional flow imposed by the cardiac pump. It has several physiological functions: diffusion of oxygen and nutrients to all tissues of the body, removal of carbon dioxide and metabolic wastes, transport of immune system cells, transport of hormones, participation in the body's thermoregulation and hemostasis. Human blood is a complex biological fluid composed of a liquid phase, plasma and suspended cells (bio-colloids)

H. Benabdelhalim (✉) · D. Brutin
Aix Marseille Univ, CNRS, IUSTI, Marseille, France
e-mail: houssine.benabdelhalim@uni-amu.fr

D. Brutin
e-mail: david.brutin@univ-amu.fr

M. Marengo and J. De Coninck (eds.), *The Surface Wettability Effect on Phase Change*,
https://doi.org/10.1007/978-3-030-82992-6_5

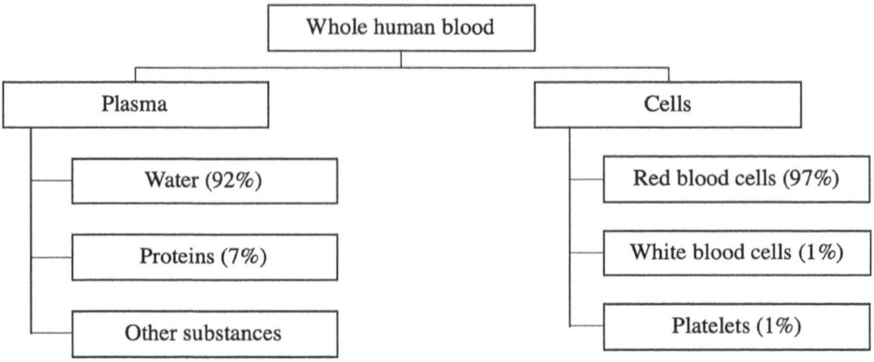

Fig. 5.1 Organizational chart showing the components of whole human blood

[39] Fig. 5.1. Plasma, the fluid in which blood cells are suspended, is an aqueous solution of components that vary widely in size.

Erythrocytes, also called red blood cells, are the most abundant blood cells. Red blood cells have a biconcave disc shape, with diameter of $7, 7 \pm 0, 7\,\mu$m. The volume proportion of red blood cells present in the blood is given by a dimensionless quantity, hematocrit. In a healthy adult subject, hematocrit ranges from 38 to 46% for a woman and 40–50% for a man. Under pathological conditions, the hematocrit may fall as low as 20% and rise to about 70%.

Leukocytes, also called white blood cells, are cells of the immune system. Due to their low physiological number in the blood, they contribute little to the mechanical properties of the blood. Platelets or thrombocytes play a major role in primary hemostasis; they are involved in coagulation. They are of little hemorheological importance because they are very small components that do not interact with each other under normal conditions.

It has been known since the 1950s that whole human blood is a non-Newtonian fluid [15], and in particular that it has marked shear-thinning properties (its viscosity decreases as the shear rate increases). Its behavior depends on hematocrit [40] and temperature [1]. These mechanical properties have different origins:

- The viscosity of blood is partly related to that of plasma, which due to the different proteins in solution is about twice as viscous as water.
- The ability of erythrocytes to form in the presence of certain molecules in solution weakly coherent aggregates in the form of rolls (rouleaux) gives the blood shear-thinning properties. Indeed, these rouleaux tend to disintegrate when the shear increases, resulting in a drop-in viscosity Fig. 5.2.
- The deformability of erythrocytes also contributes to the rheology of high shear blood Fig. 5.2, which results in a Newtonian behavior at high shear rate characterized by a plateau, a constant relative viscosity.

As shown in Fig. 5.2, with hematocrit 45% and a temperature of 37 °C, the relative viscosity of the blood as a function of shear rate reaches a plateau for values greater

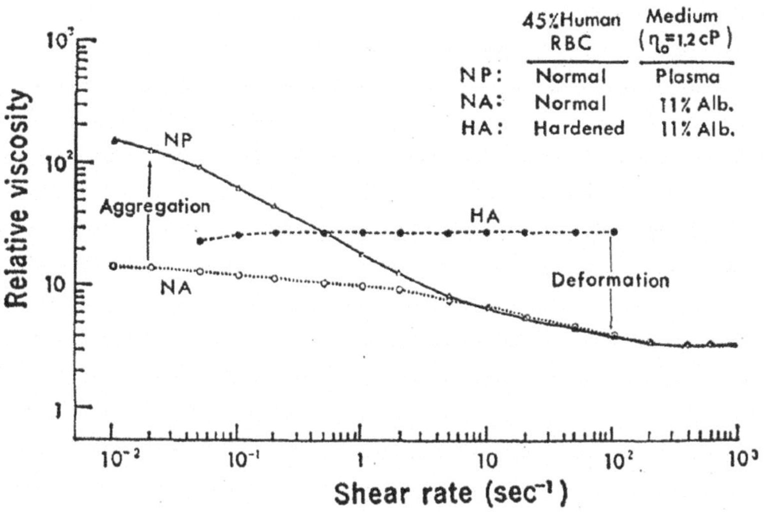

Fig. 5.2 Viscosity of whole human blood (NP curve), red blood cells suspended in 11% albumin-Ringer solution (NA curve) and hardened red blood cells in 11% albumin-Ringer solution (HA) as a function of shear rate [10]

than 200 m s^{-1}, and it measures between 4 and 5 mPa s. Also, blood has a density that varies between 1020 and 1060 kg m^3 and, this variation is due the hematocrit level, which depends on the individual. The surface tension is similar to that of water, $\gamma_{blood} = 69.8$ mN m^{-1} at 22°C, determined with the pendant drop method [6].

5.1.1 Spreading, Wetting and Evaporation of Human Blood Drops

When a drop of liquid is deposited on an ideal substrate without an initial velocity it will spread, at the equilibrium, the smaller ones have a spherical cap shape under the effect of surface tension, while the larger ones and pools are flattened by gravity. The competition between gravity and surface tension is given by the number of Bond, as follows:

$$Bo = \frac{\rho gr^2}{\gamma}$$

where, γ, the surface tension (in mN. m^{-1}) The radius of the contact line is r, ρ and h are the density and height of the drop, respectively. An important value of the number of Bond corresponds to the effects of gravity dominating those of surface tension. The value Bo = 1 corresponds to a critical size called capillary length and is written

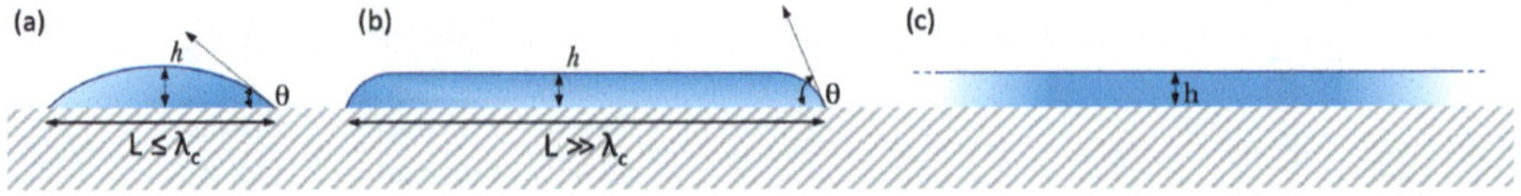

Fig. 5.3 Representation of the different possible cases depending on the radius at equilibrium **a** a droplet $r \leq l_c$, **b** a pool $r \gg l_c$, and **c** an infinite stretch of liquid [7]

as follows:

$$l_c = \sqrt{\frac{\gamma}{\rho g}}$$

To determine the relative importance of the surface tension, l_c is compared with the radius of the contact line, as shown in Fig. 5.3. In the figure, θ represents the contact angle.

The wetting of a drop of a volatile liquid is most often accompanied by evaporation. The evaporation of a drop is defined as the passage of the liquid to its vapor form. It can be seen from a molecular point of view, as the transfer of molecules from the liquid to the surrounding, while on a macroscopic scale, it is defined as a flux that crosses the interface (liquid/vapor). This phenomenon exists when the partial pressure in the gas is lower than the saturation pressure. Evaporation persists as long as the saturation pressure is not reached, and it is controlled by the diffusion of vapor into the environment. In this case, we call it purely diffusive evaporation.

When a drop evaporates in contact with a substrate, we notice the appearance of the triple line motion and the change in the local evaporation flux. The flux profile varies along with the interface and changes according to the contact angle, therefore, depending on the shape of the drop, three situations can be distinguished [7]:

– if $0° < \theta < 90°$, the evaporation flux is minimal at the top of the drop, grows with r and diverges near the triple line [13]

– if $\theta = 90°$, the evaporation flux is uniform along with the interface [18]

– if $90° < \theta < 180°$, the evaporation flux is minimal at the triple line and maximal at the top of the drop.

Different modes of evaporation exist, depending on the dynamics of the triple line and the contact angle [7]. These modes have an influence on the internal flow, hence the mass transfer. Two modes can be distinguished according to Picknett and Bexon [30]:

– the drop evaporates with a constant contact surface; the contact line is pinned, and the angle decreases over time Fig. 5.4

– the drop evaporates with a constant contact angle; the contact radius decreases over time Fig. 5.4. This mode is characterized by a recession of the contact line.

In certain cases of complex evaporation, the two modes can be alternated "Stick–slip motion". The first stage is during which the contact line is pinned to the surface, and the contact angle decreases until the receding contact angle. The second stage starts with a constant contact angle, the contact line recedes until a certain contact

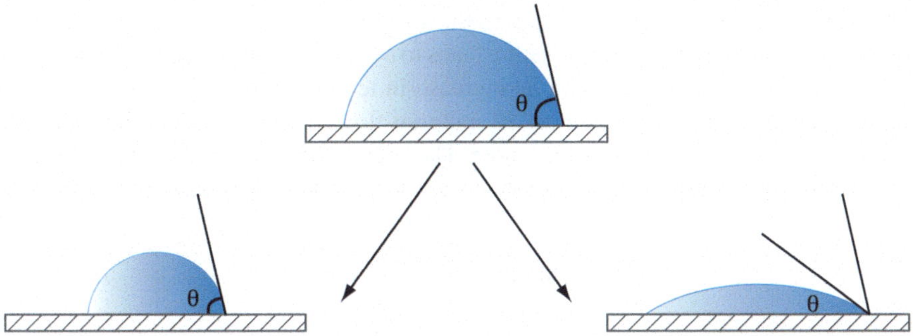

Fig. 5.4 The two evaporation modes according to the dynamics of the contact line [7]

radius is reached. Then this periodic process is repeated until equilibrium is reached. Shanhan and Sefiane give an explanation to the Stick–slip motion based on the Gibbs free energy. Also, we have the mixed-motion case, both modes happen at the same time [22].

These two dynamics described above depend on the roughness and the heterogeneity of the surface, as well as the temperature and humidity. Bridi and Vu have shown according to the wetting of the surface (Glass-Teflon) [2], the evaporation rate is linear with time in the case of a water droplet that has a contact angle $\theta < 90°$, and non-linear in the case where $\theta > 90°$.

Sobac and Brutin study the influence of substrate properties [36], for a wide range of contact angles (from 17 to 135°) and five different fluids, in the case of pinned contact line. They describe the effect of contact angle on evaporation dynamics: contact angles less than 40° have a negligible effect, unlike angles greater than 40°. Also, they give a universal relation of the evolution of the drop mass, independent of the drop size and the initial contact angle.

Evaporation gives rise to convective mechanisms within the drop [7], such as thermogravity convection and thermocapillary convection, which induce flows inside the drop.

Thermogravity convection is due to the variation in density ρ with temperature (density decreases with an increase in temperature). The presence of a vertical temperature gradient leads to an ascent of the fluid from the hottest part to the coldest part. As the fluid cools, it goes back down. This leads to a circular flow. The number of Rayleigh Ra characterizes thermogravity effects by comparing the Archimedean buoyancy force to the dissipative forces:

$$\mathrm{Ra} = \frac{\beta g L^3 \Delta T}{\nu \alpha}$$

where β is the coefficient of thermal expansion, ν is the kinematic viscosity, L is the characteristic length, and α, is the coefficient of thermal diffusion.

Thermocapillary convection is due to the variation of the surface tension γ with temperature. A difference in temperature leads to a difference in surface tension, since the surface tension decreases with an increase in temperature. This surface tension gradient generates a flow known as "Marangoni flow", which will set the liquid in motion from the hottest to the coldest area. The Marangoni Ma number characterizes the thermocapillary effects by comparing the surface force to the dissipative forces:

$$\mathrm{Ma} = \frac{-\frac{d\gamma}{dt} L \Delta T}{\rho \nu \alpha}$$

The Marangoni flow results from a surface tension gradient, which may be due to the temperature gradient as presented above, or to the local change in the composition of the dispersions (colloids or polymers), known by solutal convection [37]. Also, it can be caused by the presence of surfactants [16].

The spreading of a drop on a solid substrate can be characterized by the evolution of the radius r(t) and the apparent contact angle θ(t) over time, until the flow comes to a complete stop. The competition between forces: capillary, inertial, gravitational and viscous determine its dynamics during the different phases that characterize the flow. Spreading can be affected by temperature, humidity, surface nature and rheological behavior of fluids and their compositions. Surrounding conditions affect wetting, spreading and evaporation, since the viscosity and surface tension of the fluid are dependent on temperature, and the evaporation is a function of relative humidity (RH) and temperature (T). The contact angle and the triple line are affected by the substrate characteristic. Solid substrates are not ideal (smooth and homogeneous), they can be chemically heterogeneous and rough. Also, substrates can be porous, in this case, the spreading dynamics is largely influenced by the permeability of the pores. The rheological behavior of fluids is important for both, wetting and spreading. The fluid may exhibit Newtonian or non-Newtonian behavior: shear-thinning, shear-thickening and yield stress fluid. Fluid behavior is characterized by models that give stress and viscosity as a function of shear rate.

The dynamics of drop spread in a total wetting situation is investigated by Tanner in the case of small drops (negligible gravity effect) of a non-volatile fluid [38]. For these small drops, the dynamic results from a balance between viscous dissipation and capillary forces (viscous dissipation in the zone near the contact line). He assumed that the radius of curvature of the interface is large enough in front of the drop thickness, which allows the theory of lubrication to be applied. The drop takes the shape of a spherical cap of volume $V = (\pi/4)r^3\theta$. With r the radius of contact with the substrate, and the contact angle θ is given by Tanner's relation $\theta^3 \sim \mathrm{cte} \times \mathrm{Ca}$. The capillary number $\mathrm{Ca} = \mu U \gamma^{-1}$ represents the relative importance of effects due to viscosity and those due to capillarity. He predicted a slow growth of the drop radius over time $r(t) \sim t^{1/10}$, and a decrease in the angle of contact $\theta(t) \sim t^{-3/10}$.

Gravity plays an important role in determining the geometry of large drops and their spreading dynamics. Unlike the first case, the drops will be flattened in the central part and their curvature is localized near the edges. The spreading law is given by Lopez $r(t) \sim t^{1/8}$, in the case of a non-volatile fluid that spreads on an ideal substrate [25]. This law results from a balance between viscous dissipation in the bulk of drop and variation in potential energy. Cazabat and Stuart studied the spreading of a drop of silicone oil [9], and they observed a law of variation with an exponent $1/10$ at short times, in this case, the radius is small in regard to the capillary length $r < l_c$, from which the capillary effects remain dominant. At longer times, the radius becomes larger than the capillary length, we are moving into a regime dominated by the forces of gravity with an exponent $1/8$.

After, we briefly discussed the spreading and evaporation of a drop of a liquid, and its different modes of evaporation, and also the internal flows that can take place with the evaporation. Now we are interested in drops of human whole blood. The understanding of these phenomena in the case of a drop of blood is of crucial importance. Given its great interest in biomedical applications, specifically in disease detection, and in forensic applications by crime scene investigators.

The dynamics of spreading and evaporation of blood droplet is given by Bou-Zeid and Brutin [5]. They realize an exhaustive study on blood droplet that spread over a clean glass substrate. A blood droplet spread by following two regimes, one controlled by capillary forces and a second controlled by viscous dissipation near the contact line. A blood droplet wets the substrate keeping a spherical cap shape with a contact angle at equilibrium equal to 20°. Also, they investigated the influence of relative humidity, RH, spreading is rapid at high evaporation rates (low RH), contrary to low evaporation rates (high RH).

The first spreading regime results from a competition between capillary forces and viscous dissipation near the contact line. In Fig. 5.5, the curves obtained for different RHs merge, hence the absence of the influence of humidity. The evolution of the radius r(t) is described by a power law $r^*(t) = k\left(\frac{t}{t_d}\right)^n$, with a factor $k = 2,29 \times 10^{-2}$ and a wetting factor $n = O,65 \pm O,11$ obtained by fitting all the curves. r^* is the normalized radius, $t_d = 4\pi\mu/\gamma R_f$, is the viscous capillary characteristic time, which is used to normalize the time in this first regime.

The first regime is dominated by capillary forces and gravity is negligible Table 5.1, which gives a dynamic with a wetting factor $n = O,65 \pm O,11$. The same dynamics are observed for a fluid with a viscosity of 10 mPa s. This shows that bio-colloids do not have a remarkable effect in this first spreading regime [23].

A competition between viscous dissipation and the pressure difference (between the saturated pressure at the droplet interface and the air pressure) characterizes the spreading dynamics of the second regime Fig. 5.6. The evolution of the radius over time is given by a power law like the first regime $r^*(t) = k\left(\frac{t}{t_F}\right)^n$, with a wetting factor $n = O,19 \pm O,03$. The time was scaled by the total evaporation time $t_F = \frac{4RTm_0}{\pi \Delta P D_f D_{diff}}$, predicted by the use of the pure evaporation model as the process is limited by water diffusion.

The second regime is governed by the viscous dissipation near the contact line Table 5.1. The drop during spreading keeps the shape of a spherical cap with a

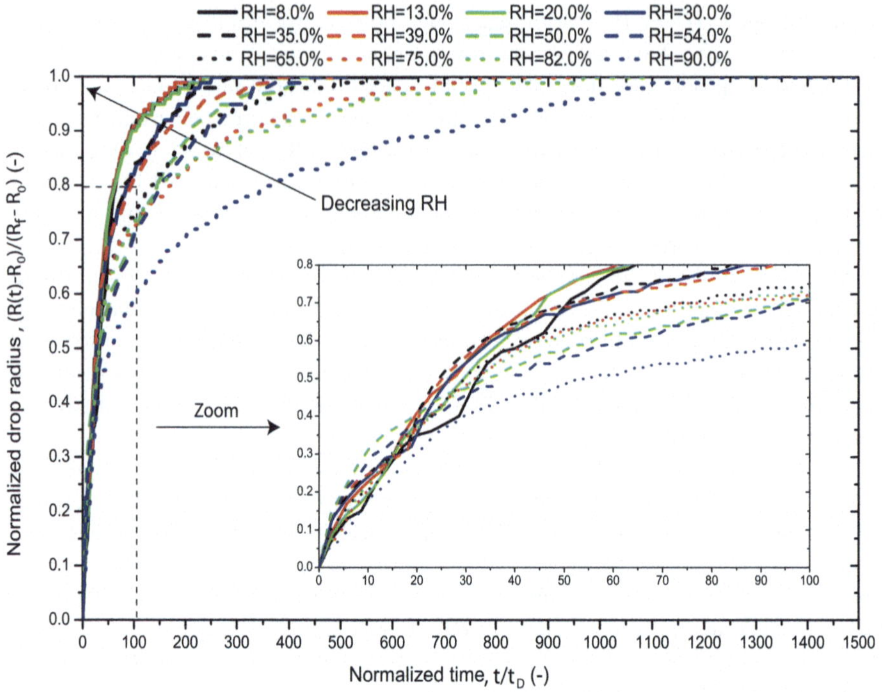

Fig. 5.5 Normalized drop radius, $(r(t) - R_0)/(R_f - R_0)$ as a function of normalized time, t/t_D, for RH varying from 8 to 90%. The inset graph is a magnification of t/t_D from 0–100 [5]

Table 5.1 Estimation of characteristic dimensionless numbers for the two regimes of blood drop spreading at R = 65% [5]

Number	Formula	Signification	Regime 1	Regime 2	Ration 1/2
R (mm)	–	Mean radius	3, 18	4, 54	0, 7
H (mm)	$R(1 - \cos(\theta))/\sin(\theta)$	Mean height	1, 05	0, 41	2, 5
U (mms^{-1})	–	Mean velocity	1, 92	0, 11	17, 4
Re	$\rho Uh/\mu$	Inertia/viscous	0,209	0, 004	44, 4
Ca	$\mu U/\gamma$	Viscous/capillary	$2, 75 \times 10^{-4}$	$1, 58 \times 10^{-5}$	17, 0
Bo	$\rho gh^2/\gamma$	Gravity/capillary	0,160	0, 024	6, 5
We	$\rho U^2h/\gamma$	Inertia/capillary	$5, 74 \times 10^{-5}$	$7, 41 \times 10^{-8}$	774
Fr	U^2/gh	Inertia/gravity	$3, 59 \times 10^{-4}$	$3, 00 \times 10^{-6}$	120
St	$\mu U/\rho gh^2$	Viscous/gravity	$1, 72 \times 10^{-3}$	$6, 38 \times 10^{-4}$	2, 7

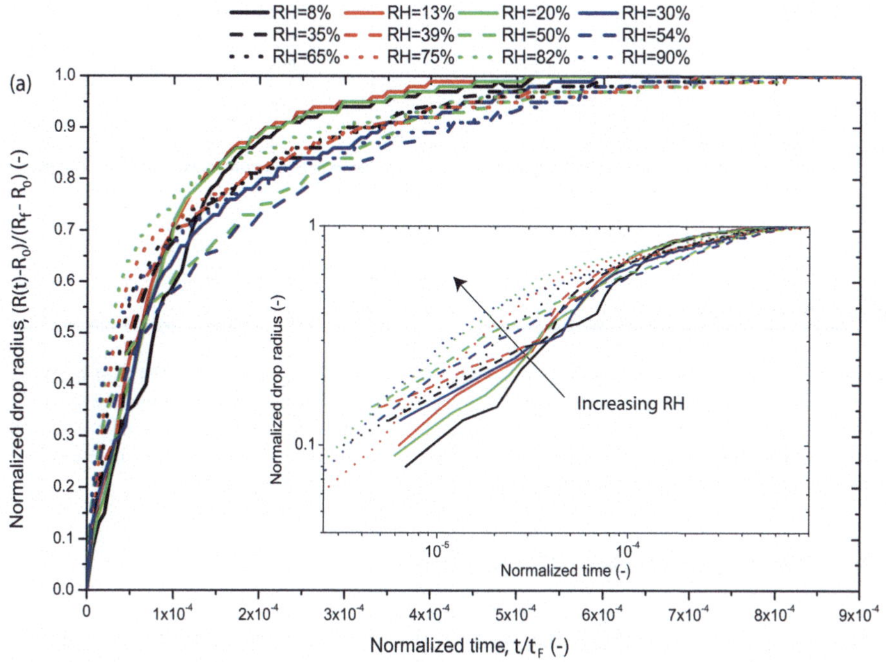

Fig. 5.6 Normalized drop radius, $(r(t) - R_0)/(R_f - R_0)$ as a function of normalized time, t/t_F, for different humidities [5]

negligible gravity effect shown by the calculation of the Bond number, less than 1 for both regimes Table 5.1.

In this experiment, the evaporation mass flow is not uniform along the surface of the drop and tends to diverge near the contact line. This will cause temperature gradients on the surface, hence a gradient of surface tension leads to a Marangoni flow inside the droplet [19]. This flow will transport the colloids to the edge near the contact line, which evaporates faster, and so the bio-colloids accumulate, forming a layer of biological deposition. The contact line moves radially with a speed of U_{cl} and then it becomes pinned. Proteins in the blood will anchor the contact line to the solid substrate.

At low relative humidity (high evaporation rate), Bou-Zeid and Brutin notice an increase in the evaporation rate near the contact line Fig. 5.7a. Unlike the case where we have a high HR Fig. 5.7b, this will intensify the transport of colloids to the contact line, resulting in an increase in flow velocity near the contact line [5].

This section presents the evaporation and spreading of blood, in the case of a blood droplet that spreads gently on a glass substrate. In which the effect of humidity on the rate of evaporation was taken into account. In the next section, we will present the final evaporation phase, which is mainly characterized by gelation, adhesion and crack formation, and by a complex pattern at the end of evaporation (total drying).

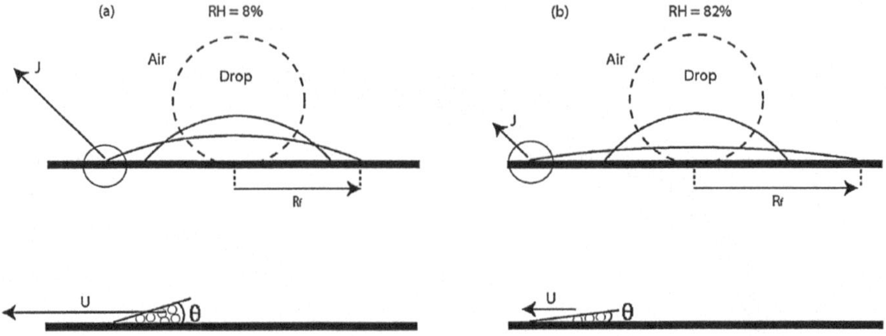

Fig. 5.7 Schematic representation of spreading drops for the case of: **a** R = 10% and **b** RH = 80%. Initially, the blood drop has a spherical shape, and at equilibrium, the drop has the shape of a spherical cap. The arrows indicate the velocity and evaporative flux of water vapor from the drop surface to the surroundings in the vicinity of the triple line. The dashed square marks the area close to the triple line, where the drop geometry can be approximated by a wedge [5]

5.1.2 Drying of Human Blood Drops

A blood drop dries on a glass substrate by following five stages until it is completely dry [6], in the end, only the non-volatile constituents, the bio-colloids, remain. Drying of the blood drop is characterized by two regimes, a first regime driven by convection, diffusion and gelation, and a second driven by diffusion through the pores of the gelled bio-colloids, with a lower evaporation rate than the first regime [35].

As described above, the presence of protein in the blood promotes contact line pinning, and the internal flow leads the bio-colloids to the edge of the drop forming a ring. This phenomenon is known as the "coffee-ring effect", observed and described experimentally by Deggen et al. [12]. Evaporation continues with an increase in the local concentration of the bio-colloidal particles ϕ_s, then the particles aggregate to form a gel at the edge of the drop [29], which propagates inwards of the droplet. This mechanism characterizes the first regime, which is clearly shown in Fig. 5.8, a ring appears at time τ_{C1} resulting in a high concentration of RBCs, and the central surface remains liquid. Then we have a transient regime that is characterized by the gelation of the center with a decrease in the rate of evaporation, as shown in Fig. 5.8, a shrinkage of the liquid part (image τ_{C2} and τ_G). And finally, a second regime in which an amount of liquid remains trapped within the gelled cellular matter and dries with a very long evaporation rate.

Figure 5.9 clearly shows the different regimes. The first regime presents a non-linear decrease compared with the evaporation of pure liquid due to the perturbations caused by gelation, with an evaporation rate equal to 5.2 μg/s. The last regime is 56 times slower than the first regime, since the liquid phase diffuses through the pores. A competition between evaporation and gelation drives this process, therefore, the authors define two distinct characteristic time. τ_G defined as gelation time, taken at the end of the sol–gel transition, and τ_D the desiccation time, which corresponds to

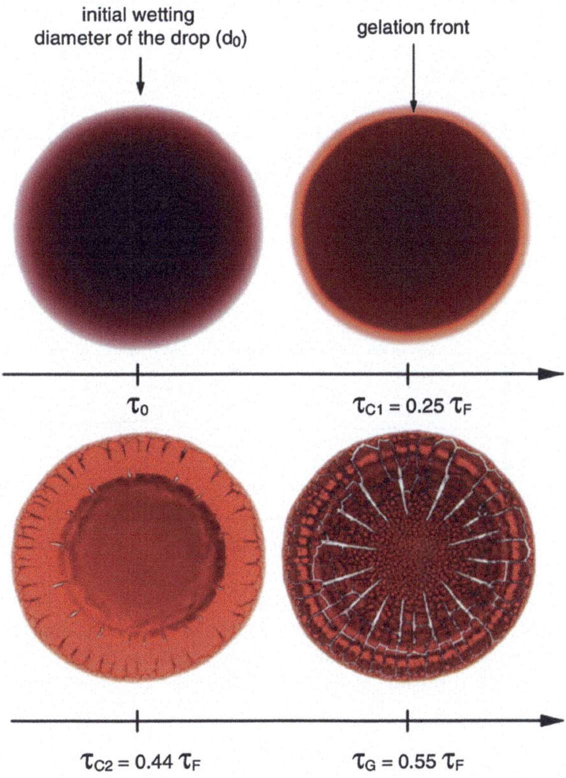

Fig. 5.8 Drying mechanism of blood droplet (drop diameter $D_0 = 8, 6$ mm, initial mass $m_0 = 17, 3$ mg, final mass $m_f = 3, 51$ mg at room temperature 22,0 °C and surrounding humidity 42%. τ_0 represent a deposit time, τ_G first time of nucleation of cracks, τ_{C2} time of nucleation of cracks at center, τ_G gelation time, τ_F total drying time [35]

the time needed for the drying of a drop, which contains solvent only, given by the relation [35]:

$$\tau_d \approx \frac{\theta_0 \rho R_0 S_0}{4\left(\frac{dm}{dt}\right)_{t=0}}$$

Developed by assuming the case of pinning contact line, and a spherical cap shape. Where ρ the density of the fluid, m the mass and V, S and R, respectively, are the volume, surface and radius at time t. Also, they develop a relation for the gelation time, which corresponds to the time of the transition between the two regimes:

$$\tau_G \approx \left(\frac{2,036}{D}\right) {m_0}^{2/3}$$

where D represents the coefficient of diffusion. In their experiment, they obtained for the first regime a coefficient equal to $5, 5 \times 10^{-7} m^2/s$, and for the last regime

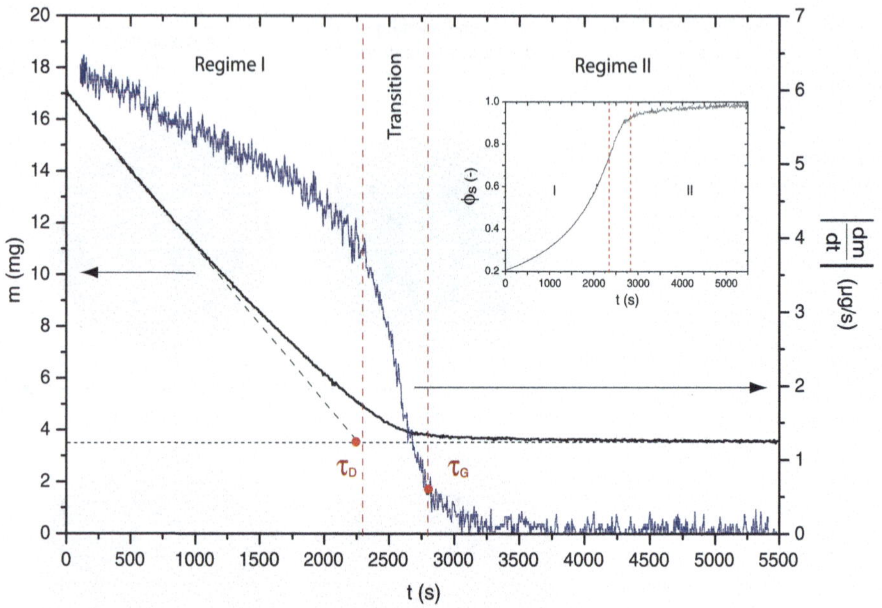

Fig. 5.9 Mass m, drying rate [dm/dt] and solid mass concentration $\varnothing_s$ as a function of time for a drying drop of blood under the same condition as Fig. 5.1. Regime 1: drying regime driven by convection, diffusion, and gelation. Regime 2: drying regime only driven by diffusion [35]

$9,8 \times 10^{-9} m^2/s$. These two values agree to the two extreme cases, pure water diffusion and diffusion through a colloidal gel [28].

The drop mass has an influence on the different regimes, which characterize the drying of the blood drop [35], the duration of the last regime becomes longer as the mass increases. Since the increase in mass leads to an intensification of the thickness of the gel deposit, as a result, more liquid is trapped and diffusion through the pores becomes longer.

The final pattern of a blood droplet disposed on a glass substrate contains cracks [6], as shown in Fig. 5.8. These are formed when the liquid diffuses through the pores of the gelled matter [22]. As described earlier, the gelation starts at the edge of the droplet and propagates toward the center. As a result, cracks also form at the edge, and during the drying process, they will spread toward the center. Sobac and Brutin observed in their experiments that the first cracks nucleate at a certain critical particle concentration of 29.6% [35].

The dynamics of crack formation during drying depends on the mechanical stresses resulting from the competition between evaporation and adhesion to the substrate [26], and the thickness of the gel layer [3]. Indeed, during evaporation, a porous matrix is formed by the aggregation of particles, so the system will be defined by the two interfaces shown in Fig. 5.10. A gel–substrate interface, where adhesion takes place, and a gel–air interface, where the solvent evaporates. A capillary pressure, P_{cap}, is created in the liquid phase caused by the presence of a menisci at the

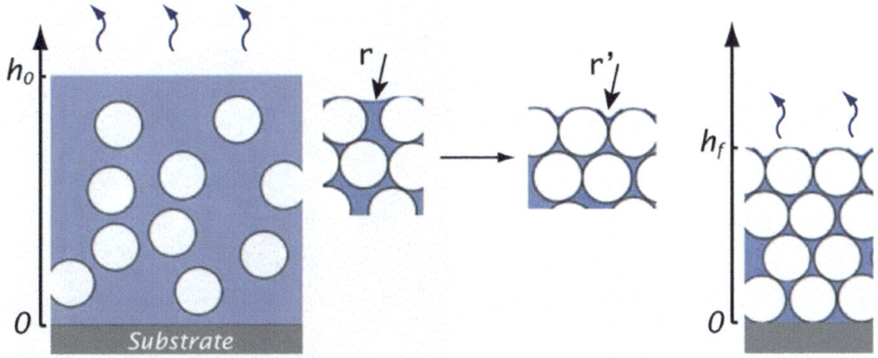

Fig. 5.10 Representation of the drying process of a colloidal system, which results in the formation of porous matrix [22] (left). Just after disposition, the system is characterized by dispersion suspension (right). A packed colloidal system takes shape during drying, known as porous matrix. r and r' denote the curvature of the menisci before and after shrinkage, respectively

gel–air interface.

$$P_{cap} = \frac{\alpha \gamma_{s,a} \cos(\theta)}{r_p}$$

where $\gamma_{s,a}$ surface tension, r_p pore radius and $\alpha \approx 10$ geometric constant. With the evaporation, the curvature of the menisci increases, thus creating a depression in the solvent. The latter has two major effects, it leads to solvent migration to the surface and also to the shrinkage of the porous matrix which is counteracted by adhesion to the substrate. As a result, a tensile stress develops [24]. During evaporation, this stress becomes important, and it relaxes by forming cracks through the release of energy.

The final pattern is influenced by the characteristics of the substrate [8] and also by the relative humidity [4], see Figs. 5.11 and 5.12 (right), respectively. Brutin et al. [8] have studied the drying of drops of blood on gold and aluminum substrates, and they showed that the thermal diffusivity of the substrate has no influence, however, substrate wettability has an important influence on the evaporation dynamics, hence on the final pattern. The initial contact angle on a glass substrate is $\theta = 15°$ with a non-uniform evaporation rate. However, it is almost equal to 90° for a gold substrate ($\theta = 91.9°$), in this case, the evaporation rate is uniform along with the interface [9], resulting in an absence of Marangoni flow. Blood remains homogeneous, and with evaporation, the particles move toward the interface by means of a radial flow that is created inside the drop. These particles compact forming a shell, which undergoes a sol–gel transition, and a glassy skin appears on the surface of the blood drop Fig. 5.11. The final pattern obtained on metallic substrates has a complex morphology, results from the existence of buckling instabilities [34, 27], and the fact that the center of the drop is composed only of air.

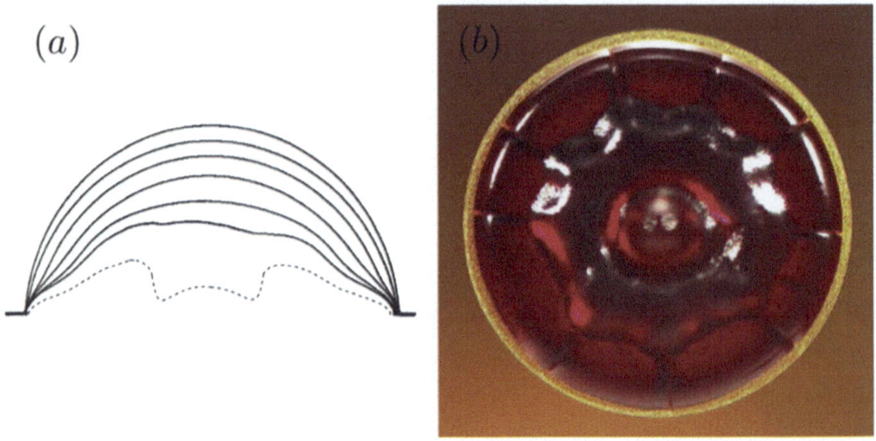

Fig. 5.11 **a** Shape evolution of blood droplet, which dries on gold substrate with an initial contact angle $\theta = 91.9°$ **b** The final pattern of blood droplet on gold substrate [34]

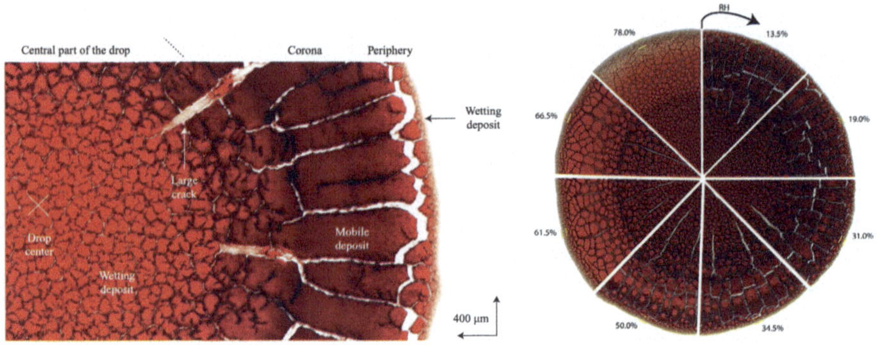

Fig. 5.12 (left) Composition of final pattern of blood droplet dries on glass substrate [6]. (right) Final pattern obtained from the drying of blood droplets with the same mass on a glass substrate under variable humidity. The fine periphery is delimited with the two yellow dashes [4]

Brutin et al. in their study on the drying of blood drops give a detailed description of the final pattern of a blood droplet [6] Fig. 5.12 (left). The pattern is mainly characterized by three distinct regions: a central part composed of small cracks, a corona formed by the accumulation of bio-colloids, which is composed of cracks and mobile plaques, and finally a well-defined border (a periphery) adhering to the glass substrate. The morphology of this pattern is highly dependent on the relative humidity RH of the surrounding Fig. 5.12b. The increase in RH (decrease in evaporation rate) changes the mechanical properties of the gel, such as the adhesion energy, which decreases according to Griffith's theory. As a result, the mobile plaques become wide with a lower adhering part, for RH values ranging from 13.5 to 50%, and for RH values above 50%, the mobile plaques are small with a most important adhering part.

This result is due to buckling instabilities [27], which will be quickly overcome by the adhesion of the gel to the substrate. Also, a change in the evaporation rate leads to a change in the internal flow, which makes the thinner periphery wider Fig. 5.12 (right). In general, for the same droplet volume, we can have several patterns depending on the humidity of the surrounding.

5.1.3 Stages of Human Blood Pool Spreading, Wetting and Drying

After a deadly blood-shedding event, blood pools are one of the important evidences that can be found at crime scenes. Pools represent the accumulation of an important quantity of blood on a substrate. From a pool of blood, the investigator or the bloodstain pattern analysts (BPAs) can make two simple conclusions by just observing, the first, if the pool is still in liquid state, the event is recent, and the second, if the pool is completely dried, the event must have taken place some hours or days ago. But these deductions are not reliable, and they can't propose them as evidence in front of the courts. Recently, courts have begun to require scientifically validated evidence for the reconstruction of crime scenes. Hence, need to understand the spreading and drying of blood pools from a physics and biomedical point of view. Several researchers investigated this topic, we can notice the pioneer works of Smith et al. about the time reconstruction of blood pools [33], the studies of Laan et al. on the morphology of drying blood pools, in which they describe the drying of blood with exhaustive manner [20], and also the works of Ramsthaler et al. and Laber and Epstein on the drying of blood pools [31] and the influence of substrate nature on coagulation [21], respectively [32].

Laan et al. have investigated the drying of blood pools, as result, a morphological characterization of the drying was given [20]. They follow the drying of several blood pools with less than 4.5ml of volume over time, on linoleum surfaces. They show that the kinetics of mass variation is similar and repeatable under different conditions, as well as morphological changes. These changes are distinguished by five stages Fig. 5.15, with different aspects related to the dominant phenomena:

(1) When a given volume of blood is deposited on a substrate, it has a dark red color, and the bio-colloids are distributed homogeneously. At the same time as the blood spreads until it reaches equilibrium, evaporation takes place and coagulation occurs, leading to a change in color to a light red. Laber and Epstein have shown that the total clotting time of blood pools depends on the substrate on which they are poured, and their volume [21]. A pool with a small volume on human skin clots rapidly compared with a pool on wood. Also, there are other parameters that may influence clotting, such as temperature and relative humidity, were not considered in their study.

(2) If the blood is ex-vivo, hemostasis will occur that is characterized by the coagulation and formation of the fibrin web, which led to the gelation. The latter

is the passage from the liquid state to the gel state. This stage begins with the gelation of the rim, after this front will propagate until the total gelation of the pool in the fourth stage. The drying of blood pools is similar to that of a gel, since the drying of both leads to a sol–gel transition.

(3) At the same time as the gel propagates toward the center of the pool, another front is formed characterized by a black color and the presence of cracks. It is defined as the drying or desiccation front. And during this stage, both fronts, drying front and gelation front, continue to propagate toward the center of the pool.

(4) At this stage, the pool is completely gelled, and evaporation will take place through the pores, and also the drying front continues to propagate toward the center. It should be noted that this front does not evolve uniformly, as can be seen in Fig. 5.13 (linoleum) and Fig. 5.14 (floor wood). Hence, the center of

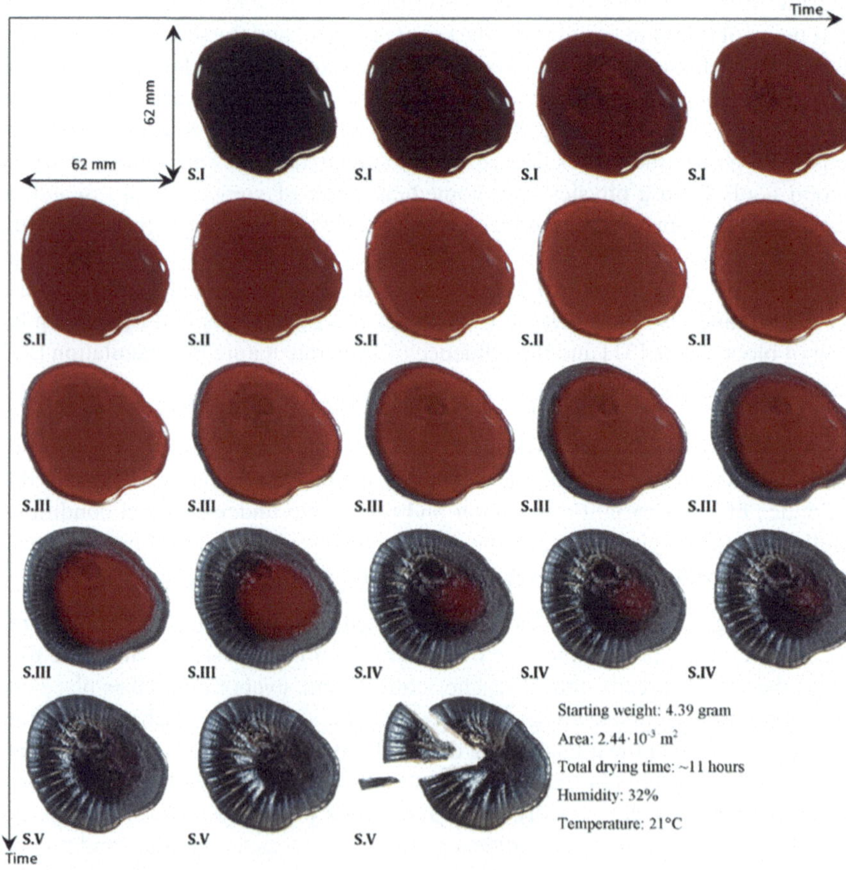

Fig. 5.13 Time lapse of a drying pool of blood from a healthy person, at 22 °C with a relative humidity of 32% [20]

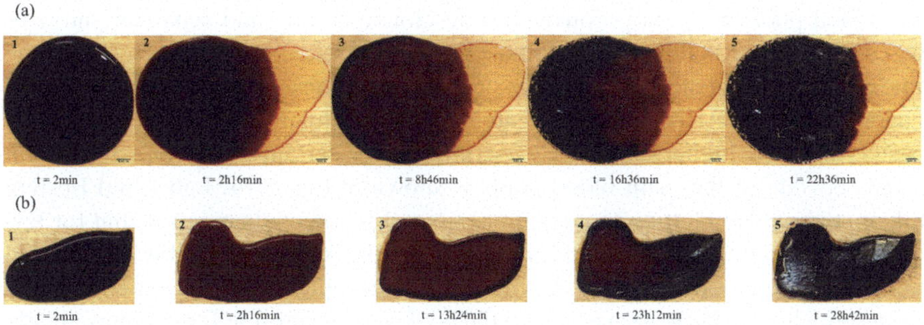

Fig. 5.14 Time laps of a drying pool of blood (m = 4,83 g ± 0.6%, Hct = 41,5%) on varnished **a** and unvarnished **b** floor wood, at 21°C with a relative humidity of 60%, showing the serum separation and the five drying stages

pools is not a geometric center, but the earlier the point where all the cracks some together.

(5) At the end, the pool has a black color, which represents a total visual drying time. And during this stage, the last volatile components continue to evaporate, and the cracks spread until they reach the center of the pool. As a result, some flakes detach and separate from the pool.

The final pattern of a completely dried pool, the total drying time and the duration of each stage are a function of the temperature, relative humidity, the nature of the substrate, the composition of the blood and the size and shape of the pool.

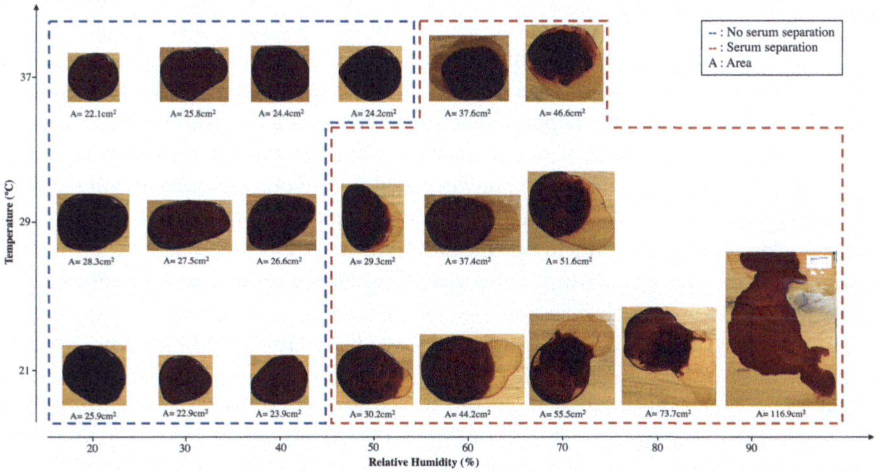

Fig. 5.15 Phase separation during the spreading and drying of 20 pools of blood with a mass of 4.83 g ± 0.6%, on varnished floor wood, under three different temperatures (21, 29 and 37 °C) and relative humidity levels ranging from 20% up to 90% with a step of 10%. The blood is taken from the same donor with a hematocrit level of 42.8% ± 2.7%

A blood pool on a non-porous substrate clots naturally and separates into two phases, liquid and cellular. The liquid phase, called serum, does not contain the same components as the plasma since some of it was consumed during coagulation. Separation is characterized by sedimentation of the cellular components due to their density, which is higher than that of serum. Under certain conditions, the serum spreads outside of the main blood pool, as shown in Fig. 5.14. Laber and Epstein in their study about clotting (coagulation) time of blood pools mention that for few pools, serum started to spread out of the edge of the main pool, and it is related to the fact that the blood is flowing down a slope. Ramsthaler et al., assumed that this serum spreading is due to the volume of the pool, and it is present for volumes greater than 10 ml. This phenomenon is encountered in crime scenes, and its understanding allows a reliable reconstruction of crime scenes, hence the need for further studies to give a more detailed scientific explanation.

Based on the fact that phase separation exists naturally and does not depend only on the volume of the pool, and assuming that the spread of the serum is due to an external factor. We choose to investigate this phenomenon, by taking a constant pool mass (volume), with varying environmental conditions and substrate types. In order to tackle the influence of blood composition on coagulation and spreading, blood is taken from the same donor in good health with hematocrit level of $Hct = 42\% \pm 2.7\%$. The blood is collected in neutral 9ml tubes, without any anticoagulant or activator, and the blood pool is created directly after collection to avoid blood clotting and to simulate a real bloody event.

Figure 5.14 shows the drying of two blood pools with the same mass ($4.83g \pm 0.6\%$) on varnished and unvarnished wood, hence a smooth and a rough surface, respectively. The two pools are created under the same conditions, a temperature of 21°C and a relative humidity of 60%. By observation, we can say that the type of surface has an important role, on the smooth surface we have a remarkable separation of the serum, however, on the rough surface, the serum spreads a little bit. In Fig. 5.14a, the serum separation started at 22 min after creating the pool, and it stopped after 3 h 50 min, the separated serum represents 24% from the total area of the pool. Also, the total area of the pool obtained at equilibrium on varnished floor wood is greater than the final area on unvarnished floor wood. So, we can put the first conclusion, the separation of serum depends on the nature of the substrate on which the pool was created. To confirm our observation, we performed other experiments, keeping the same type of substrates and modifying the environmental conditions. As results, we had the same observation.

To investigate more the serum separation, we have created 20 blood pools with the same mass on a varnished substrate, under three different temperatures (21, 27 and 37 °C), and the relative humidity was varied from 20 to 90% with a 10% step. The results obtained are shown in Fig. 5.15. The blue box represents the case without serum separation, and the red box represents the opposite case. As we can see in the figure, phase separation only occurs at high relative humidity levels. In our study, at relative humidity (RH) levels above 50%. At a constant temperature, e.g., 21 °C, this separation increases as we increase RH. Also, the increase in RH leads to an intensification of the spreading of the pool, with the same mass we have an equilibrium area of 25.9 cm^2 at 20% and 116.9 cm^2 at 90%. This is due to the fact that the increase in RH slows down evaporation, hence the main pool and serum can spread further. This effect is observed also on blood drops, which we presented in Sect. 5.2.1.

The rate of evaporation increases with an increase in the temperature. Hence, the pool dries more quickly.

As a conclusion and according to our results, the phase separation and the spreading of the pools are related to the nature of the substrate, relative humidity and temperature. Despite the presence of this phenomenon at high relative humidity levels, we have the same drying morphology with the five stages. In this study, we investigated the separation phenomenon, using blood taken from the same healthy donor, and the pools were created on non-porous substrates. Our next step will be to understand the local mechanism that triggers it and to investigate the role of porosity and hematocrit levels.

5.1.4 Drying of Human Blood Pools

In this section, we will be interested in the patented method developed by Smith et al. on the temporal reconstruction of blood pools, which can be used by investigators in crime scenes [33]. This method is based on the understanding of the dynamics of human blood drying, and the monitoring of the drying front during the third and fourth stages described above. The evaporation of a blood pools does not correspond to a one-dimensional system of mass and heat transfer as in the case of the stretch of liquid Fig. 5.3. Smith et al. have shown that in the case of pools, the triple line, the shape and the volume of the pools influence the evaporation rate. All these parameters are closely related to the advancement of the drying front. Mainly in their work, they related the evaporation rate J^* to the evolution of the drying front, therefore, they were able to determine a constant evaporation rate using only the wet surface. And by incorporating the different parameters that act in the evaporation of pools, they obtained a constant value of the diffusion coefficient D_{blood} that does not depend on the shape, and volume of the pools, D_{blood} was determined under the same external drying conditions.

The drying of pools of human whole blood corresponds to the evaporation of volatile blood components, mainly water. The decrease in mass during the drying of blood pools has the same dynamics, for any type of surfaces and also under different relative humidity level [20]. Before reaching 60% of the initial mass, the drying of a blood pools shows a linear loss of mass over time, like that of a water pool dried under the same conditions Fig. 5.16. At this stage, the liquid phase evaporates at the surface of the blood pools, without any change in the shape of the pools, hence the variation in mass corresponds to a variation in height. After this stage, the loss of mass of a pool of water continues to decrease in a linear manner until total drying, 0% of the initial mass. However, for blood pools, drying stops at 23% under dry conditions (low humidity). This value corresponds to the quantity of non-volatile components present

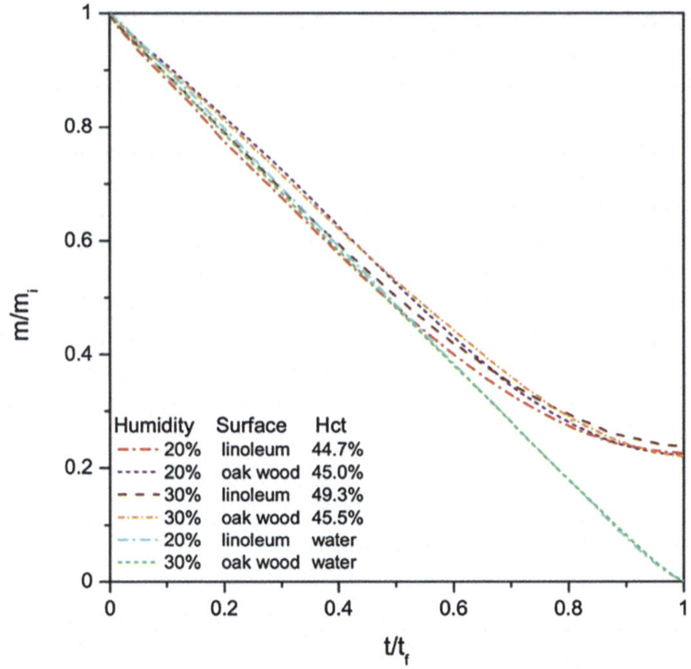

Fig. 5.16 Normalized mass as a function of normalized time of the blood and water pools. Experiments were carried out at 20 or 30% humidity, on linoleum or oak wood surfaces. The hematocrit values of the blood ranged between 44.7 and 49.3% [33]

in the blood, and it is according to hematocrit level, which explains the variations observed on the graph due to the initial hematocrit value.

The analysis of the evaporation rate J* of blood pools of different masses according to the remaining water allows to investigate deeply the loss of mass. This evaporation rate corresponds to a mass diffusion flow. Figure 5.17 shows the drying dynamics of five human blood pools of different initial masses and hematocrit level, under a temperature of 23 °C and relative humidity of 20% on the same substrate, a tile. Noting that the curve of $m = 0.30$g corresponds to a large drop, which explains the high evaporation rate, their dynamics was given in detail in Sect. 5.2.2. The evaporation dynamics of the other pools is similar to those of the alumina gels studied by Dwivedi [14]. After the creation of blood pools, the evaporation rate decreases rapidly during spreading and coagulation, until the beginning of the second stage (Gelation). Then, evaporation will take place at a constant rate characterized by a plateau until the end of the third stage (Rim desiccation). Finally, from the fourth stage, evaporation takes place in the porous medium, therefore, the evaporation rate decreases until total drying.

As can be seen in Fig. 5.17, the rate of evaporation decreases as the mass of the pool increases. A large mass results in a larger contact surface area. The phase change

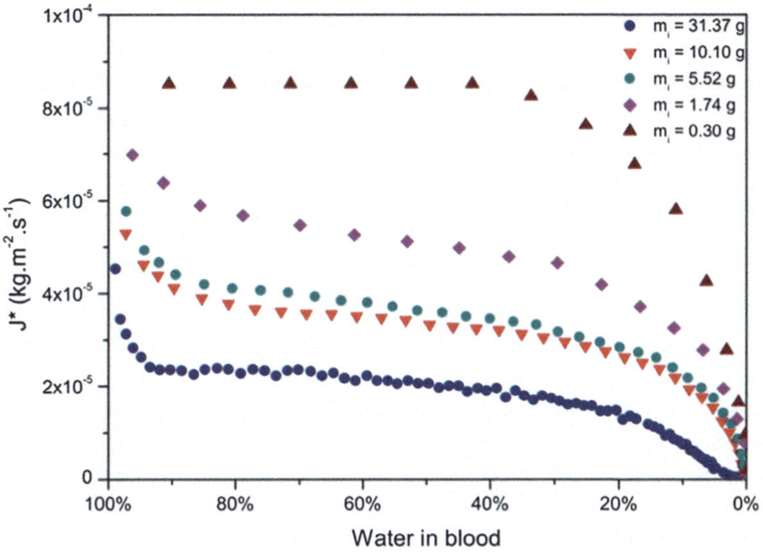

Fig. 5.17 Evaporation rates of pools having various initial masses (m_i), drying at 20% humidity, $T = 23 \pm 1$ °C, on the same surface, a tile [33]

produces humidity surrounding the pool, which slows down evaporation. Therefore, the larger the size, the more difficult it is for the aqueous phase to transform into vapor, since the surface is already saturated.

The drying of a blood pool has the same dynamics as alumina gels, and the evaporation rate can be strongly influenced by the size of the pool. Also, the latter is dependent on the shape of the pool. Smith et al. showed that the evaporation rate of 21 blood pools with the same initial mass m_i is different. accordingly, they developed a shape factor named L* to take into account the shape of the pools, and it is given by the following equation:

$$L^* = \frac{A}{hP}$$

with A, the area (in m^2), h, the height (in m) and P, the perimeter of the pool (in m). The height is estimated using the following approximation:

$$h = \frac{V}{A}$$

V is the volume (in m^3), it can be calculated with the following equation:

$$V = \frac{m}{\rho}$$

The evaporation rate is a mass flow rate per unit area, which has been calculated in the first part by using the total area. However, by using the wet area, i.e., a monitoring of the drying front by image processing software, Smith et al. were able to determine a constant evaporation rate independent of the pool size. This evaporation rate can be expressed as a function of the transfer coefficient, since the evaporation process takes place across an interface. Assuming that the blood pool is in the open air and at a temperature below the boiling point of water, and the ambient pressure is normal, the mass transfer equation can be given by [11]:

$$j^* = K_i \frac{MP_w}{RT}$$

with K_i the transfer coefficient from the liquid to the gas state (in kg m^{-1}), M the molar mass (in kg mol^{-1}), P_w the saturating vapor pressure of water at the surface (in J m^{-3}), R the universal gas constant (in J mol^{-1} k^{-1}) and T the temperature (in K). This equation shows that in the case of the evaporation of blood pools, the rate of evaporation depends on the temperature and the humidity. The saturating vapor pressure of water is given as a function of temperature by the Antoine equation [17]. The evaporation rate equation can be written as follows:

$$K_i = j^* \frac{RT}{MP_w}$$

The transfer coefficient represents the velocity of movement of a species from one phase to another. A high value of K_i implies a fast mass transfer, otherwise a slow mass transfer. The connection of the transfer coefficient with the diffusion coefficient allows to link the evaporation of a blood pool to Fick's law. In order to obtain a diffusion coefficient, Smith et al. multiplied the transfer coefficient by the Knudsen layer, L_K. The latter represents the thin evaporation layer of the vapor close to the liquid, it was used as the characteristic length, and is given by the following equation:

$$L_k = \frac{kT}{\pi d^2 P_a}$$

where k is Boltzmann's constant (in m^2 kg/s^2k), T is temperature, d is molecular diameter and P_a is atmospheric pressure.

As we said before, the drying of blood pools depends on the shape of the pool. In order to tackle its effect on the calculation of the diffusion coefficient, the shape factor is taken into account. As results, the transfer coefficient K_i of the four blood pools is weighted by the shape factor root $L^{*1/2}$ and the characteristic length L_K. The diffusion coefficients obtained for each pool are presented in Fig. 5.18 as a function of the amount of water remaining in the blood. The four curves have the same plateau with an approximately constant value equal to 1.10^{-9} m^2 s^{-1}. This value represents

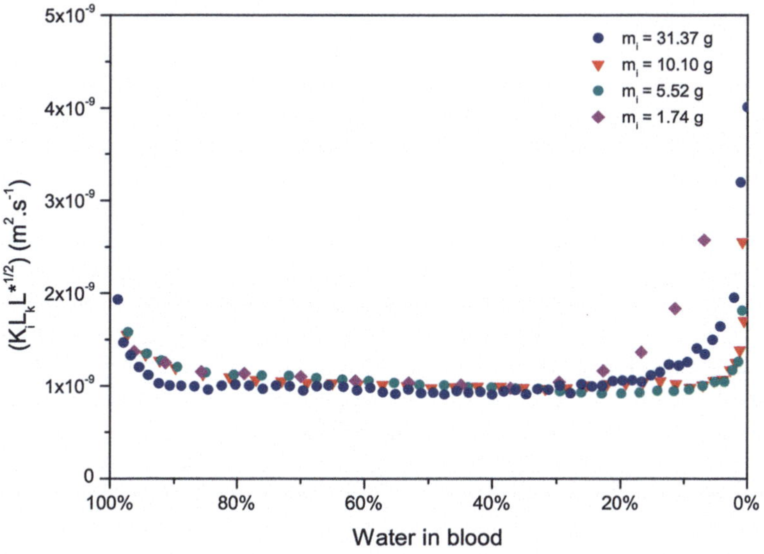

Fig. 5.18 Evaporation rate of water weighted by $(RTLkL^{*1/2})/(MP_w)$ against the percentage of water left in pools having different initial masses (m_i), drying at 20% humidity, $T = 23 \pm 1$ °C, on the same surface, a tile [33]

the diffusion coefficient of blood pools noted D_{blood}, under a temperature of 23 °C and a relative humidity of 20%. In Fig. 5.18, at the beginning and at the end of drying, the values do not follow this plateau. Directly after deposition of the pool on the substrate and before coagulation, the dynamics of the drying is the same as that of water, which explains the decrease in the evaporation rate, hence the coefficient of diffusion, during this first phase. After reaching the plateau, the drying of blood pools has the same dynamics as that of a gel. At the end of the drying process, the image processing software is unable to detect the wet surface, therefore the values of the evaporation rate are biased at the end.

This important result shows that the drying of blood pools is a complex problem, involving several parameters such as blood composition, environmental conditions, substrate nature and the shape and size of the pools. Smith et al. through their studies were able to show that the drying process of a blood pool follows the same dynamics of a gel dried under the same conditions. Also, taking into account the influence of shape and size on evaporation through shape factor and the monitoring of the wet area, they determined a blood diffusion coefficient denoted by D_{blood}. This coefficient is determined under a temperature of 23 °C and a relative humidity of 20%, however, further studies are needed to determine it under different conditions of temperature and relative humidity. Mainly, the increase in temperature leads to rapid evaporation, hence a higher diffusion coefficient. Relative humidity affects evaporation, so that its increase causes a decrease in D_{blood}. Therefore, fundamentally, we can say that shape, temperature and humidity are the key parameters to understand the drying of blood pools. Unfortunately, this is not the case since the geometry of the pool plays an important role during this process. As we know, a round pool evaporates faster than a thin, large pool, since the thin evaporation layer of the vapor near the liquid, Knudsen layer, saturates faster in the former case.

The correlation found between the diffusion coefficient D_{blood} and the transfer coefficient K_i has practical applications in crime scene investigation, and from which Smith et al. have developed a reliable and patented model for the temporal reconstruction of crime scenes, which we will discuss in detail below. Currently, bloodstain pattern analysts (BPAs) need tools that are non-destructive, reliable and based on scientifically confirmed methods, and also that can be easily brought to crime scenes. And the evolution of the drying front during the drying of blood pools is of a great interest for BPAs, as it can be followed over time by photography. Also, photography is one of the most used tools in crime scenes. Therefore, the aim of their study was to implement a method that allows to get the time in which the pool was created, from its drying morphology and by using the diffusion coefficient.

Under constant drying conditions, with a temperature of 23 °C and a relative humidity of 20%, the evaporation rate of blood pools can be calculated using the constant value $D_{blood} = 1.10^{-9}\ m^2\ s^{-1}$:

$$J^* = D_{blood} \frac{MP_w}{L_k RTL^{*0.5}}$$

L^* is obtained by: $L^* = \frac{A_i}{hP}$

where A_i is the total area of the pools, and P its perimeter, both can be calculated from a simple picture of a pool Fig. 5.13. h is the thickness, it is estimated from the calculation of the average of 30 different pools created under the same surrounding conditions, and it is equal to 1.44 ± 0.19 mm. From a picture of a pool of blood, we can have the variation of the wet area, i.e., the advancement of the drying front. So, making a relation between the latter and the variation in mass seems necessary, since, in a crime scene, we cannot have mass. Figure 5.19 shows the normalized variation in mass versus variation in wetted area, of three pools of different masses. As results, the three curves overlap, and they have the same tendency. In this figure, $A_i = A_{tot}$ and $m_i = hA_i\rho$ is the initial mass of the blood pool. The adjustment gives the following function:

$$\frac{m}{m_i} = 1 - \alpha\left[1 - \left(\frac{A}{A_i}\right)\right]^{\beta}$$

with $\alpha = 0.78$ and $\beta = 0.16$ are the fitting parameters.

Now we have the wet area as a function of mass. The time at which the pool was created can be estimated by integrating this last function in relation to the evaporation rate given by:

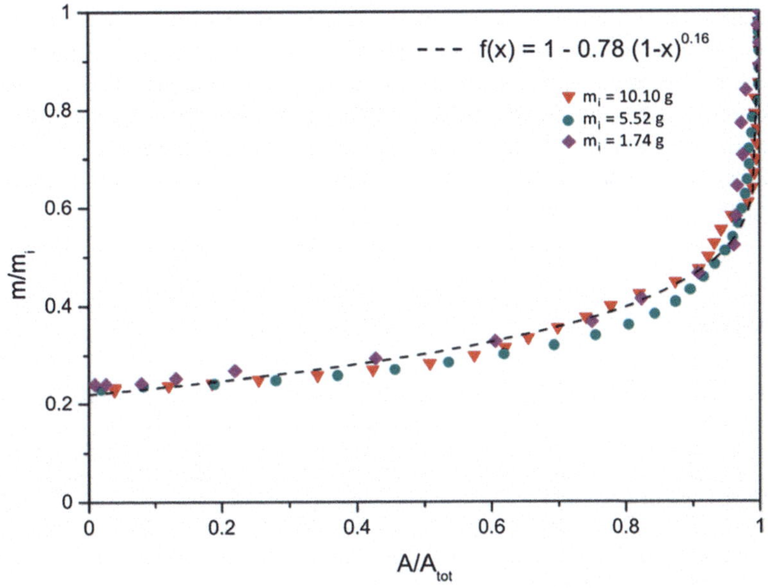

Fig. 5.19 Normalized mass as a function of normalized wet area for different pools having different initial masses (m_i), drying at 20% humidity, $T = 23 \pm 1$ °C, on the same surface, a tile [33]

$$J^* = \frac{\delta m}{\delta t A_i}$$

$\delta m = m_i - m_x$ and $\delta t = t_x - t_i$, t_x the time in which the picture was taken, and A_x the wet area at time t_x. At t_i = 0 s, the expression of t_x is given by:

$$t_x = \frac{\alpha R k_b T^2 A_0^{0,5} h^{0,5} \rho \left[1 - \left(\frac{A_t}{A_0}\right)\right]^{\beta}}{\pi d^2 M P_t^{0,5} D_{blood} P_w P_a}$$

This correlation is verified for many pools of different shapes and sizes dried under the same conditions, and it has an average error on the estimate of the creation time of the pools of 5%.

5.2 Conclusion

As evidenced in this chapter, whole human blood is a reality complex fluid that spread, wet and then dry not only depending of course on the substrate nature but also depending on the surrounding conditions. We evidenced the influence of the relative humidity on both the blood phase separation that occurs at high humidities. The perspectives of a better understanding of blood behavior on substrates are obvious for biomedical applications that need perfect reproducibility. As requested by courthouses in Europe and the USA, a better knowledge is also needed in forensic sciences in order to avoid justifying a criminal decision on the BPA guess or own experience. This is why, in 2015, the Bloodstain Pattern Analysis discipline has been included as a branch of NIST in order to gain more scientific rigor.

References

1. Barbee, James H. "The Effect of Temperature on the Relative Viscosity of Human Blood." *Biorheology* 10, no. 1 (1973): 1–5. https://doi.org/10.3233/BIR-1973-10101.
2. Birdi, K.S., and D.T. Vu. "Wettability and the Evaporation Rates of Fluids from Solid Surfaces." *Journal of Adhesion Science and Technology* 7, no. 6 (January 1993): 485–93. https://doi.org/10.1163/156856193X00808.
3. Bohn, S., L. Pauchard, and Y. Couder. "Hierarchical Crack Pattern as Formed by Successive Domain Divisions." *Physical Review E* 71, no. 4 (April 28, 2005): 046214. https://doi.org/10.1103/PhysRevE.71.046214.
4. Bou-Zeid, W., and D. Brutin. "Influence of Relative Humidity on Spreading, Pattern Formation and Adhesion of a Drying Drop of Whole Blood." *Colloids and Surfaces A: Physicochemical and Engineering Aspects* 430 (August 2013): 1–7. 0.1016/j.colsurfa.2013.03.019
5. Bou-Zeid, W., and D. Brutin. "Effect of Relative Humidity on the Spreading Dynamics of Sessile Drops of Blood." *Colloids and Surfaces A: Physicochemical and Engineering Aspects* 456 (August 2014): 273–85. https://doi.org/10.1016/j.colsurfa.2014.05.004.
6. Brutin, D., B. Sobac, B. Loquet, and J. Sampol. "Pattern Formation in Drying Drops of Blood." *Journal of Fluid Mechanics* 667 (January 25, 2011): 85–95. https://doi.org/10.1017/S0022112010005070.
7. Brutin, David. *Droplet Wetting and Evaporation: From Pure to Complex Fluids*. Edited by David Brutin. Amsterdam Boston Heidelberg: Elsevier/AP, Academic Press, 2015.
8. Brutin, David, Benjamin Sobac, and Céline Nicloux. "Influence of Substrate Nature on the Evaporation of a Sessile Drop of Blood." *Journal of Heat Transfer* 134, no. 6 (June 1, 2012): 061101. https://doi.org/10.1115/1.4006033.
9. Cazabat, A. M., and M. A. Cohen Stuart. "Dynamics of Wetting: Effects of Surface Roughness." *The Journal of Physical Chemistry* 90, no. 22 (October 1986): 5845–49. https://doi.org/10.1021/j100280a075.
10. Chien, S. "Shear Dependence of Effective Cell Volume as a Determinant of Blood Viscosity." *Science* 168, no. 3934 (May 22, 1970): 977–79. https://doi.org/10.1126/science.168.3934.977.
11. Cussler, E. L. *Diffusion: Mass Transfer in Fluid Systems*. 3rd ed. Cambridge; New York: Cambridge University Press, 2009.
12. Deegan, Robert D., Olgica Bakajin, Todd F. Dupont, Greg Huber, Sidney R. Nagel, and Thomas A. Witten. "Capillary Flow as the Cause of Ring Stains from Dried Liquid Drops." *Nature* 389 (1997): 827–829.
13. Deegan, Robert D., Olgica Bakajin, Todd F. Dupont, Greg Huber, Sidney R. Nagel, and Thomas A. Witten. "Contact Line Deposits in an Evaporating Drop." *Physical Review E* 62, no. 1 (July 2000): 756–65. https://doi.org/10.1103/PhysRevE.62.756.
14. Dwivedi, R. K. "Drying Behaviour of Alumina Gels." *Journal of Materials Science Letters* 5, no. 4 (1986): 373–76.
15. Errill, E. W. "Rheology of Blood." *Physiological Reviews* 49, no. 4 (October 1, 1969): 863–88. https://doi.org/10.1152/physrev.1969.49.4.863.

16. Girard, Fabien, Mickaël Antoni, Sylvain Faure, and Annie Steinchen. "Evaporation and Marangoni Driven Convection in Small Heated Water Droplets." *Langmuir* 22, no. 26 (December 2006): 11085–91. https://doi.org/10.1021/la061572l.
17. Haynes, William M. *CRC Handbook of Chemistry and Physics*, 2014.
18. Hu, Hua, and Ronald G. Larson. "Evaporation of a Sessile Droplet on a Substrate." *The Journal of Physical Chemistry B* 106, no. 6 (February 2002): 1334–44. https://doi.org/10.1021/jp0118322.
19. Hu, Hua, and Ronald G. Larson. "Marangoni Effect Reverses Coffee-Ring Depositions." *The Journal of Physical Chemistry B* 110, no. 14 (April 2006): 7090–94. https://doi.org/10.1021/jp0609232.
20. Laan, Nick, Fiona Smith, Celine Nicloux, and David Brutin. "Morphology of Drying Blood Pools." *Forensic Science International* 267 (October 2016): 104–9. https://doi.org/10.1016/j.forsciint.2016.08.005.
21. Laber, T.L., and B.P. Epstein. "Substrate Effects on the Clotting Time of Human Blood" 34, no. 4 (2001): 209–14. https://doi.org/10.1080/00085030.2001.10757531.
22. Lazarus, V., and L. Pauchard. "From Craquelures to Spiral Crack Patterns: Influence of Layer Thickness on the Crack Patterns Induced by Desiccation." *Soft Matter* 7, no. 6 (2011).
23. Legendre, Dominique, and Marco Maglio. "Numerical Simulation of Spreading Drops." *Colloids and Surfaces A: Physicochemical and Engineering Aspects* 432 (September 2013): 29–37. https://doi.org/10.1016/j.colsurfa.2013.04.046.
24. Lei, H., L. F. Francis, W. W. Gerberich, and L. E. Scriven. "Stress Development in Drying Coatings after Solidification." *AIChE Journal* 48, no. 3 (March 2002): 437–51. https://doi.org/10.1002/aic.690480304.
25. Lopez, Jaime, Clarence A Miller, and Eli Ruckenstein. "Spreading Kinetics of Liquid Drops on Solids." *Journal of Colloid and Interface Science* 56, no. 3 (September 1976): 460–68. https://doi.org/10.1016/0021-9797(76)90111-9.
26. Parisse, F., and C. Allain. "Shape Changes of Colloidal Suspension Droplets during Drying." *Journal de Physique II* 6, no. 7 (July 1996): 1111–19. https://doi.org/10.1051/jp2:1996119.
27. Pauchard, L. "Patterns Caused by Buckle-Driven Delamination in Desiccated Colloidal Gels." *Europhysics Letters* 74, no. 1 (2006): 188.
28. Pauchard, L., and C. Allain. "Stable and Unstable Surface Evolution during the Drying of a Polymer Solution Drop." *Physical Review E* 68, no. 5 (November 7, 2003): 052801. https://doi.org/10.1103/PhysRevE.68.052801.
29. Pauchard, L., F. Parisse, and C. Allain. "Influence of Salt Content on Crack Patterns Formed through Colloidal Suspension Desiccation." *Physical Review E* 59, no. 3 (1999): 3737. https://doi.org/10.1103/PhysRevE.59.3737.
30. Picknett, R.G, and R Bexon. "The Evaporation of Sessile or Pendant Drops in Still Air." *Journal of Colloid and Interface Science* 61, no. 2 (September 1977): 336–550. https://doi.org/10.1016/0021-9797(77)90396-4.
31. Ramsthaler, Frank, J. Schlote, C. Wagner, J. Fiscina, and M. Kettner. "The Ring Phenomenon of Diluted Blood Droplets." *International Journal of Legal Medicine* 130 (2016): 731–36.
32. Shanahan, M. E. R., and K. sefiane. "Kinetics of Triple Line Motion during Evaporation." In *Contact Angle, Wettability and Adhesion*, 6:19–31, 2009.
33. Smith, F. R., C. Nicloux, and D. Brutin. "A New Forensic Tool to Date Human Blood Pools." *Scientific Reports* 10, no. 1 (2020): 1–12.
34. Sobac, B., and D. Brutin. "Desiccation of a Sessile Drop of Blood: Cracks, Folds Formation and Delamination." *Colloids and Surfaces A: Physicochemical and Engineering Aspects* 448 (April 2014): 34–44. https://doi.org/10.1016/j.colsurfa.2014.01.076.
35. Sobac, B., and D. Brutin. "Structural and Evaporative Evolutions in Desiccating Sessile Drops of Blood." *Phys. Rev. E* 84, no. 011603 (July 6, 2011). https://doi.org/10.1103/PhysRevE.84.011603.
36. Sobac, B., and D. Brutin. "Triple-Line Behavior and Wettability Controlled by Nanocoated Substrates: Influence on Sessile Drop Evaporation." *Langmuir* 27, no. 24 (December 20, 2011): 14999–7. https://doi.org/10.1021/la203681j.

37. Takhistov, Pavlo, and Hsueh-Chia Chang. "Complex Stain Morphologies." *Industrial & Engineering Chemistry Research* 41, no. 25 (2002): 6256–69. https://doi.org/10.1021/ie010788.
38. Tanner, L.H. "The Spreading of Silicone Oil Drops on Horizontal Surfaces." *Journal of Physics D: Applied Physics* 12, no. 9 (1979): 1473.
39. Thiriet, Marc. *Biology and mechanics of blood flows: Part II: Mechanics and medical aspects.* New York; London: Springer, 2008.
40. Thurston, G.B. "Viscoelasticity of Human Blood." *Biophysical Journal* 12, no. 9 (September 1972): 1205–17. https://doi.org/10.1016/S0006-3495(72)86156-3.

Chapter 6
Evaporation Effect on the Contact Angle and Contact Line Dynamics

Vadim S. Nikolayev

Abstract This chapter shows how evaporation and condensation can modify the wetting conditions. First, we review the microscopic-scale phenomena acting near the contact line: Kelvin effect, hydrodynamic slip, vapor recoil, surface forces, and interfacial resistance. Then we address the theory of liquid flow in the wedge under evaporation at partial wetting conditions, more common in practice than complete wetting. The importance of the correct formulation of the boundary conditions at the contact line is shown. Two main evaporation regimes are addressed next. First, the evaporation into a pure vapor atmosphere is considered (like in bubble growth in boiling). It is controlled by the flow in the liquid. In the presence of contact line receding, this problem is solved by asymptotic matching of the three liquid regions: (i) the microregion near the contact line controlled by the phenomena described above, (ii) the intermediate region where the surface tension competes with the viscous effects, and (iii) the macroregion controlled by the liquid bulk effects. The asymptotic matching results in an expression for the apparent contact angle that depends both on the evaporation rate and on the contact line velocity. From such an analysis, the contact line receding dynamics caused by evaporation can be found. The theory is then compared to the available experimental data. Finally, we consider another regime of wedge evaporation, that in the atmosphere of the neutral gas, controlled by the vapor diffusion in the diffusion boundary layer. As the evaporation is weaker in this case, its effect on the apparent contact angle is smaller. However, we show how it depends on the key parameters of evaporation, e.g., the boundary layer thickness.

Keywords Partial wetting · Evaporation · Diffusion · Kelvin effect · Contact line dynamics

Nomenclature

A	Hamaker constant [J], area (Sect. 6.1.2.4, Sect. 6.2.4.3) [m^2]
C	arbitrary constant

V. S. Nikolayev (✉)
Université Paris-Saclay, CEA, CNRS, SPEC, 91191 Gif-sur-Yvette, France
e-mail: vadim.nikolayev@cea.fr

M. Marengo and J. De Coninck (eds.), *The Surface Wettability Effect on Phase Change*,
https://doi.org/10.1007/978-3-030-82992-6_6

Ca	capillary number
D	diffusion coefficient of vapor in the non-condensible gas, [m^2/s]
e	Euler number $\simeq 2.71$
f	accommodation coefficient
h	liquid layer thickness [m]
I_n	modified Bessel function of the first kind of the order n
J	mass evaporation flux, [kg/(m^2·s)]
K	curvature, [m^{-1}]
k	heat conductivity [W/(m·K)]
K_n	modified Bessel function of the second kind of the order n
L	macroscopic length scale, [m]
$\mathcal{L}$	latent heat, [J/kg]
l	length [m]
ℓ	characteristic length scale [m]
M	molar mass [kg/mol]
m	mass [kg]
$\mathbf{n}$	normal vector to the interface
P	surface force energy [J/m^2]
p	pressure [Pa]
q	heat flux [W/m^2]
R	radius of drop base, [m]
r	radial distance, [m]
R^i	interfacial thermal resistance, [(K·m^2)/W]
R^i_{diff}	interfacial kinetic resistance, [s/m]
R_g	universal gas constant [J/(mol·K)]
$\mathcal{R}$	dimensionless interfacial resistance
S	spreading coefficient, [N/m]
T	temperature [K]
t	time [s]
V	volume, [m^3]
v	liquid velocity [m/s]
W	interfacial energy [J/m^2]
w	width of dewetting ridge [m]
x, y, z	cartesian coordinates, [m]

Abbreviations

CL Contact Line

Greek symbols

β	dimensionless slip length
Δ	difference
δ	modified capillary number
γ	Marangoni coefficient, [N/(m·K)]
Λ	thickness of diffusion boundary layer, [m]

μ liquid shear viscosity, [Pa·s]; chemical potential, [J/kg] (in sec. 6.1.2.2 only)
ν liquid kinematic viscosity [m^2/s]
Φ liquid 1D flux [m^2/s]
ϕ interface slope
Π disjoining pressure, [Pa]
ρ density [kg/m^3]
σ interface tension [N/m]
θ contact angle
ε dimensionless deviation from equilibrium

Superscripts

i interfacial

Subscripts

ad	adsorption
app	apparent
CL	contact line
K	Kelvin
L	liquid
micro	microscopic
S	solid heater or substrate
s	slip
sat	saturation
V	Voinov or vapor

6.1 Introduction

Evaporation and condensation phenomena are met widely both in everyday life and in various industrial processes. Some evident everyday examples are the water boiling in one's kitchen, linen drying in open air or the morning dew disappearance on the windshield of a car. As industrial examples, one can list the steam generation for turbines used in power plants and the two-phase cooling of microprocessors with heat pipes used now in every laptop computer and mobile phone. One can recall also the spray cooling in high-power heat exchangers or the drying of solvent during thin film coating based on the colloidal solutions. In all these cases, a solid surface in contact with the liquid can be partially dried so that the triple liquid-gas-solid contact lines appear.

The experimental studies of the evaporation effect on wetting are difficult and thus quite rare; for this reason, this chapter is mainly theoretical. Some experimental data are, however, discussed in Sect. 6.2.5 below.

The contact line problem is a particular case of the free-interface problem, which can be theoretically considered; thanks to the Laplace equation

$$\Delta p = K\sigma \tag{6.1}$$

that defines the local interfacial pressure jump

$$p_V - p_L = \Delta p, \tag{6.2}$$

in terms of the surface tension σ and its curvature K; the pressures at the interface on both sides of it are denoted p_V (vapor) and p_L (liquid).

Note that the physical phenomena presented in this chapter remain invariant with respect to the sign of the mass exchange so the presentation below applies to the condensation case as well.

The mass exchange rate is always limited by the slowest dynamic phenomenon. For this reason, the physics of evaporation differs depending on the gas composition. One can consider two limit cases. In the first, the gas is air (or another non-condensible gas) with a relatively small vapor density. This is a case of drying in open air in the absence of substrate heating. The slowest dynamic phenomenon is the vapor diffusion necessary to evacuate the vapor from the gas-liquid interface to the gas bulk. The diffusion occurs in a boundary layer that forms near the interface. Farther away, the convection takes over.

In the second case, the gas is a pure vapor of the evaporating liquid and the evaporation occurs because of the substrate heating. The diffusion is not relevant here and the evaporation is limited only by the heat supply rate. Evidently, the mass exchange is much stronger in the second case. This is the case, e.g., of bubble growth in boiling.

In a tiny vicinity of the contact line called microregion hereafter, the liquid forms a curved wedge. Such a geometric singularity causes a strong increase of evaporation in its vicinity, which is one of the motivations for the studies of contact lines in the presence of phase change. There is another, perhaps even more important reason for the interest in the microregion. As will be discussed in Sect. 6.1.1, the singularity of the mass transfer causes a strong curvature that leads to a large apparent contact angle (i.e., the slope of the gas-liquid interface at a macroscopically measurable scale). In other words, evaporation can lead to a non-negligible change in wetting conditions. This change is controlled by effects that act at the microscopic scale. The objective of this chapter is to show how this apparent angle can be determined and to analyze its dependence on the system parameters.

The two-dimensional geometry is considered below. This is justified by a much stronger curvature of the gas-liquid interface in the plane perpendicular to the solid than in the plane parallel to the solid.

This chapter is structured as follows. First of all, a qualitative picture explaining the evaporation impact on the contact angle is given. In the following sections, the microscopic phenomena that are important in the microregion but negligible on the macroscopic scale are considered. The main part of the chapter is devoted to two limit cases of evaporation: first, the evaporation to the pure vapor and next, the isothermal liquid drying in the atmosphere of another gas.

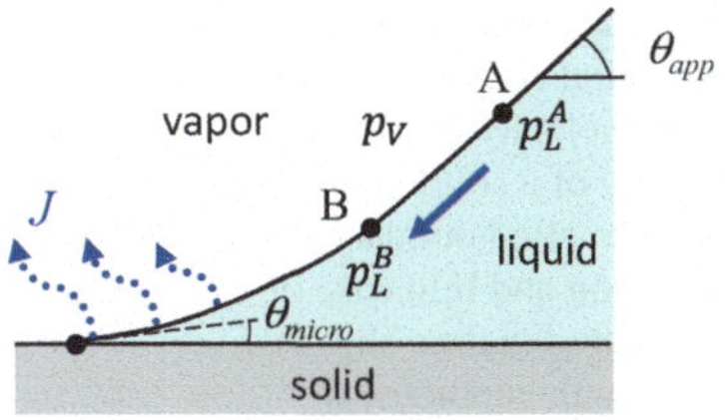

Fig. 6.1 Liquid wedge at evaporation. The fluid flow is shown by the blue arrows

6.1.1 How Evaporation Can Modify the Wetting Conditions?

First of all, one should mention that the evaporation cannot influence the static contact angle because it is controlled by the balance of intermolecular forces that are in most cases attractive and thus tend to prevent evaporation (see [45] for an extended discussion). Evaporation can only impact the apparent contact angle, i.e., that observed macroscopically. To understand this impact, let us begin with a qualitative consideration of microregion. There are two physical reasons for the difference between the apparent and microscopic contact angles. Both reasons are linked to the strong evaporation in the microregion. The first is related to the viscous pressure drop caused by the liquid flow. The second is the vapor recoil effect. We discuss here only flow-induced apparent contact angle. The vapor recoil will be discussed in Sect. 6.1.2.4.

Consider two points A and B on the vapor-liquid interface of the wedge at evaporation, with point B closer to the contact line than point A (Fig. 6.1). Evaporation is the strongest in the contact line vicinity, so there is a liquid flow toward the contact line to replenish the liquid loss. Because of the viscosity, a pressure drop appears, so there is a difference between the liquid pressures at points A and B:

$$p_L^A > p_L^B.$$

On the other hand, the vapor flow is much quicker, so the pressure p_V at the vapor side of interface is spatially homogeneous. From Eqs. (6.1–6.2), one concludes that $K^B > K^A$, so that the interfacial curvature grows toward the contact line. Note that the curvature is a rate of change of the interface slope along the interface (it is its geometrical definition). This means that the slope varies sharply near the contact line, which can cause a strong difference between the microscopic contact angle θ_{micro} and the macroscopic (apparent) contact angle θ_{app}. This effect has first been discovered by Wayner et al. [58] and studied by many other researchers. Initially, it was, however, incorrectly attributed to the impact of surface forces discussed in Sect. 6.1.2.3.

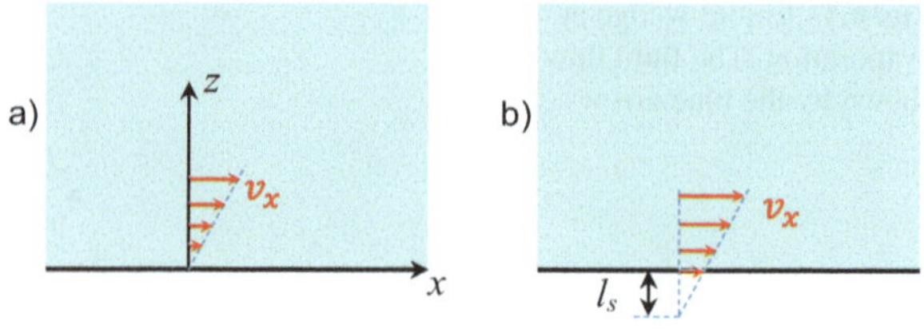

Fig. 6.2 Variation of the tangential component of the liquid velocity in the close vicinity of a solid in case of the **a** no-slip boundary condition and **b** in case of the hydrodynamic slip. The slip length geometrical definition is shown

6.1.2 Relevant Microscopic Phenomena

Evidently, near the contact line, the distance between the liquid-vapor interface and the solid is very small. This is why several effects, otherwise completely negligible, become important in this region.

6.1.2.1 Hydrodynamic Slip

For a hydrodynamic problem, one needs to define the boundary conditions, in particular, on the solid surface. One evident condition is the non-penetration of the liquid into the solid. The normal to the surface component of the liquid velocity v_z is thus zero (cf. Fig. 6.2a for the reference system). Usually, one considers that the liquid molecules stick to the solid, so the tangential to the surface component v_x of the liquid velocity is zero. This is referred to as the no-slip condition (Fig. 6.2a). When the surface is non-wettable, the statistical physics of liquids [21] shows that the liquid can slide along the solid when the tangential hydrodynamic stress is large. This phenomenon is characterized by the slip length l_s defined with the Navier boundary condition (Fig. 6.2b)

$$v_x = l_s \frac{\partial v_x}{\partial z}. \tag{6.3}$$

One needs to mention that the slip of liquid along the solid substrate is well confirmed experimentally [43], and measurements permit to determine the slip length within 10 nm accuracy [6]. The slip length can also be obtained with the molecular dynamics simulations. More detailed discussion on this phenomenon may be found in the review articles [30, 32]. The characteristic value of l_s is of the order of 20 nm. The slip effect is expected to be important at a distance $\ell_s \sim l_s/\theta_{\text{micro}}$ from the CL where θ_{micro} is a contact angle assumed to be small here. It appears in the denominator because the slip length is measured along the z axis. For small θ_{micro}, $\ell_s \gg l_s$.

6.1.2.2 Kelvin Effect

The Kelvin effect, sometimes called the Gibbs-Thomson effect [31], provides a dependence of the local temperature T_i of the vapor-liquid interface on the local interfacial pressure jump Δp. It depends on the interface curvature (cf. Eq. 6.1) and on several other factors discussed in the next sections. Consider a portion of the liquid-vapor interface that may globally be out of equilibrium. The portion is, however, assumed to be at *local* equilibrium, so the following equality holds for the chemical potentials of the phases:

$$\mu_V(p_V, T^i) = \mu_L(p_L, T^i) \tag{6.4}$$

A similar expression can be written for a flat liquid-vapor interface at equilibrium at the temperature T^i, where both liquid and vapor pressures are equal to $p_0 = p_{sat}(T_i)$

$$\mu_V(p_0, T^i) = \mu_L(p_0, T^i). \tag{6.5}$$

Let us develop now both sides of Eq. (6.4) into the Taylor series around p_0 by using the thermodynamic relation

$$\left(\frac{\partial \mu}{\partial p}\right)_T = \frac{1}{\rho},$$

where ρ is the density. With the help of Eq. (6.5), one readily obtains

$$\frac{p_V - p_0}{\rho_V} = \frac{p_L - p_0}{\rho_L}. \tag{6.6}$$

By using Eqs. (6.2, 6.6), one gets

$$p_V = p_0 - \Delta p \frac{\rho_V}{\rho_L - \rho_V}. \tag{6.7}$$

In what follows, we assume the smallness of the difference $|p_V - p_0| \ll p_0$.

For the pure vapor case, one can make use of the Clausius-Clapeyron equation

$$\left.\frac{dp}{dT}\right|_{sat} = \frac{\mathcal{L}\rho_L\rho_V}{T_{sat}(\rho_L - \rho_V)}, \tag{6.8}$$

where $\mathcal{L}$ is the latent heat and $T_{sat} = T_{sat}(p_V)$. With the relations

$$\left.\frac{dp}{dT}\right|_{sat} = \frac{p_0 - p_V}{T^i - T_{sat}}$$

and (6.7), one gets the final expression for the local equilibrium interface temperature as a function of pressure jump

$$T^i = T_{sat}\left(1 + \frac{\Delta p}{\mathcal{L}\rho_L}\right). \tag{6.9}$$

For the evaporation into the open air, one can use the ideal gas equation for the vapor

$$p_V = \rho_V^i \frac{R_g T}{M}, \tag{6.10}$$

$$p_0 = \rho_0 \frac{R_g T}{M}, \tag{6.11}$$

where ρ_0 is the saturation vapor density in the air over the flat interface at equilibrium, R_g is the gas constant and M, vapor molar weight. By using these expressions in Eq. (6.7), one finally obtains the local equilibrium interfacial vapor density ρ_V^i as a function of pressure jump under the assumption $\rho_V^i \ll \rho_L$

$$\rho_V^i = \rho_0 - \Delta p \frac{M \rho_0}{\rho_L R_g T}, \tag{6.12}$$

Note that only the local (and not global) equilibrium hypothesis was used. This means that the quantities T^i, ρ_V^i are allowed to vary along the interface while following the variation of Δp.

6.1.2.3 Surface Forces

Consider a thin liquid film of a homogeneous thickness h on a solid substrate at equilibrium (Fig. 6.3).

The liquid is surrounded by its vapor and the system is at the saturation temperature T_{sat} given by the bulk vapor pressure. The excess free energy W (appearing due to the existence of interfaces) per unit area is

$$W(h) = \sigma_{SL} + \sigma + P(h), \tag{6.13}$$

where the first two terms on the right side are the tensions of the solid-liquid and vapor-liquid interfaces, respectively. The term $P(h)$ is the energy of the surface (or thin film) forces [22] that appear because of the molecules of the solid "feel" the vapor-liquid interface. It is a differential contribution (proportional to the difference of liquid and vapor densities) of the interactions of the liquid and vapor molecules with those of the solid. Evidently, the h scale at which $P(h)$ matters is defined by the

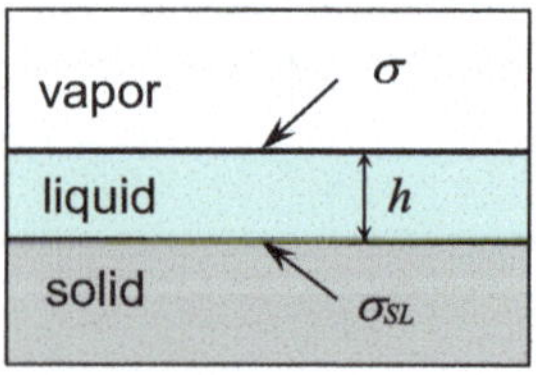

Fig. 6.3 Thin liquid film on a solid substrate

range of intermolecular interaction. Usually, this distance does not exceed several tens of nm. Obviously, $P(h \to \infty) = 0$.

The limit $h \to 0$ may be attained at partial wetting. It is evident that $W(h \to 0)$ has to be finite in this case. It has been postulated by Brochard-Wayrt et al. [8] that

$$W(h \to 0) = \sigma_{SG}, \tag{6.14}$$

which corresponds to the energy of the dry (bare) solid-gas interface. Equations (6.13, 6.14) then lead to the constraint $P(h \to 0) = S$, where

$$S = \sigma_{SG} - \sigma_{SL} - \sigma \tag{6.15}$$

is the spreading coefficient. A more general case [59, 60] $P(h \to 0) \geq S$ is adopted here. The inequality can be justified by the existence of a residual monolayer of fluid molecules at the solid surface that can modify the surface energy so $W(h \to 0) > \sigma_{SG}$.

With the account of the surface forces, the Laplace equation (6.1) becomes

$$\Delta p = K\sigma + \Pi(h), \tag{6.16}$$

where Π is called the disjoining pressure related to P via

$$\Pi = -\frac{\partial P}{\partial h}. \tag{6.17}$$

In general, the disjoining pressure includes contributions from dispersion, electrical double layers, electrostatic and structural forces [22]. Within a conventional for the contact line problem approach [45, 58], only the dispersion component (neglecting the retardation effect) is accounted for. This is justified for $h \gtrsim 10$ nm

$$\Pi(h) = A/(6\pi h^3). \tag{6.18}$$

The Hamaker constant A is positive for conventional couples of solid surfaces and fluids. For the "high-energy" metal or oxidized surfaces, its value is $A \sim 10^{-20} - 10^{-19}$ J.

Because of singularity of Eq. (6.18) at $h \to 0$, such a Π dependence usually results in a configuration where the solid is covered with a continuous (wetting or adsorption) film, of at least nanometric thickness h_{ad}, so that the vapor contact with the bare solid is nonexistent. Such a situation corresponds to the complete wetting case. When a liquid wedge configuration is forced by the macroscopic liquid shape, the wedge ends by the film (Fig. 6.4a). Another solution [17], where the contact line exists even for the complete wetting case, prevails when the evaporation rate is higher than a threshold [48], so that the film dries out (Fig. 6.4b) and $h(x) \sim \sqrt{x}$ near the contact line $x = 0$.

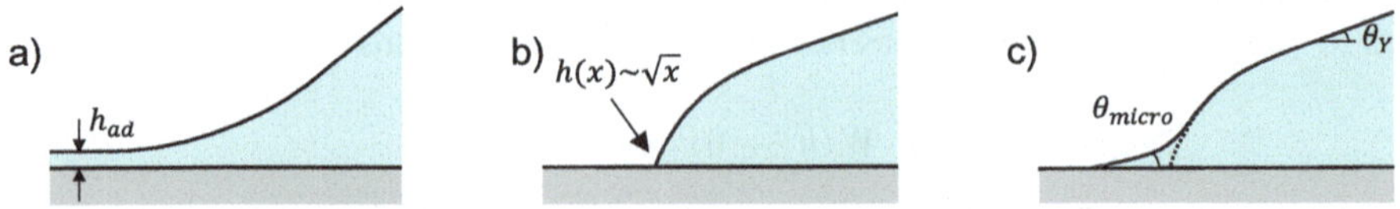

Fig. 6.4 Wedge shape in the microregion: **a** for complete wetting at equilibrium; **b** for complete wetting and strong evaporation (with the dried wetting film) [17]; **c** for the partial wetting (the dotte dline corresponds to the solution **b**)

At equilibrium, the wedge geometry appears at partial wetting (Fig. 6.4c), where $S < 0$. At a distance from the contact line, where the surface forces vanish (where $h \lesssim 100$ nm), the meniscus slope is θ_Y defined by the Young formula $\cos\theta_Y = 1 + S/\sigma$. At a smaller scale ($h \sim 1$ nm), the meniscus forms a contact angle θ_{micro} linked to θ_Y via the expression [60]

$$\cos\theta_{\text{micro}} = 1 + \frac{S - P(h \to 0)}{\sigma} = \cos\theta_Y - \frac{P(h \to 0)}{\sigma}. \tag{6.19}$$

Note that the assumption (6.14) results in $\theta_{\text{micro}} = 0$; a more general case is considered here.

To describe the partial wetting configuration at equilibrium, the disjoining pressure must be regularized at small h, e.g., [26]

$$\Pi(h) = \begin{cases} C_1 h + C_2, & h \in (0, h_m) \\ A/(6\pi h^3), & h \in (h_m, \infty) \end{cases} \tag{6.20}$$

Indeed, $\Pi(h)$ needs to be integrable at $h \to 0$; $P(h)$ would be infinite otherwise, which is nonphysical. The constant h_m is of the order of several nm and defines the position of the maximum of the $\Pi(h)$ function; $C_{1,2}$ are the constants defined from the continuity of $\Pi(h)$ at $h = h_m$ as

$$C_1 = (A - 4\pi h_m^2 P(h \to 0))/(2\pi h_m^4),$$
$$C_2 = (6h_m^2 \pi P(h \to 0) - A)/(3\pi h_m^3),$$

where $P(h \to 0)$ can be determined from θ_{micro} and θ_Y via Eq. (6.19).

At $h \geq h_m$ but still in the contact line vicinity, the interface follows the $\sqrt{x}$ law, which cross-overs to the straight wedge shapes both at $h < h_m$ and at larger h (Fig. 6.4c) but of different slopes. The slope near the contact line is θ_{micro}. Farther away, it is equal to θ_Y at equilibrium (cf. Eq. 6.19) and is different from it when the phase change occurs.

The above approach has been developed in the "local" approximation, where P is assumed to be independent of the spatial variation of h (i.e., of its spatial derivatives).

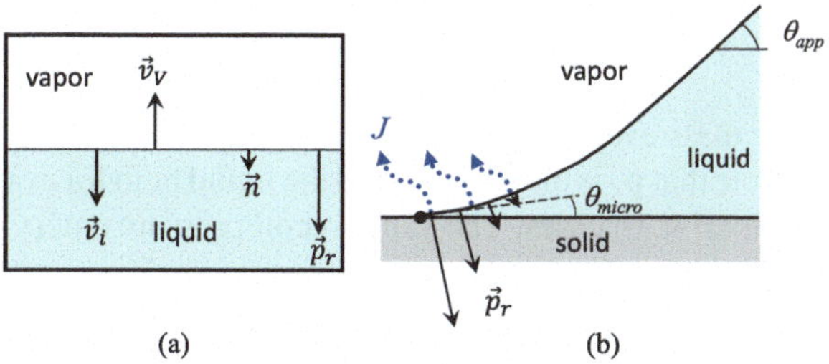

Fig. 6.5 (a) Vapor recoil force. (b) Effect of the vapor recoil on the apparent contact angle

A more adequate but also more complicated non-local approach [51] (where P is considered to be a functional of $h(x)$) gives a correction to (6.19) at large θ_Y. In particular, $\theta_{\text{micro}} \neq 0$ even for $P(h \to 0) = S$. For small angles, the results of the local and non-local approaches are essentially the same.

6.1.2.4 Vapor Recoil

Let us consider now a portion of the liquid-vapor interface of area A at evaporation conditions; see Fig. 6.5a. Every fluid molecule evaporated from the liquid interface causes a recoil force analogous to that created by the gas emitted by a rocket engine. It pushes the interface toward the liquid side in the normal direction. This force appears because the fluid necessarily expands while transforming from liquid to gas phase. Obviously, the stronger the evaporation rate, the larger is the vapor recoil force.

During the time $\mathrm{d}t$, the fluid mass $\mathrm{d}m_V$ changes state from liquid to vapor. This mass can be expressed as

$$\mathrm{d}m_V = \rho_V \mathrm{d}V_V = -\rho_L \mathrm{d}V_L, \tag{6.21}$$

where $\mathrm{d}V_L(\mathrm{d}V_V)$ is the volume change of the liquid (vapor). As a consequence, the interface displacement is $\mathbf{dl} = -\mathbf{n}\mathrm{d}V_L/A$, where $\mathbf{n}$ is the unit vector normal to the interface and directed into the liquid.

The conservation of momentum for the control volume shown in Fig. 6.5a reads

$$(\mathbf{v}_V + \mathbf{v}^i)\mathrm{d}m_V + \mathbf{p}_r \mathrm{d}t\, A = 0, \tag{6.22}$$

where $\mathbf{p}_r$ is the vapor recoil force per unit area, $\mathbf{v}^i = \mathbf{dl}/\mathrm{d}t$ is the velocity of interface and $\mathbf{v}_V = -\mathbf{n}\mathrm{d}V_V/(A\mathrm{d}t)$ is the velocity of vapor with respect to the interface. By using this expression together with Eq. (6.21), the Eq. (6.22) can be rewritten as

$$\mathbf{p}_r = \mathbf{n}J^2\left(\frac{1}{\rho_V} - \frac{1}{\rho_L}\right), \tag{6.23}$$

where

$$J = \frac{1}{A}\frac{\mathrm{d}m_V}{\mathrm{d}t} \tag{6.24}$$

is the mass evaporation flux.

Note that $\mathbf{p}_r$ is directed toward the liquid both for evaporation ($J > 0$) and condensation ($J < 0$) cases. The vapor recoil pressure enters the pressure balance equation as [38]

$$K\sigma = \Delta p + p_r, \tag{6.25}$$

where $p_r = |\mathbf{p}_r|$. In addition to the Kelvin term (6.9), a vapor recoil term appears in the interface temperature expression [4].

$$T^i = T_{sat}\left[\left(1 + \frac{\Delta p}{\mathcal{L}\rho_L}\right) + \frac{J^2}{2\mathcal{L}}\left(\frac{1}{\rho_V^2} - \frac{1}{\rho_L^2}\right)\right]. \tag{6.26}$$

The vapor recoil effect causes an increase in the apparent contact angle just like the viscous pressure drop discussed in Sect. 6.1.1. Indeed, the vapor recoil is the largest near the contact line (where J is very large, cf. Fig. 6.5b). As the curvature K increases with p_r, according to Eq. (6.25), the slope changes strongly in the contact line vicinity [38], which causes a difference between θ_{app} and θ_{micro}, just like in the pressure drop caused effect described in Sect. 6.1.1. The relative contribution of this effect is especially strong near the liquid-vapor critical point [39], where it causes the boiling crisis: the apparent contact angle growth induces the growth of dry area under the bubbles, which, in turn, triggers the complete heater dewetting.

6.1.2.5 Interfacial Kinetic Resistance

Up to now, we have considered the interface at a local equilibrium. This assumption holds when phase change rates are smaller than some generally quite high value. Even for modest average evaporation rates, high local evaporation rates can be attained in the contact line vicinity.

The non-equilibrium effects can be analyzed with the Schrage molecular-kinetic theory of evaporation [9] by using the ideal gas assumption for the vapor. According to this theory, the statistical distribution of the speeds of molecules is the Maxwell distribution. If the average vapor velocity is zero and the vapor is at temperature T_V, the mass flux perpendicular to a plane is

$$p_V\sqrt{\frac{M}{2\pi R_g T_V}}.$$

When the vapor flows with a velocity v_V perpendicularly to a plane, it is evident that the mass fluxes J^+ along $\mathbf{v}_V$ and J^- against to $\mathbf{v}_V$ should differ and depend on v_V because the velocity distribution becomes asymmetric. The expressions for them read

$$J^+ = \Gamma(a) p_V \sqrt{\frac{M}{2\pi R_g T_V}}, \tag{6.27}$$

$$J^- = \Gamma(-a) p_V \sqrt{\frac{M}{2\pi R_g T_V}}, \tag{6.28}$$

where

$$\Gamma(a) = \exp(-a^2) + a\sqrt{\pi}[1 + \text{erf}(a)] \simeq 1 + a\sqrt{\pi}$$

for a small

$$a = \frac{v_V}{\sqrt{2R_g T_V/M}} = \frac{J}{p_V}\sqrt{\frac{R_g T_V}{2M}}.$$

The latter equality is obtained by using the vapor mass flux expression $J = v_V \rho_V$ and the ideal gas equation.

Consider now the liquid-vapor interface at evaporation. It is assumed that a part of molecules that come to the interface can be reflected by it, so that the actual incoming flux arriving back to the liquid is smaller than J^-

$$J_L = f J^- = f\Gamma(-a) p_V \sqrt{\frac{M}{2\pi R_g T_V}}, \tag{6.29}$$

where f is called the evaporation coefficient. At equilibrium ($J = 0$), the incoming flux would be equal to the outgoing flux

$$J_V = f p_0 \sqrt{\frac{M}{2\pi R_g T^i}}. \tag{6.30}$$

The case $J > 0$ differs from the case $J = 0$ only by a smaller value of J_L (given by Eq. 6.29) while J_V remains to be given by Eq. (6.30). Evidently, $T^i = T_{sat}(p_0)$. By writing $J = J_L + J_V$ one obtains

$$J = \frac{2f}{2-f}\sqrt{\frac{M}{2\pi R_g}}\left(\frac{p_0}{\sqrt{T^i}} - \frac{p_V}{\sqrt{T_V}}\right). \tag{6.31}$$

The temperature is assumed to be continuous ($T^i = T_V$) so the small interfacial temperature jump [16] that may exist in some particular cases (like strong evaporation into vacuum) is neglected, so that Eq. (6.31) becomes finally

$$J = \frac{2f}{2-f}\sqrt{\frac{M}{2\pi R_g T^i}}(p_0 - p_V). \tag{6.32}$$

For the (isothermal) evaporation into the open air, one can again use Eqs. (6.10–6.11) to obtain

$$\rho_V^i = \rho_0 - J R_{diff}^i, \tag{6.33}$$

where ρ_V^i is the interfacial vapor density; R_{diff}^i is called the interfacial kinetic resistance and is defined as

$$R_{diff}^i = \frac{2-f}{2f}\sqrt{\frac{2\pi M}{R_g T_{sat}}}. \tag{6.34}$$

The value of evaporation coefficient can be set to unity which is verified experimentally [33].

For the pure vapor case, thanks to Eq. (6.8), one can linearize Eq. (6.32) by writing

$$p_0 - p_V = \frac{\mathcal{L}\rho_L\rho_V}{T_{sat}(\rho_L - \rho_V)}(T^i - T_{sat}),$$

that results in the expression

$$T^i = T_{sat} + R^i J \mathcal{L}, \tag{6.35}$$

where R^i is the interfacial thermal resistance:

$$R^i = \frac{2-f}{2f}\frac{T_{sat}\sqrt{2\pi R_g T_{sat}/M}(\rho_L - \rho_V)}{\mathcal{L}^2\rho_L\rho_V}. \tag{6.36}$$

By combining the above expression with Eq. (6.26), one can write the final expression for the local temperature of the vapor-liquid interface

$$T^i = T_{sat}\left[\left(1 + \frac{\Delta p}{\mathcal{L}\rho_L}\right) + \frac{J^2}{2\mathcal{L}}\left(\frac{1}{\rho_V^2} - \frac{1}{\rho_L^2}\right)\right] + R^i J \mathcal{L}. \tag{6.37}$$

as a function of Δp and J that can also vary along the interface.

One should note that R^i is typically is a tiny quantity, so that its effect is notable only at extremely high heat fluxes that can occur in the microregion.

6.1.3 *Liquid Flow in the Wedge*

The full Navier–Stokes equations in a wedge with a free curved boundary (i.e., the gas-liquid interface) are quite complex to solve. In particular, no analytical approach

Fig. 6.6 Geometry of the general contact line problem. The chosen direction of the normal to the interface is shown

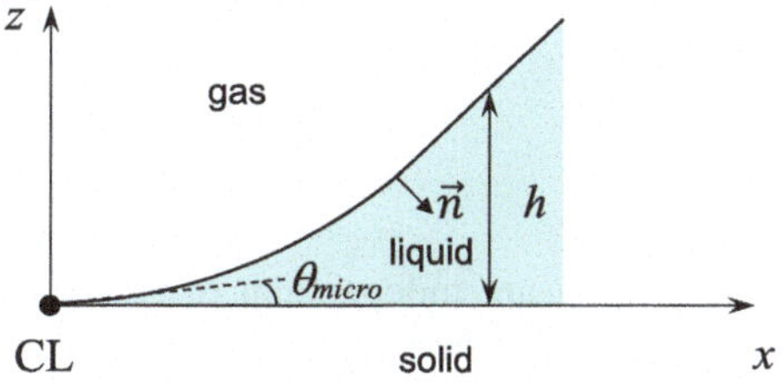

is possible in a general case, so that the only possible approach is numerical. However, the numerical algorithms describing such a problem are still insufficiently developed. The rare existing studies [1, 55] concern mainly the moving contact line problem with no phase change. Fortunately, there is a powerful tool for such studies: the lubrication approximation developed independently by [40] and [50] to study the thin film hydrodynamics. It is based on the smallness of the Reynolds number (because of the small film thickness). The inertial terms in the Navier–Stokes equations are thus unimportant and the hydrodynamics can be described by the Stokes equations in 3D,

$$\begin{aligned} \nabla p_L &= \Delta \mathbf{v}, \\ \nabla \cdot \mathbf{v} &= 0, \end{aligned} \tag{6.38}$$

where $\mathbf{v}$ is the liquid velocity and $\nabla = (\partial/\partial x, \partial/\partial y, \partial/\partial z)$ is the 3D differential operator. As described in the following sections, these equations can be further simplified by considering the liquid layer in the thin film approximation.

Note the essential role of the viscosity in the contact line region (Sect. 6.1.1). For this reason, the theoretical approaches based on the inviscid flow approximation are not relevant and will not be considered here.

In general, the contact line problem should be considered as transient: evaporation causes the progressive drying of the substrate. However, a stationary regime where the contact line is pinned by a defect on the solid is equally possible: the liquid is supplied to the microregion from the liquid bulk to compensate exactly the evaporation losses.

To consider the apparent contact angle, a 2D geometry in the $x - z$ plane (Fig. 6.6) is sufficient. For thin fluid layers, the fluid is supposed to move mainly along x axis, i.e., $v_x \gg v_z$, where $\mathbf{v} = (v_x, 0, v_z)$. In addition, the v_x variation across the layer is assumed to be much larger than along it: $\partial v_x/\partial z \gg \partial v_x/\partial x$. The Stokes equations then reduce to

$$\frac{\partial p_L}{\partial x} = \mu \frac{\partial^2 v_x}{\partial z^2}, \tag{6.39}$$

$$\frac{\partial p_L}{\partial z} = 0. \tag{6.40}$$

By taking the z derivative of (6.39) and using (6.40), one arrives at the equation $\partial^3 v_x/\partial z^3 = 0$, the solution of which is

$$v_x = C_1 + C_2 z + C_3 z^2, \tag{6.41}$$

where $C_{1,2,3}$ are independent of z. They are to be determined from the boundary conditions. The first of them defines the tangential stress at the free vapor-liquid interface $z = h(x)$ to be equal to the surface tension gradient induced (Marangoni) stress

$$\mu \frac{\partial v_x}{\partial z} = \frac{\partial \sigma}{\partial x}. \tag{6.42}$$

The volume flux Φ flowing through the film at a given position x

$$\Phi = \int_0^h v_x(z)\mathrm{d}z \tag{6.43}$$

serves as the second equation. The third condition is given by Eq. (6.3). The back substitution of the solution into Eq. (6.39) written at the vapor-liquid interface results in

$$\mu\Phi = \frac{\partial \sigma}{\partial x}\left(\frac{h^2}{2} + hl_s\right) + \left(\frac{h^3}{3} + h^2 l_s\right)\frac{\partial \Delta p}{\partial x}, \tag{6.44}$$

where the equality $\partial \Delta p/\partial x = -\partial p_L/\partial x$ has been used. It assumes the constant gas pressure which can be justified by the small gas viscosity and density with respect to those of the liquid, so that the gas pressure gradient is much smaller.

By using the fluid mass conservation, Φ can also be expressed via the component $v_n = \mathbf{v} \cdot \mathbf{n}$ of the liquid velocity normal to the vapor-liquid interface (positive when directed along $\mathbf{n}$, see Fig. 6.6)

$$\Phi \simeq \int_{x_{CL}}^{x} v_n(x)\mathrm{d}x. \tag{6.45}$$

Eq. (6.45) can thus be rewritten as

$$v_n = \frac{\partial \Phi}{\partial x}, \tag{6.46}$$

where v_n is related to the evaporation flux J (6.24) at the interface via the mass conservation law

$$J = (v^i - v_n)\rho_L. \tag{6.47}$$

The normal interface velocity

$$v^i = -\frac{\partial h}{\partial t}\left[1+\left(\frac{\partial h}{\partial x}\right)^2\right]^{-1/2} \simeq -\frac{\partial h}{\partial t}, \tag{6.48}$$

is considered to be positive if directed inside the liquid (as the vector **n** in Fig. 6.6). Here the small-slope approximation is applied.

By injecting (6.44, 6.47), and (6.48) into (6.46), one arrives finally at the expression [37]

$$\frac{\partial}{\partial x}\left\{\frac{1}{\mu}\left[h\left(\frac{h}{2}+l_s\right)\frac{\partial \sigma}{\partial x}+h^2\left(\frac{h}{3}+l_s\right)\frac{\partial \Delta p}{\partial x}\right]\right\} = -\mu\left(\frac{\partial h}{\partial t}+\frac{J}{\rho_L}\right). \tag{6.49}$$

In the same, small-slope approximation, the curvature

$$K = \frac{\partial^2 h}{\partial x^2}\left[1+\left(\frac{\partial h}{\partial x}\right)^2\right]^{-3/2} \tag{6.50}$$

can be approximated as $K \simeq \partial^2 h/\partial x^2$, and Eq. (6.25) reduces to

$$\sigma\frac{\partial^2 h}{\partial x^2} = \Delta p + p_r. \tag{6.51}$$

If both the slip, Marangoni, and vapor recoil effects are neglected, Eq. (6.49) takes a simpler form

$$\sigma\frac{\partial}{\partial x}\left(\frac{h^3}{3}\frac{\partial^3 h}{\partial x^3}\right) = -\mu\left(\frac{\partial h}{\partial t}+\frac{J}{\rho_L}\right). \tag{6.52}$$

The theory can be generalized to the 3D case by writing

$$\sigma\nabla\cdot\left[\frac{h^3}{3}\nabla(\nabla^2 h)\right] = -\mu\left(\frac{\partial h}{\partial t}+\frac{J}{\rho_L}\right), \tag{6.53}$$

where $\nabla = (\partial/\partial x, \partial/\partial y)$ is the 2D differential operator and $h = h(x, y, t)$.

One can distinguish several cases related to the CL motion. When the wedge is immobile and stationary, $\partial h/\partial t = 0$. On the other hand, the CL motion is convenient to consider in the reference where CL is immobile, in which, instead of the $\partial h/\partial t$ term, one has

$$\frac{\partial h}{\partial t} - v_{CL}\frac{\partial h}{\partial x}, \tag{6.54}$$

where the CL velocity v_{CL} is assumed to be positive at liquid receding. The $\partial h/\partial t$ term describes now only the slope change. For a large liquid volume, this term is much smaller with respect to the second, at least in the CL vicinity. This is why in the next section it is neglected.

6.1.4 Boundary Conditions

Equation (6.49) is the fourth order differential equation. To solve it, one needs four boundary conditions. In the partial wetting case considered here, two of them are geometrical constraints defined at the CL

$$\begin{aligned} h(x \to 0) &= 0, \\ \partial h/\partial x|_{x\to 0} &= \theta_{\text{micro}}. \end{aligned} \tag{6.55}$$

They are sufficient to describe the macroscopic liquid shape. However, in the microscopic and intermediate regions (i.e., for the wedge geometry), other conditions need to be imposed. The third boundary condition is given by the matching of the pressure to its macroregion value, i.e., at $x \to \infty$. The interfacial pressure jump in the macroregion is expected to be much smaller than in other regions because of much smaller K and J. For this reason, the condition

$$\Delta p(x \to \infty) = 0 \tag{6.56}$$

is applied.

The fourth boundary condition is not straightforward. It comes from the regularity of all the variables at the contact line. Indeed, from the physical point of view, the measurable quantities cannot be infinite. It can be proven mathematically (cf. the online supplementary material to the work of [28]) that a regular solution indeed exists for the case of the pure vapor.

The fourth boundary condition is specific to each mode of evaporation. However, some general features can be determined, in particular, the J asymptotics. We start from Eq. (6.49), in which $\partial h/\partial t$ is replaced by $-v_{CL}\partial h/\partial x$ as discussed in Sect. 6.1.3. To comply the evaporation in the open air with no substrate heating, the Marangoni term is omitted for this derivation (a more complete derivation can be found in Sect. 6.2.1 below). One can integrate Eq. (6.49) (with the substitution of from 0 to x and tend x to 0 by assuming the finiteness of $J(x \to 0)$. One readily obtains

$$x\frac{\partial \Delta p}{\partial x} = \frac{\mu}{l_s\theta_{\text{micro}}}\left[v_{CL} - \frac{J}{\rho_L\theta_{\text{micro}}}\right]. \tag{6.57}$$

If the pressure does not diverge at CL (i.e., Δp saturates at $x \to 0$), the left-hand side is zero, so

$$J(x \to 0) = \theta_{\text{micro}}\rho_L v_{CL}. \tag{6.58}$$

In the important particular case of immobile contact line, $J(x \to 0) = 0$.

Equation (6.58) shows that a regular solution can be obtained at the contact line, in spite of the no-slip boundary condition imposed at the solid-liquid interface: at the microscopic scale, the interface advances ($v_{CL} < 0$) by the curvature driven condensation ($J < 0$) or recedes ($v_{CL} > 0$) by evaporation ($J > 0$). Equation (6.58) is easy to interpret: in the reference of the fluid wedge, the whole amount of the fluid flow created by the entrainment by the moving substrate is spent as a mass exchange at the liquid-vapor interface.

Consider now the case of the isothermal CL motion with no overall mass exchange. When the contact line advances ($v_{CL} < 0$), the mass flux at the CL is negative, so that condensation occurs in the vicinity of the CL. It means that $J(s)$ should change sign farther away from the CL, so that evaporation occurs there. It should exactly compensate the condensed mass. A possibility of such an effect was already discussed [44, 49]. Such an effect is discussed in detail in [29].

As the fourth boundary condition, one can use the finiteness of pressure at the CL. Equation (6.57) suggests a form [25, 37]

$$x \frac{\partial \Delta p}{\partial x}\bigg|_{x \to 0} = 0. \tag{6.59}$$

Note the role of the Kelvin effect that is responsible for the explicit form of the fourth boundary condition. If the Kelvin effect is neglected, the pressure jump at the contact line cannot be identified explicitly, thus requiring application of asymptotic methods of solution instead of a generally simpler numerical treatment. The finiteness of pressure and of evaporation flux at the CL clearly shows that the Kelvin effect alone is sufficient to regularize all the physical quantities.

6.2 Evaporation into Pure Vapor

Now that all the basic "bricks" of microscale phenomena are explained, one can consider two most important regimes of evaporation listed in Sect. 6.1. First, the evaporation of liquid into the atmosphere of its pure vapor is discussed. It is an important case met in numerous industrial applications oriented to cooling, like growth of bubbles in boiling, heat pipes, and sessile drop evaporation during the spray cooling.

Near the contact line, the interfacial temperature T^i can vary along the interface according to Eq. (6.37), so the Marangoni stress

$$\frac{\partial \sigma}{\partial x} = -\gamma \frac{\partial T^i}{\partial x}, \tag{6.60}$$

can be important. Here $\gamma = -\mathrm{d}\sigma/\mathrm{d}T$ is positive for pure fluids.

A large amount of work has been done by many researchers to understand this regime. One can cite the works [2, 19, 34, 35, 45, 48, 54] as the main milestones.

Since the vapor heat conductivity is much smaller than that of the liquid, the heat flux into the vapor can be neglected, and the interfacial energy balance reads

$$q_L^i = \mathcal{L}J. \tag{6.61}$$

Now, one can see the mass evaporation flux is completely defined by the heat transfer in the liquid in this regime: the liquid heat flux is spent to compensate the latent heat of vaporization.

Similarly to the above hydrodynamics approach, one can use the thin film approximation to simplify the heat transfer problem in the contact line vicinity. First, the thermal inertia of the thin film is small and can be neglected so the problem to solve is stationary. Second, the heat convection can be neglected at small scales with respect to the heat conduction, so the temperature distribution in the liquid obeys the equation

$$\nabla^2 T_L = 0. \tag{6.62}$$

The boundary conditions are the fixed temperature both on the solid (where the superheating ΔT is fixed) and on the free interface:

$$\begin{aligned} T_L(z=0) &= T_{sat} + \Delta T, \\ T_L(z=h) &= T^i. \end{aligned} \tag{6.63}$$

One can use the small-slope approximation to solve this problem. The solution that satisfies the boundary conditions (6.63) results in the heat flux

$$q_L^i = k_L \frac{T_{sat} + \Delta T - T^i}{h}. \tag{6.64}$$

at the free interface.

By comparing Eqs. (6.61) and (6.64), one can see that the problem is local in this regime: the mass flux at coordinate x depends on the substrate and interfacial temperatures at the same point.

By combining Eq. (6.64) with Eqs. (6.37, 6.61), one obtains the flux

$$J = \frac{\Delta T - \Delta p T_{sat}/(\mathcal{L}\rho_L)}{\mathcal{L}(R^i + h/k_L)} - \frac{T_{sat}}{2\mathcal{L}^4\rho_V^2}\frac{(\Delta T - \Delta p T_{sat}/(\mathcal{L}\rho_L))^2}{(R^i + h/k_L)^3} \tag{6.65}$$

under an assumption that the vapor recoil term in Eq. (6.37) is small (which is true in most cases).

All the terms of the fourth order set of differential equations (6.49, 6.51) for the unknown functions $h(x)$ and $\Delta p(x)$ are now completely defined.

6.2.1 Fourth Boundary Condition

To provide the temperature continuity, the interfacial temperature should be equal to that of the solid at the contact line. Equation (6.37) thus results in

$$\Delta p(x \to 0) = \frac{\mathcal{L}\rho_L}{T_{sat}}(\Delta T - R^i J(x \to 0)\mathcal{L}) - \frac{J(x \to 0)^2 \rho_L}{2}\left(\frac{1}{\rho_V^2} - \frac{1}{\rho_L^2}\right). \tag{6.66}$$

Similarly to Sect. 6.1.4, to obtain $J(x \to 0)$, Eq. (6.49) (now containing the Marangoni term) is integrated from 0 to x, where x is infinitesimally small. This results in

$$-\gamma\frac{\partial T^i}{\partial x} + \theta_{\text{micro}} x \frac{\partial \Delta p}{\partial x} = \frac{\mu}{l_s \theta_{\text{micro}}}\left[\theta_{\text{micro}} v_{CL} - \frac{J)}{\rho_L}\right]. \tag{6.67}$$

The T^i derivative can be obtained by developing Eq. (6.64) into the Taylor series around $h = 0$

$$\frac{\partial T^i}{\partial x} = -\frac{\mathcal{L}\theta_{\text{micro}}}{k_L} J, \tag{6.68}$$

where Eq. (6.61) is accounted for. The substitution into Eq. (6.67) results in the expression

$$x\frac{\partial \Delta p}{\partial x} = \frac{\mu}{l_s \theta_{\text{micro}}}\left[v_{CL} - J\left(\frac{1}{\theta_{\text{micro}}\rho_L} + \frac{l_s \mathcal{L}\theta_{\text{micro}}}{\mu k_L}\gamma\right)\right]. \tag{6.69}$$

By injecting the value of J from Eq. (6.65) into Eq. (6.69), one obtains a differential equation for Δp. Its solution [25] is quite straightforward and, for this reason, is not detailed here. From it, one obtains explicitly that Δp saturates at $x \to 0$, so that the Eq. (6.59) indeed holds. Note that such behavior is different from the CL dynamics [18] for nonvolatile liquids, where the pressure boundary condition cannot be written and the limit (6.59) is nonzero.

Finally, Eqs. (6.59, 6.69) result in

$$J(x \to 0) = v_{CL}\left(\frac{1}{\theta_{\text{micro}}\rho_L} + \frac{l_s \mathcal{L}\theta_{\text{micro}}}{\mu k_L}\gamma\right)^{-1}. \tag{6.70}$$

The pressure boundary condition is easily obtained by substitution of Eq. (6.70) into Eq. (6.66).

In the particular case of the immobile contact line, $J(x \to 0)$ and

$$\Delta p(x \to 0) = \frac{\mathcal{L}\rho_L \Delta T}{T_{sat}}. \tag{6.71}$$

This formula is similar to the complete wetting case [47]. In such a situation, J also vanishes at the left domain border (which corresponds to the flat film at $x \to -\infty$, cf.

Fig. 6.4a). In this case, the boundary condition exactly coincides with the condition (6.71) written, however, for $x \to -\infty$. Note that for $R^i = 0$ and when the vapor recoil effect is neglected, Eq. (6.71) describes also the moving CL case, cf. Eq. (6.66).

The evaporation flux (6.70) is proportional to v_{CL}. When the contact line advances ($v_{CL} < 0$), the heat flux at the CL becomes negative. It means that $J(x)$ changes sign and condensation occurs in an extremely small vicinity of the CL.

As the fourth boundary condition, one can use either the pressure derivative (6.59) or directly the pressure value (6.66) with the substitution of the mass flux (6.70).

6.2.2 *Asymptotic Analysis for Immobile Contact Line*

The asymptotic analysis is helpful to provide some general idea of the behavior of the apparent contact angle because it results in analytical expressions. They are possible to obtain for some simple problem statements by neglecting secondary physical effects.

6.2.2.1 Partial Wetting and Small Superheating

One of the main goals of a microregion model is the apparent contact angle. It turns out to be possible to obtain an analytic result for its variation with ΔT when it is small enough. This result is possible for the "minimal complexity" problem, while accounting only for the Kelvin effect, which is auto-sufficient to relax the contact line singularity. The set of equations to be solved is the following:

$$\sigma \frac{\partial^2 h}{\partial x^2} = \Delta p, \tag{6.72a}$$

$$\frac{\partial}{\partial x}\left(\frac{h^3}{3}\frac{\partial \Delta p}{\partial x}\right) = \mu \frac{k_L T_{sat}}{\mathcal{L}^2 \rho_L^2} \frac{\Delta p - \mathcal{L}\rho_L \Delta T / T_{sat}}{h} \tag{6.72b}$$

with the boundary conditions (6.55, 6.56, 6.71). The scales $\sigma/(\mathcal{L}\rho_L)$, $\mathcal{L}\rho_L$ and T_{sat} are used to make the lengths, the pressure jump and the superheating dimensionless, respectively. The dimensionless counterparts of Eq. (6.72) read

$$\frac{\partial^2 \tilde{h}}{\partial \tilde{x}^2} = \Delta \tilde{p}, \quad \frac{\partial}{\partial \tilde{x}}\left(\frac{\tilde{h}^3}{3}\frac{\partial \Delta \tilde{p}}{\partial \tilde{x}}\right) = N \frac{\Delta \tilde{p} - \varepsilon}{\tilde{h}}, \tag{6.73a}$$

$$\Delta \tilde{p}(\tilde{x} \to \infty) = 0, \tag{6.73b}$$

$$\tilde{h}(\tilde{x} = 0) = 0, \quad \Delta \tilde{p}(\tilde{x} = 0) = \varepsilon, \tag{6.73c}$$

$$\partial \tilde{h}/\partial \tilde{x}|_{\tilde{x}=0} = \theta_{\text{micro}}, \tag{6.73d}$$

where $N = \mu k_L T_{sat}/\sigma^2$ and $\varepsilon = \Delta T / T_{sat}$.

We seek a solution by expanding the pressure jump $\Delta\tilde{p}$ and shape $\tilde{h}$ in a regular perturbation series in ε [25]:

$$\Delta\tilde{p} = \tilde{p}_0 + \varepsilon\tilde{p}_1 + \mathcal{O}(\varepsilon^2), \quad \tilde{h} = \tilde{h}_0 + \varepsilon\tilde{h}_1 + \mathcal{O}(\varepsilon^2). \tag{6.74}$$

One needs to substitute (6.74) into the set (6.73) and collect terms of the same order in ε. In the zeroth order, one obtains $\tilde{p}_0 = 0$ and $\tilde{h}_0 = \theta_{\text{micro}}\tilde{x}$.

The equation for $\tilde{p}_1$

$$\tilde{h}_0 \frac{\partial}{\partial\tilde{x}}\left(\tilde{h}_0^3 \frac{\partial\tilde{p}_1}{\partial\tilde{x}}\right) = \alpha^2(\tilde{p}_1 - 1), \tag{6.75a}$$

$$\tilde{p}_1(\tilde{x} = 0) = 1, \tag{6.75b}$$

$$\tilde{p}_1(\tilde{x} \to \infty) = 0, \tag{6.75c}$$

where $\alpha = \sqrt{3N}/\theta_{\text{micro}}^2$, does not contain $\tilde{h}_1$. One can easily obtain its solution (e.g., with the Wolfram MATHEMATICA software) in terms of the modified Bessel functions I_1, K_1 of the first order

$$\tilde{p}_1(\tilde{x}) = 1 + C_1 \frac{K_1(\alpha/\tilde{x})}{\tilde{x}} + C_2 \frac{I_1(\alpha/\tilde{x})}{\tilde{x}}. \tag{6.76}$$

where C_1, C_2 are integration constants. Since $I_1(\cdot)$ diverges at infinity, the boundary condition (6.75b) requires $C_2 = 0$. Since $K_1(z \to 0) \sim z^{-1}$, the second constant $C_1 = -\alpha$ is determined from (6.75c); the solution for $\Delta\tilde{p}(\tilde{x})$ is thus

$$\Delta\tilde{p} = \varepsilon - \varepsilon\frac{\alpha}{\tilde{x}} K_1\left(\frac{\alpha}{\tilde{x}}\right) + \mathcal{O}(\varepsilon^2). \tag{6.77}$$

The condition (6.75b) turns out to be satisfied automatically since $K_1(z \to \infty) \to 0$ exponentially. According to Eqs. (6.73a), the slope at infinity can be found by integrating $\Delta\tilde{p}$,

$$\theta_{\text{app}} = \theta_{\text{micro}} + \alpha\varepsilon \int_0^\infty \left[\frac{1}{z^2} - \frac{1}{z}K_1(z)\right] \mathrm{d}z = \theta_{\text{micro}} + \frac{\alpha\varepsilon\pi}{2}.$$

By returning to the dimensional variables, one obtains the final expression for the apparent contact angle in terms of the microscopic contact angle within the first order approximation:

$$\theta_{\text{app}} = \theta_{\text{micro}} + \frac{\pi\Delta T}{2\sigma\theta_{\text{micro}}^2}\sqrt{\frac{3\mu k_L}{T_{sat}}}. \tag{6.78}$$

One can see that θ_{app} varies linearly with the superheating. Note that the slope diverges for $\theta_{\text{micro}} \to 0$, i.e. the complete wetting, which is a well-known behavior [47]. The asymptote (6.78) is shown in Fig. 6.10c below.

A similar result can be obtained in the framework of another approach where the Kelvin effect is neglected but both the hydrodynamic slip (Sect. 6.1.2.1) and interfacial thermal resistance (Sect. 6.1.2.5) are accounted for [24]

$$\theta_{\text{app}} = \theta_{\text{micro}} + \frac{3\mu\Delta T}{\rho_L \mathcal{L}\sigma\theta_{\text{micro}}^3 R^i} f\left(\frac{l_s}{R^i k_L}\right),$$

where $f(\cdot)$ is a function of the order one. Note that the simultaneous introduction of both these effects is necessary to relax the contact line singularity. It is even possible to extend this result for larger ΔT [3, 19]

$$\theta_{\text{app}}^4 = \theta_{\text{micro}}^4 + \frac{12\mu\Delta T}{\rho_L \mathcal{L}\sigma R^i} \log\left(\frac{e\theta_{\text{micro}} l_s}{\theta_{\text{app}} R^i k_L}\right). \tag{6.79}$$

However, a practical implementation of this model is more complex than the approach based on the Kelvin effect.

6.2.2.2 Partial Wetting and Large Superheating

We consider again the set (6.72) but use a different scaling of variables [28]. To find it, one introduces $\theta_{\text{app}}\ell_K$ as a reference value for h, where ℓ_K is a reference value for x. By balancing two Δp-containing terms in Eq. (6.72b) (the third term is not meaningful as it can be easily eliminated by shifting of Δp by $\Delta p(x \to 0)$), one gets

$$\ell_K = \frac{\sqrt{3\mu k_L T_{\text{sat}}}}{\theta_{\text{app}}^2 \rho_L \mathcal{L}}. \tag{6.80}$$

We, therefore, make the solution dimensionless using

$$h = \theta_{\text{app}}\ell_K \tilde{h}(\tilde{x}), \quad \tilde{x} = x/\ell_K.$$

The dimensionless governing equations have the same form as (6.73a–6.73c), where N is now replaced by 1, and the superheating parameter becomes

$$\varepsilon = \frac{\sqrt{3\mu k_L T_{\text{sat}}}}{\sigma\theta_{\text{app}}^3} \frac{\Delta T}{T_{\text{sat}}} \tag{6.81}$$

The boundary condition (6.73d) is replaced by

$$\partial\tilde{h}/\partial\tilde{x}|_{\tilde{x}=0} = \theta_{\text{micro}}/\theta_{\text{app}}. \tag{6.82}$$

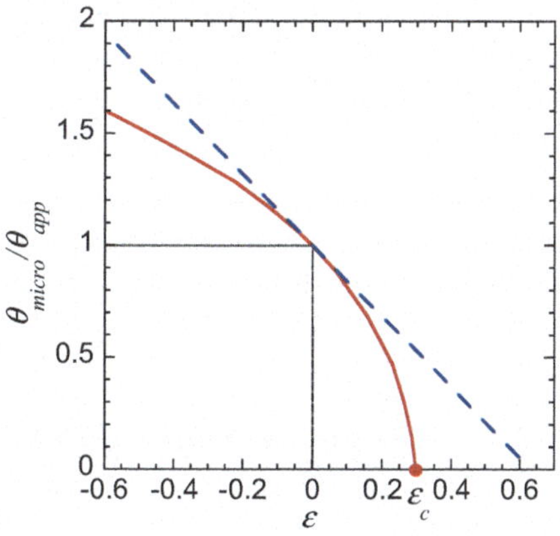

Fig. 6.7 Ratio of the microscopic and apparent contact angles as a function of the superheating parameter. Numerical solution: solid line; asymptotic expansion (6.84): dashed line

Like in the previous section, one can proceed by expanding both $\Delta\tilde{p}$ and $\tilde{h}$ in a regular perturbation series in ε.

One mentions that, with the present scaling, the boundary condition (6.73b) is equivalent to $\tilde{h}'(\tilde{x} \to \infty) = 1$. Therefore, by using this expression to define a background solution, the asymptotic expansions are

$$\Delta\tilde{p} = \varepsilon\tilde{p}_1 + \mathcal{O}(\varepsilon^2), \quad \tilde{h} = \tilde{x} + \varepsilon\tilde{h}_1 + \mathcal{O}(\varepsilon^2). \tag{6.83}$$

The solution is exactly like (6.77) but with $\alpha = 1$. It can be integrated from ∞ to 0, which results in

$$\frac{\theta_{\text{micro}}}{\theta_{\text{app}}} = 1 - \frac{\pi}{2}\varepsilon + \mathcal{O}(\varepsilon^2), \tag{6.84}$$

shown as a dashed line in Fig. 6.7.

The most important feature of the curve $\theta_{\text{micro}}/\theta_{\text{app}}(\varepsilon)$ is the existence of a terminal point $\varepsilon_c \simeq 0.297$; note that the linear approximation (6.84) overestimates ε_c by a factor $\simeq 2$. When ε approaches ε_c, $\theta_{\text{app}} \gg \theta_{\text{micro}}$, which means that the terminal point can be interpreted as corresponding to the large ΔT asymptotics. If one fixes $\varepsilon = \varepsilon_c$ in Eq. (6.81), the asymptotic expression for large ΔT follows:

$$\theta_{\text{app}} = \left(\frac{\sqrt{3\mu k_L T_{sat}}}{\varepsilon_c \sigma}\,\frac{\Delta T}{T_{sat}}\right)^{1/3}. \tag{6.85}$$

It is natural that at large ΔT, θ_{app} becomes independent of θ_{micro}, i.e. of its value at $\Delta T = 0$. This asymptote is shown in Fig. 6.10c below.

From Fig. 6.7, the terminal point corresponds to $\theta_{\text{micro}} = 0$, which means that Eq. (6.85) is the *exact* solution corresponding to this case. Since θ_{app} is an increasing function of θ_{micro} at a fixed ΔT, Eq. (6.85) provides the precise lower bound for θ_{app} for a given ΔT.

It should be stressed that the specific asymptotic expressions for θ_{app} may be invalid when other microscopic effects are accounted for. However, the general tendency suggested by the asymptotic formulas remains valid. Among such general features, one can list

- fast growth with ΔT for small ΔT and much weaker growth at large ΔT,
- strong dependence on θ_{micro} for small ΔT and a weaker dependence for large ΔT,
- increase of the difference $\theta_{app} - \theta_{micro}$ with μ and decrease with σ.

6.2.3 Parametric Study of the Apparent Contact Angle

In addition to what was mentioned above, in this section we discuss the impact of other parameters on the apparent contact angle for the case of the immobile CL. In general, all the fluid parameters impact the apparent contact angle. But some influence it stronger than the others. The following study has been conducted to show their impact and explain what it is.

The influence of the two most important parameters, namely, the solid superheating ΔT and the microscopic contact angle θ_{micro} is shown in Fig. 6.8. These curves have been calculated with the above model for the fluid parameters assumed constant. The situation shown in Fig. 6.8 is quite common: θ_{app} grows with both θ_{micro} and ΔT.

In the following, we discuss the impact of various microscopic-scale physical phenomena on the behavior of various parameters in the microregion, including the most important of them, the apparent contact angle. Unless mentioned specifically, the numerical calculations discussed here [23] are performed for water at 10 MPa,

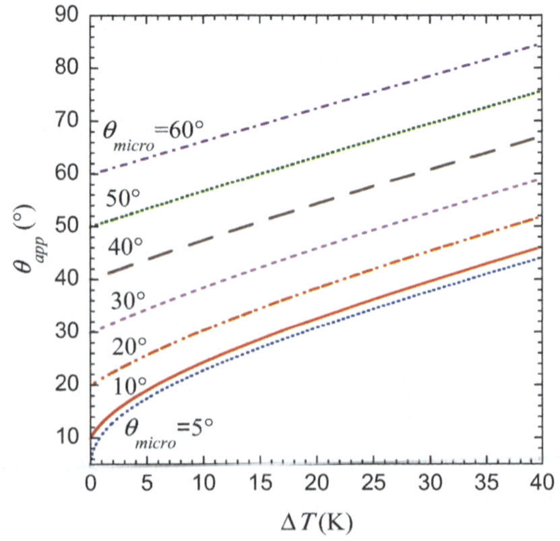

Fig. 6.8 Variation of the apparent contact angle with ΔT for different microscopic contact angle θ_{micro} for water at 1 atm and $l_s = 10$ nm

$\theta_Y = 15°$, $l_s = 10$ nm. We use the Eq. (6.72) as a reference and add various effects to study their impact.

6.2.3.1 Impact of Surface Forces for Partial Wetting

The surface forces can be introduced into the partial wetting theory as explained in Sect. 6.1.2.3, with the regularized at $h \to 0$ disjoining pressure (6.20). Note that, as mentioned in Sect. 6.1.2.3, the disjoining pressure leads to the difference between θ_{micro}, i.e., the slope at a scale smaller than that of the surface forces, and θ_Y, i.e., the static equilibrium contact angle at a larger (but still nanometer) scale.

The numerical calculations discussed here [26] use the following values related to the surface forces: $\theta_{\text{micro}} = 1°$, $h_m = 1$ nm and $A = 3.7 \cdot 10^{-20}$ J.

Examples of the computed shape of the liquid-vapor interface with and without accounting for the disjoining pressure effect are shown in Fig. 6.9. One can see that the macroscopic shapes in Fig. 6.9b for the cases $\Pi = 0$ and $\Pi \neq 0$ are very close. The curves corresponding to $\Pi = 0$ and $\Pi \neq 0$ are indistinguishable at this scale for $\Delta T = 0$. The impact of the surface forces is visible (Fig. 6.9a) only on a scale comparable to the characteristic scale (cf. Sect. 6.2.2.2)

$$\ell_K^{\text{micro}} = \frac{\sqrt{3\mu k_L T_{\text{sat}}}}{\theta_{\text{micro}}^2 \rho_L \mathcal{L}}, \tag{6.86}$$

which is $\simeq 7.2$ nm here. The wedge shape follows the trend schematized in Fig. 6.4. One can check that the variation of the parameters A, h_m and θ_{micro} at a fixed θ_Y do

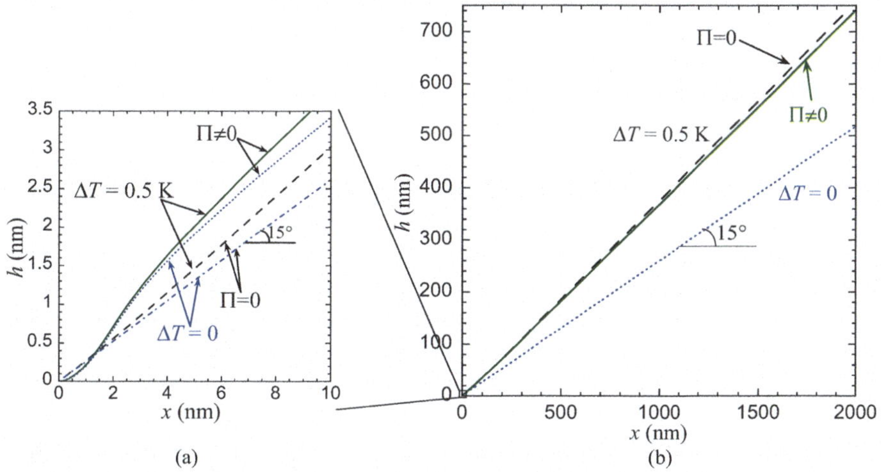

Fig. 6.9 Wedge shape computed for $\theta_Y = 15°$ with and without the impact of surface forces. **a** Close nanometric vicinity of the contact line. **b** Large scale view [23]

not impact the apparent contact angle either [23]. For this reason, in the remaining part of this chapter, we neglect the surface forces, so that $\theta_{\text{micro}} = \theta_Y$ according to Eq. (6.19).

6.2.3.2 Impact of the Slip Length

When using the scaling based on ℓ_K^{micro} (6.86), the governing equation including the slip length can be reduced to the following form by using an approach like in Sect. 6.2.2.2

$$(\tilde{h} + \mathcal{R})\frac{\partial}{\partial\tilde{x}}\left[\left(\beta\tilde{h}^2 + \tilde{h}^3\right)\frac{\partial^3\tilde{h}}{\partial\tilde{x}^3}\right] = \frac{\partial^2\tilde{h}}{\partial\tilde{x}^2} - \varepsilon^{\text{micro}},$$

where the slip parameter is

$$\beta = \frac{3l_s}{(\theta_{\text{micro}}\ell_K^{\text{micro}})}.$$

$$\varepsilon^{\text{micro}} = \frac{\sqrt{3\mu k_L T_{\text{sat}}}}{\sigma\theta_{\text{micro}}^3}\frac{\Delta T}{T_{\text{sat}}}$$

is the dimensionless superheating, and

$$\mathcal{R} = \frac{l_R}{\theta_{\text{micro}}\ell_K^{\text{micro}}}. \tag{6.87}$$

is the dimensionless interfacial resistance (cf. Sect. 6.1.2.5). The scale $l_R = R^i k_L$ is an equivalent thickness of the liquid layer of the same thermal resistance as the interface. As an example, $l_R = 2$ nm for water at 10 MPa, with R^i given by Eq. (6.36). The constraint $\mathcal{R} = 0$ is artificially imposed here to show the impact of the slip length alone; the impact of R^i is considered in the next section. Consequently, $\varepsilon_{\text{micro}}$ and β are the only parameters defining the behavior of the system.

One can argue that the scales considered in Fig. 6.10a are too small to be physically reasonable within the continuum mechanics approach. Resolving such small scales is, however, necessary to find a correct solution at a larger x.

The hydrodynamic slip is well-known to remove the CL singularity for moving contact line (in the absence of the Kelvin effect). In case of its implementation for the evaporation with immobile CL [19, 24], similarly to the moving CL case, one obtains the logarithmically divergent $\Delta p(x \to 0)$. While such a divergence is integrable (it does not case the divergence of integral properties like viscous energy dissipation), this is still nonphysical as the pressure, being a measurable quantity, cannot be infinite. The Kelvin effect causes the finite pressure at CL, which is more coherent from the physical viewpoint. With l_s increase, Δp decays slower along the wedge (Fig. 6.10a), so that in the limit $\beta \to \infty$, the dependence $\Delta p(x)$ approaches a straight line in the semi-logarithmic scale (indicating the logarithmic pressure divergence)

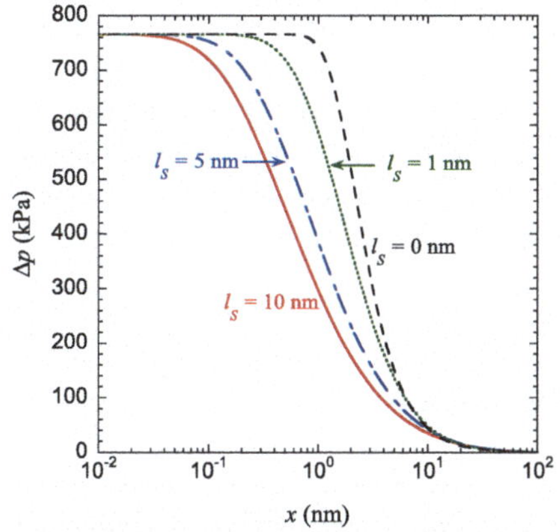

(a) Pressure jump spatial variation for water at 10 MPa, $\Delta T = 0.5$ K, $\theta_{micro} = 15°$.

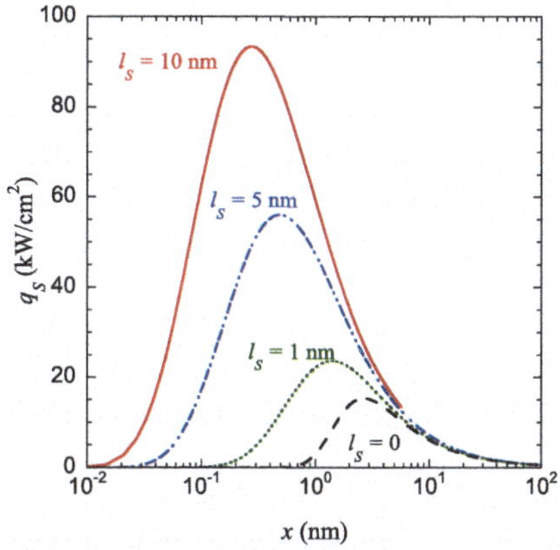

(b) Heat flux spatial variation for water at 10 MPa, $\Delta T = 0.5$ K, $\theta_{micro} = 15°$.

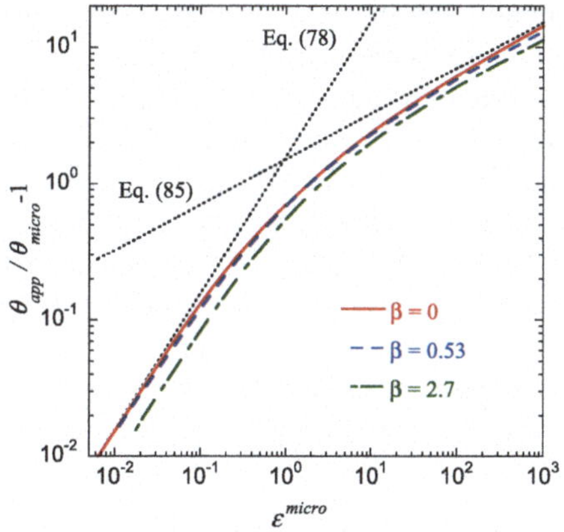

(c) θ_{app} as a function of the dimensionless superheating ε^{micro} for different β. The asymptotic solutions (78, 85) are also shown.

Fig. 6.10 Pressure and heat flux spatial variations and the apparent contact angle as a function of superheating for different slip lengths; $\mathcal{R} = 0$ [23]

and the pressure saturation at $x \to 0$ eventually disappears in the limit $\beta \to \infty$. For a fixed x, the pressure jump thus reduces with the l_s increase. Since θ_{app} is proportional to $\int_0^\infty \Delta p(x)dx$, it decreases with l_s (Fig. 6.10c).

The liquid-vapor interface heat flux variation along the liquid wedge (Fig. 6.10b) shows huge heat fluxes localized, however, in a tiny CL vicinity. The maximum of the local heat flux increases with the slip length. This is due to the fact that pressure jump (and thus the interface temperature) both saturate at a smaller scale, where the liquid thickness (and thus the conductive thermal resistance) are smaller so the flux is larger. However, such flux increase is not strong enough to cause the pressure increase. In Fig. 6.10c, the curves $\theta_{app}/\theta_{micro} - 1$ vs. ε^{micro} are compared for three values of $\beta = 0$, 0.53 and 2.7. The slopes of the curves for small and large ε_{micro} are close to those for the case $\beta = 0$ described by Eqs. (6.78, 6.85). It is evident that the apparent contact angle decreases with the slip length, however, the impact of the slip length is weak for $\beta < 1$. This is an expected behavior as the crossover to the regime controlled by the slip rather than Kelvin effect is expected at $\beta \sim 1$, cf. Fig. 6.11c below.

6.2.3.3 Impact of the Interface Thermal Resistance

The impact of the interface thermal resistance R^i (see Sect. 6.1.2.5) is studied here. Figure 6.11a shows the interfacial heat flux variation along the liquid wedge for four different l_R values. The length scale at which the flux maximum is attained remains nearly unaffected by R^i. As expected, the interfacial resistance causes heat flux reduction. Consequently, the pressure drop becomes smaller, which leads to a smaller apparent contact angle (see Fig. 6.11b, c).

In Fig. 6.11b, the variation of the parameter $\theta_{app}/\theta_{micro} - 1$ with the dimensionless superheating ε^{micro} is shown for three values of $\mathcal{R} = 0, 0.01$, and 1 (which correspond to $l_R = 0$, 1 nm and 10 nm for water at 10 MPa and $\theta_{micro} = 5°$). It is evident that the apparent contact angle decreases with $\mathcal{R}$, however, the variation is weak. Similarly to Fig. 6.10c, the curves for $\mathcal{R} \neq 0$ look shifted with respect to the curve $\mathcal{R} = 0$. The slopes for small and large ε_{micro} seem to be the same as for the case $R^i = 0$. As for simultaneous account of R^i and l_s, they both lead to the θ_{app} reduction (Fig. 6.11c). An especially fast θ_{app} decrease starts from $\beta = 1$.

6.2.4 *Simultaneous Contact Line Motion and Evaporation*

The contact line receding is quite common at evaporation; its impact on the apparent contact angle thus needs to be investigated. The moving contact line problem is more complex than its static counterpart. The difficulty comes from the flow caused by the contact line motion that is not limited to close contact line vicinity. This becomes evident if one considers a liquid wedge moving along a solid, or rather solid moving with respect to the liquid wedge (which is equivalent). Because of the viscous forces, the liquid is drugged by the moving solid not only near the contact line, but all along the wedge. Thanks to the change of liquid thickness, the hydrodynamic stress caused by this motion appears to be large only in some vicinity of the contact line, which is however larger than the vicinity where the evaporation effects are important. One can show that, in general, three length scales (Fig. 6.12) form an hierarchy. At the smallest scale, the mass exchange defines the interfacial curvature and the impact of the contact line motion is negligible. Therefore, this region can be assimilated to the microregion considered for the immobile contact line (Fig. 6.12a). Its characteristic size will be called the Voinov length $\ell_V \sim 100$ nm for the reasons

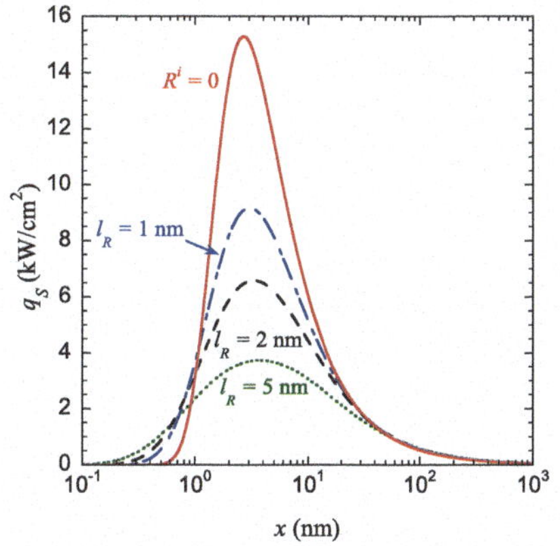

(a) Heat flux spatial variation for $l_s = 0$ and different l_R.

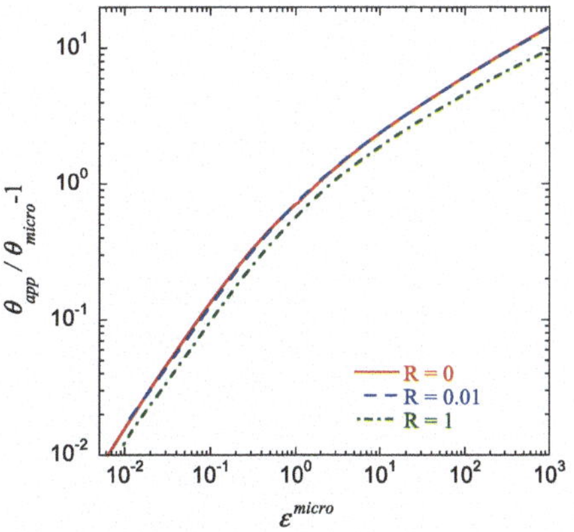

(b) θ_{app} as a function of superheating for $l_s = 0$ and different $\mathcal{R}$.

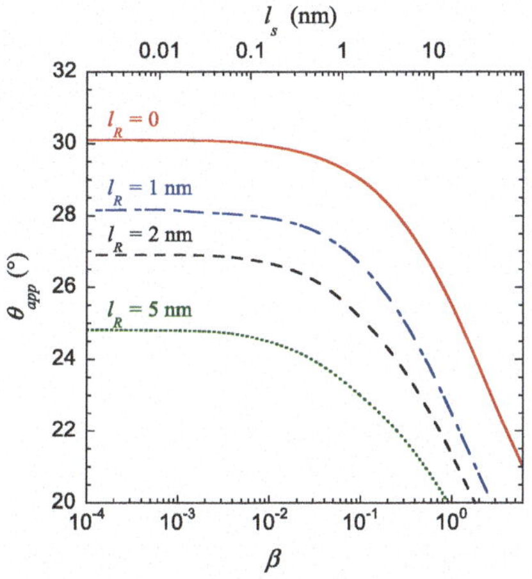

(c) θ_{app} as a function of the slip parameter for different R^i.

Fig. 6.11 Impact of the interfacial kinetic resistance on the microregion parameters for water at 10 MPa and $\Delta T = 0.5$ K [23]

that will be evident later on. The macroscopic length scale L is governed by the capillarity and gravity forces, and possibly the macroscopic fluid motion around the drop or the bubble (Fig. 6.12c). Typically, $L \sim 1$ mm, but it depends on the problem. In between these two scales lies an intermediate region (Fig. 6.12b) where the impact of mass exchange is negligible and the interfacial profile is defined by the flow caused by the contact line motion; the relevant phenomena here are the surface tension and viscosity.

Equation (6.49) can describe all the above scales (of course, within the limits of applicability of the lubrication approximation). If the gravity or other external forces

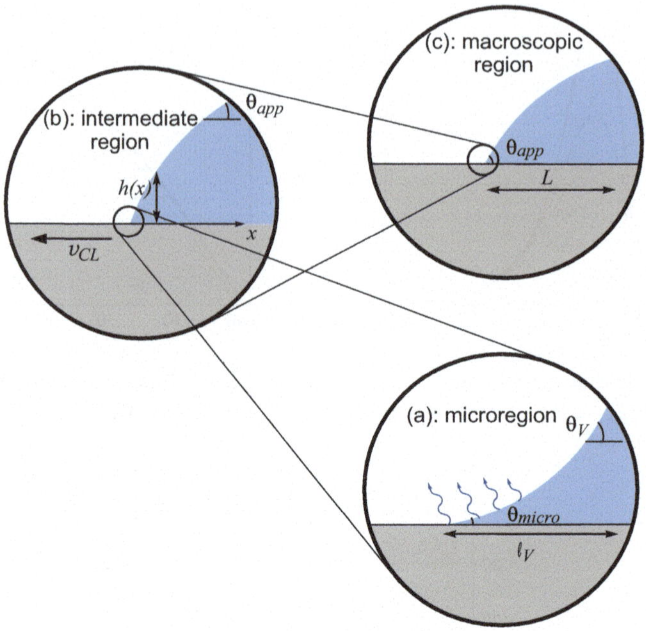

Fig. 6.12 Hierarchy of length scales in the moving CL problem with phase change

are important, they should be included in the equation. There are two possible ways of solving such a problem. In principle, one can calculate it straightforwardly by numerics. However, this is not easy, as one needs to resolve both small and large scales, which requires the application of heterogeneous adaptive grids that often cause numerical instabilities for these nonlinear differential equations. Alternatively, thanks to the strong scale difference between small and large scales, one can apply a multi-scale approach. It consists of using a mathematical method called asymptotic matching two times. First, one couples the micro and intermediate regions and then intermediate and macro regions to obtain the full solution.

6.2.4.1 Matching of Microscopic and Intermediate Regions

The impact of CL motion will be considered with the "minimal complexity" set of equations, where among all the microscopic phenomena, only the Kelvin effect is accounted for [28]. This means that the fluid flow in the wedge is described by Eq. (6.52) in the reference, where CL is immobile; the $\partial h/\partial t$ term is replaced by two terms (6.54). As the macroregion is not considered, the transient term can be neglected and only the v_{CL}-containing term remains:

$$\frac{\partial}{\partial x}\left(\frac{h^3}{3}\frac{\partial \Delta p}{\partial x}\right) = \mu v_{CL}\frac{\partial h}{\partial x} + \mu\frac{k_L T_{sat}}{\mathcal{L}^2\rho_L^2}\frac{\Delta p - \mathcal{L}\rho_L \Delta T/T_{sat}}{h}, \tag{6.88}$$

where Δp is defined by Eq. (6.72a), with the boundary conditions (6.55, 6.56, 6.71). While comparing two terms in the r.h.s. of Eq. (6.88), one can see that, at large h, the mass exchange term is small so the contact line motion term defines the wedge curvature. Inversely, near the contact line, the contact line motion term is smaller. This kind of behavior is typical for the multi-scale problems.

The asymptotic matching consists of solving two coupled problems. The first is in the "inner" region (microregion here), and the second, in the "outer" region (intermediate region here). As discussed above, in the inner region, the problem (6.72) should be solved (i.e., Eq. (6.88) with no v_{CL} term), while in the outer, the equation

$$\sigma\frac{\partial}{\partial x}\left(\frac{h^3}{3}\frac{\partial^3 h}{\partial x^3}\right) = \mu v_{CL}\frac{\partial h}{\partial x} \tag{6.89}$$

satisfying the boundary condition (6.56). This latter problem describes the moving contact line with no phase change [17]. According to the asymptotic matching strategy, the behavior at $x \to \infty$ in the inner region should match to $x \to 0$ in the outer. They should be found during the solution.

After integrating once, Eq. (6.89) reduces to

$$\sigma\frac{\partial^3 h}{\partial x^3} = \frac{3\mu v_{CL}}{h^2}. \tag{6.90}$$

Such a problem has an asymptotic (Cox-Voinov) solution first found by [57] far from the contact line:

$$\left(\frac{\partial h}{\partial x}\right)^3 = \theta_V^3 - 9Ca\log\frac{x}{\ell_V} \tag{6.91}$$

where $Ca = \mu v_{CL}/\sigma$ is the capillary number. Two parameters appear here as integration constants: the Voinov length ℓ_V and the Voinov angle θ_V. Evidently, θ_V is equivalent to θ_{app} of Sect. 6.2.2 because it is the apparent contact angle for $Ca = 0$, i.e. caused by evaporation only. We introduce it to distinguish from θ_{app} that accounts now for the contact line motion, cf. Fig. 6.12. Equation (6.91) is valid for all $Ca < 0$ at liquid advancing. For the receding case ($Ca > 0$), the situation is more complex; Eq. (6.91) is valid when the second term on the right-hand side is sufficiently smaller than the first (below the dynamic wetting transition [13, 52]) because otherwise this solution cannot be matched to the macroscopic region.

The scaling

$$\ell_K = \frac{\sqrt{3\mu k_L T_{sat}}}{\theta_V^2\rho_L\mathcal{L}}, \qquad h = \theta_V\ell_K\tilde{h}(\tilde{x}), \quad \tilde{x} = x/\ell_K, \tag{6.92}$$

is similar to that defined in Eq. (6.80) (with θ_V equivalent to θ_{app} of Sect. 6.2.2.2). The dimensionless governing equation reads

$$(\tilde{h}'''\tilde{h}^3)' = \delta\,\tilde{h}' + \frac{\tilde{h}'' - \varepsilon}{\tilde{h}}, \tag{6.93}$$

where

$$\varepsilon = \frac{\sqrt{3\mu k_L T_{sat}}}{\sigma\theta_V^3}\frac{\Delta T}{T_{sat}}$$

and

$$\delta = \frac{3Ca}{\theta_V^3}.$$

To solve this problem, we use an expansion

$$\tilde{h} = \tilde{h}_0 + \delta\,\tilde{h}_\delta + O(\delta^2).$$

It is injected into Eq. (6.93) and the equations for $\tilde{h}_0$ and for $\tilde{h}_\delta$ are written separately. The equation for $\tilde{h}_0$ corresponds to Eq. (6.72) (and to Eq. (6.88) without the first term in the r.h.s.), and, more generally, to the case of the immobile contact line considered in Sect. 6.2.2. As we saw this theory results in θ_V (formerly, θ_{app}) as a function of ΔT or its dimensionless counterpart ε. θ_V is the slope $\partial h_0/\partial x$ at $x \to \infty$ within the inner region. Within the scaling (6.92), $\tilde{h}_0'(\tilde{x} \to \infty) = 1$.

Consider now the first order, i.e., the outer region equation valid at large $\tilde{x}$ (so the second term in the r.h.s. of Eq. (6.93) can be abandoned)

$$\tilde{h}_\delta'''\tilde{h}_0^2 = \tilde{h}_\delta. \tag{6.94}$$

According to the asymptotic matching paradigm, the outer region "sees" only the $x \to \infty$ behavior $\tilde{h}_0 \simeq \tilde{x}$ in the inner region, so Eq. (6.94) can be easily solved: $h_\delta' = -\log(\tilde{x}\ell_K/\ell_V)$. A yet undetermined dimensional "Voinov" length ℓ_V plays role of an integration constant. The full expansion

$$\tilde{h}' = 1 - \delta\log(\tilde{x}\ell_K/\ell_V) \tag{6.95}$$

presents a linearized version of Eq. (6.91). The Voinov length needs to be determined as a result of the rigorous matching procedure that is easier to make numerically.

The full solution is shown in Fig. 6.13a together with the inner and outer solutions. It is now evident from Fig. 6.13b that the slope given by the inner problem saturates at $x \to \infty$ to the Voinov contact angle θ_V. The curves are shown in a semi-logarithmic scale, in which the outer asymptotic solution (6.95) is a straight line. It can be obtained as a fit to the full solution at large $\tilde{x}$. The abscissa of its intersection with the line $\tilde{h}' = 1$ gives ℓ_V/ℓ_K, cf. Eq. (6.95).

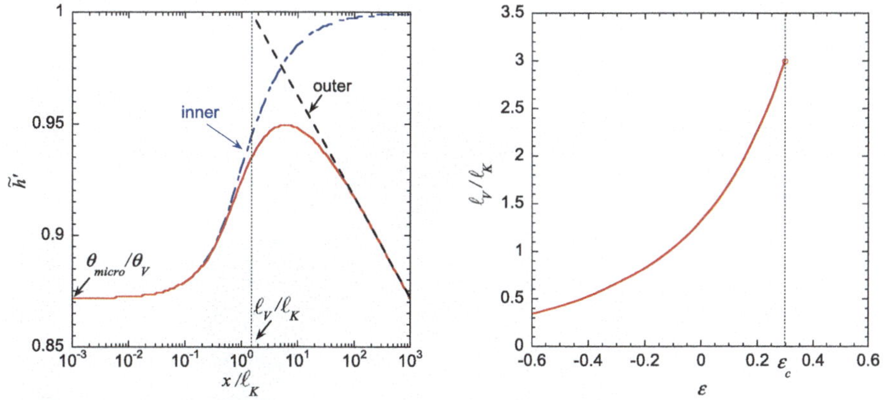

(a) Solution for $\varepsilon = 0.08$ and $\delta = -0.07$. The dash dotted line corresponds to the inner solution $\tilde{h}_0'$. The dashed line corresponds to the outer solution (95).

(b) The Voinov length ℓ_V as a function of the dimensionless superheating of the solid.

Fig. 6.13 Matching of the micro and intermediate regions [28]

Evidently, the solution of Eq. (6.93) depends on ε, so does ℓ_V. This dependence is shown in Fig. 6.13.

The multi-scale approach developed here for the microregion model based on the Kelvin effect can be generalized to any microregion description. As shown above (cf. [52] for a review), the hydrodynamic flow generated at the intermediate length scale is independent of the specific microregion effects discussed in Sect. 6.1. In fact, they influence the intermediate region only through two parameters, the Voinov length and Voinov angle. The specific value of ℓ_V depends on the microregion model. For example, for the hydrodynamic slip model in the isothermal case, $\ell_V = 3l_s/(e\,\theta_V)$ [13]. The complete wetting case is different because ℓ_V depends on the CL velocity: $\ell_V = 0.63(A/6\pi\sigma)^{1/2}/Ca^{2/3}$ [42].

Instead of using the above asymptotic matching approach, the set of Eqs. (6.49, 6.51, 6.65) with the boundary conditions (6.55, 6.56, 6.66, 6.70) can be solved numerically for the system of interest. However, as Eq. (6.91) shows, the slope depends (although weakly) on the scale at which it is defined. As a macroscopic scale, one can use a fixed right boundary x_{max} of the integration domain, so that the boundary condition (6.56) is imposed at $x = x_{max}$. An example of such a calculation is shown in Fig. 6.14. It can be checked [26] that it obeys the Cox-Voinov law.

6.2.4.2 Matching to Macroscopic Region: Drop Retraction

The Cox-Voinov law (6.91) is an extremely important result of matching of the microscopic and intermediate regions (cf. Fig. 6.12). However, a definition of the apparent contact angle as a slope at a coordinate x is still vague as it depends (although

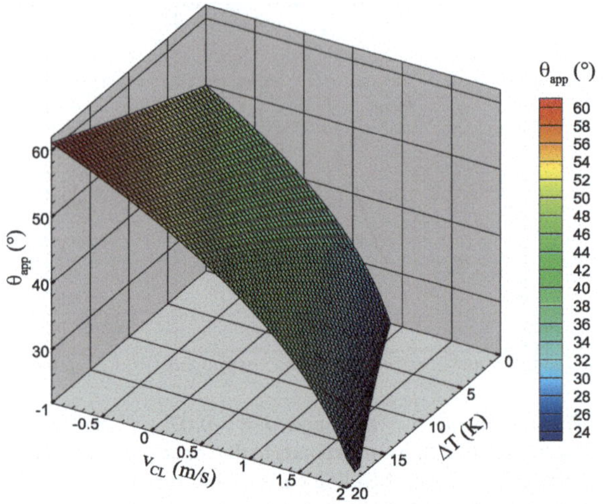

Fig. 6.14 θ_{app} as a function of the wall superheating ΔT and the CL velocity ($v_{CL} > 0$ at receding) calculated for water at atmospheric pressure; $x_{max} = 10\,\mu$m and $\theta_{\text{micro}} = 40°$

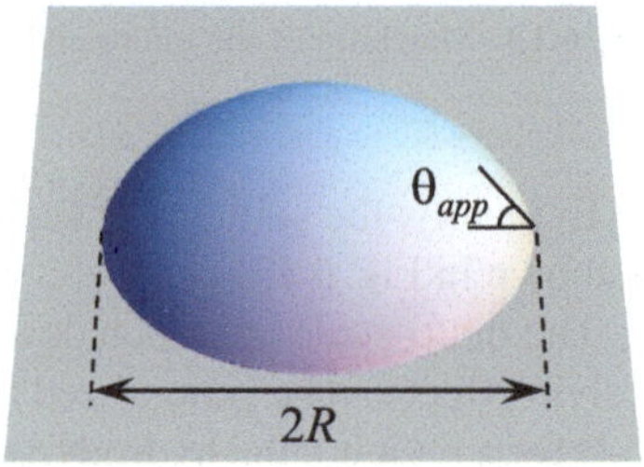

Fig. 6.15 Sessile drop on a flat substrate

weakly) on an unknown variable x. For practical applications, θ_{app} should thus be determined at a macroscopic scale. This means that the second matching procedure, between intermediate and macroscopic regions, should be applied. We discuss next an example of such a matching for the case of the drop geometry. It was first done by [42] for the complete wetting case, with no mass exchange. We generalize it here for any microregion model and show that it can be applied for the evaporation case (under an assumption).

Consider a drop on a flat substrate (Fig. 6.15), small enough so the gravity does not impact its shape. Let it be out of equilibrium slowly spreading or retracting to gain an equilibrium shape, possibly due to the evaporation effect in the microregion. In the case of cylindrical symmetry, $h = h(r, t)$ (where $r = \sqrt{x^2 + y^2}$), the evaporation loss in the macroregion is neglected, and Eq. (6.53) reduces to

$$-3\frac{\mu}{\sigma}\frac{\partial h}{\partial t} = \frac{1}{r}\frac{\partial}{\partial r}\left\{rh^3\frac{\partial}{\partial r}\left[\frac{1}{r}\frac{\partial}{\partial r}\left(r\frac{\partial h}{\partial r}\right)\right]\right\}. \tag{6.96}$$

As previously, h is expanded for small Ca

$$h = h_0 + Cah_1 + O(Ca^2). \tag{6.97}$$

$Ca = \mu v_{CL}/\sigma$ is defined by using

$$v_{CL} = -\mathrm{d}R/\mathrm{d}t \tag{6.98}$$

in agreement with the adopted sign convention, where R is the radius of CL. As the drop dynamics is driven by the CL motion, it is evident that in the zeroth order, Eq. (6.96) reduces to

$$\frac{1}{r}\frac{\partial}{\partial r}\left(r\frac{\partial h_0}{\partial r}\right) = \text{const},$$

thus describing the spherical cap in a small-slope approximation

$$h_0 = \frac{2V}{\pi R^2}\left(1 - \frac{r^2}{R^2}\right). \tag{6.99}$$

The drop volume

$$V = 2\pi\int_0^R hr\mathrm{d}r \tag{6.100}$$

is assumed to remain constant throughout CL motion, which means that the evaporation losses are small with respect to the drop mass. The apparent contact angle can now be clearly defined from this spherical cap shape as

$$\theta_{\text{app}} = -\left.\frac{\partial h_0}{\partial r}\right|_{r=R} = \frac{4V}{\pi R^3}. \tag{6.101}$$

One mentions that to the first order in Ca,

$$-\frac{\mu}{\sigma}\frac{\partial h}{\partial t} = Ca\frac{\partial h_0}{\partial R},$$

and Eq. (6.96) reduces to

$$-3\theta_{\text{app}}\left(1 - 2\frac{r^2}{R^2}\right) = \frac{1}{r}\frac{\partial}{\partial r}\left\{rh_0^3\frac{\partial}{\partial r}\left[\frac{1}{r}\frac{\partial}{\partial r}\left(r\frac{\partial h_1}{\partial r}\right)\right]\right\}. \tag{6.102}$$

Instead of using the method of [42], it is much simpler to integrate it directly by applying the boundary condition $h_1(r = R) = 0$ at the contact line, the symmetry conditions

$$\partial h_1/\partial r = 0,$$
$$\partial^3 h_1/\partial r^3 = 0$$

at $r = 0$ and the mass conservation that follows from Eq. (6.100):

$$\int_0^R h_1 r \mathrm{d}r = 0.$$

The result for the interface slope is

$$\frac{\partial h_1}{\partial r} = \frac{3R}{r\theta_{\text{app}}^2}\left[2\frac{r^2}{R^2} + \log\left(1 - \frac{r^2}{R^2}\right)\right]. \tag{6.103}$$

For matching of this solution to the intermediate region, one needs to find its asymptotics at $x \simeq R - r \to 0$

$$\left(\frac{\partial h}{\partial x}\right)^3 = \theta_{\text{app}}^3 - 9Ca \log\left(\frac{2e^2 x}{R}\right). \tag{6.104}$$

By equalizing this expression to Eq. (6.91) (which is the $x \to \infty$ asymptotics coming from the intermediate region), one obtains the final expression

$$\theta_{\text{app}}^3 = \theta_V^3 - 9Ca \log\left(\frac{L}{\ell_V}\right), \tag{6.105}$$

for the apparent contact angle, where the macroscopic scale $L = R/(2e^2)$ is now uniquely defined. Equation (6.105) represents the final result of the multi-scale approach for the drop geometry; θ_{app} is now uniquely defined as a function of the system parameters. One can understand at last the definition of the apparent contact angle as the interface slope obtained by extrapolation of the macroscopic interface shape to the contact line. It is precisely the contact angle commonly measured in experiments.

Note that Eq. (6.105) can be used to describe the drop dynamics. The most well-known example is the drop spreading in the complete wetting case with no mass exchange, where $\theta_V = \theta_{\text{micro}} = 0$. By using the definitions (6.98, 6.101), one obtains from Eq. (6.105) the scaling

$$\frac{\mathrm{d}R}{\mathrm{d}t} \sim R^{-9},$$

that results in the Tanner law [17] $R \sim t^{1/10}$.

Equation (6.105) clearly shows how the microregion impacts the macroscopic behavior. Two parameters come from the microregion, the Voinov angle θ_V and length ℓ_V. When $\Delta T = 0$ (no mass exchange), they are constant; $\theta_V = \theta_{micro}$. When

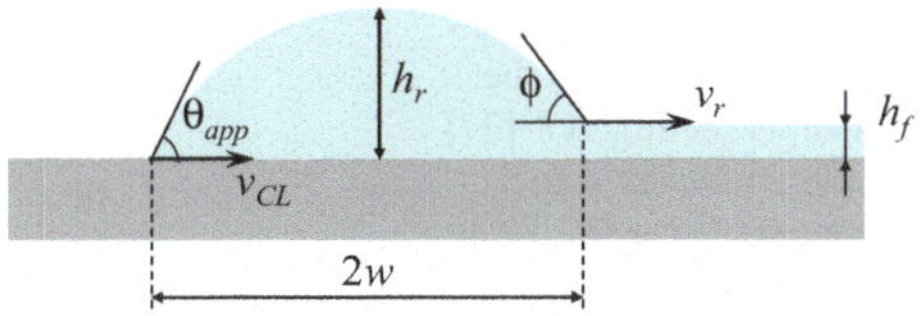

Fig. 6.16 2D Dewetting ridge

$\Delta T \neq 0$, both these parameters depend on ΔT. The dependence $\theta_V(\Delta T)$ is the most important, cf. Sect. 6.2.3. An example for the dependence of $\ell_V(\Delta T)$ provided by the Kelvin effect is shown in Fig. 6.13b.

6.2.4.3 Dewetting

In this section, we briefly consider the dewetting phenomenon. Imagine a classical (isothermal) case of a liquid film with a straight contact line that forms on a flat surface, e.g., during a receding of a liquid plug from a capillary slot. If the equilibrium contact angle is large, the CL moves, thus leading to the reduction of the film interface energy. However, the liquid cannot flow into the film because of the high viscous friction. The receding liquid forms thus a ridge (called also rim) at the film edge (Fig. 6.16). As the evaporation impacts the apparent contact angle, it is expected to impact the macroscopic ridge dynamics discussed next.

The first dewetting theory [7] was based on the phenomenological approach to the contact line motion [17]. For uniformity, it is presented here by using the above hydrodynamic theory that results in

$$\theta_{\text{app}}^3 = \theta_V^3 - \frac{v_{CL}}{v^*}, \tag{6.106}$$

where v^* is a constant (logarithmic) term. The theory is developed in the approximation of the small $\theta_{\text{app}} \ll 1$, so the slope ϕ of the rear edge of the ridge (Fig. 6.16) is equally small. It obeys a law similar to (6.106) but accounting for the fact that the equilibrium ϕ value is zero

$$\phi^3 = \frac{v_r}{v^*}, \tag{6.107}$$

where v_r is the velocity of the rear ridge edge. As mentioned above, all the liquid from the film of the thickness h_f is gathered in the ridge so its area A (in the plane of Fig. 6.16) growth obeys the law $\dot{A} = v_r h_f$. Consider the late stages of the ridge growth, where its height $h_r \gg h_f$. The area is defined as $A \sim h_r w$, where w is the ridge half-width, so $\dot{A} \sim h_r \dot{w}$ for slow variation of θ_{app} because $h_r \sim w\theta_{\text{app}}/4$. This means that

$$\frac{\dot{w}}{v_r} \sim \frac{h_f}{h_r} \ll 1,$$

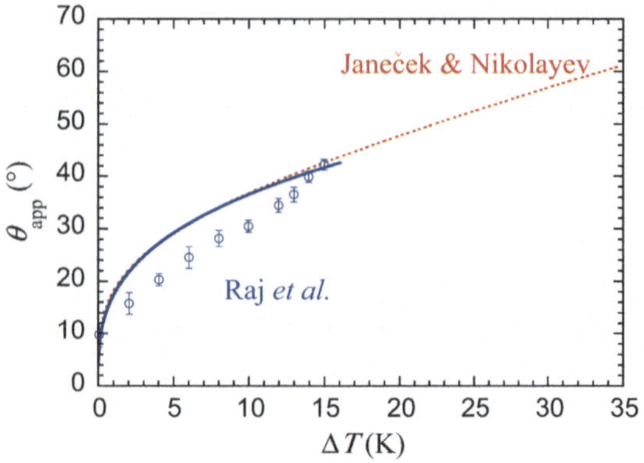

Fig. 6.17 The calculated dependency of θ_V on the wall superheating ΔT [27] calculated for the static contact angle for FC-72 at 0.04 MPa and $\theta_{micr} = 8°$ (dotted line) compared to the theoretical results for complete wetting [47], solid line. Their experimental data (where $\theta = 8°$) are shown with circles

which justifies that $v_{CL} = v_r - 2\dot{w} \simeq v_r$. Since $h_f \ll h_r$, $\theta_{\text{app}} \simeq \phi$. With two latter equalities, Eqs. (6.106, 6.107) reduce to $\theta_{\text{app}}^3 \simeq \theta_V^3/2$ and

$$v_{CL} \simeq \theta_V^3 \frac{v^*}{2}. \tag{6.108}$$

Equation (6.108) is the central result of this phenomenological approach, which shows the $v_{CL} \propto \theta_V^3$ proportionality but is not able to predict the prefactor.

To obtain the prefactor, one needs a more sophisticated multi-scale theoretical approach [53] based on the asymptotic matching technique similar to that described for the drop case (Sect. 6.2.4.2). An additional complication of the dewetting ridge geometry is its dissymmetry: one needs to match the ridge to the film of a finite (typically $\sim 50\,\mu$m) thickness. The matching is possible to do asymptotically under the above assumption $h_f \ll h_r$. The asymptotic model of [53] has been initially developed for the microregion model based on the hydrodynamic slip. Its main asymptotic result can easily be generalized to any microregion description as

$$Ca = \frac{\theta_V^3}{9}\left[\log\left(aCa^{1/3}\frac{L^2}{\ell_V h_f}\right)\right]^{-1}, \tag{6.109}$$

where $a \approx 1.094$ is a matching constant and $L = 2w/e$ is the characteristic macroscopic scale for the ridge case. Equation (6.109) defines the CL velocity. One can mention the explicit expression for the logarithmic prefactor. The apparent contact angle obeys Eq. (6.105), where the above L value should be used.

6.2.5 Comparison with Experimental Data

Some typical dependencies of θ_V on the wall superheating are shown in Fig. 6.17.

Note a high value of the apparent contact angle for a superheating of several tens of Kelvin characteristic of the boiling conditions. The agreement with the experimental data of [47] is reasonably good. The discrepancy between the experimental and theoretical data can probably be attributed to the neglect of the CL velocity (i.e., of the second term in Eq. (6.105)). The CL was moving in the experiment; the motion was not quantified by the authors and thus cannot be properly modeled.

The simultaneous phase change and CL motion was experimentally observed by [15] in a closed transparent sapphire capillary slot (Hele-Shaw cell) filled with pure ethanol. The sapphire substrate was heated by the electric current passed through a transparent indium-tin oxide film deposited on the back sapphire surface. A 60 μm thick ethanol film was laid down by a receding liquid. While the ethanol wets completely the sapphire at equilibrium, the dewetting phenomenon was observed at evaporation (Fig. 6.18a). The ridge shape was well-approximated by a circular arc (except at late times where gravity played a role), from which θ_{app} and w were obtained.

The θ_V values (Fig. 6.18b) were calculated via Eq. (6.105) with $L = 2w/e$. According to estimations, $\ell_s \gg \ell_K$ for the ethanol, where the slip length $l_s = 20$ nm was used. For this reason, the hydrodynamic slip prevails over the Kelvin effect in the microregion and $\ell_V = \ell_s$ was assumed. The line in Fig. 6.18b is a theoretical curve similar to those of Fig. 6.8 calculated for the same parameters (including the same l_s). The agreement is very good, which shows the validity of the above theory. Figure 6.18b shows that the θ_V values can be high in spite of nearly zero equilibrium contact angle. Such high θ_V values are at the origin of the solid wall dewetting.

Once a liquid film is deposited by the moving liquid meniscus, its CL is expected to recede because of two phenomena. First, according to the theory discussed above, the contact angle grows, and the substrate dewetting occurs. Second, the liquid in the CL vicinity vaporizes; there is no flow in the film so its length decreases; in other words, CL recedes because of the evaporative mass loss. Let us consider now the relative contribution of evaporative mass loss at the CL. Figure 6.18c presents the experimental CL receding velocity together with the result of Eq. (6.109), which describes only the first (dewetting) contribution. One can see that, in this particular experiment, CL recedes mainly because of the contact angle growth via dewetting; the evaporative mass loss is responsible only for 14% of the CL velocity.

6.3 Diffusion-Controlled Evaporation

It has been shown experimentally [56] that the increase of apparent contact angle with evaporation rate exists in the case of isothermal drying in the atmosphere of a neutral gas. One can compare this case to the regime of evaporation to the pure vapor atmosphere induced by heating considered above. The theoretical description of drying is more complex as the mass exchange is controlled by the phenomena in the gas domain, which is generally much larger than the liquid domain that controls the pure vapor case so natural or forced convection is often important. Another reason for the increased complexity is the non-locality of evaporation flux: the evaporation

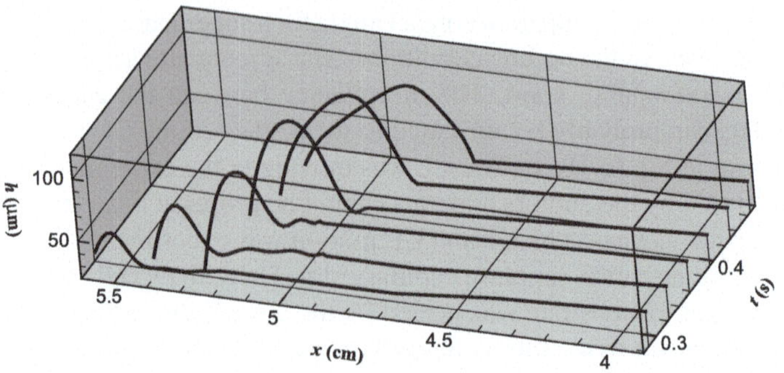

(a) The dewetting ridge dynamics caused by evaporation. The ridge shape is obtained with the optical methods.

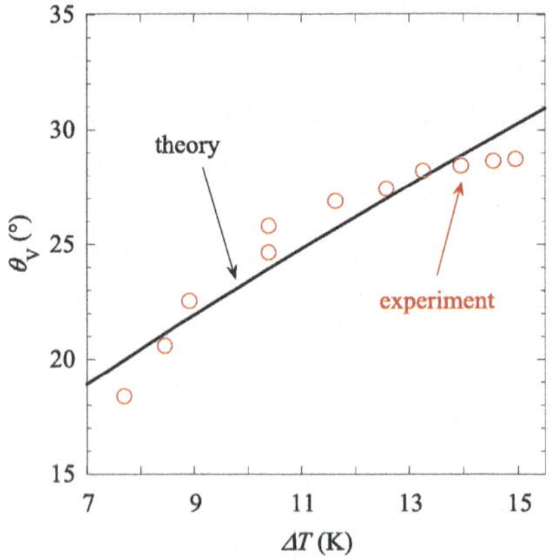

(b) The Voinov angle θ_V recalculated from experimental θ_{app} data versus wall superheating ΔT and the theoretical model.

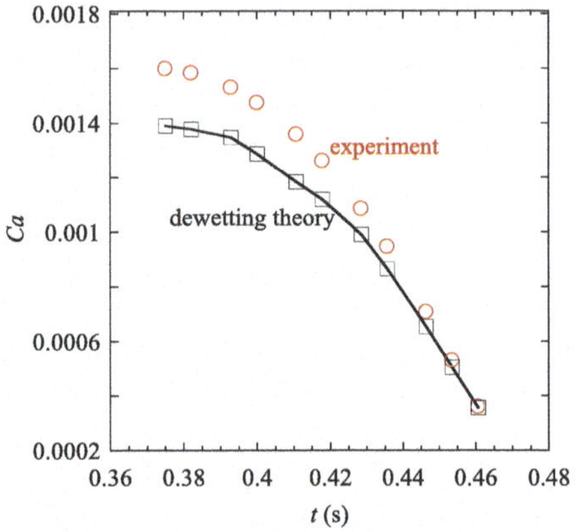

(c) Experimental data for the dimensionless CL velocity compared to the theoretical values given by Eq. (109).

Fig. 6.18 Evaporation-caused dewetting phenomenon. The dewetting theory results compared to experimental data on the ethanol film on sapphire [15]

rate at one point of the interface depends not only on the local vapor density, but on its distribution over the whole liquid-vapor interface. As the mass exchange rates are generally much smaller at drying, the effect on the contact angle is expected to be much weaker than in the pure vapor case. For this reason, we consider this regime more briefly than the pure vapor case.

There are many theoretical approaches to the drying description. Early approaches neglect the impact of evaporation on the interface shape (and thus on the impact on contact angle). One of the first important steps was achieved by [10] who have obtained the stationary diffusion problem solution in the half-space above a spherical-cap-shaped sessile liquid droplet posed on a solid substrate with a fixed contact angle. The success of this solution is due to the integrability of the resulting local evaporation flux so that the total evaporation rate can be calculated and compared to the experiment. The integrability of the background solution for the mass flux (i.e., a weaker divergence) is another reason for the weaker effect of evaporation on the apparent contact angle in this regime comparing to the regime discussed in Sect. 6.2.

In earlier theoretical approaches [5, 20, 41, 46], the evaporation flux distribution along the interface was imposed independently of the interface shape. The full coupling of the problems in the vapor and liquid phases through the Kelvin effect is studied in more recent works [11, 12, 14, 29, 36]. The Kelvin effect is important as it makes the evaporation flux to be finite at the contact line, cf. Eq. (6.58). It also causes the impact of evaporation on the apparent contact angle.

6.3.1 Problem Statement

Consider a diffusion boundary layer of thickness Λ above the thin liquid wedge-like film. The value of Λ depends on the gas dynamics in the remaining part of the space (defined by the natural or forced convection). The film is placed on a flat and homogeneous substrate in a situation of partial wetting. Its contact line is pinned; the problem of contact line motion under saturation conditions has been solved by Janeček et al. [29]. The atmosphere of a non-condensable inert gas surrounds the substrate and the condensation or evaporation mass exchange with it is controlled by the vapor transfer in the gas. The gas supersaturation with vapor causes condensation onto the film, and under-saturation causes its evaporation. The fluid is assumed to be isothermal, which may be justified when the substrate has a good thermal conductivity and is maintained at room temperature. For this reason, the temperature T is equivalent to T_{sat} used above.

First, some assumptions about the characteristic scale hierarchy need to be made. One can identify a macroscopic length scale L of the liquid phase. It is related to the macroscopic-level interface curvature, e.g., the interface curvature radius. Another macroscopic length scale concerns the gas phase description. It is the diffusion boundary layer thickness Λ. It is millimetric (for natural convection) or smaller, for forced convection. For simplicity, we consider the diffusion boundary layer of homogeneous thickness (band-like). The idea is to find a stationary solution for the vapor diffusion that can be representative of real microscopic situation. For this, we assume that

$$L \gg \Lambda \gg h \tag{6.110}$$

where h is the wedge thickness, see Fig. 6.19a, b. It is evident that the inequality (6.110) is violated far enough from the wedge apex (contact line) for any wedge slope. However, if the slope is small, the region of validity of such a geometry is large. As for the pure vapor case, the hydrodynamic singularity related to the evaporation flow is solved at the microscopic scale that we call ℓ_K which is typically nanometric (Fig. 6.19d). An intermediate region (Fig. 6.19c) is needed to match micro and macro regions. We will see later that, generally, in this case, two intermediate regions need to be introduced.

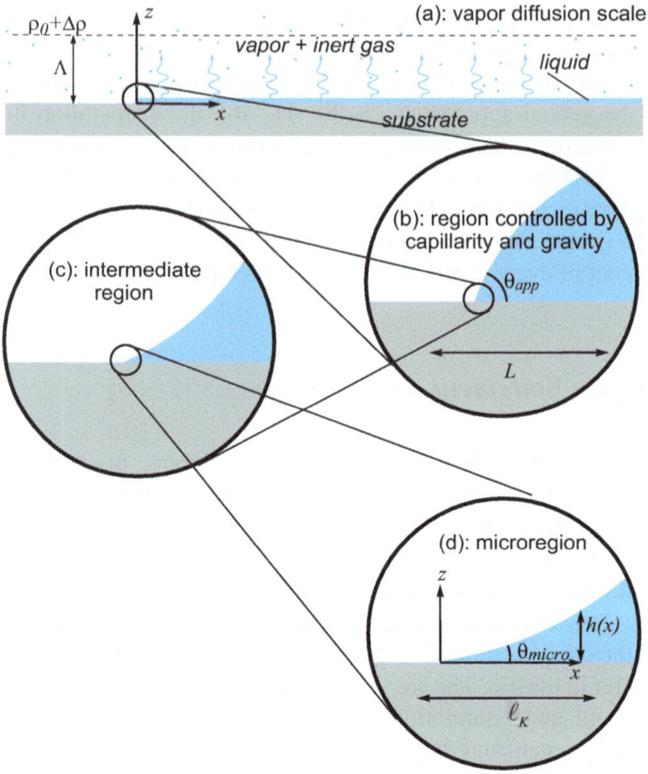

Fig. 6.19 Hierarchy of scales considered in the article and geometries for the vapor diffusion (**a**) and hydrodynamic (**b-d**) problems. Note that the radius of curvature in (**b**) is assumed to be much larger than the diffusive boundary layer width Λ shown in (**a**)

The equation governing the liquid flow in the wedge is the stationary version of Eq. (6.52):

$$\frac{\partial}{\partial x}\left(\frac{h^3}{3}\frac{\partial \Delta p}{\partial x}\right) = -\mu \frac{J}{\rho_L}, \text{ where } \Delta p = \sigma K = \sigma \frac{\partial^2 h}{\partial x^2}. \tag{6.111}$$

The interfacial mass flux J is controlled by the vapor diffusion

$$J = -D \frac{\partial \rho_V}{\partial z}\bigg|_{z=0}, \tag{6.112}$$

where $\rho_V = \rho_V(x, z)$ now varies in space and D is the coefficient of the vapor diffusion in the ambient gas. The vapor diffusion equation reads

$$\frac{\partial^2 \rho_V}{\partial x^2} + \frac{\partial^2 \rho_V}{\partial z^2} = 0. \tag{6.113}$$

The boundary conditions for the liquid phase are discussed in Sect. 6.1.4: $h(x \to 0) = 0$ and $\partial h/\partial x|_{x\to 0} = \theta_{\text{micro}}$ at the CL, the vanishing curvature for from CL, $K(x \to \infty) = 0$, and the regularity of all hydrodynamic quantities at the CL, which is equivalent to $J(x \to 0) = 0$.

Within the small wedge slope approximation, the liquid-gas interface seen from the large scale of the gas atmosphere is assumed to coincide with the line $z = 0$, $x \geq 0$, as shown in Fig. 6.19a. This assumption is valid for a liquid height much lower than the boundary layer thickness, i.e., for $x \ll \Lambda/\theta_{\text{micro}}$. As θ_{micro} is a small angle, the model is valid over a distance to the contact line much larger than Λ. The boundary conditions for the diffusion equation are defined at the upper and lower boundaries of the diffusion layer. At the upper boundary $z = \Lambda$

$$\rho_V(x, z = \Lambda) = \rho_0 + \Delta\rho, \tag{6.114}$$

where ρ_0 is the vapor density at thermodynamic equilibrium for a flat liquid-gas interface, $\Delta\rho$ is the deviation from the equilibrium vapor density; $\Delta\rho < 0$ corresponds to evaporation. The vapor cannot penetrate into the solid boundary ($z = 0$ and $x < 0$)

$$\frac{\partial \rho_V}{\partial z} = 0. \tag{6.115}$$

The vapor density at the liquid-gas interface can deviate from ρ_0 because of two reasons: (i) the interfacial pressure jump (Kelvin effect) and (ii) the interfacial kinetic resistance R^i_{diff} defined by Eq. (6.34). Equations (6.12, 6.33) can thus be combined, so the boundary condition

$$\rho_V(x, z = 0) = \rho_0 - \Delta p \frac{M\rho_0}{\rho_L R_g T} - J R^i_{diff}, \tag{6.116}$$

is imposed for $x \geq 0$. In the following sections, we focus on the diffusion controlled regime; the impact of the kinetic interfacial resistance can be found in [12].

6.3.2 *Kelvin Effect and Dimensionless Formulation*

The microregion size (Fig. 6.19) for the case of the singularity relaxation with the Kelvin effect can be easily obtained with a scaling analysis [29] that results in

$$\ell_K = \frac{1}{\rho_L}\sqrt{\frac{3\mu M \rho_0 D}{\theta^3_{\text{micro}} R_g T}}. \tag{6.117}$$

For the analysis below, the dimensionless abscissa is $\tilde{x} = x/\ell_K$. The liquid height is scaled as $\tilde{h} = h/(\theta_{\text{micro}}\ell_K)$, and the dimensionless flux, as $\tilde{J} = 3\mu J/(\theta_{\text{micro}}^4 \sigma \rho_L)$. The dimensionless lubrication equation reads

$$\frac{\partial}{\partial \tilde{x}}\left(\tilde{h}^3 \frac{\partial^3 \tilde{h}}{\partial \tilde{x}^3}\right) = -\tilde{J}. \tag{6.118}$$

Density deviation is reduced as $\tilde{\rho} = (\rho_V - \rho_0)/C$, with

$$C = \frac{\rho_L \sigma \theta_{\text{micro}}^4 \ell_K}{3\mu D}. \tag{6.119}$$

The z and Λ variables are reduced with ℓ_K and the dimensionless boundary conditions for the diffusion problem for $\tilde{z} = 0,\ \tilde{x} \geq 0$ are

$$\frac{\partial \tilde{\rho}}{\partial \tilde{z}} = -\tilde{J}, \tag{6.120}$$

$$\tilde{\rho} = -\tilde{h}''. \tag{6.121}$$

The trivial boundary condition (6.115) is valid for $\tilde{z} = 0,\ \tilde{x} < 0$.

6.3.3 *Weak Evaporation Approximation*

The deviation from equilibrium $\varepsilon = \Delta\rho/C$ is assumed to be small; the variables are expanded in a regular perturbation series. At the zero order corresponding to the straight wedge in thermodynamic equilibrium, $\tilde{h}_0 = \tilde{x}$, $\tilde{J}_0 = 0$ and $\tilde{\rho}_0 = 0$;

$$\tilde{h} = \tilde{x} + \varepsilon \tilde{h}_1 + O(\varepsilon^2), \quad \tilde{J} = \varepsilon \tilde{J}_1 + O(\varepsilon^2), \quad \tilde{\rho} = \varepsilon \tilde{\rho}_1 + O(\varepsilon^2).$$

Note that ε is negative during evaporation and positive during condensation, so $\tilde{J}_1$ and $\tilde{h}_1$ are always negative.

The first order problem is described by the fluid flow equation

$$\frac{\partial}{\partial \tilde{x}}\left(\tilde{x}^3 \frac{\partial^3 \tilde{h}_1}{\partial \tilde{x}^3}\right) = -\tilde{J}_1, \tag{6.122}$$

with the boundary conditions $\tilde{J}_1 = \tilde{h}_1 = \tilde{h}_1' = 0$ at $\tilde{x} = 0$ and $\tilde{h}_1'' = 0$ at $\tilde{x} \to \infty$. The diffusion part of the first order problem remains as above.

There is no analytical solution and the problem should be solved numerically. A direct numerical solution of the problem would be complicated as a nonlinear 1D fluid flow equation needs to be coupled to the 2D vapor diffusion problem in a stripe.

For this reason, it is more convenient to solve the 2D problem analytically by using the boundary integral method. It consists of reducing the 2D differential equation (6.113) to the integral over the domain boundary by using the Green function [12]. By redefining the density variable as $\tilde{\rho}_1 - 1$, all the boundary conditions except those at $\tilde{z} = 0, \tilde{x} \geq 0$ become trivial; only the integral over this portion involves the boundary condition (6.120). The interfacial density $\tilde{\rho}_1(\tilde{x}, \tilde{z} = 0)$ over it can be expressed as a function of the mass flux as

$$\tilde{\rho}_1(\tilde{x}, \tilde{z} = 0) = 1 - \int_0^\infty G_0(\tilde{x} - \tilde{x}')\tilde{J}_1(\tilde{x}')\mathrm{d}\tilde{x}', \tag{6.123}$$

where

$$G_0(\tilde{x}) = \frac{1}{\pi} \log \left| \frac{\exp\left(\frac{\pi \tilde{x}}{2\tilde{\Lambda}}\right) - 1}{\exp\left(\frac{\pi \tilde{x}}{2\tilde{\Lambda}}\right) + 1} \right|$$

is the Green function for the stripe geometry.

By combining Eqs. (6.121–6.123), one gets the governing integral equation

$$\tilde{x}^3 \int_0^\infty \frac{\partial G_0(\tilde{x} - \tilde{x}')}{\partial \tilde{x}} \tilde{J}_1(\tilde{x}')\partial \tilde{x}' = - \int_0^{\tilde{x}} \tilde{J}_1(\tilde{x}')\partial \tilde{x}' \tag{6.124}$$

that can be solved numerically [12]. Once $\tilde{J}_1$ is known, the slope can be computed with Eq. (6.122).

6.3.4 Impact of the Thickness of Diffusion Boundary Layer

One of the most important parameters that impacts the evaporation rate and thus the apparent contact angle is the thickness Λ of the diffusion boundary layer; its impact is studied here.

Because of the presence of the length scale $\Lambda \gg \ell_K$, one identifies two intermediate regimes of different asymptotic behavior. The first intermediate region ranges from ℓ_K to Λ, while the second goes from Λ to L. As we do not discuss here the coupling to the macroregion, L is considered to be infinitely large.

Within both intermediate regions, the Kelvin effect can be neglected, which means that ρ_V at the interface is constant and equal to the saturation density ρ_0; $\tilde{\rho} = 0$. This is a "Deegan-like" [10] diffusion problem in the 2D boundary layer geometry that admits an analytical solution [12]:

$$\tilde{J}_1(\tilde{x}) = -\frac{1}{\tilde{\Lambda}\sqrt{2}}\sqrt{1+\coth\left(\frac{\pi\tilde{x}}{2\tilde{\Lambda}}\right)}. \tag{6.125}$$

Unfortunately, the fluid flow part of the problem, Eq. (6.122), cannot be solved analytically. For this reason, the asymptotic solutions are considered. The asymptotic expressions for the flux are different for two intermediate regions. The limit $\tilde{x} \ll \tilde{\Lambda}$

$$\tilde{J}_1(\tilde{x}) \simeq -\frac{1}{\sqrt{\pi\tilde{\Lambda}\tilde{x}}}. \tag{6.126}$$

describes the first intermediate region, while the limit $\tilde{x} \gg \tilde{\Lambda}$

$$\tilde{J}_1(\tilde{x}) \simeq -\frac{1}{\tilde{\Lambda}}, \tag{6.127}$$

describes the second intermediate region. The scaling (6.126) was originally obtained [10] for the thin 3D axisymmetric drop and infinite boundary layer. However, the present solution is different because the geometry is 2D, so the asymptotic solutions are always Λ-dependent (the solution for the stationary diffusion problem in 2D does not exist in the infinite domain). Instead of Λ, the solution of Deegan et al. [10] is controlled by the drop diameter. The flux diverges at the contact line, as expected: it cannot describe the microregion dominated by the Kelvin effect.

One can get the curvature and the slope after successive integrations of Eq. (6.122) with the above asymptotic expressions for $\tilde{J}_1$ and by applying the boundary conditions to determine some of the integration constants [12].

Consider the first intermediate region $1 \ll \tilde{x} \ll \tilde{\Lambda}$. The asymptotic solution for the curvature is

$$\frac{\partial^2\tilde{h}_1}{\partial\tilde{x}^2} \simeq -\frac{4}{3}\frac{1}{\sqrt{\pi\tilde{\Lambda}\tilde{x}^3}}. \tag{6.128}$$

while the slope is

$$\frac{\partial\tilde{h}_1}{\partial\tilde{x}} \simeq \frac{8}{3}\frac{1}{\sqrt{\pi\tilde{\Lambda}}}\left(\frac{1}{\sqrt{\tilde{x}}} - \frac{1}{\sqrt{C_1}}\right), \tag{6.129}$$

where C_1 corresponds to the dimensionless size of the microregion, so $C_1 \sim 1$ is expected. For the second intermediate region $\tilde{\Lambda} \ll \tilde{x} \ll L$, by integrating Eq. (6.127), one gets

$$\frac{\partial^2\tilde{h}_1}{\partial\tilde{x}^2} \simeq -\frac{1}{\tilde{\Lambda}\tilde{x}} \tag{6.130}$$

for the curvature and

$$\frac{\partial\tilde{h}_1}{\partial\tilde{x}} \simeq -\frac{1}{\tilde{\Lambda}}\left[\log\left(\frac{\tilde{x}}{\tilde{\Lambda}}\right) + C_2\right], \tag{6.131}$$

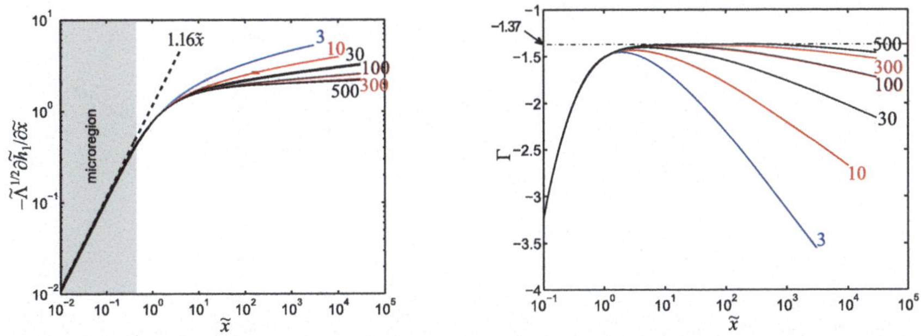

(a) The microregion slope behavior. The dashed line is the microregion asymptote.

(b) Solid lines are $\Gamma(\tilde{x})$ curves; dash-dotted line is the first intermediate region asymptote.

Fig. 6.20 Matching of microregion and first intermediate region [12] for different $\tilde{\Lambda}$

for the slope, with an integration constant C_2. At $\tilde{\Lambda} \gg C_1$, the slopes (6.129) and (6.131) should match at $\tilde{x} \sim \tilde{\Lambda}$. This matching results in

$$C_2 = C_3\sqrt{\tilde{\Lambda}/C_1} \tag{6.132}$$

with a constant $C_3 \sim 8/(3\sqrt{\pi}) \simeq 1.5$. Its exact value and those of the other constants can be determined by matching to the microregion model.

The microregion asymptotic solution cannot be obtained analytically. Instead, the constants C_1 and C_2 that enter Eqs. (6.129, 6.131) are determined by solving numerically Eq. (6.124), which is valid in microscopic and both intermediate regions.

The numerics shows that in the microregion, $\partial\tilde{h}_1/\partial\tilde{x} \sim -\tilde{x}/\sqrt{\tilde{\Lambda}}$ (Fig. 6.20a). The $\tilde{\Lambda}^{-1/2}$ scaling is important to match the asymptotics (6.129) in the first intermediate region. The departure from the linear behavior corresponds to the end of microregion.

The first intermediate region is described by Eq. (6.129) that involves a characteristic length C_1. It can be obtained numerically by plotting (Fig. 6.20b) the quantity

$$\Gamma = 3/8(\pi\tilde{\Lambda})^{1/2}\frac{\partial\tilde{h}_1}{\partial\tilde{x}} - \tilde{x}^{-1/2}$$

that is expected to be constant ($\Gamma = -C_1^{-1/2}$) in the first intermediate region, cf. Eq. (6.129). Figure 6.20b shows the $\tilde{x}$-independent region only for $\tilde{\Lambda} \gg 1$, which is exactly the criterion of existence of the first intermediate region. From the numerical value $\Gamma \simeq -1.37$, one gets $C_1 \simeq 0.53$, so that the precise microregion size is $0.53\ell_K$.

To analyze the large-scale behavior (in the second intermediate region), one needs to renormalize the slope data of Fig. 6.20. It is done in Fig. 6.21a. Note that the microregion behavior cannot be seen on these curves (the scale is too large).

A logarithmic variation of $\partial\tilde{h}_1/\partial\tilde{x}$ is observed for $x > \Lambda$, as predicted by Eq. (6.131). The integration constant $C_2 = -\tilde{\Lambda}\frac{\partial\tilde{h}_1}{\partial\tilde{x}}(\tilde{x} = \tilde{\Lambda})$ is obtained by fitting

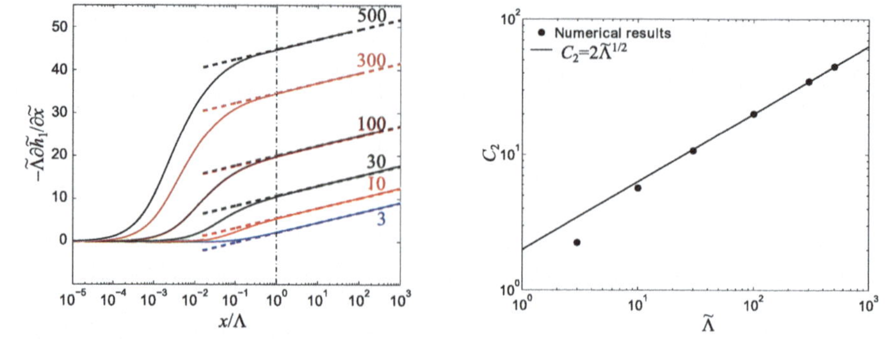

(a) Renormalized interface slope for different values of $\tilde{\Lambda}$; solid lines: numerical results; dashed lines: asymptotes given by Eq. (131).

(b) Matching constant C_2 as a function of $\tilde{\Lambda}$.

Fig. 6.21 Behavior of the interface slope in the intermediate regions: matching of the first and second intermediate regions [12]

Eq. (6.131) to the numerical results (see dashed lines in Fig. 6.21a) and taking the intersection of each fit with the dash-dotted line $\tilde{x} = \tilde{\Lambda}$. The parameter C_2 is given in Fig. 6.21b as a function of $\tilde{\Lambda}$. For $\tilde{\Lambda} \gg 1$, the law (6.132) indeed holds, $C_2 = \chi\sqrt{\tilde{\Lambda}}$, where $\chi = C_3/\sqrt{C_1} \simeq 2$ is a constant. By using $C_1 \simeq 0.53$ obtained above, one gets $C_3 \simeq 1.46$.

6.3.5 Apparent Contact Angle

The main result of such a model is the behavior of the interface slope observed at the macroscopic length scale L. According to Eq. (6.131), the slope depends on the distance x from the CL

$$\frac{\partial h}{\partial x} \simeq \theta_{\text{micro}} - \frac{3\mu D\Delta\rho}{\rho_L\sigma\theta_{\text{micro}}^3\Lambda}\left[\chi\sqrt{\frac{\Lambda}{\ell_K}} + \log\left(\frac{x}{\Lambda}\right)\right]. \tag{6.133}$$

This expression means that the apparent contact angle that should be obtained as a result of the matching to the macroregion (Sect. 6.2.4.2), logarithmically depends on the macroscopic scale L (i.e., the interfacial curvature radius), similarly to the moving contact line problem.

Suppose that the logarithmic term in Eq. (6.133) can be neglected. One then obtains an approximate expression of the slope that is independent of the scale L and can thus be associated with the apparent contact angle

$$\theta_{\text{app}} \simeq \theta_{\text{micro}} - \frac{3\chi}{\sqrt{\ell_K}}\frac{\mu D}{\rho_L\sigma\theta_{\text{micro}}^3}\frac{\Delta\rho}{\sqrt{\Lambda}}. \tag{6.134}$$

On the other hand, Eq. (6.134) can also be obtained by tending $\tilde{x} \to \infty$ in Eq. (6.129), i.e., within the first intermediate region because it is the "in-between" result. In this case, one gets instead of $3\chi \simeq 6$ the coefficient $8/\sqrt{\pi C_1} \simeq 6.2$. Two values are very close, which is not surprising because both intermediate regions are matched (in the asymptotic sense) for $\tilde{x} \to \infty$ for the first and $\tilde{x} \to \tilde{\Lambda}$ for the second. The agreement shows simply that the matching has been performed correctly.

Note that Eq. (6.134) is the result of the intermediate regions where microscopic effects are negligible. Therefore, its form is independent of the microscopic singularity relaxation mechanism. Similarly to the pure vapor case and moving CL, the microscopic details impact only the microregion length scale ℓ_K.

Equation (6.134) is very close to the result of Berteloot et al. [5] (one needs to drop the independent Cox-Voinov term in their Eq. (13) since they considered the moving CL case. The boundary layer thickness Λ is included in their parameter J_0. Unlike the approach presented here, Berteloot et al. [5] introduced a microscopic scale phenomenologically, and found the numerical prefactor $3\chi = 8$. In the present approach, the scales are matched rigorously, and a more precise value of $3\chi \simeq 6$ is obtained. Note that Berteloot et al. [5] considered only the first intermediate region $x \ll \Lambda$. The approach described here reveals another logarithmic term that was implicitly neglected by them.

6.4 Conclusions

While not impacting the contact angles at the nanoscale, evaporation causes an increase of the apparent (experimentally measurable at the macroscale) contact angle that leads to a change of the wetting properties (both dynamic and static) as if this change were induced by a surface modification. The control of this phenomenon is possible via the control of the local evaporation rate in a tiny vicinity of the contact line by choosing the degree of non-equilibrium. The specific way of control depends on the phase change regime. Two limiting regimes have been considered above. Strong evaporation (and, accordingly, a strong change of the wetting properties) can be achieved at evaporation into the atmosphere of pure vapor. In this case, the evaporation rate is controlled by the superheating ΔT of the heater with respect to the saturation temperature for the system pressure, like e.g., in boiling. The contact angle depends on many system parameters but the most important dependence is on ΔT and on the static contact angle (i.e. on that at $\Delta T = 0$). Another limiting regime, that of slow evaporation, is achieved at isothermal drying in the under-saturated atmosphere of a non-condensible gas. It is controlled by the vapor diffusion within the diffusion boundary layer. The main control parameters here are the under-saturation $\Delta\rho$ (with respect to the saturation vapor density for the system temperature), the thickness of the diffusion boundary layer (defined mainly by the externally imposed flow velocity), and of course, the static contact angle.

Because of many reasons, it is difficult to measure the contact angles at evaporation conditions. Some of these reasons are the continuous variation of the contact angle, the optical aberrations in the presence of thermal gradients, and the necessity of simultaneous thermal control and optical observations.

It is, however, possible to predict the contact angles theoretically. The hydrodynamic flow caused by evaporation induces a contact line singularity similar to that caused by the contact line motion. For this reason, the contact angle calculation is a delicate issue and calculation error can be high if the problem is approached incorrectly. Like in the contact line motion problem, several approaches are possible to relax the singularity. At complete wetting, the continuous wetting film approach is possible so the actual triple contact is absent. In a much more common case of partial wetting, one can use a model based on the simultaneous action of two effects: the hydrodynamic slip and the kinetic resistance. Another approach consists in the account of the Kelvin effect that is self-sufficient to relax the singularity. The Kelvin effect can be used alone or in any combination with other effects. The models based on the Kelvin effect present an evident advantage of providing the finite values of the evaporation flux and liquid pressure at the contact line. These values can be defined *a priori* and serve as boundary conditions for calculations.

A multi-scale approach is a powerful tool for an accurate and numerically efficient contact angle calculation that is necessary to describe the evaporation (in some cases) for the static and always to describe the contact line motion that is often caused by evaporation. By using the multi-scale approach, one can explicitly define all the parameters (both microscopic and macroscopic) to find the apparent contact angle that is a function of the specific geometry of wetting (drop, bubble, tube, ridge, etc.).

Acknowledgements The author would like to express his gratitude to all his former and present Ph.D. students and postdocs, especially V. Janeček and L. Fourgeaud, without whose contribution this work would not be possible. The author would like to thank the European Space Agency for the support through the MAP TOPDESS and CNES for the support in the framework of the GdR MFA.

References

1. Afkhami S, Buongiorno J, Guion A, Popinet S, Saade Y, Scardovelli R, Zaleski S (2018) Transition in a numerical model of contact line dynamics and forced dewetting. J Comput Phys 374:1061–1093, https://doi.org/10.1016/j.jcp.2018.06.078
2. Anderson DM, Davis SH (1994) Local fluid and heat flow near contact lines. J Fluid Mech 268:231–265, https://doi.org/10.1017/S0022112094001333
3. Anderson DM, Janeček V (2018) Comment on L. M. Hocking, "On contact angles in evaporating liquids" [Phys. Fluids 7, 2950–2955 (1995)]. Phys Fluids 30(7):079101, https://doi.org/10.1063/1.5041445
4. Anderson DM, Cermelli P, Fried E, Gurtin ME, McFadden GB (2007) General dynamical sharp-interface conditions for phase transformations in viscous heat-conducting fluids. J Fluid Mech 581:323–370, https://doi.org/10.1017/S0022112007005587
5. Berteloot G, Pham CT, Daerr A, Lequeux F, Limat L (2008) Evaporation-induced flow near a contact line: Consequences on coating and contact angle. Europhys Lett 83(1):14003, https://doi.org/10.1209/0295-5075/83/14003

6. Bouzigues CI, Tabeling P, Bocquet L (2008) Nanofluidics in the Debye layer at hydrophilic and hydrophobic surfaces. Phys Rev Lett 101(11):114503, https://doi.org/10.1103/PhysRevLett.101.114503
7. Brochard-Wyart F, de Meglio JM, Quéré D (1987) Démouillage. Etude du retrait d'un film de liquide non mouillant déposé sur un plan ou une fibre. C R Acad Sci, Ser II 304:553–558
8. Brochard-Wyart F, Di Meglio JM, Quere D, de Gennes PG (1991) Spreading of nonvolatile liquids in a continuum picture. Langmuir 7(2):335–338, https://doi.org/10.1021/la00050a023
9. Carey VP (1992) Liquid-Vapor Phase Change Phenomena. Hemisphere, Washington D.C.
10. Deegan RD, Bakajin O, Dupont TF, Huber G, Nagel SR, Witten TA (2000) Contact line deposits in an evaporating drop. Phys Rev E 62(1):756–765, https://doi.org/10.1103/PhysRevE.62.756
11. Doumenc F, Guerrier B (2011) A model coupling the liquid and gas phases for a totally wetting evaporative meniscus. Eur Phys J Special Topics 197:281–293, https://doi.org/10.1140/epjst/e2011-01470-7
12. Doumenc F, Janecek V, Nikolayev VS (2018) Thin wedge evaporation/condensation controlled by the vapor dynamics in the atmosphere. Eur Phys J E 41(12):147, https://doi.org/10.1140/epje/i2018-11758-8
13. Eggers J (2005) Contact line motion for partially wetting fluids. Phys Rev E 72(6):061605, https://doi.org/10.1103/PhysRevE.72.061605
14. Eggers J, Pismen LM (2010) Nonlocal description of evaporating drops. Phys Fluids 22(11):112101, https://doi.org/10.1063/1.3491133
15. Fourgeaud L, Ercolani E, Duplat J, Gully P, Nikolayev VS (2016) Evaporation-driven dewetting of a liquid film. Phys Rev Fluids 1(4):041901, https://doi.org/10.1103/PhysRevFluids.1.041901
16. Gatapova EY, Graur IA, Kabov OA, Aniskin VM, Filipenko MA, Sharipov F, Tadrist L (2017) The temperature jump at water-air interface during evaporation. Int J Heat Mass Transfer 104:800–812, https://doi.org/10.1016/j.ijheatmasstransfer.2016.08.111
17. de Gennes PG (1985) Wetting: statics and dynamics. Rev Mod Phys 57:827–863
18. Hocking LM (1983) The spreading of a thin drop by gravity and capillarity. Q J Mechanics Appl Math 36(1):55–69, https://doi.org/10.1093/qjmam/36.1.55
19. Hocking LM (1995) On contact angles in evaporating liquids. Phys Fluids 7:2950–2955, https://doi.org/10.1063/1.868672
20. Hu H, Larson R (2002) Evaporation of a sessile droplet on a substrate. J Phys Chem B 106(6):1334–1344, https://doi.org/10.1021/jp0118322
21. Huang DM, Sendner C, Horinek D, Netz RR, Bocquet L (2008) Water slippage versus contact angle: A quasiuniversal relationship. Phys Rev Lett 101(22):226101, https://doi.org/10.1103/PhysRevLett.101.226101
22. Israelachvili JN (1992) Intermolecular and Surface Forces, 2nd edn. Academic Press, London
23. Janeček V (2012) Evaporation at microscopic scale and at high heat flux. PhD thesis, Université Pierre et Marie Curie Paris 6, http://tel.archives-ouvertes.fr/tel-00782517
24. Janeček V, Anderson DM (2016) Microregion model of a contact line including evaporation, kinetics and slip length. Interfacial Phenomena and Heat Transfer 4(2-3):93–107, https://doi.org/10.1615/InterfacPhenomHeatTransfer.2017017202
25. Janeček V, Nikolayev VS (2012) Contact line singularity at partial wetting during evaporation driven by substrate heating. Europhys Lett 100(1):14003, https://doi.org/10.1209/0295-5075/100/14003
26. Janeček V, Nikolayev VS (2013) Apparent-contact-angle model at partial wetting and evaporation: impact of surface forces. Phys Rev E 87(1):012404, https://doi.org/10.1103/PhysRevE.87.012404
27. Janeček V, Nikolayev VS (2014) Triggering the boiling crisis: a study of the dry spot spreading mechanism. Interfacial Phenomena and Heat Transfer 2(4):363–383, https://doi.org/10.1615/InterfacPhenomHeatTransfer.2015012273
28. Janeček V, Andreotti B, Pražák D, Bárta T, Nikolayev VS (2013) Moving contact line of a volatile fluid. Phys Rev E 88(6):060404, https://doi.org/10.1103/PhysRevE.88.060404

29. Janeček VV, Doumenc F, Guerrier B, Nikolayev VS (2015) Can hydrodynamic contact line paradox be solved by evaporation-condensation? J Colloid Interface Sci 460:329–338, https://doi.org/10.1016/j.jcis.2015.08.062
30. Koplik J, Banavar JR (1995) Continuum deductions from molecular hydrodynamics. Annu Rev Fluid Mech 27:257–292, https://doi.org/10.1146/annurev.fl.27.010195.001353
31. Landau LD, Lifshitz EM (1980) Statistical physics. Pergamon Press, Oxford
32. Lauga E, Brenner MP, Stone HA (2007) Microfluidics: The no-slip boundary condition. In: Tropea C, Yarin A, Foss J (eds) Springer Handbook of Experimental Fluid Dynamics, Springer, New York, chap 19, pp 1217–1240
33. Marek R, Straub J (2001) The origin of thermocapillary convection in subcooled nucleate pool boiling. Int J Heat Mass Transfer 44:619–632, https://doi.org/10.1016/S0017-9310(00)00124-1
34. Moosman S, Homsy GM (1980) Evaporating menisci of wetting fluids. J Colloid Interface Sci 73(1):212–223, https://doi.org/10.1016/0021-9797(80)90138-1
35. Morris SJS (2001) Contact angles for evaporating liquids predicted and compared with existing experiments. J Fluid Mech 432:1–30, https://doi.org/10.1017/S0022112000003074
36. Morris SJS (2014) On the contact region of a diffusion-limited evaporating drop: a local analysis. J Fluid Mech 739:308–337, https://doi.org/10.1017/jfm.2013.577
37. Nikolayev VS (2010) Dynamics of the triple contact line on a nonisothermal heater at partial wetting. Phys Fluids 22(8):082105, https://doi.org/10.1063/1.3483558
38. Nikolayev VS, Beysens DA (1999) Boiling crisis and non-equilibrium drying transition. Europhys Lett 47(3):345–351, https://doi.org/10.1209/epl/i1999-00395-x
39. Nikolayev VS, Chatain D, Garrabos Y, Beysens D (2006) Experimental evidence of the vapor recoil mechanism in the boiling crisis. Phys Rev Lett 97:184503, https://doi.org/10.1103/PhysRevLett.97.184503
40. Petroff NP (1883) Friction in machines and the effect of lubricant. Inzhenernyj Zhurnal, Sankt-Peterburg 1-4:71–140, 228–279, 377–436 and 535–564, published in four issues, in Russian
41. Pham CT, Berteloot G, Lequeux F, Limat L (2010) Dynamics of complete wetting liquid under evaporation. Europhys Lett 92(5):54005, https://doi.org/10.1209/0295-5075/92/54005
42. Pismen LM, Eggers J (2008) Solvability condition for the moving contact line. Phys Rev E 78(5):056304, https://doi.org/10.1103/PhysRevE.78.056304
43. Pit R, Hervet H, Léger L (2000) Direct experimental evidence of slip in hexadecane: solid interfaces. Phys Rev Lett 85:980–983, https://doi.org/10.1103/PhysRevLett.85.980
44. Pomeau Y (2000) Representation of the moving contact line in the equations of fluid mechanics. C R Acad Sci, Ser IIb 238:411–416, https://doi.org/10.1016/S1620-7742(00)00043-X, (in French)
45. Potash M, Wayner PC (1972) Evaporation from a two-dimensional extended meniscus. Int J Heat Mass Transfer 15(10):1851–1863, https://doi.org/10.1016/0017-9310(72)90058-0
46. Poulard C, Guéna G, Cazabat AM, Boudaoud A, Ben Amar M (2005) Rescaling the dynamics of evaporating drops. Langmuir 21(18):8226–8233, https://doi.org/10.1021/la050406v
47. Raj R, Kunkelmann C, Stephan P, Plawsky J, Kim J (2012) Contact line behavior for a highly wetting fluid under superheated conditions. Int J Heat Mass Transfer 55(9-10):2664–2675, https://doi.org/10.1016/j.ijheatmasstransfer.2011.12.026
48. Rednikov AY, Colinet P (2011) Truncated versus extended microfilms at a vapor-liquid contact line on a heated substrate. Langmuir 27(5):1758–1769, https://doi.org/10.1021/la102065c
49. Reyes R, Wayner PC Jr (1996) A Kelvin–Clapeyron adsorption model for spreading on a heated plate. J Heat Transfer 118(4):822–830, https://doi.org/10.1115/1.2822576
50. Reynolds O (1886) On the theory of lubrication and its application to Mr Beauchamp Tower's experiments, including an experimental determination of viscosity of olive oil. Phil Trans R Soc Lond 177:157–234, https://doi.org/10.1098/rstl.1886.0005
51. Snoeijer JH, Andreotti B (2008) A microscopic view on contact angle selection. Phys Fluids 20(5):057101, https://doi.org/10.1063/1.2913675
52. Snoeijer JH, Andreotti B (2013) Moving contact lines: Scales, regimes, and dynamical transitions. Annu Rev Fluid Mech 45(1):269–292, https://doi.org/10.1146/annurev-fluid-011212-140734

53. Snoeijer JH, Eggers J (2010) Asymptotic analysis of the dewetting rim. Phys Rev E 82(5):056314, 10.1103/PhysRevE.82.056314
54. Stephan P, Hammer J (1994) A new model for nucleate boiling heat transfer. Heat Mass Transfer 30(2):119–125, https://doi.org/10.1007/BF00715018
55. Sui Y, Ding H, Spelt PD (2014) Numerical simulations of flows with moving contact lines. Annu Rev Fluid Mech 46(1):97–119, https://doi.org/10.1146/annurev-fluid-010313-141338
56. Tsoumpas Y, Dehaeck S, Galvagno M, Rednikov A, Ottevaere H, Thiele U, Colinet P (2014) Nonequilibrium Gibbs criterion for completely wetting volatile liquids. Langmuir 30(40):11847–11852, https://doi.org/10.1021/la502708f
57. Voinov O (1976) Hydrodynamics of wetting. Fluid Dyn 11(5):714–721, https://doi.org/10.1007/BF01012963
58. Wayner PC, Kao YK, LaCroix LV (1976) The interline heat-transfer coefficient of an evaporating wetting film. Int J Heat Mass Transfer 19:487–492, https://doi.org/10.1016/0017-9310(76)90161-7
59. Yeh EK, Newman J, Radke CJ (1999a) Equilibrium configurations of liquid droplets on solid surfaces under the influence of thin-film forces: Part I. Thermodynamics. Colloids Surf, A 156(1-3):137–144, https://doi.org/10.1016/S0927-7757(99)00065-5
60. Yeh EK, Newman J, Radke CJ (1999b) Equilibrium configurations of liquid droplets on solid surfaces under the influence of thin-film forces: Part II. Shape calculations. Colloids Surf, A 156(1-3):525–546, https://doi.org/10.1016/S0927-7757(99)00110-7

Chapter 7
Leidenfrost Effect and Surface Wettability

Prashant Agrawal and Glen McHale

Abstract The Leidenfrost effect is a case of thin-film boiling where a drop of liquid levitates on a surface heated to temperatures significantly higher than the liquid's boiling point. When the drop contacts this superheated surface, a thin film of vapor (typically around 100 microns) forms instantaneously between the surface and the drop. The vapor layer supports the weight of the drop and thermally shields it from immediate evaporation. Due to the absence of direct contact between the drop and the surface, the Leidenfrost effect represents the case of a perfectly hydrophobic surface. In this chapter, we discuss the effect of surface wettability on the onset of this thin-film boiling state. We discuss passive methods, such as surface texturing, and active methods, such as using external fields to alter and control the transition to the Leidenfrost effect. The absence of a contact line provides extremely high mobility to these levitating drops and virtually eliminates friction. We discuss how this reduced friction can, in one case, reduce viscous drag on solid objects and, in another case, by introducing an asymmetry in the vapor flow, induce self-propulsion of levitating drops.

7.1 Introduction

Consider a drop, with a radius smaller than the capillary length, on a perfectly smooth surface (Fig. 7.1a). The drop assumes a spherical cap shape with an equilibrium contact angle θ_e to minimize the total surface free energy. θ_e is defined by the balance of three interfacial energies: solid–liquid (σ_{sl}), liquid–vapor (σ_{lv}) and solid–vapor (σ_{sv}), and through energy minimization θ_e is given by Young's law as:

P. Agrawal
Smart Materials and Surfaces Laboratory, Faculty of Engineering & Environment, Northumbria University, Newcastle upon Tyne NE1 8ST, UK

G. McHale (✉)
Institute of Multiscale Thermofluids, School of Engineering, University of Edinburgh, Edinburgh EH9 3FB, UK
e-mail: glen.mchale@ed.ac.uk

M. Marengo and J. De Coninck (eds.), *The Surface Wettability Effect on Phase Change*,
https://doi.org/10.1007/978-3-030-82992-6_7

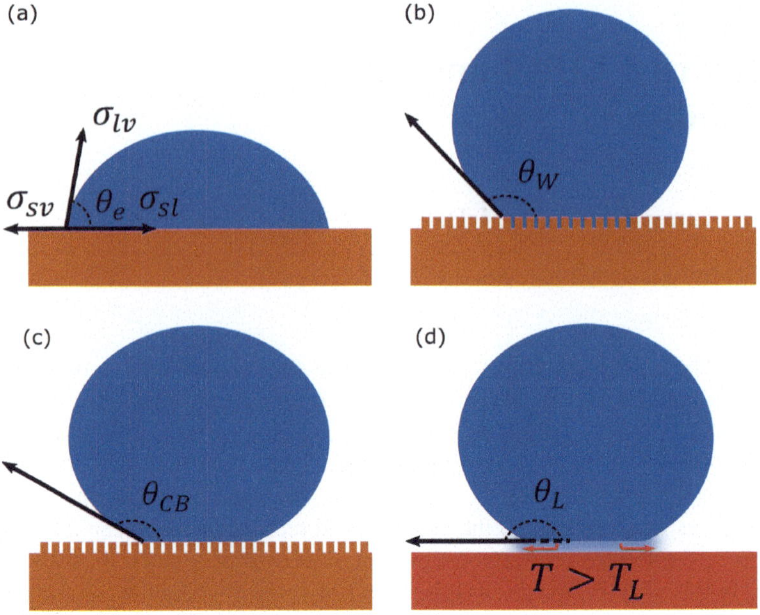

Fig. 7.1 Drop states on a surface: **a** Equilibrium contact angle on an ideal smooth surface with equilibrium contact angle θ_e; **b** Drop in a Wenzel state on a rough surface with apparent contact angle θ_W, idealized as a surface with pillars; **c** Drop in a Cassie-Baxter state on a surface with pillars apparent contact angle θ_{CB}; **d** Drop in a Leidenfrost state with apparent contact angle θ_L over a superheated surface undergoing continuous phase change

$$\cos\theta_e = \frac{\sigma_{sv} - \sigma_{sl}}{\sigma_{lv}}. \tag{7.1}$$

A low θ_e implies that a liquid has a higher affinity to adhere to and wet the surface. However, real surfaces are rough where the contact line undergoes a stick and slip motion. Surfaces are characterized by an advancing (θ_{adv}) and a receding (θ_{rec}) contact angle and the difference between these angles is termed as the contact angle hysteresis of a surface. These rough surfaces can be modeled as a surface patterned with pillars (Fig. 7.1), where the drop can wet the surface in one of two configurations or a combination of them. In the first configuration (Fig. 7.1b), the liquid penetrates the pillared surface and is in complete contact with the solid surface. This is called the Wenzel state and the apparent contact angle of the drop (θ_W) is given by $\cos\theta_W = r_s\cos\theta_e$, where r_s is the roughness ratio defined as the ratio of area of the rough surface to the area of the flat surface [1]. As $r_s > 1$, the Wenzel equation suggests that the roughness makes an intrinsically hydrophilic surface (θ_e <90°) more hydrophilic ($\theta_W < \theta_e$), and, similarly, an intrinsically hydrophobic surface (θ_e >90°) more hydrophobic ($\theta_W > \theta_e$). Although a droplet in a Wenzel state appears more hydrophobic, these surfaces have a high contact angle hysteresis and, therefore, are not very water repellent in the sense that the droplet is not easily removed from the surface by tilting it at an angle to the horizontal.

In the second configuration (Fig. 7.1c), the liquid rests on top of the pillars with air pockets between the liquid and the base surface. This is the Cassie-Baxter state and the (cosine of the) apparent contact angle (θ_{CB}) is given by a surface area weighted average of $\cos\theta_e$ and $\cos 180°$, i.e., $\cos\theta_{CB} = \phi_s \cos\theta_e - (1-\phi_s)$, where ϕ_s is the solid surface area fraction and $(1 - \phi_s)$ is the air surface area fraction. Due to minimal contact with the solid surface, a drop in a Cassie-Baxter state has a high contact angle and a low contact angle hysteresis [2]. From the Cassie-Baxter equation, we can infer that as the contact area between the drop and the pillars decreases, the apparent contact angle increases, i.e., as $\phi_s \rightarrow 0$, $\theta_{CB} \rightarrow 180°$. The extreme case of superhydrophobicity will be a zero contact between the pillars and the drop. This situation can be realized by continuously injecting vapor between the drop and the surface such that it supports the drop's weight (Fig. 7.1d). Here, due to zero contact with the surface, the surface is in a perfect water-repellent state, i.e., $\theta_L = 180°$ and has no contact angle hysteresis. The critical factor in achieving this state of perfect hydrophobicity is sustaining the layer of air/vapor between the liquid and the surface. This can be achieved by either pumping air through a porous material [3–6] or using thin-film boiling via the Leidenfrost effect [7] (Fig. 7.1d).

The onset of the Leidenfrost regime can be identified by looking at drop evaporation times on a heated substrate at different temperatures (Fig. 7.2a). A drop below the saturation temperature on a substrate undergoes slow evaporation. The rate of evaporation rapidly increases as the temperature of the substrate approaches the saturation temperature. Above the saturation temperature, bubbles form at the

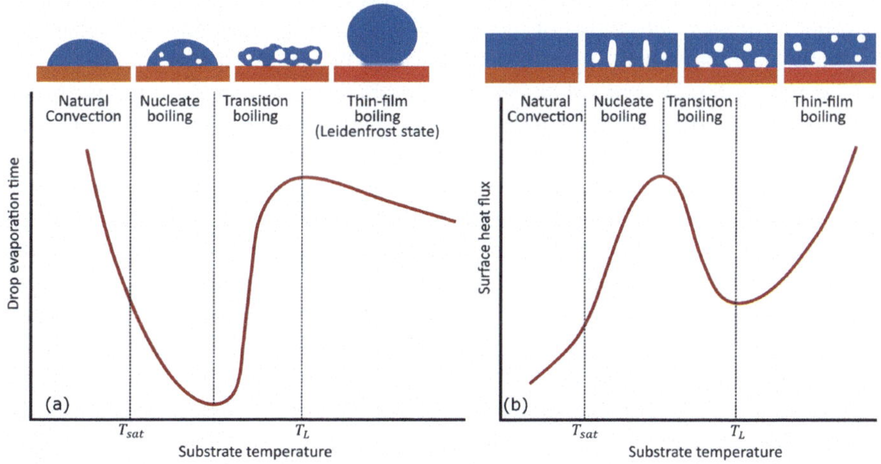

Fig. 7.2 Methods of identifying Leidenfrost temperature: **a** Measuring the drop evaporation lifetime for a sessile drop on a heated surface; **b** Measuring the surface heat flux in pool boiling experiments

liquid–solid interface, which results in a rapid phase change of the liquid, significantly reducing the drop lifetime. This nucleate boiling regime also corresponds to an increased heat flux from the substrate as the rapid phase change of the liquid extracts significant latent heat from the surface. On further increasing the temperature, the boiling enters a transition regime, where the heat flux drops as the increased bubble formation starts partially shielding the drop from the superheated surface. As a result, the drop lifetime may see a slight increase in this brief transition regime. At higher temperatures, the heat flux from the surface is sufficient to form a constant layer of vapor between the drop and the substrate. In this film-boiling regime, the drop is thermally and physically shielded from the superheated substrate, which eliminates bubble nucleation and increases the drop's lifetime significantly. The temperature at which this sudden increase in the drop's lifetime is observed is identified as the Leidenfrost temperature. Above the Leidenfrost temperature, the drop lifetime starts decreasing again.

Another method of determining the Leidenfrost temperature of a liquid-surface combination is through pool boiling (Fig. 7.2b). A pool of liquid on a heated substrate undergoes a similar phase change transition with temperature, where the different boiling phases are identified by the heat flux from the surface. Below the saturation temperature, the heat flux increases slowly with the substrate temperature. As bubble formation starts in the nucleate boiling regime, the heat flux increases rapidly, reaching a maximum critical heat flux. In the transition boiling regime that follows next, the bubble nucleation starts partially shielding the substrate as bubble growth resists liquid re-wetting, which limits heat transfer. As the Leidenfrost regime sets in, the heat flux drops significantly, due to complete thermal insulation from the vapor layer, resulting in a minimum in the heat flux from the surface.

The Leidenfrost effect is not just limited to (evaporating) liquid drops levitating on solid substrates but is also observed with sublimating ices on solid substrates [8]. Moreover, the underlying substrate need not be a rigid material, as levitation of evaporating liquids and solids can also be obtained on non-evaporating heated liquid baths [9, 10]. Another subset of the Leidenfrost phenomenon is the inverse Leidenfrost effect, where the substrate undergoes a phase change to continuously levitate a non-evaporating liquid drop [11, 12] or a solid object [13]. Therefore, the levitating objects can be evaporating liquids like water [14], acetone [9], benzene [15], nitrogen [16], oxygen [17], non-evaporating liquids like silicone oil [12, 18] and solids like dry-ice undergoing sublimation [10, 19], elastic hydrogels [20] or non-evaporating rigid blocks [13]. The substrates can be rigid solids like copper, steel, brass, Pyrex, silicon, sapphire glass, evaporating solids like dry-ice [21], non-evaporating liquids like silicone oil [9, 22, 23], molten metals [24] and evaporating liquids like water [16, 25] and liquid nitrogen [11, 12, 18, 26]. Figure 7.3 summarizes the different configurations of the levitating object and the substrate in a Leidenfrost state. In this chapter, we will discuss the classical Leidenfrost state of an evaporating liquid over a heated rigid solid substrate.

In the following sections, we first briefly discuss the dynamics of a sessile liquid drop on a heated solid surface in the Leidenfrost state. We discuss drop shape and hydrodynamic factors that govern the shape of the vapor layer separating the drop

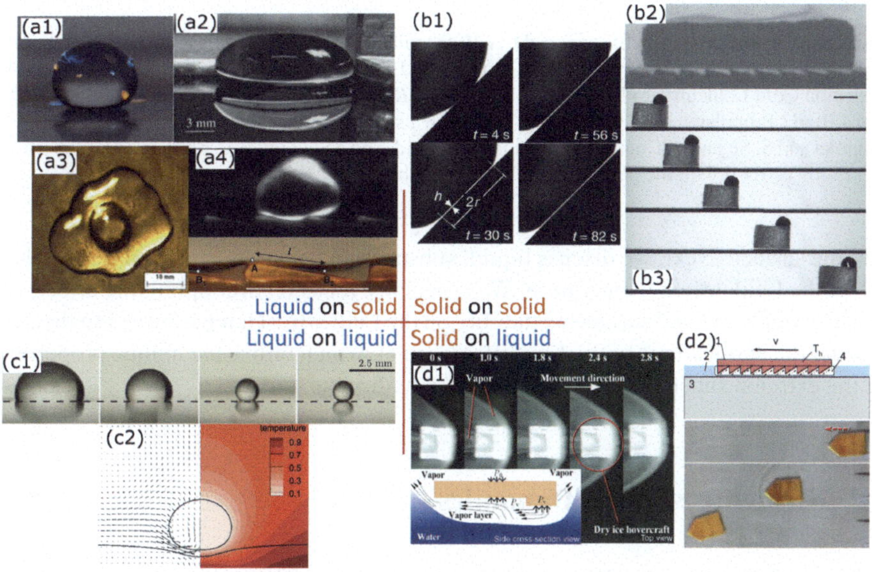

Fig. 7.3 Different types of Leidenfrost phenomena: **a1** A drop with radius $R < l_c$ on a slightly curved stainless steel surface at 300 °C; **a2** A gravity flattened water puddle $R > l_c$ on a silicon surface [27] (Reprinted from [A. Biance, C. Clanet and D. Quéré, "Leidenfrost drops", *Physics of Fluids* 15, 1632–1637 (2003)] with the permission of AIP Publishing); **a3** Vapor dome in a water puddle of volume 3 mL [28] (Reprinted from [G. Paul, I. Manna, P. K. Das and P. K. Das., "Formation, Growth, and Eruption Cycle of Vapor Domes beneath a Liquid Puddle during Leidenfrost Phenomena." *Applied Physics Letters* 103, 084101 (2013)] with the permission of AIP Publishing); **a4** R134a droplet on a ratchet surface. Magnified view of the liquid–vapor interface of water on the brass surface at 460 °C [29] (Reprinted figure with permission from [H. Linke, B. J. Alemán, L. D. Melling, M. J. Taormina, M. J. Francis, C. C. Dow-Hygelund, V. Narayanan, R. P. Taylor, and A. Stout, "Self-Propelled Leidenfrost Droplets." *Physical Review Letters* 96, 154502 (2006)] Copyright 2006 by the American Physical Society); **b1** Time sequence of contact of a hydrogel sphere on a surface at 215 °C [30] (Reprinted figure with permission from [S. Waitukaitis, K. Harth and M. van Hecke, "From Bouncing to Floating: The Leidenfrost Effect with Hydrogel Spheres." *Physical Review Letters* 121, 048001, 2018] Copyright 2018 by the American Physical Society); **b2** A dry-ice platelet levitating and self-propelling on a ratchet surface at 450 °C [19] (Reprinted from [G. Dupeux, T. Baier, V. Bacot, S. Hardt, C. Clanet and D. Quéré, "Self-Propelling Uneven Leidenfrost Solids." *Physics of Fluids* 25, 051704 (2013)] with the permission of AIP Publishing); **b3** A dry-ice platelet on a flat surface (320 °C) with a small weight placed asymmetrically. The mass tilts the platelet slightly such that the ice self-propels [19] (Reprinted from [G. Dupeux, T. Baier, V. Bacot, S. Hardt, C. Clanet and D. Quéré, "Self-Propelling Uneven Leidenfrost Solids." *Physics of Fluids* 25, 051704 (2013)] with the permission of AIP Publishing); **c1** Acetone drops on a water bath (70 °C) [25] (Reprinted from [S. D. Janssens, S. Koizumi and E. Fried, "Behavior of Self-Propelled Acetone Droplets in a Leidenfrost State on Liquid Substrates." *Physics of Fluids* 29, 032103 (2017)] with the permission of AIP Publishing); **c2** Depiction of velocity vectors (left) and temperature (right) for a liquid levitating on a liquid [31] (Reprinted from *International Journal of Heat and Mass Transfer*, 128, L. Qiao, Z. Zeng, H. Xie, H. Liu, and L. Zhang, "Modeling Leidenfrost Drops over Heated Liquid Substrates", 1296–1306, Copyright (2019), with permission

Fig. 7.3 (continued) from Elsevier); **d1** Self-propulsion of a dry-ice hovercraft on water [10] (Adapted from [M. Shi, X. Ji, S. Feng, Q. Yang, T. J. Lu, and F. Xu, "Self-Propelled Hovercraft Based on Cold Leidenfrost Phenomenon." *Scientific Reports* 6, 28574 (2016)], licensed under CC BY 4.0); **d2** Propulsion of a 'Leidenfrost glider' on a shallow water layer [13] (Reprinted from [H. Sugioka and S. Segawa, "Controllable Leidenfrost Glider on a Shallow Water Layer." *AIP Advances* 8, 115209 (2018)] with the permission of AIP Publishing)

and the surface. Next, we discuss liquid, substrate and environmental properties that affect the Leidenfrost temperature. We especially focus on decoupling the effects of surface roughness and surface wettability on the Leidenfrost temperature. Finally, we discuss the applications of the Leidenfrost effect, which utilize the unique properties of the extreme superhydrophobicity/perfect hydrophobicity.

7.2 Leidenfrost Drop Dynamics

A Leidenfrost drop is an extreme case of superhydrophobicity as it is in a completely non-wetting state on a surface. An idealized Leidenfrost drop can be imagined as a sessile drop resting on a surface with θ_e =180°. However, for large drops gravity distorts their shape, which is critically important in dictating heat transfer dynamics, as it affects vapor layer thickness, drop evaporation rates and substrate temperature distribution.

Leidenfrost drops with a radius R smaller than the capillary length l_c ($l_c = \sqrt{\sigma_{lv}/\rho_l g}$, where g is the acceleration due to gravity and ρ_l is the density of liquid) assume a quasi-spherical shape on the surface, where the liquid–vapor interface near the surface is flattened, with a contact radius R_c, due to gravity (Fig. 7.3a1). This drop shape is analogous to liquid marbles [32], where there is a gain in surface energy ($\sigma_{lv} R_c^4/R^2$) due to the lowering of the drop's gravitational potential energy ($\rho_l g R^3 \delta$) by a length scale δ. Equating these two energy differences provides a typical length scale for the contact radius: $R_c \sim R^2/l_c$. Drops with $R > l_c$ form disk-like puddles (Fig. 7.3a2), where a balance of Laplace pressure ($2\sigma_{lv}/\text{h}$) and hydrostatic pressure ($\rho_l \text{gh}^2/2$) result in a puddle height $\text{h} = 2l_c$ [27]. This approximation of puddle height is not valid for values of $R \sim l_c$ as the Laplace pressure from the drop's equatorial curvature squeezes the drop and results in a slightly larger h [32]. In the case of puddles, the contact radius is equal to the puddle's radius: $R_c = R$.

The phase change of the liquid due to the heat transfer from the substrate feeds the vapor film, which supports the drop's weight. Given the typical scale of vapor layer thickness $l_v \sim 100\mu\text{m}$ and substrate temperatures less than 1000 °C, thermal conduction through the vapor layer is the primary mode of heat transfer. Radiative heat flux from the heated surface, given by $\sigma_S(T_s^4 - T_{sat}^4)$, where σ_S is the Stefan's constant, T_s is the substrate temperature and T_{sat} is the saturation temperature of the liquid, can be ignored as its magnitude is comparable to thermal conduction only above 1000 °C [33]. Assuming that most of the heat transferred from the surface

($\propto \pi R_c^2 \mathrm{k_s} \Delta T / l_v$, $\mathrm{k_s}$ is the thermal conductivity of the substrate) is used as the latent heat of evaporation ($\propto \pi R_c^2 \rho_v \mathrm{L} v_0$, where L is the latent heat of evaporation of the liquid and ρ_v is the density of vapor), typical evaporation flux from the liquid–vapor interface (v_0) can be written as:

$$v_0 = \frac{\mathrm{k_s} \Delta T}{l_v \rho_v \mathrm{L}}. \tag{7.2}$$

Temperature measurements of the drop have shown that its temperature is close to the liquid's boiling point [27], therefore, $\Delta T = T_s - T_{sat}$. This vapor flows out radially between the liquid–vapor interface and the surface. Using the lubrication approximation and assuming a radially symmetric Poiseuille flow, the radial velocity of the flow (v_r) can be written as [34]:

$$v_r = \frac{z(z - l_v)}{2\nu \rho_v} \frac{dp_v}{dr}, \tag{7.3}$$

where ν is the kinematic viscosity of the vapour, z is the vertical coordinate, r is the radial coordinate and p_v is the pressure in the vapor layer. Using the continuity equation ($\nabla \cdot \boldsymbol{v} = 0$) with the boundary condition of vapor influx $v_z = -v_0$ at $z = l_v$, the governing equation for pressure distribution in the vapor layer is obtained as:

$$\frac{1}{r} \frac{d}{dr} \left(r l_v^3 \frac{dp_v}{dr} \right) = -12 \nu \rho_v v_0. \tag{7.4}$$

To obtain estimates for the vapor layer thickness, let us assume a flat liquid–vapor interface, i.e., a constant l_v. Consequently, on integrating Eq. (7.4) from $r = 0$ to $r = R_c$, the radial pressure distribution in the vapor layer is obtained as:

$$p_v(r) = p_{atm} + \frac{3\nu \rho_v v_0}{l_v^3} \left(R_c^2 - r^2 \right). \tag{7.5}$$

Equation (7.5) integrated over the drop's contact radius will be equal to the drop's weight, which in conjunction with Eq. (7.2) provides the vapor layer thickness. For drops with $R > l_c$ [27, 34]:

$$l_v = \left[\frac{3\nu \mathrm{k_s} \Delta T R^2}{4 \mathrm{L} \rho_l g l_c} \right]^{1/4}. \tag{7.6}$$

Equation (7.2) is a simplified 1D model for the heat flux that assumes that the heat transfer and evaporation occur from the contact area (πR_c^2) only. This assumption is valid for drops with $R > l_c$, however, for drops with $R < l_c$ evaporation occurs from the entire surface of the drop. In this case, the typical scale of the temperature gradient is modified to $\Delta T / R$ and, therefore, the heat transferred from the surface is proportional to $\pi R^2 \mathrm{k_s} \Delta T / R$. Therefore, for drops with $R < l_c$, the vapor layer thickness is obtained as [27]:

$$l_v = \left[\frac{\nu k_s \Delta T \rho_l g R^4}{L \sigma_{lv}^2} \right]^{1/3}. \quad (7.7)$$

From Eqs. (7.6) and (7.7) it is seen that the vapor film thickness decreases with a decrease in the drop radius. Therefore, as a drop evaporates its vapor layer thickness continuously decreases with time.

Equation (7.5) provides additional information regarding the morphology of the liquid–vapor interface near the surface. The pressure field underneath the drop is maximum at the center and radially decreases toward its edge. This high pressure at the center can overcome the surface tension of the liquid and distort the liquid–vapor interface to form a pocket of vapor (Fig. 7.4a). We can infer from Eq. (7.5) that larger drops will have a higher pressure at the center, therefore, a bigger vapor pocket (Fig. 7.4b). In large puddles, these vapor pockets become unstable and rise within the liquid as a bubble (Fig. 7.4c, d) and burst at the upper liquid–air interface (sometimes termed as a 'chimney effect') [28]. Although Eqs. (7.6) and (7.7) assume a flat liquid–vapor interface, the theoretical estimates provide a good approximation to the scale of mass fluxes [27, 35] and average vapor layer thicknesses measured experimentally via X-ray imaging [36, 37], optical interference [38] and equivalent capacitance [39]. Nevertheless, the accurate shape of the vapor layer and the whole drop can also be obtained numerically (Fig. 7.4e) [40, 41].

The above-described models assume a static system where we do not consider dynamic flow inside the drop. The flow inside the drop could originate because of natural convection, viscous drag from the escaping vapor or Marangoni flows due to temperature differences between the top of the drop and the liquid–vapor interface [33]. However, the typical scale of these flows ($\sim$10 cm s^{-1}) is not significant, compared to surface tension forces, to alter the shape of the drop or heat transfer dynamics. However, these internal flows affect a drop's movement on the surface. Large drops have axisymmetric counter-rotating toroidal internal flows and mostly remain stably positioned, apart from minor fluctuations. However, small, milli-metric-sized droplets have a single asymmetric toroidal flow which causes the droplet to roll on a perfectly level surface in random directions like a 'wheel'[42]

To initialize and sustain a vapor film underneath a Leidenfrost drop, the drop requires a continuous supply of latent heat from the surface. This heat transfer introduces a local cooling inside the substrate, as seen in Fig. 7.4f [43]. In situations where the residence time of a drop on the surface is very short, for example in drop impact, some surfaces maintain isothermal conditions [44, 45], while some surfaces show some local cooling [46, 47]. However, for higher drop residence times, this local cooling becomes significant, which can be as strong as 80 °C for low thermally conductive materials [43]. Using simulations van Limbeek et al. showed that the temperature field in the substrate depends on the drop size and the liquid–vapor interface geometry, as seen in Fig. 7.4g, h [43]. In the limit of infinite thermal conductivity, the substrates maintain isothermal conditions.

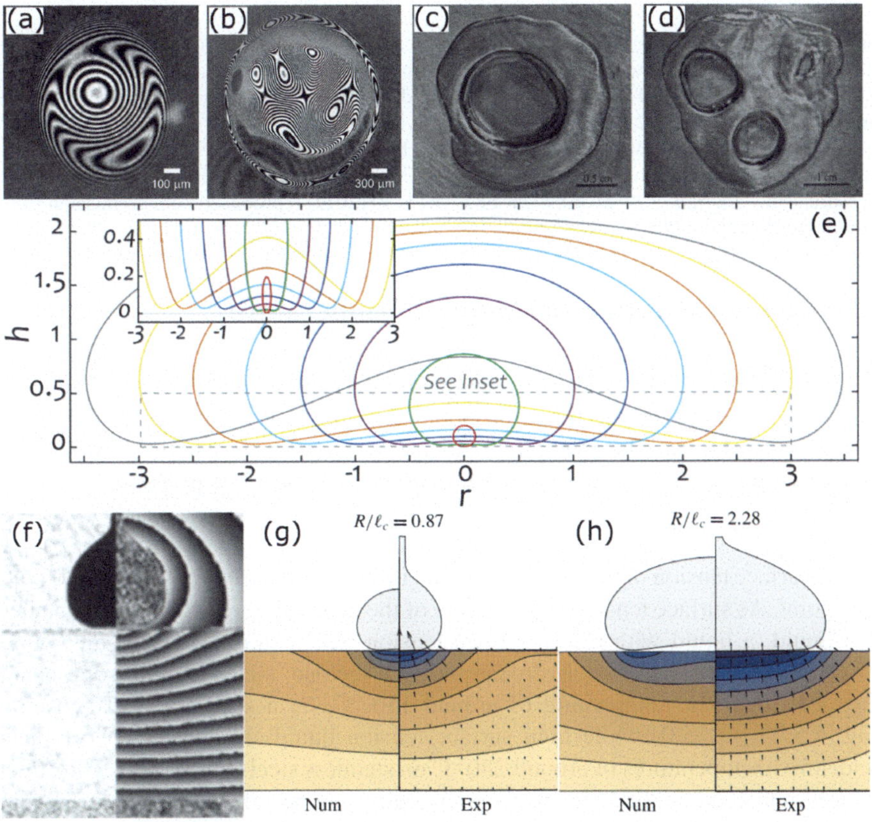

Fig. 7.4 Vapor layer geometry underneath a Leidenfrost drop and its effects: **a** Interference patterns for a water drop (top view) at the surface temperature of 245 °C. Drop size $r_{max} = 1.2$ mm [38] (Reprinted figure with permission from [J. C. Burton, A. L. Sharpe, R. C.A. Van Der Veen, A. Franco, and S. R. Nagel, "Geometry of the Vapor Layer under a Leidenfrost Drop." *Physical Review Letters* 109, 074301 (2012)] Copyright 2012 by the American Physical Society); **b** Interference patterns for a drop with $r_{max} = 2.7$ mm [38] (Reprinted figure with permission from [J. C. Burton, A. L. Sharpe, R. C.A. Van Der Veen, A. Franco, and S. R. Nagel, "Geometry of the Vapor Layer under a Leidenfrost Drop." *Physical Review Letters* 109, 074301 (2012)] Copyright 2012 by the American Physical Society) **c**, **d** Puddles of water on Duralumin substrate at 300 °C (top view). One or several bubbles can be observed in the puddle, depending on its size. The bubbles rise in the liquid and burst at the upper liquid-vapor interface [27] (Reprinted from [A. Biance, C. Clanet and D. Quéré, "Leidenfrost drops", *Physics of Fluids* 15, 1632–1637 (2003)] with the permission of AIP Publishing); **e** Numerically calculated Leidenfrost drop shapes for water drops of size $R_{\max} = 0.1l_c$, $0.5l_c$, $1l_c$, $1.5l_c$, $2l_c$, $2.5l_c$, $3l_c$, and $3.5l_c$, where l_c is the capillary length of water [40] (Reprinted figure with permission from [B. Sobac, A. Rednikov, S. Dorbolo, and P. Colinet, "Leidenfrost Effect: Accurate Drop Shape Modeling and Refined Scaling Laws." *Physical Review E - Statistical, Nonlinear, and Soft Matter Physics* 90, 053011 (2014)] Copyright 2014 by the American Physical Society); **f** Interferometric images of the reconstructed temperature field for an ethanol drop on a quartz plate with $T_s = 330$ °C [43] (M. A. J. van Limbeek, M. H. K. Schaarsberg, B. Sobac, A. Rednikov, C. Sun, P. Colinet, and D. Lohse, "Leidenfrost Drops Cooling Surfaces:

Fig. 7.4 (continued) Theory and Interferometric Measurement." *Journal of Fluid Mechanics* 827: 614–39 (2017) reproduced with permission from Cambridge University Press); **g** Temperature field in a quartz plate (T_s = 330 °C) underneath ethanol drops, numerical (left) vs experiments (right) for drop size (**g**) R = 0.87 l_c and (**h**) R = 2.28 l_c. Arrows indicate the relative magnitude and direction of heat flux [43] (M. A. J. van Limbeek, M. H. K. Schaarsberg, B. Sobac, A. Rednikov, C. Sun, P. Colinet, and D. Lohse, "Leidenfrost Drops Cooling Surfaces: Theory and Interferometric Measurement." *Journal of Fluid Mechanics* 827: 614–39 (2017) reproduced with permission from Cambridge University Press)

7.3 Factors Affecting Leidenfrost Temperature

As discussed before, the Leidenfrost temperature is determined when a stable vapor film is present between the liquid drop and the superheated surface. The stability of this vapor film, and therefore, the Leidenfrost temperature, can be controlled by altering liquid properties, environmental conditions and surface properties.

a. Liquid properties

The surface tension of the deposited liquid is important in determining the Leidenfrost point. As surface tension is a measure of the cohesive force between molecules of a liquid, a liquid with a low surface tension will require a lower temperature for phase change. Therefore, hydrocarbons like acetone, methanol and isopropanol have a Leidenfrost temperature of around 150 °C, on a stainless steel substrate (Fig. 7.5a1) [48, 53], while high surface tension liquids like water have typical Leidenfrost temperatures of around 250 °C on stainless steel (Fig. 7.5a1). The effect of surface tension on the Leidenfrost temperature can be extended to impure liquid drops, for example, water drops contaminated with salts or surfactants. The presence of salts increases the surface tension of water, therefore resulting in increased Leidenfrost temperatures [54], whereas ionic surfactants like sodium dodecyl sulfate (SDS) and hexadecylcetyltrimethylammonium bromide (CTAB) reduce the surface tension of water resulting in reduced Leidenfrost temperatures on different surfaces (Fig. 7.5a3) [49, 55].

The deposition method of the drop also affects the Leidenfrost temperature. For a drop gently placed on a surface, the vapor pressure has to support only the weight of the drop or the gravitational potential energy of the drop mass. However, if the drop is released at a certain distance from a surface or strikes the surface with a non-zero velocity, the vapor pressure has to account for the additional kinetic energy of the drop to ensure a non-contact deposition. As a result, with an increase in velocity of the impacting drop the Leidenfrost temperature also increases [56, 57]. In the case of such impacting drops, the Leidenfrost temperature is referred to as the dynamic Leidenfrost temperature and is an important quantity for characterizing surfaces for applications in spray cooling and spray combustion [37, 58].

The initial temperature of the liquid drop (T_{li}), or liquid sub-cooling, before deposition on the surface also affects the Leidenfrost temperature. The temperature of a drop in a Leidenfrost state is close to the saturation point of the liquid (T_{sat}) [27].

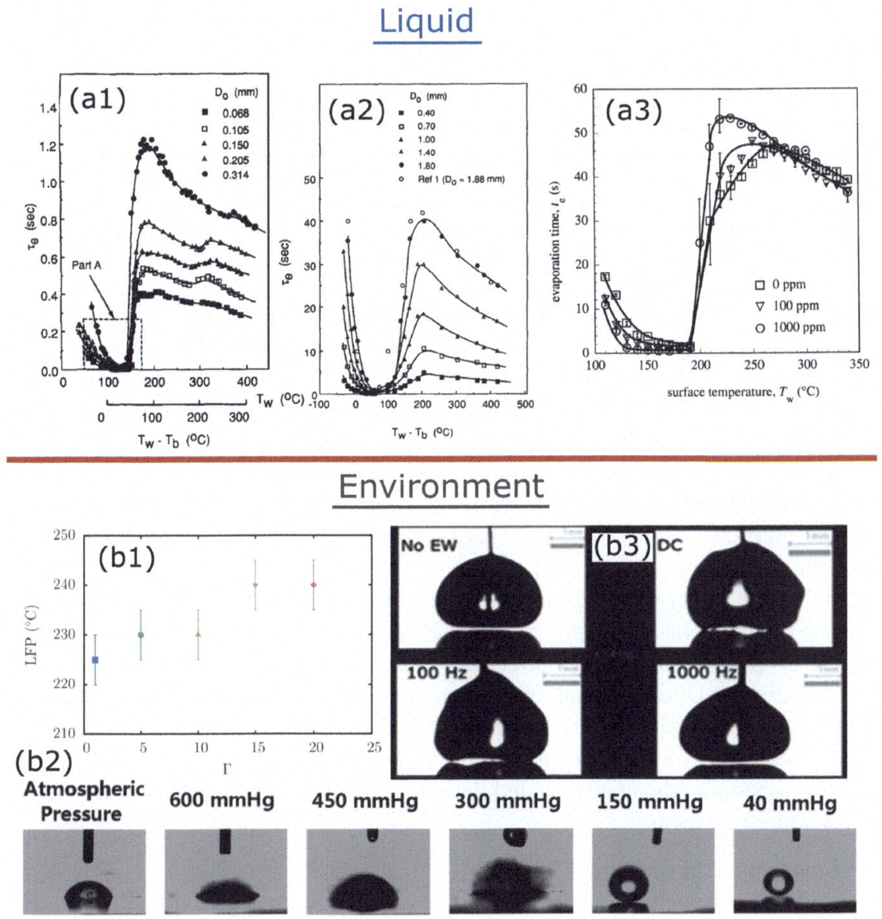

Fig. 7.5 Different factors affecting Leidenfrost temperature: **a1** Evaporation rates of a heptane drop of different sizes (D_0) on stainless steel plate [48] (Reprinted from *International Journal of Heat and Mass Transfer* 34(7), T.Y. Xiong and M. C. Yuen, "Evaporation of a Liquid Droplet on a Hot Plate", 1881–1894, Copyright (1991), with permission from Elsevier); **a2** Evaporation rates of a water drop of different sizes (D_0) on stainless steel plate [48] (Reprinted from *International Journal of Heat and Mass Transfer* 34(7), T.Y. Xiong and M. C. Yuen, "Evaporation of a Liquid Droplet on a Hot Plate", 1881–1894, Copyright (1991), with permission from Elsevier); **a3** Measured drop evaporation times for water drops with different concentrations of surfactant sodium dodecyl sulfate (SDS) on a stainless steel surface at different temperatures [49] (Reprinted from [Y.M. Qiao and S. Chandra, "Experiments on Adding a Surfactant to Water Drops Boiling on a Hot Surface." *Proceedings of the Royal Society A: Mathematical, Physical and Engineering Sciences* 453 (1959): 673–89] with permission from The Royal Society (U.K.)); **b1** Variation of the Leidenfrost temperature with non-dimensional acceleration due to gravity Γ, non-dimensionalized by $g = 9.81\ \mathrm{m\ s^{-2}}$ [50] (Reprinted figure with permission from [L. Maquet, M. Brandenbourger, B. Sobac, A.-L. Biance, P. Colinet and S. Dorbolo "Leidenfrost drops: Effect of gravity", EPL, 110, 24001 (2015) with permission from IOP publishing); **b2** Water drops deposited on a heated polished aluminium surface

Fig. 7.5 (continued) ($T_s = 105$ °C) at different surrounding pressures [51] (Reprinted figure with permission from [D. Orejon, K. Sefiane and Y. Takata, "Effect of Ambient Pressure on Leidenfrost Temperature." *Physical Review E - Statistical, Nonlinear, and Soft Matter Physics*, 90 (5): 1–6 (2014)] Copyright 2014 by the American Physical Society); **b3** Effect of the frequency of an AC signal on suppression of the Leidenfrost state of isopropanol drops on an anodized aluminum plate at 80 V (for both AC and DC) [52] (Reprinted from [O. Ozkan, A. Shahriari and V. Bahadur, "Electrostatic Suppression of the Leidenfrost State Using AC Electric Fields." *Applied Physics Letters* 111, 141608 (2017)] with the permission of AIP publishing)

Therefore, a sub-cooled liquid drop will require an initial additional specific heat input proportional to $C_p(T_{sat} - T_{li})$, where C_p is the specific heat capacity of the liquid, over the latent heat of vaporization. This additional heat has to be supplied rapidly by the substrate till the drop reaches a steady-state temperature of T_{sat}, while it is levitating. For substrates with high thermal conductivity, transfer of this additional energy occurs almost instantaneously, and therefore, liquid sub-cooling does not significantly impact the Leidenfrost temperature [15]. However, for substrates with low thermal conductivity, like glass, liquid sub-cooling can increase Leidenfrost temperatures by over 200 °C [59].

Finally, the effect of the drop volume on the Leidenfrost temperature has been found to be inconclusive. Experiments by Baumeister et al. on flat surfaces suggest that the drop volume does not affect the Leidenfrost temperature [60]. However, recent studies by Duursma et al. on microtextured surfaces show a significant effect of drop size on the Leidenfrost temperature [61]. Additional experiments and simulations are required to provide insight into the dynamics of drops on these microtextured surfaces, especially the geometry and the role of the liquid–vapor interface on these microtextured surfaces.

b. Environmental conditions

The Leidenfrost temperature can be controlled through external fields. For example, increasing the gravitational field increases the Leidenfrost temperature as the weight of the drop (or hydrostatic pressure) increases (Fig. 7.5b1). Therefore, a higher temperature is required to produce sufficient vapor to balance this increased drop weight [50]. Similarly, the Leidenfrost temperature can also be altered by changing the surrounding pressure around the drop [62, 63]. By lowering the surrounding pressure, the saturation temperature of the liquid decreases. As a result, the evaporation rate increases and the pressure required to initiate and sustain levitation is obtained at a lower Leidenfrost temperature (Fig. 7.5b2) [51, 64].

Low-frequency vibrations of the order 100 Hz can suppress the Leidenfrost effect at a surface acceleration of 100g. At surface temperatures just above the Leidenfrost temperature, vibrations increase the frequency of intermittent contact of the drop with the heated surface forcing a transition from film boiling to contact boiling [65].

Electric fields also provide another method of suppressing the Leidenfrost effect. When a potential difference is applied across a slightly charged drop and the surface, the liquid–vapor interface and the surface act like two plates of a capacitor separated by an insulator vapor layer [66]. This configuration is similar to electrowetting, where an electric field is used to modify the wetting properties of a surface [67]. In electrowetting, the presence of charges at the interface increases surface wetting allowing control over the contact angle of the liquid. In the case of Leidenfrost drops, electrohydrodynamic forces acting on the liquid–vapor interface introduce instabilities that lead to direct contact with the surface leading to contact boiling (Fig. 7.5b3) [52, 66, 68].

c. Surface properties

The Leidenfrost state sets in after the transition boiling regime where the liquid tries to re-wet the surface. Once the drop enters the Leidenfrost regime, the requirement for sustaining levitation on a surface is the continuous production of a vapor layer with sufficient pressure to balance the weight of the drop. Considering the onset and sustenance of the Leidenfrost state, the following surface properties affect the Leidenfrost point of a liquid drop on a surface: (1) Surface thermal conductivity, (2) Surface roughness or porosity and (3) Surface wettability or wickability. Table 1 summarizes key experimental works where the effect of surface properties on the Leidenfrost temperature was investigated.

In many cases these three properties are interdependent, however, with recent progress in manufacturing techniques these properties can be de-coupled and their effects on the Leidenfrost point can be isolated. The types of surfaces that can be manufactured are bare substrates, nanoporous sheets, micro-structures on surfaces, micro-pillars on surfaces and surfaces with different wettabilities.

I. Thermal conductivity

Let us consider an extremely smooth surface with infinite thermal conductivity heated to the Leidenfrost temperature of a liquid T_L. As mentioned in Sect. 7.2, when a drop is deposited on this surface at a given temperature, the surface is cooled locally underneath the drop. Due to its infinite thermal conductivity, the substrate remains isothermal at T_L, i.e., zero temperature gradients. This also implies that the liquid requires a local surface temperature of T_L for levitation. Let us consider another extremely smooth surface but with finite thermal conductivity. Here, due to local cooling by the drop and the substrate's finite thermal inertia, there will be local temperature gradients in the substrate. Therefore, to obtain a local surface temperature of T_L for maintaining levitation, the substrate has to be heated to a higher temperature, which practically implies a higher Leidenfrost temperature. For example, stainless steel has a lower thermal conductivity than copper and, therefore, shows higher Leidenfrost temperatures [69]. The formation of oxide layers on copper decreases the transfer of heat from the surface and shows slightly higher Leidenfrost temperatures than polished aluminum [15]. The difference in Leidenfrost temperatures is more pronounced with a larger difference in thermal conductivities. For example, the Leidenfrost temperature of the water on Pyrex glass is about 515 °C, while on stainless steel of the same roughness, it is about 305 °C [59]. However, these studies do not account for the effect of parameters like surface wettability on the reported Leidenfrost temperatures. As such some studies also report no significant effect of thermal conductivity on the Leidenfrost temperature despite an order of magnitude difference in thermal diffusivities [70].

II. Surface roughness

A real surface can be idealized as a surface with pillars, where the height and spacing between the pillars determine the roughness of the surface. For a drop levitating on these pillars, the height of the pillars increases the effective vapor layer thickness. The spacing between the pillars, or pillar density, also allows an easy passage for the vapor to leak to the atmosphere. These two effects can reduce the vapor pressure sustaining the drop weight. The pillars can also act as protrusions, which can increase the frequency of intermittent contacts with the surface, requiring higher Leidenfrost temperatures. The dominating mechanism which affects the Leidenfrost temperature depends on the scale of roughness, primarily the pillar spacing. It is important to mention here that surface roughness also affects surface wettability. In this discussion, we will consider experimental findings where the effects of surface wettability are not significant or have been isolated from surface roughness. The effect of surface roughness from surface wettability can be isolated by either using low surface tension liquids [53, 59, 74], or using machined surfaces and materials with similar wettabilities [45, 75, 76].

Two competing effects determine the Leidenfrost temperature for drops on pillared surfaces (Fig. 7.6a). When a drop is placed on a heated pillared surface, the drop forms a meniscus with the pillars, which induce a surface adhering capillary pressure, $P_c = 2\sigma_{lv}\cos\theta/l_p$, where l_p is the scale of pillar spacing. This capillary adhesion is resisted by the vapor pressure generated due to liquid phase change, which flows out radially underneath the drop. Assuming a Poiseuille flow, this vapor flow (typical velocity V) is resisted by shear losses due to the vapor layer thickness (pressure drop scales as $\nu V/l_v^2$) and shear flow between the pillar spacing (pressure drop scales as $\nu V/l_p^2$). Therefore, although surfaces with a smaller pillar spacing increase capillary adhesion, the vapor pressure underneath the drop also increases as the resistance to flow increases. For surfaces with typical pillar spacing $\leq$100 μm, higher Leidenfrost temperatures are obtained due to a decrease in vapor pressure with increasing pillar spacing. Kwon et al. fabricated pillars (width 10 μm and height 10 μm) with pillar spacings varying from 3.3 to 100 μm [71]. These surfaces are superhydrophilic and showed an increase in the Leidenfrost temperature from 300 °C to 370 °C (Fig. 7.6b). Duursma et al. reached the same conclusion experimentally, but attributed the increase in Leidenfrost temperature with pillar spacing to a reduced effective area for heat transfer from the surface to the drop, supported by numerical simulations (Fig. 7.6c) [61]. The effect of vapor leakage on increased Leidenfrost temperatures is also seen in levitation on stainless steel meshes [73]. An increase in separation between wires increases the size of the air gap in the meshes that reduces the vapor pressure necessary for levitation, and thus results in higher Leidenfrost temperatures (Fig. 7.6d). Porous materials with increasing porosity also show an increased Leidenfrost temperature [74]. Reduced pressures due to vapor leakage increase the Leidenfrost temperature for wetting methanol drops. The role

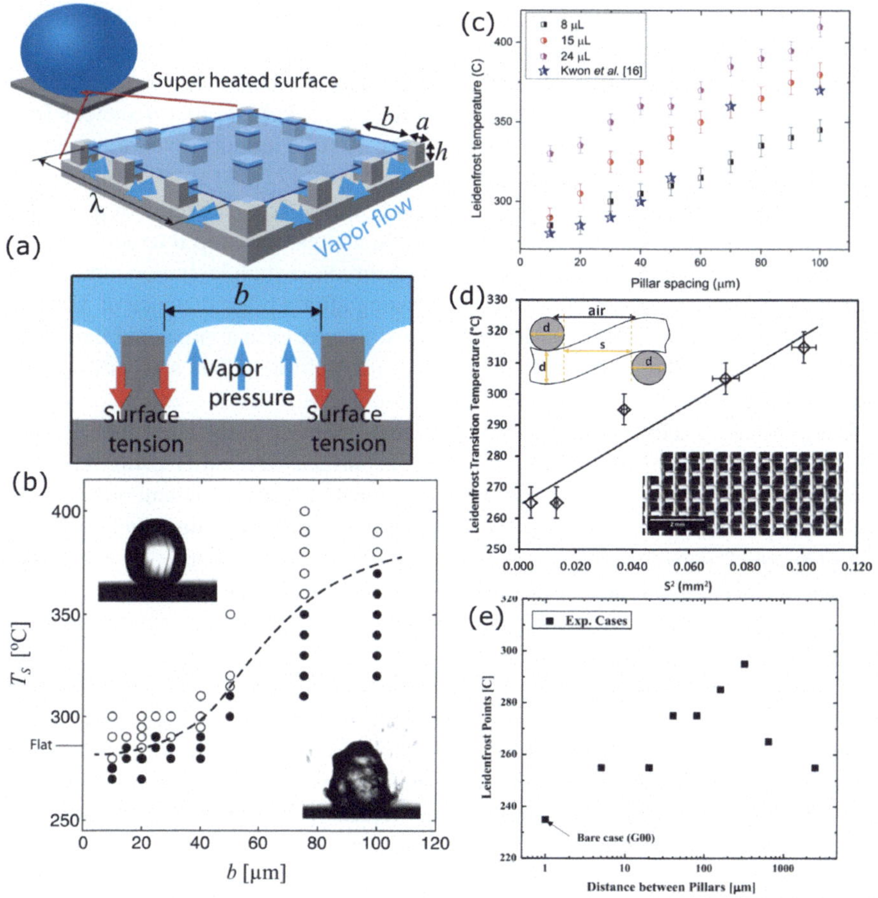

Fig. 7.6 Effect of surface roughness on Leidenfrost temperature: **a** Schematic representation of a drop on a micro-pillared surface. Capillary adhesion on the pillared surface is countered by the escape of vapor between the pillars [71] (Reprinted from [H. M. Kwon, J. C. Bird and K. K. Varanasi, "Increasing Leidenfrost Point Using Micro-Nano Hierarchical Surface Structures." *Applied Physics Letters* 103, 201601 (2013)] with the permission of AIP Publishing); **b** Transition to Leidenfrost state on surfaces with different pillar spacing. The dashed line is a guide to the eye [71] (Reprinted from [H. M. Kwon, J. C. Bird and K. K. Varanasi, "Increasing Leidenfrost Point Using Micro-Nano Hierarchical Surface Structures." *Applied Physics Letters* 103, 201601 (2013)] with the permission of AIP Publishing); **c** Leidenfrost temperature on micro-pillared Si surfaces with varying pillar spacing for different drop volumes [61] (Reprinted from G. Duursma, R. Kennedy, K. Sefiane, and Y. Yu, "Leidenfrost Droplets on Microstructured Surfaces." *Heat Transfer Engineering* 37 (13–14): 1190–1200 (2016) with permission from Taylor & Francis Ltd, http://www.tandfonline.com)); **d** Leidenfrost transition temperature with open area s^2. Insets show a depiction of a plain weave mesh and an image of a #50 stainless steel mesh at bottom [73] (Reprinted from *Materials Letters*, 176, N.R. Geraldi, G. McHale, B.B. Xu, G.G. Wells, L.E. Dodd, D. Wood, and M. I. Newton, "Leidenfrost Transition Temperature for Stainless Steel Meshes", 205–208, Copyright (2016), with permission from Elsevier); **e** Leidenfrost temperature on pillared surfaces with varying pillar spacing [72] (Reprinted from *International Journal of Heat and Mass Transfer*, 139, S.H. Kim, G. Lee, H.M. Kim and M.H. Kim, "Leidenfrost Point and Droplet Dynamics on Heated Micropillar Array Surface.", 1–9, Copyright (2019), with permission from Elsevier)

of thermal conductivity at higher porosities is also speculated to contribute to this increase in Leidenfrost temperature [74].

However, a monotonic increase in the pillar distance does not increase the Leidenfrost temperature monotonically (Fig. 7.6e). Kim et al. varied pillar distances from 5 μm to 2560 μm (height 20 μm and diameter 20 μm) and observed a maximum Leidenfrost temperature at a pillar distance of 320 μm [72]. At smaller pillar spacings, the increase in Leidenfrost temperature was attributed to the reduction in vapor pressure due to vapor leakage. At larger pillar spacings, the drop interacts with the base substrate and generates enough vapor pressure to sustain levitation, and, therefore, the Leidenfrost temperature starts decreasing after a pillar distance of 320 μm. At a pillar distance of 2560 μm, the Leidenfrost temperature approached that of the polished surface as the drop hardly interacted with multiple pillars and sees only the base substrate.

Using micro-machining and surface coatings, Kim et al. isolated the effect of surface wettability and surface roughness [45]. Surface wettability using Au, SiO_2 and nanoporous coatings on Si wafers was controlled across flat smooth surfaces and micro-pillared surfaces. At a pillar spacing of 500 μm, the pillars (height 15 μm and diameter 5 μm) act as protrusions and increase the frequency of intermitted contact of the liquid–vapor interface with the surface. As a result, all surface coatings with micro-pillars have a higher Leidenfrost temperature than those on flat surfaces, irrespective of the surface wettability.

Here we have not discussed the Leidenfrost temperature on surfaces with nanoscale roughness as in all cases we observed that the wettability of the surface changes significantly on the introduction of a nanoscale coating or roughness. In such cases, the wettability of the surface dominated the increase in Leidenfrost temperature [77]. The role of surface roughness in increasing intermittent contacts cannot be ruled out to some extent.

III. Surface wettability

As discussed in Sect. 7.1, the Leidenfrost regime sets in after intense heterogeneous bubble nucleation on the surface. Surface cavities are believed to act as activation sites where bubble nucleation starts [83]. Bubbles from such multiple nucleation sites grow, merge and detach from the surface depending on the surface superheat. After bubble departure, the liquid re-wets these nucleation sites and the process repeats. Therefore, bubble nucleation and the tendency of the liquid to re-wet the surface after bubble departure, i.e., surface wettability [84], is crucial in affecting the Leidenfrost temperature.

The surface wettability of an ideal smooth surface is governed by its surface energy that is an intrinsic material property. Experiments on extremely smooth surfaces have shown that the Leidenfrost point increases with increasing wettability. For example, experiments by Nagai et al. with water on sapphire glass (contact angle $\approx 90°$) and stainless steel (contact angle $\approx 60°$) showed a lower Leidenfrost temperature for sapphire due to a lower surface wettability [53]. Similar observations with water on polished aluminum (contact angle $\approx 78°$) and PTFE coating on aluminum (contact angle $\approx 122°$) were reported (Fig. 7.7a) [78].

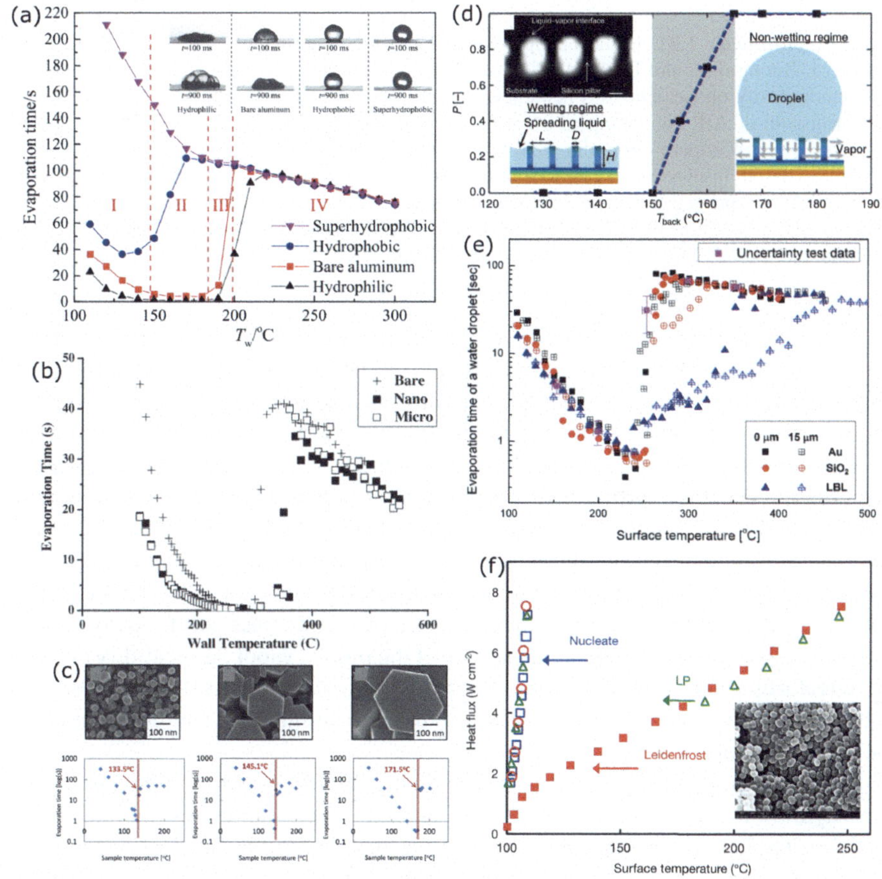

Fig. 7.7 Effect of surface wettability on Leidenfrost temperature: **a** Evaporation times of water drops on the surfaces with various wettabilities. Inset shows drop boiling on surfaces at a temperature T_s = 130 °C: hydrophilic (θ_e = 10°), bare aluminum (θ_e = 78°), hydrophobic (θ_e = 122°) and superhydrophobic (θ_e = 157°) [78] (Reprinted from *Applied Thermal Engineering*, 167, Q. Ma, X. Wu, T. Li and F. Chu., "Droplet Boiling on Heated Surfaces with Various Wettabilities", 114703, Copyright (2020), with permission from Elsevier); **b** Evaporation time on different oxidized zirconium-textured surfaces [79] (Reprinted from *Nuclear Engineering and Design*, 278. S.H. Kim, H.S. Ahn, J. Kim, M.H. Kim and H.S. Park, "Experimental Study of Water Droplets on Over-Heated Nano/Microstructured Zirconium Surfaces" 367–76, Copyright (2014), with permission from Elsevier); **c** Evaporation time of droplets on surfaces with ZnO rods of different sizes: 38.5 ± 10.4 nm, 133.1 ± 60.7 nm, 257.9 ± 100.4 nm [80] (Reprinted from *Materials Chemistry and Physics*, 217, T. Kano, T. Isobe, S. Matsushita and A. Nakajima, "Hydrophobicity and Leidenfrost Point of ZnO Nanorod Array Combined with Nanoscale Roughness on the Topmost Surface", 192–98, Copyright (2018), with permission from Elsevier); **d** Non-wetting probability of a drop on a micro-pillared superhydrophilic surface as a function of substrate temperature. The inset shows the contact between the pillars and the drop [81] (Reprinted by permission from Springer Nature: Springer, *Nature Communications*, "Non-Wetting Droplets on Hot Superhydrophilic Surfaces", S. Adera, R. Raj, R. Enright and E.N. Wang, 2013); **e** Drop evaporation times for different surface temperatures.

Fig. 7.7 (continued) LBL is the nanoporous SiO_2 surface [45] (Reprinted from [H. Kim, B. Truong, J. Buongiorno and L. W. Hu, "On the Effect of Surface Roughness Height, Wettability, and Nanoporosity on Leidenfrost Phenomena." *Applied Physics Letters* 98, 083121 (2011)] with the permission of AIP Publishing); **f** Heat flux on the surface of an immersion heater with four different coatings: superhydrophilic (SHL), hydrophilic (HL), hydrophobic (HB) and superhydrophobic (SHB). Inset shows the SEM image of the superhydrophobic nanoparticle coating, scale is 1 μm [82] (Reprinted by permission from Springer Nature: Nature, *Nature*, "Stabilization of Leidenfrost Vapour Layer by Textured Superhydrophobic Surfaces", I.U. Vakarelski, N. A. Patankar, J.O. Marston, D.Y.C. Chan and S.T. Thoroddsen, 2012)

As mentioned in Sect. 7.3 c II, on real surfaces, surface wettability depends on the surface roughness. Introducing micro-scale roughness via surface oxidation [79], micro-pillars [45], laser irradiation [76] can increase the wettability of a surface and therefore demonstrate a higher Leidenfrost temperature (Fig. 7.7b). Similarly, micro-scale roughness in the form of pillars and holes [85], coated nanorods [80] and ribs [75] can decrease surface wettability and show a lower Leidenfrost temperature (Fig. 7.7c). However, there is a unique exception to this observation. Pillared surfaces demonstrating superhydrophilicity at room temperature showed remarkable non-wetting properties at lower temperatures than on flat surfaces [81]. This effect is not due to the Leidenfrost phenomenon as there is a distinct solid–liquid contact (Fig. 7.7d). Due to the arrangement and size of the pillars, vapor permeability through these features is very low. Therefore, vapor leakage at these relatively low superheats is low, which increases the vapor pressure underneath the drop and does not allow the drop to collapse on the surface texture.

Nanoscale roughness can push surface wettability to its extremities, i.e., create superhydrophilic and superhydrophobic surfaces. The nanopores on the rough surface act as heterogeneous bubble nucleation sites. The bubble nucleation temperature for a nanopore cavity (scale r_p) can be estimated by equating the Young–Laplace pressure across the bubble interface ($\Delta p = 2\sigma_{lv}/r_p$) with the Clausius–Clapeyron equation, which characterizes the phase transition of a liquid ($\frac{dp}{dT} = \frac{\mathrm{L}}{T\Delta v}$, where Δv is the specific volume change) [83]:

$$T_{\text{nucleation}} = T_{\text{sat}} exp\left(\frac{2\sigma_{lv}\Delta v}{r_p \mathrm{L}}\right). \tag{7.8}$$

From Eq. (7.8), the nucleation temperature of bubbles for water on a 20 nm pore size is about 218 °C [45], which implies that bubbles nucleate easily on nanoporous surfaces compared to flat surfaces. However, material wettability determines whether this bubble nucleation is observed as nucleate boiling or as thin-film boiling. For example, with intrinsically hydrophilic materials, nanoscale roughness in the form of nanowires [86], nanotubes [87], nanoparticles [78] and nanostructures [79] makes surfaces superhydrophilic and, therefore, significantly increases the Leidenfrost temperatures (Fig. 7.7e). This significantly increased wettability is attributed to a higher wickability of the surface due to increased capillary suction pressures in the surface nanopores. Combining macro-scale features with nanostructures is also seen to increase the Leidenfrost temperature drastically. For example, in the experiments by Kim et al., micro-pillars coated with nanoparticles showed significantly higher Leidenfrost temperatures than flat surfaces coated with the same nanoparticles despite similar wettabilities [45]. Though increased intermitted contact, due to the micro-pillars, is credited to this observation, a relatively higher surface wickability could also be the reason behind the increased Leidenfrost temperatures. For example, hierarchical micro/nanoporous mesh-like surface textures demonstrate no Leidenfrost limit (up to approximately 600 °C) due to a combined effect of vapor leakage and high surface wickability [88, 89].

Similarly, nanostructures coated with intrinsically hydrophobic materials make a surface superhydrophobic. For example, salinized nanoparticles and nanocones on a surface increase the static contact angle of the surface to about 165° (advancing) and 160° (receding). These nanopores on the surface trap a layer of air, termed as a plastron [90]. Given the intrinsic low wettability of the coating material, water is unable to displace this layer of air, unless external pressure is applied [91, 92], or via diffusion over time [93, 94]. As a result, a drop rests in a Cassie-Baxter state on the surface with reduced contact with the surface texture. This reduced contact area also results in a reduced heat transfer rate. Therefore, on these superhydrophobic surfaces, the nucleate boiling and transition boiling regimes are hardly observed above the saturation temperature. A drop's evaporation rate steadily reduces with surface temperature (Fig. 7.7a, f) and the drop undergoes a smooth transition to the Leidenfrost state [82]. As the drop evaporation rate does not give an indication of the onset of the Leidenfrost state, interferometric measurements of a drop's base indicate that a complete vapor layer is formed at temperatures just above the saturation temperature (around 130 °C). The drop's mobility also increases significantly at this temperature, indicating a transition to the Leidenfrost state [95].

7.4 Applications

The presence of a stable vapor layer between the solid and the liquid provides lubricating properties between the two phases. As a result of the physical and chemical properties of this vapor layer, several applications of the Leidenfrost effect have been explored recently (Fig. 7.8).

a. Drop self-propulsion

For a drop in the Leidenfrost state over a surface, the surface acts as a zero contact-angle hysteresis surface. The vapor layer acts as a lubricant and provides extreme mobility to the levitating drop. As a result, the drop can be propelled using small forces. These forces can take the form of externally applied fields such as surface tilt-induced gravity, electric [107] and magnetic [17] fields, or propulsion can be achieved passively by introducing an asymmetry in the flow of the vapor on which the drop levitates (Fig. 7.8a1, a2, a3). First demonstrated by Linke et al. [29], Leidenfrost

Table 1 Summary of experimental observations on Leidenfrost temperature variation with surface properties

Citation	Substrate	Roughness	Apparent contact angle	Leidenfrost temperature	Key finding/Parameter studied
Gottfried et al. [96] Baumeister and Simon [60]	Polished stainless steel plate	–	≈87° (water)	≈250 °C	Leidenfrost temperature changes with different liquids and liquid composition
		–	≈26° (butanol)	≈218 °C	
		–	≈23° (ethanol)	≈180 °C	
		–	≈88° (5% NaCl solution)	≈250 °C	
Forrest et al. [97]	Al 6061, as received with solvent clean	Oxide thickness: 51 Å Hydrocarbon thickness: 54 Å	≈45°	–	Surface wetting decreases due to higher hydrocarbon length, therefore, CHF increases. Oxide layer thickness also changes but its isolated effect needs to be investigated
	Al 6061, chemical etch	Oxide thickness: 4.5 nm Hydrocarbon thickness: 0.48 nm	≈32°	–	
	Al 6061, mild contamination	Oxide thickness: 3.6 nm Hydrocarbon thickness: 9.4 nm	≈82°	–	

(continued)

Table 1 (continued)

Citation	Substrate	Roughness	Apparent contact angle	Leidenfrost temperature	Key finding/Parameter studied
	Al 6061, mild contamination followed by solvent clean	Oxide thickness: 3.7 nm Hydrocarbon thickness: 1.3 nm	≈45°	–	
	Zirconium	Oxide thickness: 5.7 nm Hydrocarbon thickness: 0.85 nm	≈95°	–	
	Chemical etch zirconium	Oxide thickness: 7.7 nm Hydrocarbon thickness: 17 nm	≈48°	–	
Chabicovsky et al. [98]	Austenitic stainless steel	–	–	≈590 °C	Increased Leidenfrost temperature with increased oxide layer thickness. Changes in surface wettability are not reported
	Oxidized stainless steel Thermal conductivity of oxide layer 0.2 W m^{-1} K^{-1}	Oxide thickness: 50 μm	–	≈840 °C	
Nagai and Nishio [53]	Sapphire Thermal conductivity ≈ 24 Wm^{-1} K^{-1}	R_a: 3 nm R_z: 24 nm	≈90° (water) ≈0° (acetone, ethanol, cyclohexane, R113)	≈200 °C (water) ≈140 °C (acetone) ≈170 °C (ethanol) ≈160 °C (cyclohexane) ≈120 °C (R113)	Leidenfrost temperature for acetone, cyclohexane, ethanol, R113 are same on sapphire and steel. This implies that for similar wetting surfaces, with similar thermal diffusivity, Leidenfrost temperature does not depend on surface roughness. For water sapphire has much lower Leidenfrost temperature because of lower wetting

(continued)

Table 1 (continued)

Citation	Substrate	Roughness	Apparent contact angle	Leidenfrost temperature	Key finding/Parameter studied
	Stainless steel Thermal conductivity ≈ 19 Wm^{-1} K^{-1}	R_a: 39 nm R_z: 386 nm	≈60° (water) ≈0° (acetone, ethanol, cyclohexane, R113)	≈250 °C (water) ≈150 °C (acetone) ≈170 °C (ethanol) ≈170 °C (cyclohexane) ≈120 °C (R113)	
Emmerson [70]	Stainless steel Thermal diffusivity ≈ 15 mm^2/s	–	–	≈282 °C	Thermal diffusivity of the surfaces does not affect the Leidenfrost temperature. Changes in surface wettability and roughness are not reported
	Monel Thermal diffusivity ≈ 21 mm^2/s	–	–	≈316 °C	
	Brass Thermal diffusivity ≈ 166 mm^2/s	–	–	≈284 °C	
Bernardin and Mudawar [15]	Aluminum polished	Average roughness: 97 nm	–	≈171 °C (water) ≈135 °C (acetone)	Increase in surface roughness increases the Leidenfrost temperature Changes in surface wettability due to roughness are not reported. The formation of oxide layer on copper can be a reason for relatively increased Leidenfrost temperature
	Aluminum particle blasted	Average roughness: 970 nm	–	≈250 °C (water) ≈155 °C (acetone)	
	Aluminum rough sanded	Average roughness: 2960 nm	–	≈263 °C (water) ≈160 °C (acetone)	

(continued)

Table 1 (continued)

Citation	Substrate	Roughness	Apparent contact angle	Leidenfrost temperature	Key finding/Parameter studied
	Silver polished	–	–	≈176 °C (water)	
	Nickel polished	–	–	≈173 °C (water)	
	Copper polished	–	–	≈198 °C (water)	
Baumeister et al. [59]	Pyrex glass	RMS: 76–101 nm	–	≈515 °C (water at 100 °C) >700 °C (water at 26 °C) ≈260 °C (ethanol at 100 °C) ≈360 °C (ethanol at 26 °C)	Effect of liquid sub-cooling: Leidenfrost temperature on low thermally conductive surfaces is higher for sub-cooled liquid Effect of material type: Leidenfrost temperature changes with material type. Changes in wettability not reported Effect of surface roughness: Leidenfrost temperature increases with surface roughness. Changes in wettability not reported
	Stainless steel	RMS: 76–101 nm	–	≈305 °C (water at 100 °C) ≈305 °C (water at 26 °C) ≈190 °C (ethanol at 100 °C) ≈190 °C (ethanol at 26 °C)	

(continued)

Table 1 (continued)

Citation	Substrate	Roughness	Apparent contact angle	Leidenfrost temperature	Key finding/Parameter studied
	Brass	RMS: 76–101 nm	–	≈235 °C (water at 100 °C) ≈235 °C (water at 26 °C)	
	Aluminum	RMS: 76–101 nm	–	≈235 °C (water at 100 °C) ≈235 °C (water at 26 °C) ≈155 °C (ethanol at 100 °C) ≈155 °C (ethanol at 26 °C)	
	Aluminum	RMS: 635 nm	–	≈265 °C (water at 100 °C)	
Lee et al. [99]	Porous zirconia. Porosity: 30% Thermal conductivity ≈ 1.25 Wm^{-1} K^{-1}	R_a=0.12 μm	≈0° (water) ≈0° (ethanol)	≈274 °C (ethanol) ≈550 °C (water)	Analytical model disproves the effect of vapor leakage Thermal properties and capillary wicking in the surface affect the Leidenfrost temperature
	Porous zirconia. Porosity: 20% Thermal conductivity ≈ 2 Wm^{-1} K^{-1}	R_a=0.21 μm	≈0° (water) ≈0° (ethanol)	≈267 °C (ethanol) ≈505 °C (water)	

(continued)

Table 1 (continued)

Citation	Substrate	Roughness	Apparent contact angle	Leidenfrost temperature	Key finding/Parameter studied
	Porous zirconia. Porosity: 8.3% Thermal conductivity ≈ 2.4 Wm^{-1} K^{-1}	R_a=0.18 μm	≈52° (water) ≈0° (ethanol)	≈255 °C (ethanol) ≈488 °C (water)	
	Porous zirconia. Porosity: 0.7% Thermal conductivity ≈ 2.9 Wm^{-1} K^{-1}	R_a=0.14 μm	≈84° (water) ≈0° (ethanol)	≈244 °C (ethanol) ≈422 °C (water)	
Auliano et al. [86]	Silicon	~ 4 nm	Static: 43°	≈225 °C	Increased Leidenfrost temperature with increased surface wettability
	Silicon nanowire	Diameter: 221 nm Height: 3.6 nm Pitch: 628 nm	Static: 4°	≈300 °C	
Kim et al. [87]	Bare zirconium	–	Static: 43°	≈300 °C	Increased Leidenfrost temperature with increased surface wettability
	Zirconium nanotube	Diameter: 20 nm Height: 2.52 nm Pitch: 628 nm	Static: 4.2°	≈370 °C	
Liu and Craig [100]	SiO_2-coated GaAs	RMS: 3.8 nm	≈13°–40°	≈280 °C	Increased Leidenfrost temperature with increased surface wettability
	Au	RMS: 5.3 nm	≈49°–77°	≈280 °C	
	Fluorinated SiO_2-coated GaAs	RMS: 4.7 nm	≈90°–121°	≈190 °C	

(continued)

Table 1 (continued)

Citation	Substrate	Roughness	Apparent contact angle	Leidenfrost temperature	Key finding/Parameter studied
Ma et al. [78]	Hydrophilic silica nanoparticles on aluminum	R_a< 30 nm	≈10°	≈220 °C	Increased Leidenfrost temperature with increased surface wettability
	Bare aluminum	R_a< 30 nm	≈78°	≈200 °C	
	PTFE coating on aluminum	R_a< 30 nm	≈122°	≈170 °C	
	Hydrophobic silica nanoparticles on aluminum	$R_a \approx 350$ nm	≈157°	Cannot determine	
Lee et al. [77]	Polished zirconium surface	R_a: 0.36 μm	<7°	≈314 °C	In general, increased surface wettability increases Leidenfrost temperature. On superhydrophilic surfaces wettability has stronger influence than surface roughness
	Zirconium surface with micro-roughness	R_a: 1.9 μm	<11°	≈342 °C	
	Zirconium surface with multiscale textures	R_a: 2.4 μm	≈0°	≈465 °C	
Kim et al. [79]	Bare Zirconium alloy	R_a: 0.15 μm	≈49°	≈300 °C	Increased Leidenfrost temperature with increased surface wettability. High Leidenfrost temperature on nanostructured surfaces attributed to enhanced wetting and vapor leakage
	Microstructured surface	R_a: 0.32 μm	≈5°	≈340 °C	
	Nanostructured surface	–	≈0°	≈370 °C	
Kim et al. [45]	Smooth Si with Au layer	R_a < 0.5 nm	≈ 83°	≈264 °C	Increased Leidenfrost temperature with increased surface wettability. Nanoporosity escalates the Leidenfrost temperature significantly due to low cavity nucleation temperature. Micro-pillars assist in increasing intermittent contact of liquid with the surface
	Smooth Si with SiO_2 layer	R_a < 0.5 nm	≈19°	≈274 °C	
	Smooth Si with nanoporous SiO_2 layer	Particle diameter: 23 nm	≈0°	≈359 °C	
	Pillared Si with Au layer	Pillar height: 15 μm, Pitch: 500 μm Pillar diameter: 5 μm	≈83°	≈290 °C	

(continued)

Table 1 (continued)

Citation	Substrate	Roughness	Apparent contact angle	Leidenfrost temperature	Key finding/Parameter studied
	Pillared Si with SiO_2 layer	Pillar height: 15 μm, Pitch: 500 μm Pillar diameter: 5 μm	≈19°	≈325 °C	
	Pillared Si with nanoporous SiO_2 layer	Pillar height: 15 μm, Pitch: 500 μm Pillar diameter: 5 μm	≈0°	≈453 °C	
Kruse et al. [76]	Polished steel surface	–	≈80°	≈280 °C	An increase in the Leidenfrost temperature is typically attributed to a reduction of the contact angle, an increase in the surface roughness, or an increase in the nanoporosity. Substantial capillary wicking due to intermittent contact with the surface is the main reason
	Steel irradiated with laser irradiated to create mounds: below surface growth mounds	R_{rms}: 4.4 μm Average height: 15 μm Structure distance: 11.3 μm	≈12°	≈316 °C	
	Steel irradiated with laser irradiated to create mounds: below surface growth mounds	R_{rms}: 5.7 μm Average height: 20 μm Structure distance: 11.7 μm	≈5°	≈340 °C	

(continued)

Table 1 (continued)

Citation	Substrate	Roughness	Apparent contact angle	Leidenfrost temperature	Key finding/Parameter studied
	Steel irradiated with laser irradiated to create mounds: above surface growth mounds	R_{rms}: 4.5 μm Average height: 15 μm Structure distance: 14.1 μm	≈0°	≈360 °C	
	Steel irradiated with laser irradiated to create mounds: above surface growth mounds	R_{rms}: 6.0 μm Average height: 14 μm Structure distance: 21 μm	≈0°	≈405 °C	
	Steel irradiated with laser irradiated to create mounds: nanostructure covered pyramids	R_{rms}: 5.4 μm Average height: 154 μm Structure distance: 24.5 μm	≈15°	≈455 °C	
Adera et al. [81]	Silicon	–	≈38–42°	≈220 °C	Drops are in a Cassie-Baxter state and in contact with the surface pillars. Minimum surface interaction and sufficient pillar spacing to sustain vapor pressure to support drop weight
	Pillared Silicon substrates	Pillar diameter: 3.1–9.1 μm Pillar height: 15–27.5 μm Pillar spacing: 10–30 μm	≈0°	≈170 °C	

(continued)

Table 1 (continued)

Citation	Substrate	Roughness	Apparent contact angle	Leidenfrost temperature	Key finding/Parameter studied
Kwon et al. [71]	Smooth Silicon	–	≈0°	≈300 °C	Leidenfrost temperature increases with increasing pillar spacing. Vapor leakage at large pilar spacings is attributed to the increased Leidenfrost temperatures. Increased wickability with nanostructures increases Leidenfrost temperature
	Pillared silicon surface	Post width: 10 μm Post height: 10 μm Pillar spacing: 3.3–100 μm	≈0°	≈300–370 °C	
	Pillared silicon surface with nanocoatings		≈0°	≈400 °C	
Del Cerro et al. [85]	Stainless steel with pillars Pillar spacing 17 μm	Height: 10–26 μm	≈115°	≈130–150 °C	Sufficient vapor pressure generated to support drop weight and minimal contact with surface increases drop lifetime
	Stainless steel with holes Holes spacing 17 μm	Depth: 3–9 μm	≈115°	≈160–180 °C	
	Plain stainless steel	–	–	≈240–260 °C	
Duursma et al. [61]	Silicon with different square pillar	Pillar spacing: 10–100 μm Height 10 μm	6–24°	≈280–330 °C (8 μl) ≈290–375 °C (15 μl) ≈325–400 °C (24 μl)	Leidenfrost temperature depends on drop volume and increases with pillar spacing. Decrease in heat transfer area to the drop is the main mechanism
Hays et al. [75]	Teflon-coated silicon substrates with a micro rib pattern. Rib widths range from 2 to 20 μm, heights range from 15 to 20 μm, and spacing between the ribs ranges from 20 to 38 μm	Cavity fraction: 0	≈109°–125°	–	Qualitative observations show that an increased cavity fraction reduces Leidenfrost temperature
		Cavity fraction: 0.5	≈127°–140° (longitudinal) ≈121°–168° (transverse)	–	
		Cavity fraction: 0.8	≈140°–150° (longitudinal) ≈133°–168° (transverse)	–	

(continued)

Table 1 (continued)

Citation	Substrate	Roughness	Apparent contact angle	Leidenfrost temperature	Key finding/Parameter studied
		Cavity fraction: 0.95	≈146°–158° (longitudinal) ≈144°–168° (transverse)	–	
Park and Kim [101]	Silicon Pillar diameter: 9 μm Pillar height: 22 μm	Smooth	–	≈275 °C (water) ≈150 °C (ethanol)	For water drops, Leidenfrost temperature was higher on P60 and P120 surfaces compared with the smooth surface, while it was lower on P15 and P30 surfaces For ethanol, Leidenfrost temperature increased with increasing pillar distance. Changes in surface wettability not considered
		Pillar pitch: 15 μm	–	≈250 °C (water) ≈165 °C (ethanol)	
		Pillar pitch: 30 μm	–	≈260 °C (water) ≈155 °C (ethanol)	
		Pillar pitch: 60 μm	–	≈350 °C (water) ≈175 °C (ethanol)	
		Pillar pitch: 120 μm	–	≈475 °C (water) ≈180 °C (ethanol)	

(continued)

Table 1 (continued)

Citation	Substrate	Roughness	Apparent contact angle	Leidenfrost temperature	Key finding/Parameter studied
Kim et al. [72]	Photoresist. All surfaces have height: 20 μm and diameter: 20 μm	Polished	–	≈235 °C	Increased pillar spacing first increases Leidenfrost temperature and then decreases. Reduction of supporting vapor pressure at 320 μm causes highest Leidenfrost temperature. At higher pillar spacing the drop can also interact with the base of the substrate Changes in surface wettability due to pillar spacing are not considered
		Pillar distance: 5 μm	–	≈255 °C	
		Pillar distance: 20 μm	–	≈255 °C	
		Pillar distance: 40 μm	–	≈275 °C	
		Pillar distance: 80 μm	–	≈275 °C	
		Pillar distance: 160 μm	–	≈285 °C	
		Pillar distance: 320 μm	–	≈295 °C	
		Pillar distance: 640 μm	–	≈265 °C	
		Pillar distance: 2560 μm	–	≈255 °C	

(continued)

Table 1 (continued)

Citation	Substrate	Roughness	Apparent contact angle	Leidenfrost temperature	Key finding/Parameter studied
Geraldi et al. [73]	Stainless steel mesh gauge #50	Wire diameter (d): 0.193 mm Wire separation (s): 0.317 mm Solid fraction ($d/(s+d)$): 0.38	≈0°	≈315 °C	Leakage of vapor through pores increases Leidenfrost temperature
	Stainless steel mesh gauge #60	Wire diameter (d): 0.145 mm Wire separation (s): 0.27 mm Solid fraction ($d/(s+d)$): 0.35	≈0°	≈305 °C	
	Stainless steel mesh gauge #98	Wire diameter (d): 0.08 mm Wire separation (s): 0.19 mm Solid fraction ($d/(s+d)$): 0.29	≈0°	≈295 °C	

(continued)

Table 1 (continued)

Citation	Substrate	Roughness	Apparent contact angle	Leidenfrost temperature	Key finding/Parameter studied
	Stainless steel mesh gauge #150	Wire diameter (d): 0.07 mm Wire separation (s): 0.12 mm Solid fraction ($d/(s+d)$): 0.36	≈0°	≈ 265 °C	
	Stainless steel mesh gauge #250	Wire diameter (d): 0.04 mm Wire separation (s): 0.07 mm Solid fraction ($d/(s+d)$): 0.38	≈0°	≈265 °C	
Kano et al. [80]	ZnO nanorods on Si coated with hydrophobic coating, additional structures on nanorods All hydrophobic 110–150	Rod diameter: 38.5 μm	≈90°	≈180 °C	Increased Leidenfrost temperature with increased surface wettability
		Rod diameter: 57.5 μm	≈100°	≈170 °C	
		Rod diameter: 133.1 μm	≈95°	≈160 °C	
	Plain Si substrate with hydrophobic coating	–	≈120°	≈150 °C	

(continued)

Table 1 (continued)

Citation	Substrate	Roughness	Apparent contact angle	Leidenfrost temperature	Key finding/Parameter studied
Avedisian and Koplik [74]	Stainless steel	–	≈0° (methanol)	≈170 °C	Increased Leidenfrost temperature with increasing surface porosity. Attributed to vapor leakage. Thermal conductivity is also speculated but not proved
	Ceramic alumina: porosity 10% 20, 40%	–	≈0° (methanol)	≈297 °C	
	Ceramic alumina: porosity 20%	–	≈0° (methanol)	≈372 °C	
	Ceramic alumina: porosity 40%	–	≈0° (methanol)	>427 °C	
Bourrianne et al. [95]	Silicon	–	≈16°–42°	≈210 °C	Surface geometry for superhydrophobic surfaces does not affect transition to Leidenfrost regime
	Silicon and silicon with nanocones coated with nanoparticles (Glaco)	–	≈160°–165°	Cannot determine	
Vakarelski et al. [82]	Stainless steel sphere cleaned with organic reagents	$R_a < 60$ nm	<30 °C	≈420 °C	Increased Leidenfrost temperature with increased surface wettability. Smooth transition to Leidenfrost regime on superhydrophobic surfaces
	Stainless steel sphere salinized with trichloro(1H,1H,2H,2H-perfluorooctyl)silane	$R_a < 60$ nm	≈100 °C	≈170 °C	
	Stainless steel sphere treated with commercial superhydrophobic coating	Typical spacing of structure: 0.1–2 μm, depth of 50–300 nm	>160 °C	Cannot determine	
	Stainless steel sphere was plasma cleaned	$R_a < 60$ nm	<10 °C	>600 °C	

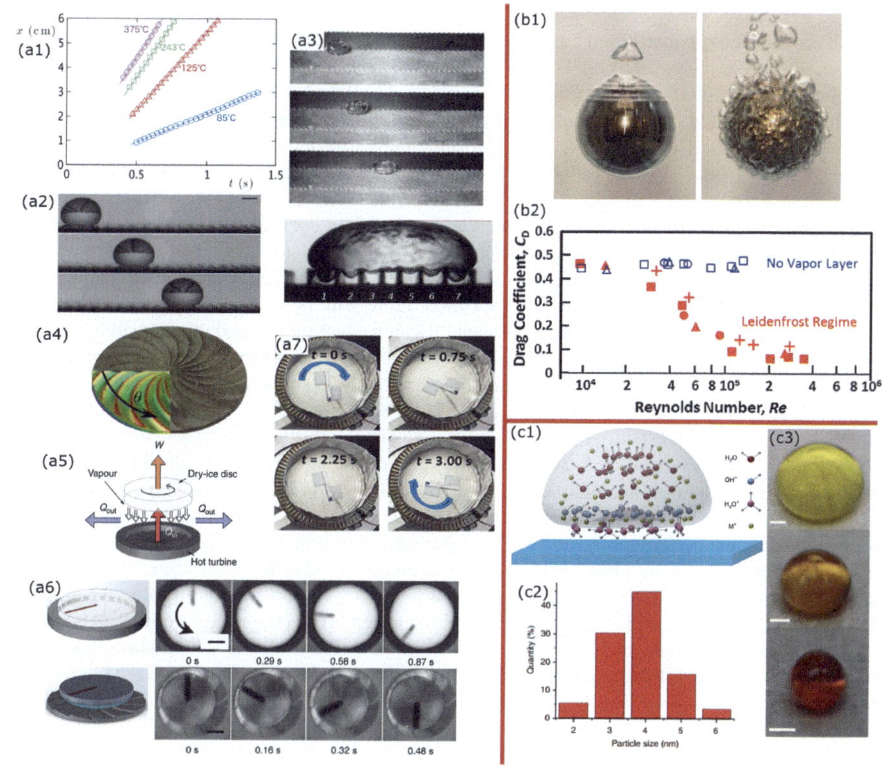

Fig. 7.8 Applications of Leidenfrost effect: **a1** Position (x) of water drops self-propelling on a hot superhydrophobic ratchet vs time (t) at different temperatures [102] (Reprinted from [G. Dupeux, P. Bourrianne, Q. Magdelaine, C. Clanet and D. Quéré, "Propulsion on a Superhydrophobic Ratchet." *Scientific Reports*, 4, 5280 (2014)], licensed under CC-BYNC-ND 4.0); **a2** Side view of a drop of volume 60 mL propelling on a ratchet at 85°C. The bar represents 2 mm and successive photos are 0.28 s apart [102] (Reprinted from [G. Dupeux, P. Bourrianne, Q. Magdelaine, C. Clanet and D. Quéré, "Propulsion on a Superhydrophobic Ratchet." *Scientific Reports*, 4, 5280 (2014)], licensed under CC-BYNC-ND 4.0); **a3** Self-propulsion of an acetone drop (volume 200 μL at $T_s = 400$ °C. Image also shows the drop on the crenelated cross-section of the herringbone structure [103] (Reprinted figure with permission from [D. Soto, G. Lagubeau, C. Clanet, and D. Quere, "Surfing on a Herringbone." *Physical Review Fluids*, 1, 013902 (2016)] Copyright 2016 by the American Physical Society); **a4** CNC machined turbine-inspired substrate (Reprinted from [G.G. Wells, R. Ledesma-Aguilar, G. McHale and K. Sefiane. 2015. "A Sublimation Heat Engine" *Nature Communications*, 6, 9390 (2015)], licensed under CC-BY 4.0); **a5** Concept of a Leidenfrost engine operating between the substrate temperature T_s and ambient temperature [34] (Reprinted from [G.G. Wells, R. Ledesma-Aguilar, G. McHale and K. Sefiane. 2015. "A Sublimation Heat Engine" *Nature Communications*, 6, 9390 (2015)], licensed under CC-BY 4.0); **a6** Time sequence of a disc of dry-ice with radius 2 cm (top panel) and a drop of water supporting a metal plate (bottom panel) rotating on the turbine-inspired surface at $T_s = 500$ °C. The panels show a sequence over time, showing the rotation of the disc [34] (Reprinted from [G.G. Wells, R. Ledesma-Aguilar, G. McHale and K. Sefiane. 2015. "A Sublimation Heat Engine" *Nature Communications*, 6, 9390 (2015)], licensed under CC-BY 4.0); **a7** Snapshot of a rotating

Fig. 7.8 (continued) paper-based Leidenfrost rotor [104] (Reprinted from [H. Xu, A. Thissandier, R. Zhao, P. Tao, C. Song, J. Wu, W. Shang, and T. Deng, "Self-Propelled Rotation of Paper-Based Leidenfrost Rotor." *Applied Physics Letters* 114, 113703 (2019)] with the permission of AIP Publishing); **b1** Heated sphere (15 mm) in a fluorinated liquid at temperature above the Leidenfrost temperature (left panel) and below the Leidenfrost temperature (right panel) [105] (Reprinted figure with permission from [I.U. Vakarelski, J.O. Marston, D.Y.C. Chan and S.T. Thoroddsen, "Drag Reduction by Leidenfrost Vapor Layers." *Physical Review Letters*, 106, 214501 (2011)] Copyright 2011 by the American Physical Society); **b2** Variation of drag coefficient with Reynolds number for different spheres: Open (blue) data points are at 25 °C for steel (square), tungsten carbide (triangles) and agate (circles). Solid (red) points are spheres heated to 200 °C. Heated ascending steel spheres are denoted by red crosses [105] (Reprinted figure with permission from [I.U. Vakarelski, J.O. Marston, D.Y.C. Chan and S.T. Thoroddsen, "Drag Reduction by Leidenfrost Vapor Layers." *Physical Review Letters*, 106, 214501 (2011)] Copyright 2011 by the American Physical Society); **c1** Representation of self-ionization of water in a Leidenfrost state [106] (Reprinted from [R. Abdelaziz, D. Disci-Zayed, M.K. Hedayati, J.H. Pöhls, A.U. Zillohu, B. Erkartal, V.S.K. Chakravadhanula, V. Duppel, L. Kienle and M. Elbahri, "Green Chemistry and Nanofabrication in a Levitated Leidenfrost Drop." *Nature Communications*, 4, 2400 (2013)], licensed under CC-BY-NC-ND 3.0); **c2** Particle size distribution of the gold nanoparticles [106] (Reprinted from [R. Abdelaziz, D. Disci-Zayed, M.K. Hedayati, J.H. Pöhls, A.U. Zillohu, B. Erkartal, V.S.K. Chakravadhanula, V. Duppel, L. Kienle and M. Elbahri, "Green Chemistry and Nanofabrication in a Levitated Leidenfrost Drop." *Nature Communications*, 4, 2400 (2013)], licensed under CC-BY-NC-ND 3.0); **c3** Formation of gold nanoparticles without any reducing agent in a Leidenfrost drop from gold salts. As the drop evaporates the color of particles changes from yellow to red [106] (Reprinted from [R. Abdelaziz, D. Disci-Zayed, M.K. Hedayati, J.H. Pöhls, A.U. Zillohu, B. Erkartal, V.S.K. Chakravadhanula, V. Duppel, L. Kienle and M. Elbahri, "Green Chemistry and Nanofabrication in a Levitated Leidenfrost Drop." *Nature Communications*, 4, 2400 (2013)], licensed under CC-BY-NC-ND 3.0)

drops undergo self-propulsion when placed on ratchet-like heated textured surfaces. Several mechanisms have been proposed to explain this self-propulsion, such as jet thrust, Marangoni flows [108], thermal creep-induced drag force [109] and drag due to vapor rectification [8, 29]. However, vapor rectification-induced drag force has been numerically [8, 25, 110, 111], analytically [34, 42, 103, 112] and experimentally [25, 34, 42, 103] proven to be the main mechanism for self-propulsion of levitating liquids and solids. As discussed in Sect. 7.2, on a flat surface the vapor flow is radially symmetric. The surface asymmetry rectifies the flow of the exiting vapor in a preferential direction. This redirected vapor produces a viscous drag on the levitating liquid (or solid) and propels it in a specific direction, achieving terminal speeds of the order of 100 mms^{-1} [113–115]. Apart from ratchet-textured surfaces [110, 116–122], self-propulsion is obtained on non-parallel structures [123], nano and macro-textured substrates [124–127], herringbone-like [103, 128] and herringbone-ratchet-like [113] textured surfaces. Additionally, superhydrophobic coatings can be used to lower the temperature at which self-propulsion is observed [35, 102]. In this 'cold' Leidenfrost regime, although the drop is not levitating, it encounters a very low pinning force from the surface. As such, the vapor flow at temperatures even below the saturation temperature is sufficient to propel the drop (Fig. 7.8a1).

An asymmetry in the flow of vapor can also be introduced by an asymmetric mass distribution of the levitating object. For example, asymmetrically placed blocks of dry-ice self-propel when placed on a heated surface [25]. A similar self-propulsion is observed in millimetric-sized drops on flat surfaces. Due to the nearly radially symmetric vapor flow, large drops have symmetric counter-rotating internal flows. However, as the drop size reduces, this flow symmetry is disrupted, and a single internal rotating flow remains. This rolling drop entrains the surrounding air asymmetrically, which tilts its base and leads to self-propulsion in a precise but randomly determined direction [42].

The above-mentioned concepts of linear self-propulsion can also be extended to rotate levitating liquids and solids. For example, turbine-inspired textured substrates (Fig. 7.8a4) can rotate liquids and solids and transfer this torque to surface tension coupled non-volatile solids (Fig. 7.8a5, a6) [34, 35]. Similarly, asymmetric mass distribution can be used to rotate levitating dry-ice blocks [25] and water-fed paper rotor (Fig. 7.8a7) [104].

The concept of self-propulsion can find potential applications in drop and material transport in microfluidics systems. Recently, this self-propulsion mechanism has shown potential utility in thermochemical biomass reactors to transport cellulose-based particles passively at 750 °C for pyrolysis, without the use of active components like injectors and conveyors [120]. Recent work on selective heating of substrates with micro-patterned heating elements [129] has opened opportunities for precise control of drop motion without the need for macro-textured features [130]. The technique provides a precise electrical control of substrate heating while using thermally induced vapor rectification for drop propulsion. Additionally, these self-propelling drops function like a heat engine, converting temperature differences into mechanical motion (Fig. 7.8a5). The low-friction characteristic of this self-propulsion can lead to the development of micro-scale engines, where material wear due to solid friction is a significant problem. At macro-scales such 'Leidenfrost engines' can also be used in extreme gravity, temperature and pressure environments, such as mining, space and planetary exploration, due to their reduced structural complexity [34, 35]. As self-propulsion of both solids and liquids can be achieved, such engines can use ices of H_2O, CH_4 and CO_2 on other planetary bodies for thermal energy harvesting for early human explorers.

b. Drag reduction

Hydrodynamic drag due to liquid flow over a surface comprises of two components: wake drag due to normal stresses on the body and skin friction due to shear stresses because of fluid viscosity. For example, a solid sphere moving in a liquid at $Re < 0.5$ has a drag coefficient given by $C_D = 64/Re$, due to a no-slip boundary condition with the liquid. The drag coefficient on a superhydrophobic sphere reduces as the continuous liquid–solid interface is replaced partially by a liquid–vapor interface (due to the plastron), which introduces a slip boundary to the flow [131–136]. As an extreme case, if the liquid–solid interface is replaced completely by a liquid–vapor interface then the drag coefficient reduces by 33%, showing a significant reduction in the skin friction on the sphere [131]. Such a continuous vapor layer can be sustained by heating the sphere above the Leidenfrost temperature (Fig. 7.8b1). These Leidenfrost vapors layers can reduce the drag by almost an order of magnitude, compared to cold spheres [105, 137, 138]. Even at sub-Leidenfrost temperatures, the collapse of the plastron can be limited by supplying a low superheat ($T_s - T_{sat} \leq 25$ °C) and an 80–90% reduction in drag can be obtained compared to an unheated superhydrophobic surface for turbulent flows $26{,}100 \leq Re \leq 52{,}000$ (Fig. 7.8b2) [139, 140]. Here, superhydrophobic coatings on such spheres increase the stability of the vapor film between the liquid and the sphere surface, either as a fully formed Leidenfrost vapor film or as a plastron [82]. This allows better control and study of heat transfer properties as the surfaces are devoid of different boiling transitions and associated heat fluxes.

iii. Drop reactor

The above applications of Leidenfrost drops exploited the physical attributes of the vapor film. Another recent application scope utilizes chemical attributes of the Leidenfrost drops. Experiments by Gilbert and Shaw demonstrated that water undergoes self-ionization, into hydroxyl (H_3O^+) and hydronium ions (OH^-), and charge separation due to a fast rupture of the liquid–gas interface [141]. This process is enhanced by the presence of salts or extreme pressure and temperature [142, 143]. Similar charge separation is observed at the liquid–vapor interface of a Leidenfrost drop [106]. The hydronium ions are carried away by the vapor while the drop is left with the hydroxyl ions (Fig. 7.8c1). Thus, a Leidenfrost drop acts as a chemical reactor, wherein the excess hydroxyl ions act as reducing agents. Consequently, metal cations in corresponding salt solutions can be reduced to metal atoms without the addition of any reducing agents. The atoms can then coalesce to form clusters and nanoparticles. For example, palladium nanoparticles can be synthesized using a Pd precursor aqueous solution, charcoal supports and a hot plate [144]. Similarly, plasmonic gold nanoparticles (Fig. 7.8c2, c3), zinc-oxide (ZnO) nanoparticles, copper-oxide (CuO) nanorods and platinum nanoparticles can be synthesized using their corresponding salt solutions. Moreover, by using additional reducing agents like citric acid or sodium hydroxide solutions, nanoporous superhydrophilic metals can be fabricated [106, 145].

These Leidenfrost reactors can also be used to create nanoscale coatings [106]. For example, high-performance indium oxide (In_2O_3) and zinc-oxide (ZnO) thin-film transistors were prepared as an alternative to traditional spray pyrolysis techniques [146]. The process is akin to a chemical vapor deposition procedure where the deposition occurs out of the vapor film and the crystal formation occurs on the substrate [147]. Additionally, Leidenfrost drops can initiate colloidal assembly during drop evaporation, where a Leidenfrost drop acts as a confining geometry. Silica particles, along with carbon black nanoparticles when inserted in a Leidenfrost drop, form microgranules after the drop dries out. These microgranules comprise packed silica particles and demonstrate photonic properties in the full visible range of light, with potential applications in developing color pigments [148].

7.5 Summary

The onset of the Leidenfrost state of a liquid droplet on a superheated surface is directly linked to the surface wettability. In principle, surface wettability depends on the physical and chemical nature of the surface, i.e. surface roughness (physical nature) and intrinsic surface wettability (chemical nature of an extremely smooth surface). Isolating the effect of this physical and chemical nature of the surface is extremely challenging, but critical to understand and control the onset of the Leidenfrost regime for applications in metal forming, heat exchangers, microfluidics and drag reduction. The consensus is that an increase in surface wettability increases the Leidenfrost temperature. Here, multi-scale superhydrophilic surfaces provide the highest Leidenfrost temperatures as a combined result of frequent intermittent contacts with the surface and high capillary wicking. On the other end of the spectrum, superhydrophobic surfaces decrease Leidenfrost temperatures, due to low surface adhesion.

Acknowledgement The authors would like to thank funding from EPSRC grant EP/P005896/1.

References

1. R. N. Wenzel, “Resistance of solid surfaces to wetting by water,” Ind. Eng. Chem., vol. 28, no. 8, pp. 988–994, 1936.
2. A. B. D. Cassie and S. Baxter, “Wettability of porous surfaces,” Trans. Faraday Soc., vol. 40, no. 5, pp. 546–551, 1944.
3. M. A. Goldshtik, V. M. Khanin, and V. G. Ligai, “A liquid drop on an air cushion as an analogue of leidenfrost boiling,” J. Fluid Mech., vol. 166, pp. 1–20, 1986.
4. W. Bouwhuis, K. G. Winkels, I. R. Peters, P. Brunet, D. Van Der Meer, and J. H. Snoeijer, “Oscillating and star-shaped drops levitated by an airflow,” *Phys. Rev. E - Stat. Nonlinear, Soft Matter Phys.*, vol. 88, no. 023017, 2013.
5. H. De Maleprade *et al.*, “Air-propelled, herringbone-textured platelets,” *Phys. Rev. Fluids*, vol. 3, no. 104101, 2018.
6. D. Soto, H. De Maleprade, C. Clanet, and D. Quéré, “Air-levitated platelets: From take off to motion,” J. Fluid Mech., vol. 814, pp. 535–546, 2017.
7. J. G. Leidenfrost, “On the fixation of water in diverse fire,” Int. J. Heat Mass Transf., vol. 9, no. 11, pp. 1153–1166, 1966.
8. T. Baier, G. Dupeux, S. Herbert, S. Hardt, and D. Quéré, “Propulsion mechanisms for Leidenfrost solids on ratchets,” *Phys. Rev. E - Stat. Nonlinear, Soft Matter Phys.*, vol. 87, no. 021001 (R), 2013.
9. Y. Iida and T. Takashima, “A study of liquid-liquid Leidenfrost film boiling,” JSME Int. J., vol. 31, no. 4, pp. 727–733, 1988.
10. M. Shi, X. Ji, S. Feng, Q. Yang, T. J. Lu, and F. Xu, “Self-Propelled Hovercraft Based on Cold Leidenfrost Phenomenon,” *Sci. Rep.*, vol. 6, no. 28574, 2016.
11. M. Adda-Bedia, S. Kumar, F. Lechenault, S. Moulinet, M. Schillaci, and D. Vella, “Inverse Leidenfrost Effect: Levitating Drops on Liquid Nitrogen,” Langmuir, vol. 32, no. 17, pp. 4179–4188, 2016.

12. A. Gauthier, C. Diddens, R. Proville, D. Lohse, and D. van der Meer, "Self-propulsion of inverse Leidenfrost drops on a cryogenic bath," Proc. Natl. Acad. Sci. U. S. A., vol. 116, no. 4, pp. 1174–1179, 2019.
13. H. Sugioka and S. Segawa, "Controllable Leidenfrost glider on a shallow water layer," *AIP Adv.*, vol. 8, no. 115209, 2018.
14. B. S. Gottfried and K. J. Bell, "Film boiling of spheroidal droplets: Leidenfrost Phenomenon," *Ind. Eng. Chem. Fundam.*, vol. 5, no. 4, pp. 561–568, 1966.
15. J. D. Bernardin and I. Mudawar, "The Leidenfrost Point: Experimental Study and Assessment of Existing Models," *J. Heat Transfer*, vol. 121, no. 4, pp. 894–903, 1999.
16. M. Le Merrer, C. Clanet, D. Queŕeá, E. Raphaël, and F. Chevy, "Wave drag on floating bodies," Proc. Natl. Acad. Sci. U. S. A., vol. 108, no. 37, pp. 15064–15068, 2011.
17. K. Piroird, C. Clanet, and D. Quéré, "Magnetic control of Leidenfrost drops," *Phys. Rev. E - Stat. Nonlinear, Soft Matter Phys.*, vol. 85, no. 5, pp. 10–13, 2012.
18. A. Gauthier, D. van der Meer, J. H. Snoeijer, and G. Lajoinie, "Capillary orbits," *Nat. Commun.*, vol. 10, no. 3947, 2019.
19. G. Dupeux, T. Baier, V. Bacot, S. Hardt, C. Clanet, and D. Quéré, "Self-propelling uneven Leidenfrost solids," *Phys. Fluids*, vol. 25, no. 051704, 2013.
20. S. R. Waitukaitis, A. Zuiderwijk, A. Souslov, C. Coulais, and M. Van Hecke, "Coupling the Leidenfrost effect and elastic deformations to power sustained bouncing," Nat. Phys., vol. 13, no. 11, pp. 1095–1099, 2017.
21. A. Milionis, C. Antonini, S. Jung, A. Nelson, T. M. Schutzius, and D. Poulikakos, "Contactless Transport and Mixing of Liquids on Self-Sustained Sublimating Coatings," *Langmuir*, vol. 33, no. 8, pp. 1799–1809, 2017.
22. E. Mogilevskiy, "Levitation of a nonboiling droplet over hot liquid bath," *Phys. Fluids*, vol. 32, no. 012114, 2020.
23. M. A. Alchalabi, N. Kouraytem, E. Q. Li, and S. T. Thoroddsen, "Vortex-induced vapor explosion during drop impact on a superheated pool," *Exp. Therm. Fluid Sci.*, vol. 87, pp. 60–68, 2017.
24. Y. Ding and J. Liu, "Dynamic interactions of Leidenfrost droplets on liquid metal surface," *Appl. Phys. Lett.*, vol. 109, no. 121904, 2016.
25. S. D. Janssens, S. Koizumi, and E. Fried, "Behavior of self-propelled acetone droplets in a Leidenfrost state on liquid substrates," *Phys. Fluids*, vol. 29, no. 032103, 2017.
26. Y. S. Song *et al.*, "Vitrification and levitation of a liquid droplet on liquid nitrogen," *Proc. Natl. Acad. Sci. U. S. A.*, vol. 107, no. 10, pp. 4596–4600, 2010.
27. A. L. Biance, C. Clanet, and D. Quéré, "Leidenfrost drops," *Phys. Fluids*, vol. 15, no. 6, pp. 1632–1637, 2003.
28. G. Paul, I. Manna, P. Kumar Das, and P. K. Das, "Formation, growth, and eruption cycle of vapor domes beneath a liquid puddle during Leidenfrost phenomena," *Appl. Phys. Lett.*, vol. 103, no. 084101, 2013.
29. H. Linke *et al.*, "Self-propelled leidenfrost droplets," *Phys. Rev. Lett.*, vol. 96, no. 154502, 2006.
30. S. Waitukaitis, K. Harth, and M. van Hecke, "From Bouncing to Floating: The Leidenfrost Effect with Hydrogel Spheres," *Phys. Rev. Lett.*, vol. 121, no. 048001, 2018.
31. L. Qiao, Z. Zeng, H. Xie, H. Liu, and L. Zhang, "Modeling Leidenfrost drops over heated liquid substrates," *Int. J. Heat Mass Transf.*, vol. 128, pp. 1296–1306, 2019.
32. P. Aussillous and D. Quéré, "Properties of liquid marbles," *Proc. R. Soc. A Math. Phys. Eng. Sci.*, vol. 462, no. 2067, pp. 973–999, 2006.
33. D. Quéré, "Leidenfrost Dynamics," *Annu. Rev. Fluid Mech.*, vol. 45, pp. 197–215, 2013.
34. G. G. Wells, R. Ledesma-Aguilar, G. McHale, and K. Sefiane, "A sublimation heat engine," *Nat. Commun.*, vol. 6, no. 6390, 2015.
35. P. Agrawal *et al.*, "Leidenfrost heat engine: Sustained rotation of levitating rotors on turbine-inspired substrates," *Appl. Energy*, vol. 240, pp. 399–408, 2019.
36. P. R. Jones *et al.*, "High-speed X-ray imaging of the Leidenfrost collapse," *Sci. Rep.*, vol. 9, no. 1598, 2019.

37. G. C. Lee *et al.*, "Measurement of the vapor layer under a dynamic Leidenfrost drop," *Int. J. Heat Mass Transf.*, vol. 124, pp. 1163–1171, 2018.
38. J. C. Burton, A. L. Sharpe, R. C. A. Van Der Veen, A. Franco, and S. R. Nagel, "Geometry of the vapor layer under a Leidenfrost drop," *Phys. Rev. Lett.*, vol. 109, no. 074301, 2012.
39. T. Roques-Carmes, A. Domps, P. Marchal, and L. Marchal-Heussler, "Equivalent capacitive thickness of the vapor layer below Leidenfrost drops," *Exp. Fluids*, vol. 59, no. 115, 2018.
40. B. Sobac, A. Rednikov, S. Dorbolo, and P. Colinet, "Leidenfrost effect: Accurate drop shape modeling and refined scaling laws," *Phys. Rev. E - Stat. Nonlinear, Soft Matter Phys.*, vol. 90, no. 053011, 2014.
41. C. Cai, I. Mudawar, H. Liu, and C. Si, "Theoretical Leidenfrost point (LFP) model for sessile droplet," *Int. J. Heat Mass Transf.*, vol. 146, p. 118802, 2020.
42. A. Bouillant, T. Mouterde, P. Bourrianne, A. Lagarde, C. Clanet, and D. Quéré, "Leidenfrost wheels," *Nat. Phys.*, vol. 14, pp. 1188–1192, 2018.
43. M. A. J. Van Limbeek et al., "Leidenfrost drops cooling surfaces: Theory and interferometric measurement," J. Fluid Mech., vol. 827, pp. 614–639, 2017.
44. T. Tran, H. J. J. Staat, A. Prosperetti, C. Sun, and D. Lohse, "Drop impact on superheated surfaces," *Phys. Rev. Lett.*, vol. 108, no. 036101, 2012.
45. H. Kim, B. Truong, J. Buongiorno, and L. W. Hu, "On the effect of surface roughness height, wettability, and nanoporosity on leidenfrost phenomena," *Appl. Phys. Lett.*, vol. 98, no. 083121, 2011.
46. M. A. J. Van Limbeek, M. Shirota, P. Sleutel, C. Sun, A. Prosperetti, and D. Lohse, "Vapour cooling of poorly conducting hot substrates increases the dynamic Leidenfrost temperature," Int. J. Heat Mass Transf., vol. 97, pp. 101–109, 2016.
47. T. Tran *et al.*, "Droplet impact on superheated micro-structured surfaces," *Soft Matter*, vol. 9, no. 12, pp. 3272–3282, 2013.
48. T. Y. Xiong and M. C. Yuen, "Evaporation of a liquid droplet on a hot plate," *Int. J. Heat Mass Transf.*, vol. 34, no. 7, pp. 1881–1894, 1991.
49. Y. M. Qiao and S. Chandra, "Experiments on adding a surfactant to water drops boiling on a hot surface," *Proc. R. Soc. A Math. Phys. Eng. Sci.*, vol. 453, no. 1959, pp. 673–689, 1997.
50. L. Maquet, M. Brandenbourger, B. Sobac, A. L. Biance, P. Colinet, and S. Dorbolo, "Leidenfrost drops: Effect of gravity," *Epl*, vol. 110, no. 24001, 2015.
51. D. Orejon, K. Sefiane, and Y. Takata, "Effect of ambient pressure on Leidenfrost temperature," *Phys. Rev. E - Stat. Nonlinear, Soft Matter Phys.*, vol. 90, no. 5, pp. 1–6, 2014.
52. O. Ozkan, A. Shahriari, and V. Bahadur, "Electrostatic suppression of the Leidenfrost state using AC electric fields," *Appl. Phys. Lett.*, vol. 111, no. 141608, 2017.
53. N. Nagai and S. Nishio, "Leidenfrost temperature on an extremely smooth surface," *Exp. Therm. Fluid Sci.*, vol. 12, no. 3, pp. 373–379, 1996.
54. C. K. Huang and V. P. Carey, "The effects of dissolved salt on the Leidenfrost transition," *Int. J. Heat Mass Transf.*, vol. 50, pp. 269–282, 2007.
55. P. Zhang, B. Peng, X. Yang, J. Wang, and L. Jiang, "Regulating Droplet Dynamic Wetting Behaviors Using Surfactant Additives on High-Temperature Surfaces," *Adv. Mater. Interfaces*, vol. 2000501, pp. 1–8, 2020.
56. N. Hatta, H. Pujimoto, K. Kinoshita, and H. Takuda, "Experimental study of deformation mechanism of a water droplet impinging on hot metallic surfaces above the leidenfrost temperature," *J. Fluids Eng. Trans. ASME*, vol. 119, no. 3, pp. 692–699, 1997.
57. M. Shirota, M. A. J. Van Limbeek, C. Sun, A. Prosperetti, and D. Lohse, "Dynamic Leidenfrost Effect: Relevant Time and Length Scales," *Phys. Rev. Lett.*, vol. 116, no. 064501, 2016.
58. S. C. Yao and K. Y. Cai, "The dynamics and leidenfrost temperature of drops impacting on a hot surface at small angles," *Exp. Therm. Fluid Sci.*, vol. 1, no. 4, pp. 363–371, 1988.
59. K. J. Baumeister, R. E. Henry, and F. F. Simon, "Role of the surface in the measurement of the Leidenfrost temperature," *Augment. Convect. Heat mass Transf. ASME Winter Annu. Meet. New York*, 1970.
60. K. J. Baumeister and F. F. Simon, "Leidenfrost temperature—Its correlation for liquid metals, cryogens, hydrocarbons, and water," *J. Heat Transfer*, vol. 95, no. 2, pp. 166–173, 1973.

61. G. Duursma, R. Kennedy, K. Sefiane, and Y. Yu, "Leidenfrost Droplets on Microstructured Surfaces," *Heat Transf. Eng.*, vol. 37, no. 13–14, pp. 1190–1200, 2016.
62. A. Sharon and S. G. Bankoff, "Destabilization of leidenfrost boiling by a sudden rise of ambient pressure," *Chem. Eng. Sci.*, vol. 37, no. 8, pp. 1173–1179, 1982.
63. G. S. Emmerson and C. W. Snoek, "The effect of pressure on the leidenfrost point of discrete drops of water and freon on a brass surface," *Int. J. Heat Mass Transf.*, vol. 21, no. 8, pp. 1081–1086, 1978.
64. F. Celestini, T. Frisch, and Y. Pomeau, "Room temperature water Leidenfrost droplets," *Soft Matter*, vol. 9, no. 40, pp. 9535–9538, 2013.
65. B. T. Ng, Y. M. Hung, and M. K. Tan, "Suppression of the Leidenfrost effect via low frequency vibrations," *Soft Matter*, vol. 11, no. 4, pp. 775–784, 2015.
66. K. Takano, I. Tanasawa, and S. Nishio, "Active enhancement of evaporation of a liquid drop on a hot solid surface using a static electric field," *Int. J. Heat Mass Transf.*, vol. 37, no. 1, pp. 65–71, 1994.
67. B. Berge, "Électrocapillarité et mouillage de films isolants par l'eau," *C. R. Acad. Sci. Paris*, vol. 317, Série, pp. 157–163, 1993.
68. F. Celestini and G. Kirstetter, "Effect of an electric field on a Leidenfrost droplet," *Soft Matter*, vol. 8, no. 22, pp. 5992–5995, 2012.
69. V. E. Nakoryakov, S. Y. Misyura, and S. L. Elistratov, "The behavior of water droplets on the heated surface," *Int. J. Heat Mass Transf.*, vol. 55, no. 23–24, pp. 6609–6617, 2012.
70. G. S. Emmerson, "The effect of pressure and surface material on the Leidenfrost point of discrete drops of water," *Int. J. Heat Mass Transf.*, vol. 18, no. 3, pp. 381–386, 1975.
71. H. M. Kwon, J. C. Bird, and K. K. Varanasi, "Increasing Leidenfrost point using micro-nano hierarchical surface structures," *Appl. Phys. Lett.*, vol. 103, no. 201601, 2013.
72. S. H. Kim, G. Lee, H. M. Kim, and M. H. Kim, "Leidenfrost point and droplet dynamics on heated micropillar array surface," *Int. J. Heat Mass Transf.*, vol. 139, pp. 1–9, 2019.
73. N. R. Geraldi *et al.*, "Leidenfrost transition temperature for stainless steel meshes," *Mater. Lett.*, vol. 176, pp. 205–208, 2016.
74. C. T. Avedisian and J. Koplik, "Leidenfrost boiling of methanol droplets on hot porous/ceramic surfaces," *Int. J. Heat Mass Transf.*, vol. 30, no. 2, pp. 379–393, 1987.
75. R. Hays, D. Maynes, and J. Crockett, "Thermal transport to droplets on heated superhydrophobic substrates," *Int. J. Heat Mass Transf.*, vol. 98, pp. 70–80, 2016.
76. C. Kruse *et al.*, "Extraordinary shifts of the leidenfrost temperature from multiscale micro/nanostructured surfaces," *Langmuir*, vol. 29, no. 31, pp. 9798–9806, 2013.
77. G. C. Lee, J. young Kang, H. S. Park, K. Moriyama, S. H. Kim, and M. H. Kim, "Induced liquid-solid contact via micro/nano multiscale texture on a surface and its effect on the Leidenfrost temperature," *Exp. Therm. Fluid Sci.*, vol. 84, pp. 156–164, 2017.
78. Q. Ma, X. Wu, T. Li, and F. Chu, "Droplet boiling on heated surfaces with various wettabilities," *Appl. Therm. Eng.*, vol. 167, no. 114703, 2020.
79. S. H. Kim, H. S. Ahn, J. Kim, M. H. Kim, and H. S. Park, "Experimental study of water droplets on over-heated nano/microstructured zirconium surfaces," *Nucl. Eng. Des.*, vol. 278, pp. 367–376, 2014.
80. T. Kano, T. Isobe, S. Matsushita, and A. Nakajima, "Hydrophobicity and Leidenfrost point of ZnO nanorod array combined with nanoscale roughness on the topmost surface," *Mater. Chem. Phys.*, vol. 217, pp. 192–198, 2018.
81. S. Adera, R. Raj, R. Enright, and E. N. Wang, "Non-wetting droplets on hot superhydrophilic surfaces," *Nat. Commun.*, vol. 4, no. 2518, 2013.
82. I. U. Vakarelski, N. A. Patankar, J. O. Marston, D. Y. C. Chan, and S. T. Thoroddsen, "Stabilization of Leidenfrost vapour layer by textured superhydrophobic surfaces," *Nature*, vol. 489, no. 7415, pp. 274–277, 2012.
83. J. D. Bernardin and I. Mudawar, "A Cavity Activation and Bubble Growth Model of the Leidenfrost Point," *J. Heat Transfer*, vol. 124, no. 5, pp. 864–874, 2002.
84. E. Teodori, T. Valente, I. Malavasi, A. S. Moita, M. Marengo, and A. L. N. Moreira, "Effect of extreme wetting scenarios on pool boiling conditions," *Appl. Therm. Eng.*, vol. 115, pp. 1424–1437, 2017.

85. D. A. Del Cerro, Á. G. Marín, G. R. B. E. Römer, B. Pathiraj, D. Lohse, and A. J. Huis In 'T Veld, "Leidenfrost point reduction on micropatterned metallic surfaces," *Langmuir*, vol. 28, no. 42, pp. 15106–15110, 2012.
86. M. Auliano, D. Auliano, M. Fernandino, P. Zhang, and C. A. Dorao, "Water droplet dynamics on a heated nanowire surface," *Appl. Phys. Lett.*, vol. 113, no. 253703, 2018.
87. S. H. Kim, H. Seon Ahn, J. Kim, M. Kaviany, and M. Hwan Kim, "Dynamics of water droplet on a heated nanotubes surface," *Appl. Phys. Lett.*, vol. 102, no. 233901, 2013.
88. S. M. Sajadi, P. Irajizad, V. Kashyap, N. Farokhnia, and H. Ghasemi, "Surfaces for high heat dissipation with no Leidenfrost limit," *Appl. Phys. Lett.*, vol. 111, no. 021605, 2017.
89. N. Farokhnia, S. M. Sajadi, P. Irajizad, and H. Ghasemi, "Decoupled Hierarchical Structures for Suppression of Leidenfrost Phenomenon," *Langmuir*, vol. 33, no. 10, pp. 2541–2550, 2017.
90. N. J. Shirtcliffe, G. McHale, M. I. Newton, C. C. Perry, and F. B. Pyatt, "Plastron properties of a superhydrophobic surface," *Appl. Phys. Lett.*, vol. 89, no. 104106, 2006.
91. S. Srinivasan *et al.*, "Quantification of feather structure, wettability and resistance to liquid penetration," *J. R. Soc. Interface*, vol. 11, no. 96, 2014.
92. N. J. Shirtcliffe, G. McHale, M. I. Newton, and Y. Zhang, "Superhydrophobic copper tubes with possible flow enhancement and drag reduction," *ACS Appl. Mater. Interfaces*, vol. 1, no. 6, pp. 1316–1323, 2009.
93. R. Poetes, K. Holtzmann, K. Franze, and U. Steiner, "Metastable underwater superhydrophobicity," *Phys. Rev. Lett.*, vol. 105, no. 166104, 2010.
94. P. Lv, Y. Xue, Y. Shi, H. Lin, and H. Duan, "Metastable states and wetting transition of submerged superhydrophobic structures," *Phys. Rev. Lett.*, vol. 112, no. 196101, 2014.
95. P. Bourrianne, C. Lv, and D. Quéré, "The cold Leidenfrost regime," *Sci. Adv.*, vol. 5, 2019.
96. B. S. Gottfried, C. J. Lee, and K. J. Bell, "The leidenfrost phenomenon: film boiling of liquid droplets on a flat plate," *Int. J. Heat Mass Transf.*, vol. 9, no. 11, pp. 1167–1188, 1966.
97. E. Forrest, R. Schulze, C. Liu, and D. Dombrowski, "Influence of surface contamination on the wettability of heat transfer surfaces," *Int. J. Heat Mass Transf.*, vol. 91, pp. 311–317, 2015.
98. M. Chabičovský, M. Hnízdil, A. A. Tseng, and M. Raudenský, "Effects of oxide layer on Leidenfrost temperature during spray cooling of steel at high temperatures," *Int. J. Heat Mass Transf.*, vol. 88, pp. 236–246, 2015.
99. G. C. Lee, S. H. Kim, J. young Kang, M. H. Kim, and H. J. Jo, "Leidenfrost temperature on porous wick surfaces: Decoupling the effects of the capillary wicking and thermal properties," *Int. J. Heat Mass Transf.*, vol. 145, no. 118809, 2019.
100. G. Liu and V. S. J. Craig, "Macroscopically flat and smooth superhydrophobic surfaces: Heating induced wetting transitions up to the Leidenfrost temperature," *Faraday Discuss.*, vol. 146, pp. 141–151, 2010.
101. J. Park and D. E. Kim, "Droplet dynamics on superheated surfaces with circular micropillars," *Int. J. Heat Mass Transf.*, vol. 142, no. 118459, 2019.
102. G. Dupeux, P. Bourrianne, Q. Magdelaine, C. Clanet, and D. Quéré, "Propulsion on a superhydrophobic ratchet," *Sci. Rep.*, vol. 4, no. 5280, 2014.
103. D. Soto, G. Lagubeau, C. Clanet, and D. Quere, "Surfing on a herringbone," *Phys. Rev. Fluids*, vol. 1, no. 013902, 2016.
104. H. Xu *et al.*, "Self-propelled rotation of paper-based Leidenfrost rotor," *Appl. Phys. Lett.*, vol. 114, no. 113703, 2019.
105. I. U. Vakarelski, J. O. Marston, D. Y. C. Chan, and S. T. Thoroddsen, "Drag reduction by leidenfrost vapor layers," *Phys. Rev. Lett.*, vol. 106, no. 214501, 2011.
106. R. Abdelaziz *et al.*, "Green chemistry and nanofabrication in a levitated Leidenfrost drop," *Nat. Commun.*, vol. 4, no. 2400, 2013.
107. S. Wildeman and C. Sun, "Electric field makes Leidenfrost droplets take a leap," *Soft Matter*, vol. 12, pp. 9622–9632, 2016.
108. M. Mrinal, X. Wang, and C. Luo, "Self-rotation-induced propulsion of a Leidenfrost drop on a ratchet," *Langmuir*, vol. 33, no. 25, pp. 6307–6313, 2017.

109. A. Würger, "Leidenfrost gas ratchets driven by thermal creep," *Phys. Rev. Lett.*, vol. 107, no. 164502, 2011.
110. T. R. Cousins, R. E. Goldstein, J. W. Jaworski, and A. I. Pesci, "A ratchet trap for Leidenfrost drops," *J. Fluid Mech.*, vol. 696, pp. 215–227, 2012.
111. Q. Li, Q. J. Kang, M. M. Francois, and A. J. Hu, "Lattice Boltzmann modeling of self-propelled Leidenfrost droplets on ratchet surfaces," *Soft Matter*, vol. 12, pp. 302–312, 2015.
112. G. Dupeux, M. Le Merrer, G. Lagubeau, C. Clanet, S. Hardt, and D. Quéré, "Viscous mechanism for Leidenfrost propulsion on a ratchet," *EPL (Europhysics Lett.*, vol. 96, no. 58001, 2011.
113. L. E. Dodd *et al.*, "Low-Friction Self-Centering Droplet Propulsion and Transport Using a Leidenfrost Herringbone-Ratchet Structure," *Phys. Rev. Appl.*, vol. 11, no. 034063, 2019.
114. J. T. Ok, J. Choi, E. Brown, and S. Park, "Effect of different fluids on rectified motion of Leidenfrost droplets on micro/sub-micron ratchets," *Microelectron. Eng.*, vol. 158, pp. 130–134, 2016.
115. J. M. Arter, D. J. Cleaver, K. Takashina, and A. T. Rhead, "Self-propelling Leidenfrost droplets on a variable topography surface," *Appl. Phys. Lett.*, vol. 113, no. 243704, 2018.
116. G. Lagubeau, M. Le Merrer, C. Clanet, and D. Quéré, "Leidenfrost on a ratchet," Nat. Phys., vol. 7, no. 5, pp. 395–398, 2011.
117. R. L. Agapov *et al.*, "Length scale of Leidenfrost ratchet switches droplet directionality," *Nanoscale*, vol. 6, no. 15, pp. 9293–9299, 2014.
118. A. Cole, B. Jury, and K. Takashina, "A Leidenfrost Thermostat," *J. Heat Transfer*, vol. 137, no. 034502, 2015.
119. J. Li *et al.*, "Rectification of Mobile Leidenfrost Droplets by Planar Ratchets," *Small*, vol. 1901751, 2020.
120. C. Krumm *et al.*, "Micro-ratcheted surfaces for a heat engine biomass conveyor," *Energy Environ. Sci.*, vol. 9, no. 5, pp. 1645–1649, 2016.
121. Z. hai Jia, M. yao Chen, and H. tao Zhu, "Reversible self-propelled Leidenfrost droplets on ratchet surfaces," *Appl. Phys. Lett.*, vol. 110, no. 091603, 2017.
122. A. Grounds, R. Still, and K. Takashina, "Enhanced droplet control by transition boiling," *Sci. Rep.*, vol. 2, no. 720, 2012.
123. C. Luo, M. Mrinal, and X. Wang, "Self-propulsion of Leidenfrost Drops between Non-Parallel Structures," *Sci. Rep.*, vol. 7, no. 12108, 2017.
124. C. Kruse *et al.*, "Self-propelled droplets on heated surfaces with angled self-assembled micro/nanostructures," *Microfluid. Nanofluidics*, vol. 18, pp. 1417–1424, 2015.
125. M. yao Chen, Z. hai Jia, T. Zhang, and Y. yuan Fei, "Self-propulsion of Leidenfrost droplets on micropillared hot surfaces with gradient wettability," *Appl. Surf. Sci.*, vol. 433, pp. 336–340, 2018.
126. C. Liu, J. Ju, J. Ma, Y. Zheng, and L. Jiang, "Directional drop transport achieved on high-temperature anisotropic wetting surfaces," *Adv. Mater.*, vol. 26, no. 35, pp. 6086–6091, 2014.
127. P. Zhang, B. Peng, J. Wang, and L. Jiang, "Bioinspired Self-Propulsion of Water Droplets at the Convergence of Janus-Textured Heated Substrates," *Adv. Funct. Mater.*, vol. 1904535, 2019.
128. G. Dupeux, M. Le Merrer, C. Clanet, and D. Quéré, "Trapping Leidenfrost drops with crenelations," *Phys. Rev. Lett.*, vol. 107, no. 114503, 2011.
129. L. E. Dodd *et al.*, "Low friction droplet transportation on a substrate with a selective Leidenfrost effect," *ACS Appl. Mater. Interfaces*, vol. 8, no. 34, pp. 22658–22663, 2016.
130. L. E. Dodd *et al.*, "Planar selective Leidenfrost propulsion without physically structured substrates or walls," *Appl. Phys. Lett.*, vol. 117, no. 081601, 2020.
131. G. McHale, N. J. Shirtcliffe, C. R. Evans, and M. I. Newton, "Terminal velocity and drag reduction measurements on superhydrophobic spheres," *Appl. Phys. Lett.*, vol. 94, no. 064104, 2009.
132. B. R. K. Gruncell, N. D. Sandham, and G. McHale, "Simulations of laminar flow past a superhydrophobic sphere with drag reduction and separation delay," *Phys. Fluids*, vol. 25, no. 043601, 2013.

133. I. U. Vakarelski, D. Y. C. C. Chan, J. O. Marston, and S. T. Thoroddsen, "Dynamic air layer on textured superhydrophobic surfaces," *Langmuir*, vol. 29, no. 35, pp. 11074–11081, 2013.
134. G. McHale, M. R. Flynn, and M. I. Newton, "Plastron induced drag reduction and increased slip on a superhydrophobic sphere," *Soft Matter*, vol. 7, pp. 10100–10107, 2011.
135. A. Busse, N. D. Sandham, G. McHale, and M. I. Newton, "Change in drag, apparent slip and optimum air layer thickness for laminar flow over an idealised superhydrophobic surface," *J. Fluid Mech.*, vol. 727, pp. 488–508, 2013.
136. G. McHale, M. I. Newton, and N. J. Shirtcliffe, "Immersed superhydrophobic surfaces: Gas exchange, slip and drag reduction properties," *Soft Matter*, vol. 6, no. 4, pp. 714–719, 2010.
137. A. Jetly, I. U. Vakarelski, Z. Yang, and S. T. Thoroddsen, "Giant drag reduction on Leidenfrost spheres evaluated from extended free-fall trajectories," *Exp. Therm. Fluid Sci.*, vol. 102, no. October 2018, pp. 181–188, 2019.
138. I. U. Vakarelski, J. D. Berry, D. Y. C. Chan, and S. T. Thoroddsen, "Leidenfrost Vapor Layers Reduce Drag without the Crisis in High Viscosity Liquids," *Phys. Rev. Lett.*, vol. 117, no. 114503, 2016.
139. D. Saranadhi, D. Chen, J. A. Kleingartner, S. Srinivasan, R. E. Cohen, and G. H. Mckinley, "Sustained drag reduction in a turbulent flow using a low-Temperature Leidenfrost surface," *Sci. Adv.*, vol. 2, no. 1600686, pp. 1–9, 2016.
140. I. U. Vakarelski, D. Y. C. C. Chan, and S. T. Thoroddsen, "Leidenfrost vapour layer moderation of the drag crisis and trajectories of superhydrophobic and hydrophilic spheres falling in water," *Soft Matter*, vol. 10, no. 31, pp. 5662–5668, 2014.
141. H. W. Gilbert and P. E. Shaw, "Electrical charges arising at a liquid-gas interface," *Proc. Phys. Soc. London*, vol. 37, no. 1, pp. 195–214, 1924.
142. R. Leberman and A. K. Soper, "Effect of high salt concentrations on water structure," *Nature*, vol. 378, no. 23, pp. 364–366, 1995.
143. P. Postorino, R. H. Tromp, M. A. Ricci, A. K. Soper, and G. W. Neilson, "The interatomic structure of water at supercritical temperatures," *Nature*, vol. 366, no. 6456, pp. 668–670, 1993.
144. D. W. Lee, M. H. Jin, Y. J. Lee, J. H. Park, C. B. Lee, and J. S. Park, "Reducing-Agent-Free Instant Synthesis of Carbon-Supported Pd Catalysts in a Green Leidenfrost Droplet Reactor and Catalytic Activity in Formic Acid Dehydrogenation," *Sci. Rep.*, vol. 6, no. 26474, 2016.
145. D. W. Lee *et al.*, "Straightforward Synthesis of Metal Nanoparticles and Hierarchical Porous Metals Assisted by Partial Film Boiling Phenomena," *Chem. Mater.*, vol. 27, no. 15, pp. 5151–5160, 2015.
146. I. Isakov *et al.*, "Exploring the Leidenfrost Effect for the Deposition of High-Quality In2O3 Layers via Spray Pyrolysis at Low Temperatures and Their Application in High Electron Mobility Transistors," *Adv. Funct. Mater.*, vol. 1606407, pp. 1–9, 2017.
147. M. Ortel and V. Wagner, "Leidenfrost temperature related CVD-like growth mechanism in ZnO-TFTs deposited by pulsed spray pyrolysis," *J. Cryst. Growth*, vol. 363, pp. 185–189, 2013.
148. C. H. Lim, H. Kang, and S. H. Kim, "Colloidal assembly in leidenfrost drops for noniridescent structural color pigments," *Langmuir*, vol. 30, no. 28, pp. 8350–8356, 2014.

Chapter 8
On the Development of Icephobic Surfaces: Bridging Experiments and Simulations

Irene Tagliaro, Alessio Cerpelloni, Vasileios-Martin Nikiforidis, Rohit Pillai, and Carlo Antonini

Keywords Anti-icing · Deicing · Icephobic · Surfaces · Molecular dynamics

8.1 Introduction

Ice formation on surfaces is a common phenomenon occurring in the presence of water at temperatures below the freezing point, and can negatively impact many aspects of our lives. Atmospheric icing, which forms due to the natural presence of water as small liquid drops or vapor in the air, can cause damage to ground transportation, airplanes, power lines, and communication systems (e.g., telephone and cable television line operations) and other man-made structures and devices.

For example, ice accretion on aircraft (Fig. 8.1) can affect the proper functioning of aerodynamic surfaces, increasing drag, and increasing stall speed (the slowest speed an aircraft can maintain level flight) [1], as well as probes and sensors, leading, e.g., to incorrect engine power settings.

A recent accident caused by in-flight ice formation occurred at Clarence Center, New York, on February 12, 2009. As a result of extended ice buildup on the aircraft wings and windshield, the aircraft crashed, resulting in the death of 50 people. Over the years, other fatal accidents have also occurred, such as that of January 9, 1997, involving the Embraer EMB-12RT, due to a formation of thin rough glaze/mixed ice coverage on the wings leading edge, and that of March 22, 1992, in Flushing, New

I. Tagliaro · A. Cerpelloni · C. Antonini
Department of Materials Science, University of Milano-Bicocca, Milan, Italy
e-mail: carlo.antonini@unimib.it

V.-M. Nikiforidis · R. Pillai (✉)
School of Engineering, University of Edinburgh, Edinburgh E9 3FB, UK
e-mail: R.Pillai@ed.ac.uk

M. Marengo and J. De Coninck (eds.), *The Surface Wettability Effect on Phase Change*,
https://doi.org/10.1007/978-3-030-82992-6_8

Fig. 8.1 Wing of an airplane with accretions of glaze ice. Title: Ohio National Guard. Author: The National Guard. *Source* https://flic.kr/p/2irysRE. License: CC BY 2.0. Description: "Airmen work on a C-130H Hercules, spraying it with deicing fluid during a snowstorm, Feb. 07, 2020, at the 179th Airlift Wing, Mansfield, Ohio."

York, in which 27 deaths occurred due to ice accumulation under the aircraft wings [2].

Icing is also a problem on the ground in cold regions, either at high latitudes or in mountain areas. Indeed, ice and snow accumulation on overhead power grid components such as conductors, ground wires, and insulators can cause power line collapse and black-outs that may last several hours; this is both a severe inconvenience for the users, and a source of economic losses for energy operators [3]. The destructive effects of icing phenomena on overhead power lines (Fig. 8.2) are attributable to the excessive ice or snow accumulation, which can lead to electrical flashover or aeroelastic instabilities: one example is galloping, a self-excited vibration of transmission line under the action of wind load, eventually causing damages to conductors or structural collapse [3]. Additionally, ice removal from power lines and wind turbines can be a dangerous operation for technicians [4].

The atmospheric icing events affecting ground structures and operations can occur in many parts of the world, including North America, Europe, Asia, and South Africa. Such events are recurrent in some countries (USA, Canada, Iceland, etc.) occasionally causing a serious socioeconomic impact. While they are less frequent in other countries, but can still cause serious damage to overhead electric lines (Sweden, Spain, South Africa, Italy, etc.) [5].

Fig. 8.2 Atmospheric icing of the overhead power line. Title: IMG_0337. Author: KOMUnews. *Source* https://flic.kr/p/4cEson. License: CC BY 2.0

For example, from January 4 to 10, 1998, a major ice storm caused catastrophic damage to power lines in eastern Canada and the northeastern United States, resulting in power losses. The storm destroyed and damaged 3,000 km of power lines, 4,000 power transformers, and 1,000 steel towers [6, 7]. More than 3 million utility customers lost power in Canada and 1 million in the northeastern United States. The estimated damages were $4.4 billion [8].

Two other examples of atmospheric icing occurred in Northern Iceland on September 10 and December 29, 2012. In both the events, two wet snowstorms accompanied by strong winds (20–40 m/s) caused an extreme wet snow load on power transmission and distribution lines. This led to the breakage of many wooden poles. During the events, in the worst cases, the inhabitants of the affected areas remained without electricity for several days [9].

In the southwestern region of Sweden, during February 1999, there were outages of the power line lasting up to 6 hours. The triggers were low temperatures and strong winds which gave rise to ice accretion, followed by flashovers. Indeed, if ice or snow accretion is followed by a period of thaw, service experience indicates that the likelihood of an electrical outage because of flashover is high [5].

8.1.1 Atmospheric Icing

Atmospheric icing on structures is commonly classified into three main categories, based on the atmospheric conditions leading to ice formation: in-cloud icing, precipitation icing, and hoar frost [5]. In-cloud icing occurs due to small micron-sized drops, typically in the range of 10–40 μm. These drops are found in clouds, as the name suggests, or in fog; the amount of liquid water mass present in the unit volume is expressed by the liquid water content (LWC), with values in the range 0.1–3 g/m^3. When their temperature reduces to below the freezing point, the drops in the cloud become supercooled, i.e., they are in a metastable liquid state (see detailed discussion in Sect. 8.2) and start freezing after impact on a solid surface (e.g., the aircraft aerodynamic surfaces). Precipitation icing phenomena are caused by larger supercooled drops in the rain (with a sub-millimeter or millimeter diameters) and snowflakes that precipitate on ground structures. Hoar frost, or simply frost, originates from water vapor because of direct deposition (vapor-to-solid) or a two-step condensation-freezing (vapor-to-liquid-to-solid) process.

An additional classification is based on the ice type and properties, spanning glaze ice, rime ice, snow (wet and dry) and frost, illustrated in Fig. 8.3. According to the technical brochure periodically released by the International Council on Large Electric Systems (CIGRE) [10], in precipitation icing, wet and dry snow and glaze ice due to freezing rain can be observed; for in-cloud icing, both glaze and rime ice can be observed. The main properties of the different ice types are listed and discussed here below.

Glaze ice (Fig. 3a) is translucent and has the highest density (up to 900 kg/m^3). Typically, it is icicle-shaped and its adhesion to surfaces is very strong. In precipitation icing, raindrops pass through a region of the atmosphere characterized by thermal inversion (change in the normal tendency of the air to cool down with altitude). Glaze ice formation originates at temperatures close to the freezing point [5, 11].

Fig. 8.3 Pictures of tree branches with different ice formation: **a** Glaze ice. Title: Glaze ice from freezing rain (29 January 2009) (Newark, Ohio, USA) 38. Author: James St. John. *Source* https://flic.kr/p/Qz1EDZ. License: CC BY 2.0. **b** Rime ice. Title: Rime ice. Author: Yellowstone National Park. *Source* https://flic.kr/p/qU7mMN. License: Public Domain 1.0. **c** Snow. Title: Snow. Author: Eneco Muino. *Source* https://flic.kr/p/r9TgJR. License: CC BY 2.0. d) Frost. Title: Frost. Author: Beni Rupp. *Source* https://flic.kr/p/NCrmXv. License: CC BY 2.0

Rime ice originates at lower temperatures below −5 °C, when supercooled drops almost instantaneously freeze upon contact with a surface (Fig. 3b). Its characteristic white color is caused by the presence of air bubbles, which remain trapped due to the rapid freezing process. Rime ice can have a variable density ranging from 150 to 700 kg/m^3. There are two types of rime: soft rime and hard rime. When the air temperature is between −5 and −10 °C and the wind speed oscillates between 5 and 15 m/s, soft rime grows (150–300 kg/m^3). If, on the other hand, the temperatures range is from −10 to −15 °C and the wind speed oscillates between 15 and 25 m/s, a denser and more compact structure (with fewer air bubbles) originates that is the hard rime (300–700 kg/m^3) [5, 11, 12].

Dry snow forms (Fig. 3c) at a temperature below freezing and accumulates on surfaces at low wind speeds. It is easy to remove from surfaces and its density ranges between 50 and 100 kg/m^3 [3, 13]. The main difference between wet and dry snow is the presence of liquid water inside the flakes (absent in dry snow), which can be as high as 40% of the total mass. Wet snow density can span from 100 to 850 kg/m^3. The high density is mostly because it forms when wind compresses the snow deposited on the surfaces, making the growth more compact. Wet snow generally forms between 0.5 and 2.0 °C. The formation process takes place when, precipitating, the snowflakes cross different layers of air having higher temperatures at lower altitudes. If the air temperature exceeds 0 °C near the ground, the snowflakes begin to melt, and liquid water appears in between ice crystals. Wet snow easily adheres to surfaces when the internal content of liquid water is between 15 and 40% [3, 10, 13] due to capillary forces.

Frost is formed by the deposition process of water vapor present in the air (Fig. 3d). It usually has a density of less than 100 kg/m^3, and is easy to remove from surfaces (due to low adhesion) and appears as a collection of small crystals [3, 10].

8.1.2 From Traditional Methods to Surface Strategies Against Icing

Regardless of the type of ice, frequent problems related to icing require effective and efficient systems to control ice formation and accretion. The systems are normally classified as anti-icing, when ice formation is prevented, or de-icing, when ice is intermittently formed and then removed.

The traditional methods involve the use of thermal energy for heating surfaces, antifreeze liquids and mechanical vibrations. However, these methods are often energetically or economically demanding [13], or may require human intervention.

As traditional methods have drawbacks, passive surfaces and coatings have attracted interest. The goal in this case is to ensure that any formed ice adheres weakly and can debond naturally due to gravity or other external forces, thereby requiring no energy input and being able to be used in conjunction with traditional methods. However, for industrial use, these surfaces require modifying the surface topography and chemistry at the micro-/nano-scale in a way that is easy to manufacture, cost-effective and durable. This remains a challenge and motivates the current chapter.

8.1.3 Existing Literature and the Goal of This Chapter

Recently, several reviews have already discussed the role of surface strategies against ice formation [14–18]. Schutzius et al. [14] analyzed the key aspects of ice formation and discussed rational strategies to develop icephobic surfaces. Amirfazli and Antonini [15] discussed the relationship between superhydrophobicity and icephobicity to highlight that different mechanisms are involved in ice formation and can be addressed to control it, underlining the strong impact of environmental conditions on the icing process. Stone [16] focused on the comparison between "classical" superhydrophobic textured surfaces and liquid-infused porous surfaces (referred to as SLIPS or LIS), where an infused liquid strongly wets and is retained within a porous material, creating a liquid layer at the interface that strongly reduces ice adhesion. Huang et al. [18] specifically looked into the potential of coatings for aerospace applications, thus discussing engineering relevant coatings on metals (e.g., aluminum), analyzing them in the framework of aerospace anti-icing and de-icing requirements, e.g., in terms of durability. Irajizad et al. [17] and Shen et al. [19] described the different techniques to evaluate the performance against ice, highlighting how discrepancies may arise from different measurements and different laboratories.

Performing a simple but revealing "Scopus" query, we found that in 2020 alone, there were more than 700 publications related to the terms "icephobic" and "anti-icing surfaces", with an impressive expansion of the field over the past 10 years (see Fig. 4a)—the use of different terminology will be discussed in Sect. 8.3. By sorting the publications as experimental or numerical, we found that approximately two-thirds of the studies use an experimental approach, and one-third a numerical approach (see Fig. 4b).

Despite the existence of such an extremely rich literature resulting from the significant efforts of scientists and engineers on understanding and tackling the problem of icing, we have noticed that a large gap still exists in the analysis and correlation between experimental and numerical approaches. The challenge lies in the inherent complexity of ice formation, a multiscale process, where nanoscale interaction mechanisms (10^{-9} m such as wetting and adhesion) on the surface with characteristic micro and nano-topography (10^{-9}–10^{-6} m) induce macroscopic effects up to the size of drops (10^{-3} m) on large structures (10^{0}–10^{2} m scale). Such a large-scale span makes it difficult to create robust and reliable models to predict ice formation and to design icephobic surfaces.

To tackle this point, we have conceived this chapter as structured in three main parts: in Sect. 8.2, we introduce the key concepts of nucleation physics, which is fundamental to the mechanism of icing. In Sect. 8.3, we discuss the different surface strategies against icing, based on some of the most recent approaches. In Sect. 8.4, we

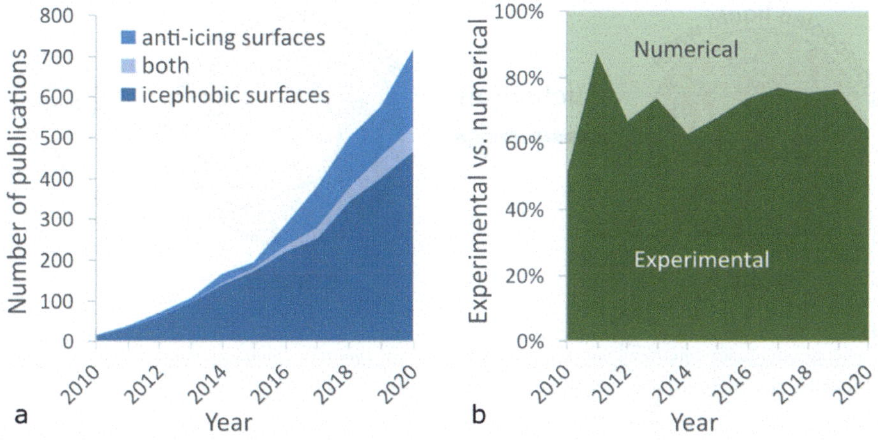

Fig. 8.4 Overview of anti-icing and icephobic surfaces research, using publication statistics as extracted from Scopus. **a** Number of published documents from 2010 to 2020, identified using as search queries ALL ("icephobic" OR "icephobicity"), ALL ("anti-icing surface"), and both. **b** Share of publications related to experimental vs. numerical approach. Publications were classified searching for the words "numerical" and "experimental" in the title, abstract, or keywords, e.g., TITLE-ABS-KEY (experimental). The query was performed on scopus.com on January 3, 2021

present numerical tools that are being developed to model ice behavior on surfaces, with the major objective of bridging the gap between theory and experiments.

8.2 Nucleation Physics: Key Concepts

The mechanism of ice nucleation is a stochastic process, influenced by various factors related to phenomena occurring at different time and spatial scales. This complicates the analysis of results from experiments, which have to be conducted under rigorous control of temperature, humidity, cooling rate, water purity, etc.

Understanding ice nucleation requires elaborating on the concept of metastability. Based on thermodynamic principles, the ice phase at 0 °C is stable, and supercooled water is in a metastable state. This means that it can persist for a long time if there is no large-scale disturbance in the system. Metastability is thus related to the transition from a less ordered phase (liquid water) with high entropy to the new phase (solid ice) with highly ordered molecules at lower entropy; entropy must be reduced for ice formation to occur. This is reflected in the existence of the energy barrier between the metastable state and the equilibrium state. To overcome the energy barrier a driving force is required, which is the chemical potential difference between the solid and liquid phase, $\Delta\mu = \mu_s - \mu_l$, where the subscripts s and l indicate the solid and liquid phase, respectively. In the case of liquid to solid phase change, the chemical potential is the difference between the equilibrium freezing temperature, at a given

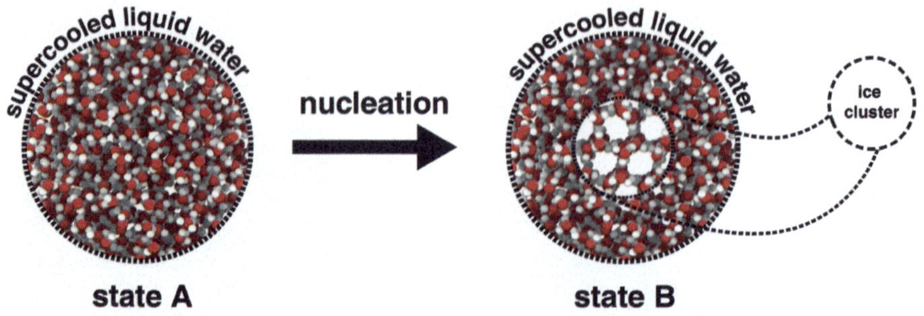

Fig. 8.5 Homogeneous nucleation explained by presenting two states of a system. In state A, the system is composed only of supercooled liquid water. In state B, a cluster of water molecules in the solid phase (ice embryo) is established within the supercooled liquid

pressure, and the effective freezing temperature at which phase change occurs, i.e., $\Delta T_s = T_f - T$.

Ice nucleation is classified into two categories: *homogeneous* and *heterogeneous* nucleation, respectively. Most natural nucleation processes are heterogeneous, because the stable phase grows on top of an external nucleation seed such as impurity or a surface; by contrast, homogeneous nucleation takes place entirely in the metastable phase (Fig. 8.5). The absence of pre-existing nucleation seed in homogeneous nucleation implies that a cluster of the stable phase must develop spontaneously before the liquid can crystallize; this process requires the overcoming of a very high free energy barrier; therefore, homogeneous nucleation is a rare event. Since the principal interest in this chapter is ice formation on solid surfaces, the focus will be on the fundamentals of heterogeneous ice nucleation. Homogeneous nucleation is only used to introduce the concept of nucleation and help develop a theoretical framework to understand nucleation, namely classical nucleation theory.

8.2.1 Classical Nucleation Theory

Classical Nucleation Theory (CNT) has found application to many areas, where the nucleation of a new phase from the parent metastable phase is observed. CNT was first introduced for drop condensation of supercooled water vapor [20]. Subsequent developments enabled its application to the crystallization of supercooled liquids [21, 22]. Although simplifying assumptions are made in CNT, it does a remarkably good job at reproducing observed trends, and has compared well to molecular dynamics simulation results [23]. According to CNT, clusters of crystalline atoms of any size are treated as macroscopic objects whose growth or collapse can be predicted using a simple set of equations, as discussed below.

8.2.1.1 Homogeneous Ice Nucleation

CNT relies on the use of two fundamental concepts, which are introduced here: the Gibbs free energy barrier of freezing, ΔG_f, and the rate of nucleation, J, associated with the kinetics of ice nuclei formation. In order to illustrate the Gibbs free energy barrier, we define two states, *A* and *B* of the same system (Fig. 8.1). State *A* comprises only supercooled liquid water (metastable phase). In contrast, in state *B*, an unstable equilibrium between the supercooled liquid water and a cluster of water molecules in the solid phase (ice embryo) is established. The Gibbs free energy represents the change in interfacial free energy as the system is transformed from state *A* to state *B*; it, therefore, can also be interpreted as the barrier needed to be overcome for freezing to occur, and is given by

$$\Delta G_f = \rho V \Delta \mu + \gamma_{iw} A$$

where ρ, V, γ_{iw}, and A are the ice embryo density, embryo volume, ice-water interfacial tension, and embryo surface area, respectively, while $\Delta \mu$ is the difference in chemical potential between the liquid and ice phases that drive the freezing process. For an embryo that is assumed to be spherical, V and A can be further simplified using well-known expressions for a sphere.

Since the solid phase (ice embryo) is more stable than that of the liquid phase, $\Delta \mu < 0$, and ΔG_f is the sum of two terms: (i) a negative volume term, driving toward phase change, which increases with the volume of ice embryo and (ii) an opposing positive surface term, related to the energy needed to produce a new water–ice interface $\gamma_{iw}(4\pi r^2)$.

Note that ΔG_f is a function of r, and by differentiating, we can find that the surface term is predominant (increasing function) up until a critical ice embryo radius $r_{crit} = 2\gamma / \rho \Delta \mu$ is reached, at which the free energy barrier reaches its maximum $\Delta G_f^* = \frac{16\pi \gamma_{iw}^3}{3(\rho \Delta \mu)^2}$. This means that an ice embryo is unstable below the critical radius and is required to surmount this critical value to be able to grow (Fig. 8.6).

Taking into account that the chemical potential $\Delta \mu$ is proportional to the latent heat of freezing, h_f, and liquid supercooling degree ΔT_s, we can write $\Delta \mu = h_f \frac{\Delta T_s}{T_f}$. As a result, the critical Gibbs free energy barrier can be expressed as

$$\Delta G_f^* = \frac{16\pi \gamma_{iw}^3}{3h_f^2 \left(\frac{\Delta T_s}{T_f} \right)^2}.$$

This equation indicates that the larger the supercooling, the easier is for the ice embryo to surmount the barrier and grow. Also, it explains why ΔT_s is a driving cause for ice nucleation.

The second important factor is the ice nucleation rate, $J[m^{-3}s^{-1}]$, which is the rate at which molecules are accumulated on an ice embryo, and consequently is an estimate of the probability of freezing. It assumes that there is no correlation between

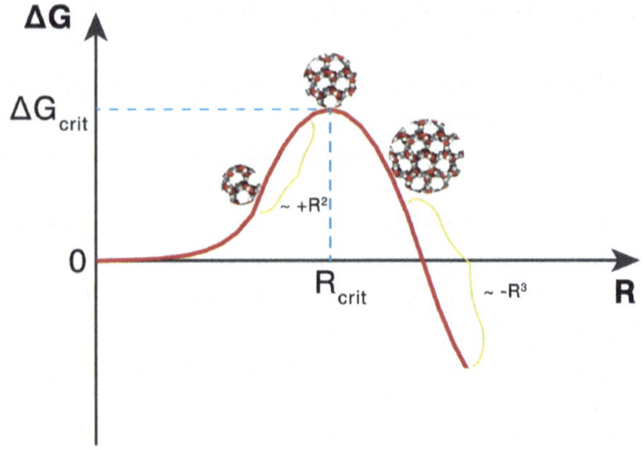

Fig. 8.6 Gibbs free energy change as a function of ice embryo radius. The ice embryo is metastable below a critical radius, corresponding to the free energy barrier, above which freezing takes place

successive events increasing or decreasing the number of molecules of ice embryo. Also, it is assumed that the most active nucleation site dominates ice nucleation; once an ice cluster overcomes the critical size, ice formation in the liquid continuous spontaneously. If these conditions are met the ice nucleation rate, which is the time-independent probability per unit time per unit volume of forming a critical nucleus, is given by the following expression:

$$J \approx \frac{k_B T}{h} \exp\left(-\frac{\Delta G_f^*}{k_B T} - \frac{\Delta g_{act}}{k_B T}\right),$$

where k_B is the Boltzmann constant, h is Planck's constant, and Δg_{act} is the activation energy barrier associated with the transfer of water molecules across the ice–water interface. Equation reveals two crucial aspects of ice nucleation. First, the ice nucleation rate is strongly (exponentially) dependent on temperature. Thus, an increase of 1 K in supercooling increases the ice nucleation rate by 2–3 orders of magnitude. Second, the two terms in the exponent indicate the existence of two energy barriers; the first term is associated with the Gibbs free energy barrier and the second term is related to Δg_{act}, an energy barrier associated with self-diffusion in water. This second barrier takes into account the reduced mobility of molecules at the interface, which delays the transport of molecules leaving the liquid phase and joining the solid phase. For low supercooling ($\Delta T_s < 30K$), Δg_{act} has a negligible effect and nucleation rate can be expressed by the more familiar expression $J \approx \frac{k_B T}{h} \exp\left(\frac{-\Delta G_f^*}{k_B T}\right)$.

For high supercooling ($\Delta T_s > 30K$), the term Δg_{act} also needs to be included, and ice nucleation is considered as needing to overcome two energy barriers. Moving toward strong supercooling and below a critical temperature (spinodal), T_{sp}, the free energy barrier for nucleation is zero, and the liquid phase changes spontaneously into a solid one in a very short time.

8.2.1.2 Heterogeneous Ice Nucleation

Heterogeneous ice nucleation, which is nucleation that occurs in the presence of foreign material, is the most common process for ice formation. It can take place at a higher temperature than that of homogeneous nucleation, because of impurities that act as nucleating agents and facilitate the formation of critical ice nuclei. From a thermodynamic standpoint, heterogeneous ice nucleation occurs due to a decrease in the energy barrier needed for freezing. Thus, the energy barrier for heterogeneous nucleation will be lower than that of the homogeneous freezing barrier, i.e.:

$$\Delta G^{*}_{f,het} < \Delta G^{*}_{f,hom}.$$

The presence of nucleation agents on a surface favour heterogeneous nucleation. These agents depend on surface characteristics, such as topography and chemical composition. Thus, all these factors should be considered when designing icephobic materials. To account for surface effects, it is usual to express the heterogeneous nucleation barrier as [24]:

$$\Delta G^{*}_{f,het} = f\left(cos(\theta_{iw}), \frac{r}{r_i^{*}}\right)\Delta G^{*}_{f,hom},$$

where f is analytically described using

$$f\left(cos(\theta_{iw}), \frac{r}{r_i^{*}}\right) = 1 + \left(\frac{1 - cos(\theta_{iw})\left(\frac{r}{r_i^{*}}\right)}{g}\right) + \left(\frac{r}{r_i^{*}}\right)^{3}\left[2 - 3\left(\frac{\frac{r}{r_i^{*}} - cos(\theta_{iw})}{g}\right) + \left(\frac{\frac{r}{r_i^{*}} - cos(\theta_{iw})}{g}\right)\right] + 3cos(\theta_{iw})\left(\frac{r}{r_i^{*}}\right)^{2}\left(\left(\frac{\frac{r}{r_i^{*}} - cos(\theta_{iw})}{g}\right) - 1\right).$$

Thus, f accounts for surface wettability and texture, through the ice-water contact angle, θ_{iw}; the surface radius of curvature, r, non-dimensionalized through the critical ice embryo radius, r_i^{*}, the function $g = \left(1 + \left(\frac{r}{r_i^{*}}\right)^{2} - 2cos(\theta_{iw})\frac{r}{r_i^{*}}\right)^{1/2}$. This equation predicts that the value of f varies from 0 to 1 and thus, $\Delta G^{*}_{f,het}$ varies from 0 to $\Delta G^{*}_{f,hom}$. This equation is the best theoretical estimate of heterogeneous nucleation and provides comprehensive information of the factors influencing both homogeneous and heterogeneous nucleation. It indicates that the icephobic performance of a surface can be modified by tuning characteristics such as topography, crystal structure, and chemical composition, in such a way to tweak the interfacial tension and contact angle. That is an essential step to analyze ice nucleation experiments involving water on the surface and to evaluate the potential icephobic performance of the surface.

8.2.1.3 Shortcomings of Classical Nucleation Theory

One of the most significant advantages of classical nucleation theory is its simplicity. However, its predictions are not always in agreement with experimental results. A significant criticism directed to CNT is that the application of bulk thermodynamics properties to microscopic systems is not accurate. For example, the theory prediction for the free energy barrier, given by the equation, varies as γ_{iw}^3. Thus, a small change of the ice-water interfacial tension, γ_{iw}, leads to significant variation of the free energy barrier. Also, according to the capillary approximation, ice embryos have a spherical shape, while in reality, this is not always true. Depending on the shape, the value of interfacial ice-water tension, γ_{iw}, assumed by CNT, can differ from the real one. Several improvements on CNT have been proposed for specific cases, in the form of corrective factors to the CNT equations; moreover, several alternative theories—phenomenological, kinetic, and microscopic—have been developed [25]. However, due to its simplicity, CNT continues to be the most popular framework for the design and analysis of the experimental and theoretical work on ice nucleation.

8.3 Latest Advances on Surface Strategies Against Icing

In the past decade, research related to the development of surfaces against ice has been booming, as clearly illustrated in Fig. 4a. The terminology is, however, not always consistent. By checking the publications in the past 10 years, the most common words are: (i) "icephobic surfaces" or "icephobicity" (72% of papers), and (ii) "anti-icing surfaces" (22%); the remaining papers (6%) used both expressions.

The absence of a standard nomenclature is common in a newly emerging field. Also, different terminologies are due to different strategies that have been investigated and developed.

Surfaces can have functional properties against ice if they reduce ice adhesion stress (typically <100 kPa) [26–29], suppress the nucleation temperature, or delay the freezing time [30–33], but also if they promote liquid water removal from the surface by drop rebound or shedding before freezing can occur [34–37]. As such, anytime the expressions "icephobicity" and "anti-icing surfaces" are used, it must be clarified which of the above functionalities is being targeted.

In accordance with the majority of studies, in this chapter, we preferentially use the term "icephobicity", which we have also selected for the chapter title. Following the analogy with classical anti-icing and de-icing systems, the term "icephobicity" is used to describe the anti-icing or de-icing properties of a surface [38–41]. The term "anti-icing surfaces" will be primarily used for surfaces that are able to prevent ice formation, whereas the term "de-icing surfaces" for those that favor ice detachment due to reduced adhesion [38, 42].

8.3.1 Ice Adhesion

To understand how ice adheres to surfaces, many parameters need to be considered. As noted by H. Sojoudi et al. [43], it is necessary to identify what triggers ice formation and the environmental conditions in which it occurs (e.g., humidity). Different types of ice may form (for example, glaze ice or rime ice, wet snow, etc.) with different adhesion and cohesion strengths [13]. It is also necessary to consider the type of surface on which the ice forms, as each surface has specific mechanical, physical, and chemical properties that influence the acting forces, the types of interactions, and therefore the degree of ice adhesion to the surface.

The interactions between any substrate and ice deposited or grown on it are governed by a combination of forces, with the prominent ones being van der Waals forces, electrostatic forces, hydrogen bonding forces, and mechanical adhesion [44–46]. Capillary forces, e.g., in cases of wet snow, may also play a role. Finally, the presence of a liquid water at the ice-substrate interface (normally referred to as quasi-liquid layer, QLL, or liquid-like layer, LLL) may be relevant, as discussed in Sect. 8.4.

In the following paragraphs, we discuss various approaches and materials to obtain icephobic surfaces.

8.3.2 Icephobic Surfaces

Icephobicity has been pursued following several different experimental approaches. Successful methods exploit surface chemical modification using molecules with low energy groups to obtain icephobic behaviors (i.e., fluorinated molecules) [26, 47–52]. Alternatively, hydrophobic polymeric surfaces have been studied as a promising solution for anti-icing applications. In principle, low permittivity, a measure of the electric polarizability, allows hydrophobic polymer surfaces to lower molecular interactions (i.e., electrostatic forces) with water. Soft deformable materials have also been demonstrated to have icephobic potential [46, 53–64] because of a reduction in ice adhesion. The presence of lubricating liquids or free molecules that can flow in the materials was found to promote the sliding of water and ice at the interface in organogels and SLIPS (superhydrophobic liquid-infused porous surfaces) [63, 65–67]. Chemical surface patterning has helped in controlling the phenomenon of condensation and subsequent ice propagation due to ice bridges [68–71]. Moreover, many approaches based on bio-inspiration have developed interesting solutions, such as the application of antifreeze proteins (AFP) [72–84]. All these approaches, schematically illustrated in Fig. 8.7, are discussed in the next sub-sections.

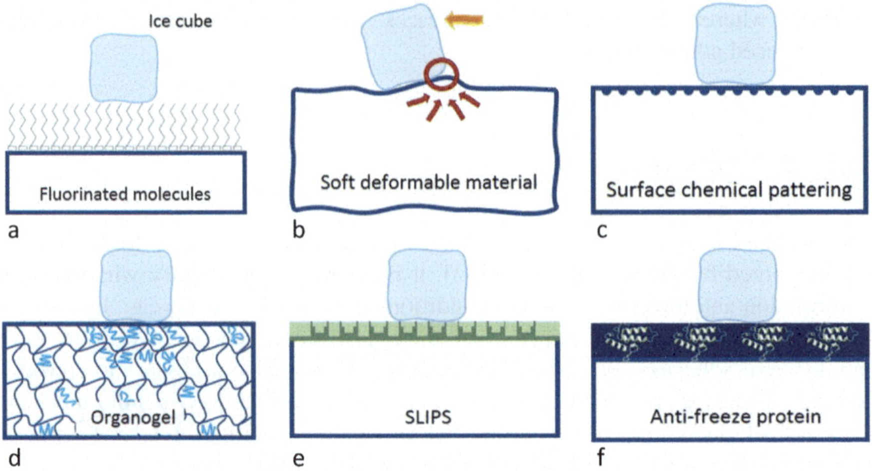

Fig. 8.7 Different approaches for icephobic applications; **a** surface modified with fluorinated molecules; **b** soft deformable material; **c** surface chemical patterning; **d** organogel; **e** SLIPS; **f** surface modified with antifreeze proteins

8.3.2.1 Fluorinated Materials

The most common and widely studied classes of anti-icing materials are fluorinated and siloxane-based polymers. Thanks to their exceptional low surface energy, due to $-CF_3$ groups, much lower than liquid water (<72 mN/m), they are natural candidates for anti-icing applications (Table 8.1).

The icephobicity of PTFE, poly(tetrafluoroethylene), for example, has been widely studied. Moreover, oxide surfaces (i.e., Al_2O_3, SiO_2, TiO_2) or metals are good substrates to host reactions between hydroxide surface groups and functionalizing molecules to obtain fluorinated materials, where the group orientation has attracted attention.

Yang et al. [48] systematically investigated the icephobic properties of a series of fluoropolymer materials including: (i) pristine PTFE plates, (ii) sandblasted PTFE

Table 8.1 Surface energy of materials (values reported from Ref. [46])

Material	Surface energy (mN m^{-1})
– CF_3 group	6
– CH_3 group	21
PTFE	20
PDMS	22
PVDF	25
Al	>100
Steel	>100

plates, (iii) PTFE coatings of different kinds. Smooth fluorinated surfaces were found to have very little ice adhesion (~100 kPa) if compared to a bare Al surface (~1500 kPa), while rough fluorinated surfaces showed a slightly worse behavior (~200 kPa). Fluorinated coatings on the other hand showed a higher ice adhesion ranging from 600 to 1300 kPa.

In other studies, anodized Al with a PTFE deposited coating, to combine low surface energy with micro-scale roughness, showed ice adhesion as low as ~140 kPa, decreasing with the increase of anodization cycles [49].

Very high morphological non-homogeneities such as micrometric Porous Superhydrophobic Polyvinylidene Fluoride (PVDF) were found to produce negligible ice accretion on wind turbine blades [50] due to the low adhesion and limited contact area between water drops and coating.

Although morphology plays a critical role, good anti-icing properties have also been obtained by orienting fluorinated functionalizing groups. By creating covalent bonds between surfaces and functionalization molecules, it is possible to grow uniform monolayers of self-assembling molecules, thus orienting fluorinated or other hydrophobic groups in the desired direction. Driving the low free energy groups toward the water interface allows maximizing the water repellence effect and, in principle, ice adhesion (Fig. 8.8).

The importance of molecule orientation has been investigated in flexible dangling fluorinated chains, with higher molecular mobility with respect to PTFE and other rigid structures. Water repellency was found to be connected to the molecular aggregation state, which is related to the polymer molecular weight dispersity [51]. Other studies proved that limiting the chain mobility through a crosslinked polymer layer underneath a thin upper layer of fluorinated molecules prevents local rearrangements, fixing the structure in the maximized state of water repellency. This approach resulted in lowering the contact angle hysteresis and ice adhesion by sixfold compared with the bare silicon and steel substates [26]. Alternatively, 1H, 1H, 2H, 2H-perfluorodecyl acrylate has been crosslinked with divinylbenzene aiming at preventing the inward orientation of fluorinated groups, thus obtaining a consistent reduction on the contact angle hysteresis, $\Delta\theta$ [52]. These functionalization approaches are based on wet chemistry or chemical vapor deposition which are sometimes difficult to scale up for industrial purposes.

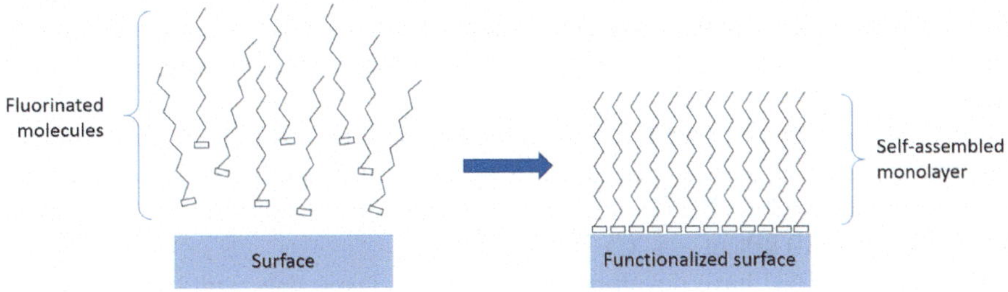

Fig. 8.8 Self-assembly of functionalizing molecules forming a monolayer on a solid surface

The more relevant papers in the field of anti-icing based on fluorinated materials date until 2015. More recently, fluorinated surfaces appear to be less appealing in comparison to other cheaper and environmentally safer alternatives, e.g., non-fluorinated siloxane-based materials.

8.3.2.2 Soft Deformable Materials

Polymers with an inorganic backbone, such as siloxane-based elastomers, have been studied as promising anti-icing solutions. Only a few examples of C-based elastomers have been reported as possible candidates for anti-icing purposes [53, 57].

The mechanism involved in elastomer icephobicity has been attributed not only to the low surface energy, related to methyl groups, but also to their viscoelastic properties [58]. Being a deformable soft material has been proved to impact on both wetting and ice adhesion [58].

Rheological properties, often determined by assessment of storage and loss moduli, are highly influenced by the crosslinking degree and by the molecular weight of polymeric chains. While the molecular weight is an intrinsic characteristic of the polymer, the crosslinking degree can be tuned by means of adjusting amounts of crosslinking agents or playing with the vulcanization kinetics.

The idea at the base of the low interaction between water and soft deformable materials (e.g., siloxane) is that, by tuning rheological-mechanical properties, not only are wetting properties modified [59], but low adhesion strength can also be achieved. This concept has been more in detail explored by means of tuning the material viscoelastic properties [60].

It has been reported that the best siloxane-based icephobic materials must have a low glass transition temperature (T_g). Lower T_g means more molecular flexibility and mobility of chains, which results in a softer material with low surface energy and thus lower ice adhesion [46].

Moreover, a low elastic modulus promotes easy detachment of ice under low applied loads. The critical shear stress of fractures has been demonstrated to decrease with the shear modulus [60].

Wang et al. [61] show that there is a dependence of ice adhesion stress on moduli and thickness of polydimethylsiloxane (PDMS). The high difference in elastic modulus between ice and PDMS leads to a concentration of stress building up on the front line, which enables easy detachment of ice from the elastomeric surface. Larger thickness corresponds to larger vertical displacements leading to increased stress.

Liu et al. [62] characterized the anti-icing behavior of soft PDMS with different elastic moduli, tuning the material elasticity by adding differently terminated siloxane molecules to the pristine PDMS. Ice adhesion stress on soft PDMS was found to be extremely low (~10 kPa at −5 °C).

Beemer et al. [63] have developed PDMS viscoelastic gels with tunable elastic modulus, by modifying the amount of different molecular weights of non-reactive trimethyl terminated PDMS, reporting ice adhesion stress of ~5 kPa.

Moreover, interface deformation and stiffness inhomogeneity of the substrate is recently being studied for boosting icephobicity. The introduction of a PDMS sponge-like structure was found to produce a dramatic reduction of ice adhesion stress [64]. Sandwich-like PDMS sponges with modulated elastic modulus result in low ice adhesion stress as low as ~1 kPa. He et al. [54] explained this behavior as a combination of nano- micro- and macro-effects, related, respectively, to surface chemistry, hierarchical surface structure, and maximization of the driving forces of crack initiation and propagation. More specifically, nanoeffects originate from the atomistic nature of the ice-solid interface, which is governed by surface chemistry. On the other hand, micro-effects are induced by the topography, creating a Cassie-state behavior and macro-effects are introduced by material sponge-like structures.

Irregular arrangements of polymer morphology were also found to be relevant in weakening ice-surface interactions. Yu et al. [55] found out that the occurrence of a microphase separation of PDMS containing polyacrylate copolymers is able to create PDMS chains aggregated on the top of the polymer surfaces, which have the effect of lowering ice adhesion.

A PDMS two-phase composite material showed very low ice adhesion (~2 kPa) [56]. The material was made of a softer PDMS micrometric phase dispersed in a continuous PDMS harder phase. The coating exploits the localization of the elastic energy at the interface to promote the ice sliding on the substrate. According to the mechanism proposed in the study, by applying a shear rate of 1 mm s^{-1}, ice is firstly detached from the soft dispersed phase, generating an elastic stress field that enables the detachment from the continuous harder phase.

More recently, other approaches to soft material wetting use the concept of poroelasticity [85]. Poroelasticity has been explained as the effect of the solvent flowing through the network, when the material undergoes compression. When crosslinked, siloxane-based polymers can be seen as a three-dimensional network of molecules, where non-crosslinked molecules can flow and migrate to the material surface [86]. Recent work has suggested that compressibility effects (i.e., poroelasticity) could play roles for swollen networks and gels. When the elastic network undergoes compression, the solvent flows through the network causing local changes to the relative concentration of solvent and the network. It has been seen that, for large drops and long timescales, wetting is dominated by the effect of poroelasticity. On the other hand, viscoelastic response dominates for small drops or short timescales. In this scenario, wetting and adhesion have been correlated to poroelasticity. As such, the effect of poroelasticity in the context of icing requires further investigation, to understand if the migration of non-crosslinked molecules at the ice-solid interface affects ice adhesion and ice detachment mechanism.

Combined siloxane-fluorinated strategies are also promising alternatives. Peng et al. [87] produced icephobic coatings of branched PDMS crosslinked with fluorinated polyhedral oligomeric silsesquioxane. The material showed low ice shear strength: the authors proposed that the patterned microstructure can facilitate stress concentration and promote microcracks development between ice and coating surface.

Recent studies based on siloxane as icephobic materials mainly focus on PDMS copolymers or composite materials, mixing and combing the above-mentioned approaches [88–90].

8.3.2.3 Organogel

In addition to elastomers, siloxane-based materials have also been used for the production of organogels. Organogels are semisolids with an organic liquid (immiscible with water) immobilized inside the crosslinked network of gelator aggregates. The network act as a slow-release reservoir of liquid to sustain the durability of the surface [63]. Organogels have recently been used for icephobic applications due to the low shear modulus and the presence of lubricating liquid which promotes the sliding of water and ice at the interface, similar to SLIPS (superhydrophobic liquid infused porous surfaces) and LIS (liquid-infused surfaces) [91–94].

Golovin et al. [65] have suggested that the insertion of small miscible polymeric chains (lubricant) in the network induces ice slippage at the interface. The author obtained ultra-low ice adhesion organogels (0.15 kPa) by changing the crosslink density of PDMS.

Wang et al. [66] have designed an organogel icephobic coating by swelling crosslinked polymer networks with liquid paraffin. Paraffin swollen crosslinked PDMS exhibits low ice adhesion stress (~2 kPa) at −30 °C and also down to −70 °C.

Organogels were also demonstrated to be a durable and reliable solution in harsh conditions [67].

8.3.2.4 Surface Chemical Patterning

Surface chemical patterning is a promising route to control a variety of phase-change phenomena, and is not limited to icing. For example, Rioboo et al. [68] showed experimentally that the position of the active nucleation sites in boiling can be controlled through chemical patterning of smooth surfaces. Nature also shows some examples of surface patterning for water collection [95]: the desert beetle Stenocara collects water from mist by a specific patterning of the overwings (elytra) on its back. A near-random array of hydrophilic bumps (~0.5 mm in diameter, ~0.5–1.5 mm apart) are located over an otherwise wax-coated, hydrophilic region. This pattern enables optimal collection and mobility of drops down the insect back to capture the drinking water.

In the context of icing, the idea of surface patterning has been used by Boreyko et al. [69, 70] to investigate and control the two-step condensation-freezing process on a smooth surface. Specifically, they focused on possible strategies to target the phenomenon of ice bridging. Ice crystals on surfaces, formed by frozen drops, harvest water from neighboring liquid drops creating ice bridges. Such bridges propagate across the surface and lead to the complete freezing of the condensate drops. This is clearly a critical cascade process: once nucleation is triggered on a part of the solid surface, ice bridges lead to freezing of the entire surface, even on hydrophobic surfaces with isolated drops.

Chemical patterning can help to control the phenomenon of condensation and subsequent ice propagation. Boreyko et al. [69] fabricated and tested surfaces with hydrophilic circles (10 μm, 20–80 μm apart) and lines on a hydrophobic background. They first found that there is an optimal spacing between hydrophilic spots, as a compromise between two competing requirements: on one hand, the need to separate the hydrophilic spots from each other, to increase the distance between condensation locations of isolated drops and reduce the ice-bridging effect; on the other hand, increase the number (and thus the density) of hydrophilic nucleation spots, to avoid undesirable and uncontrolled condensation on hydrophobic areas. For the investigated circular geometries, surfaces with a center-to-center distance four times higher than the hydrophilic spot diameter provided the best results in reducing the freezing front propagation speed (typically of the order of 10 μm/s).

Also, intentionally triggering icing at early stages may be desirable [69]: since the separation between drops decreases with increasing duration of condensation, early frozen drops are not close enough to neighbor drops and thus do not initiate the cascade ice-bridging process.

Finally, the formation of isolated ice islands can also be advantageous to promote ice shedding: as shown by icing wind tunnel tests [71], shedding of small isolated ice requires lower aerodynamic forces, when compared to the shedding of a compact ice front, as typically observed on hydrophilic surfaces.

8.3.2.5 Antifreeze Proteins

While searching for new strategies for icephobicity, we may need to turn back once again to nature and try to mimic strategies that have emerged through evolution in living organisms. One of such strategies is based on the presence of the so-called antifreeze proteins (AFPs) or ice-binding proteins (IBPs) [72, 73, 77–79]. Although the two terms are sometimes found as synonyms, strictly speaking, IBPs is the more general term to indicate proteins that bind ice and modify its structure; AFPs are a class of IBPs, used for proteins with specific antifreeze functionality.

AFPs can interact with ice crystals, modifying their shape, to control ice formation [80, 81]: specifically, AFPs functions are classified as "freeze avoidance", when crystal growth is prevented, and "freeze tolerance", when the growth of harmful large crystals at the expense of smaller ones, a process known as Ostwald ripening, is inhibited [80, 82, 83]. Such property enables organism survival even under exposure to subfreezing temperatures, which may, in absence of AFPs, lead to the organism's injury or death [79].

AFPs affect ice nucleation and growth mainly by thermal hysteresis (TH) and ice recrystallization inhibition (IRI or simply RI) [80, 82, 83]. In TH, AFPs promote a freezing point depression, with the freezing point lower than the melting point. TH activity may range from 0 to 2 °C in moderately active AFPs, and 2–13 °C in hyperactive AFPs. In IRI, AFPs act as inhibitors of the ripening effect. In our daily lives, we have all experienced ripening when storing ice-cream in the freezer for a long time: due to ripening, ice crystals grow and affect the macroscopic rheology of ice-cream, which loses its initial pleasant softness and becomes harder and harder. AFPs can either slow or stop ripening, and for this reason, AFPs can be used in the food industry to preserve the quality of ice cream [84].

Although AFPs have been studied for more than 40 years, the mechanism underpinning AFPs functions is still debated: we refer the reader to Refs. [79, 83] for a comprehensive discussion. Nonetheless, the generally accepted mechanism is called the so-called adsorption-inhibition mechanism. AFPs can bind to specific planes of a growing ice crystal, leading to ice crystal growth on a curved surface between two adjacent adsorbed proteins [74]. Based on the classical nucleation theory discussed in Sect. 8.2, this condition is energetically unfavorable and causes a depression in the freezing point, without affecting the energetics of melting; this leads to TH activity. The same mechanism may also explain IRI activity. Indeed, AFPs with TH activity also show IRI. However, there is not yet an understanding of how IRI and TH activities scale [77].

In the past few years, there have been attempts to fabricate anti-icing surfaces on solid substrates, such as aluminum [75] and glass [76]. Gwak et al. [75] used the AFP from the Antarctic marine diatom *Chaetoceros neogracile* (Cn-AFP) and bound it to an aluminum substrate, taking advantage of an Al-binding peptide (ABP). An aluminum surface coated by ABP-Cn_AFO was reported to decrease significantly the supercooling degree, before water freezes on the surface. Proteins can be easily denatured via temperature. In their study, Gwak et al. used trehalose, a disaccharide (i.e., sugar consisting of two molecules of glucose) to retain water molecules, helping proteins maintain their conformation under dehydration conditions and thus extending their lifetime, which is in the order of few days.

Along the same line, Kasahara et al. [76] fabricated a polypeptide-coated glass with polyethylene glycol. They reported that the combination of the polypeptide with polyethylene glycol led to a decrease of water supercooling on the surface, and of frost growth, compared to bare glass.

To conclude, there are still many open questions on the AFPs functions. However, we have just started to scratch the surface of the complex interaction phenomena between biomolecules and water or ice: there is certainly the opportunity to improve our understanding and thus provide a rational framework for the design of AFP-inspired anti-icing surfaces.

8.3.3 Experimental Testing for Anti-Icing Surfaces

8.3.3.1 Ice Adhesion Tests

The starting point of ice adhesion test analysis is that there is no standard and universally recognized method for uniquely measuring ice adhesion. Each ice adhesion measurement is often only comparable within the same laboratory conditions and a comparison between tests from different test rigs is not straightforward. In this regard, a measurement will more reliable when the experimental conditions are similar to real application conditions.

Ice adhesion stress (ζ_{adh}), or the area normalized tangential force required to remove ice from a surface it is adhered to, is assumed to be determined by surface-ice intermolecular interactions. At the macroscale, these interactions are quantified using the concept of *surface energy*, which typically comprises *polar* (electrostatics and hydrogen bonding) and *dispersive* (van der Waals interactions) surface energies. For a given surface, the polar and dispersive surface energies are calculated from liquid wettability, i.e., expressions relying on contact angle measurements with polar and apolar liquids, respectively. ζ_{adh} is then defined as the work per unit area needed to overcome these surface energies.

For example, in the case of tensile separation (e.g., when a surface is turned upside down and a drop detaches) the work of adhesion is $W_t = \gamma_{lv}(1 + \cos\theta_r)$, where θ_r is the receding contact angle [96]. A similar expression can be derived for shear separation (W_s), e.g., when a surface is tilted and a drop slides off, by replacing θ_r with contact angle hysteresis [97]; for *polar surfaces*, additional polar contributions are incorporated into the surface energy, but ζ_{adh} is still traditionally predicted from wettability alone for all surfaces.

Using wettability to predict ζ_{adh} assumes that, as the surface-energy characteristics of water and ice are very similar [98], their removal characteristics are likely to be highly correlated. In practice, however, this method has led to contradictory results with papers showing ζ_{adh} being best predicted by either W_s [99] or W_t [100], or neither [39]. This is because ζ_{adh} depends on fundamental intermolecular interactions between both ice-surface and ice-ice atoms. The stress response of (solid) ice is, therefore, different from (liquid) water [101]; water cannot resist shear stress but ice removal from a surface involves a complex combination of shear and tensile stresses acting at the interface [102], which needs to be considered. For example, inducing pockets of high interfacial stress can accelerate ice fracture and reduce ice adhesion [103]. Thus, accurately predicting ice adhesion requires that we look beyond simple wettability-based measures and take both the interface and the bulk ice microstructure into account. The common tests used in labs today can be classified as follows:

Horizontal shear test. In this method, water is put into containers (with predefined geometry such as small cylinders) positioned on the sample (Fig. 9a). The sample is placed in a freezer until the water is completely frozen. A lateral force is then applied to the ice block and the peak removal force is measured at the moment of detachment by means of a probe. ζ_{adh} is defined as the ratio of peak removal force to the interface area of ice.

Vertical shear test. Similar to the horizontal shear test but once the ice block has formed, the substrate is placed vertically, and the force is applied from top to bottom (Fig. 9b).

Centrifugal adhesion test. A propeller or board is coated with the icephobic material to be assessed (Fig. 9c). Ice is then made to grow on the upper part of the propeller by means of supercooled drop spray or inside test containers (as in the

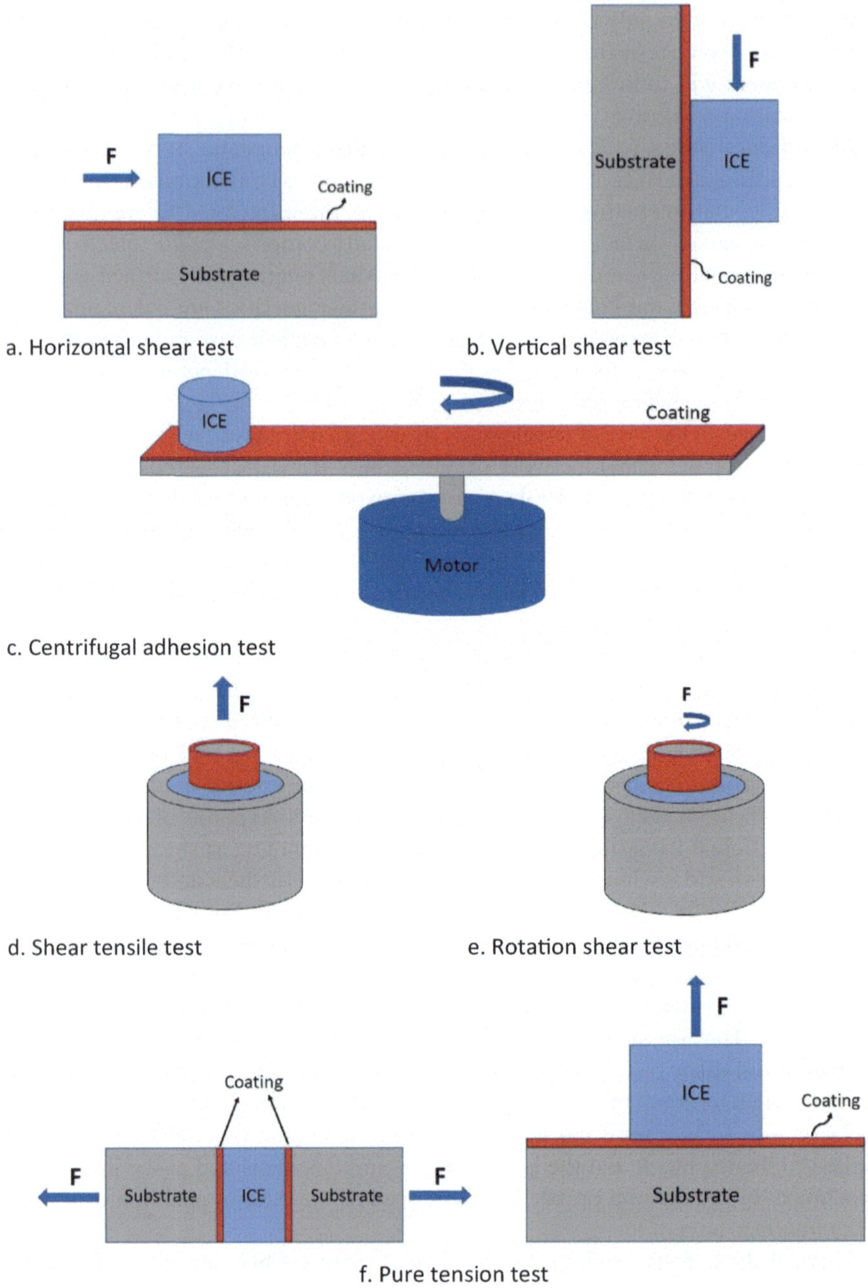

Fig. 8.9 Schematics of ice adhesion test configurations

previous methods). Finally, the motor is driven with gradually increasing rotational speed to induce a centrifugal force. The value of centrifugal force, at the moment of detachment, is divided by the ice-sample contact area to derive the ice adhesion stress.

Shear tensile test (0° cone test). Two concentric aluminum cylinders are used with a predefined space between them (Fig. 9d). First, the outer surface of the smaller cylinder is coated with icephobic material, while the inner surface of the larger cylinder remains unchanged. The smaller diameter cylinder is then inserted inside the larger one and the space between the cylinders is filled with water. The system is then stored in a freezer. Once ice has formed, a tensile force is applied to the smaller cylinder. Since the adhesion of ice to aluminum (i.e., to the inner surface of the larger cylinder) is high, the ice will only come off at the ice-coating interface of the smaller diameter cylinder. The resulting tensile force, measured at the moment of detachment, is divided by the area of the ice-substrate interface to derive the ice adhesion stress.

Rotation (or torsional) shear test. This test is analogous to the shear tensile test, however in this case the internal cylinder is not extracted but is made to rotate on itself by imparting torsion (Fig. 9e).

Pure tension test. Water is frozen between two substrates previously covered by the icephobic coating and subsequently, a gradually increasing traction force is applied to the two substrates (Fig. 9f). Alternatively, this test can be performed by freezing water on a sample previously coated with the icephobic material and applying a traction force to the ice formed.

8.3.3.2 Experimental Techniques Relevant for AFPs

As we mentioned AFPs, we briefly report here the two main methods used within the AFP community to assess the performance in icing conditions [77]. The first method is the so-called 'splat assay', in which a water drop impacts onto an ultracold metal surface from a height higher than 1 m, to attain thin wafers of ice grains; the temporal evolution of such grains is monitored to evaluate ice recrystallization inhibition. This method clearly presents some interesting and clear analogies with the study of water and molten drops on cold surfaces [104–107]. The second method is the 'sandwich assay', where multiple ice crystals are surrounded by solvent and sandwiched between two glass plates in high sucrose concentrations. Ice crystal size is monitored during a set time and analyzed to extract the rate constant for ice recrystallization. The data obtained from the two assays are not directly comparable due to differences in sample composition and assay conditions. As such, there is certainly room for improving and standardizing the tests, and possibly combining techniques used in the icing and AFP-research communities.

8.4 Numerical Simulations: Bridging the Gap Between Theory and Experiments

8.4.1 State-of-The-Art in Numerical Modelling

Even though the performance of experimental microscopic techniques is continuously improving, it is still extremely difficult to quantify the influence of a variety of factors on the ice nucleation experimentally. In theory, computational modeling can provide insight into physical phenomena, where experimental investigation remains challenging. However, as icing on surfaces is a multiscale phenomenon (nucleation occurs at the scale of a few hundred molecules and nanoseconds, whereas a supercooled drop is typically on the order of millimetres and seconds or beyond), one modeling tool cannot accurately model all spatial and temporal scales. Instead, the literature is broadly split between macroscale continuum modeling and nanoscale molecular modeling, with little in between, summarized below.

8.4.2 Continuum Modeling: Achievements, Strengths, and Limitations

Continuum modeling has been used in surface icing modeling primarily for two scenarios: (a) to study the impact and freezing of supercooled large drops, and (b) to look at the forces experienced by and stability of formed ice and predict fracture mechanics. Broadly speaking, the term "continuum modeling" can mean one of two approaches. The first kind of approach involves directly simulating particles, called Lagrangian methods. A Lagrangian method directly solves equations of motion for the trajectory of each particle, and the water drop physics is reproduced by averaging or sampling across many particles at different locations. This method can directly obtain the motion trajectory for each particle, and is additionally can reproduce the dynamics of splashing, daughter drop formation, etc., directly. The second approach which has been used for the majority of studies in the literature is the Eulerian method which treats drops as a continuous medium, and uses conservation equations (typically mass, momentum, and energy). In this case, the resulting set of partial differential equations are approximated by discretizing the computational domain into a series of subdomains to resolve the motion of the drop. Inside each subdomain, the PDE can be simplified into a set of linear algebraic equations, and the solution is then obtained by simultaneously solving them numerically. Phase change (so freezing) is then achieved by specifying turning individual subdomains from liquid to solid when certain conditions have been met. The most common Eulerian techniques are Finite Volume Method (FEM) and Finite Element Method (FEM).

There are many well-established continuum codes at this point, from NASA's LEWICE [108], FENSAP [109], the recently-developed WISE [110], etc. Each of these is usually designed with a specific application in mind (e.g., TURBICE[111] is used exclusively for wind turbine modeling). These continuum codes can be used to solve the flow fields, drop trajectory, impact, and ice accretion, and obtain final ice structures in various meteorological conditions. Therefore, they are able to estimate the role of parameters such as temperature and drop size variation on the accreted ice and show how the aerodynamics of the structure is affected by the added ice. Additionally, these codes have also been sometimes used to optimize anti-icing techniques by simulating ice fracture as thermal or electric loads are applied. Good reviews on the modeling of icing in general [112], and for specific applications such as marine icing [113] already exist. However, continuum modeling cannot take into account the role of the surface in inducing or delaying nucleation within the supercooled liquid above it, as this requires high-resolution information about the surface-ice interactions. Thus, they cannot be used to design or validate new icephobic surface designs, which is the goal of the chapter.

8.4.3 Molecular Modeling: Achievements, Strengths, and Limitations

As the intermolecular forces between the ice and surface act at the nanoscale, the molecular nature of this interface can no longer be ignored. The interfacial dynamics do not, in reality, occur on the surface as continuum models assume, but over a region less than a few nanometres thick adjacent to the surface; within this interfacial region, the molecules arrange in semi-ordered layers, existing in neither a fully liquid- nor solid-like state. The importance of this region on interfacial phenomena in general and for icing, in particular, is well-documented and is typically accessible only using molecular modeling, which involves computing the trajectory of each molecule and updating their velocities using Newton's second law when they interact with each other. Molecular Dynamics (MD) tracks individual atoms and computes interatomic interactions at every timestep, and is highly accurate in reproducing physical phenomena across the engineering sciences. MD is extremely computationally expensive to simulate even very small systems ($\sim 10^6$ atoms) for very small periods of time ($\sim 10^{-6}$ s). MD has, however, been widely used to simulate icing phenomena due to its high accuracy and ability to resolve nanoscale physics. In the remainder of this section, we focus on summarizing MD results for the three important issues discussed thus far, namely ice nucleation, surface wettability, role of interfacial region, and ice adhesion, respectively.

8.4.3.1 Ice Nucleation

Ice nucleation mechanism is influenced by several independent molecular factors such as hydrophobicity [114, 115], surface morphology [116], and charge distribution [117]. A comparative evaluation of experimental results and computer simulations of ice nucleation indicated that ice has higher nucleation rates at the surface regions with a considerable number of defects. Consequently, the objective of many MD simulations of ice nucleation is to find and quantify the relation between ice nucleation rate and the different molecular properties, such as water-surface interactions and surface morphologies. As such, early computational investigations targeted on clarifying how the surface affected the first layers of water. Pruppacher et al. [118] summarised the requirements of a good ice nucleation agent (INA). Referring to a solid surface, these were:

"(a) *Hydrophobicity or chemical bond requirement* - Given that ice stands united by hydrogen bonds, ice nucleation was predicted to be favored at surfaces having available similar hydrogen bonds; (b) *Crystallographic requirement*—The closer the crystallographic matching of the surface to ice, the better its ice-nucleating ability was expected to be (e.g., salts, such as silver iodide, AgI) [117]; (c) *Active site requirement*—Ice nucleation could also be induced by the presence of an active site. An active site could be many things, but most commonly is subdivided into (i) morphological surface inhomogeneities including corrugations, cracks, or cavities at the surface; (ii) chemical inhomogeneities in the surface, usually originated by the presence of foreign ions that are hydrophilic about the rest of the solid surface; and (iii) electrical inhomogeneities, such as sharply delimited boundaries between surface regions of different electric field sign".

Although the criteria listed above can be instrumental in identifying, by MD simulations, the surfaces promoting or inhibiting ice nucleation, in many cases these fall short. For instance, the widely accepted notion that ice nucleation agents have surfaces similar to the surface of ice has recently been questioned [119], and even amorphous surfaces have been shown to act as good ice nuclei [120]. Water on crystal surfaces nucleates almost exclusively heterogeneously, with nucleation rates depending on the surface characteristics. Extensive research work has been focused on understanding how the surface affecting the first few layers of water (on the order of a nanometre from the surface), in particular concerning the second criteria listed above, which is the crystallographic requirement or lattice "mismatch", where a surface that has a structure alike with ice and acts as a template for the crystal. Cox et al. [116] investigated the role of lattice mismatch employing model hexagonal surfaces. They noticed that, for atomically flat surfaces, a nominally zero lattice mismatch generated disordered contact layers containing smaller rings and observed hexagonal ice-like layers only for surfaces with larger lattice constants. They instead concluded that, rather than comparing the surface lattice structure to that of ice, a good match between the surface lattice constant and the averaged adjacent oxygen-oxygen distance in the first liquid layer was a better predictor surface icing characteristics. This is because the first layer of water is able to change topology and bind to the surface as a flat layer (with a fixed oxygen-oxygen distance) without needing to form hexagonal rings.

In keeping with this theme, various studies provided evidence that contrary to Pruppacher et al. [118], one surface characteristic alone is insufficient to explain a material ice-nucleating ability. Glatz et al. [121] placed hydroxyl ions in different configurations on the surface to determine how charge interacts with other surface factors and the interfacial water, to influence ice nucleation. Additionally, they used multiple-site models (which model each atom individually inside the water molecule) in their simulations and were, able to find correlations between the water orientation (which cannot be obtained using simpler single-site water models which assume the molecule to be a sphere, and thus lack rotational degrees of freedom) and the ice nucleation propensity of the surface. They concluded that, in addition to hydrogen bonding and lattice match, the precise location of the hydroxyl ions played a role in the probability of the water molecules assuming configurations similar to the bilayer structure of ice, which in turn increased the probability of ice nucleation. Fitzner et al. [122] examined the ability of genetic crystalline substrates to promote ice nucleation as a function of the hydrophobicity and the surface morphology. They indicated three conditions where a crystalline surface can promote heterogeneous ice nucleation: (a) The formation of a water overlayer that acts as an in-plane template; (b) The emergence of a contact layer buckled in an ice-like manner; and (c) nucleation on compact surfaces with very high interaction strength.

8.4.3.2 Role of Wettability

Recently Lupi et al. [123], using a coarse-grained single-site water model, found that rough amorphous surface did not enhance ice nucleation, whereas smooth graphite surfaces promoted ice nucleation. That was attributed to the fact that the graphite surface induced a layering in the density profile of water above the surface, while the rough amorphous surface did not. The same research group using the same methodology investigated the impact of hydrophilicity on the ice nucleation propensity of graphite surfaces [114]. The hydrophilicity of the surface was modified in two different ways: first by modifying the water-surface interaction strength; and secondly by introducing hydrophilic components at the surface. For temperatures around −60 °C, it was discovered that modifying the interaction potential led to enhanced ice nucleation and extent layering in water density profile above the surface, while increasing the density of hydrophilic components was adverse to both layering and nucleation ability. Lupi and Molinero concluded that the density layering in a nanoscopic interfacial region correlated well with ice nucleation efficiency, rather than the wettability. Interestingly, Li et al. [124] found a similar trend for the nucleation of microdrops for temperatures close to −40 °C. In previous work, Goertz et al. [125] demonstrated that the viscosity of the interfacial liquid layer is higher than that of bulk water, the effect being prominent on hydrophilic surfaces. As such, an increase in viscosity on hydrophilic surfaces decreases the diffusion of water molecules at the interface, causing an increase of the activation energy barrier Δg_{act}

(which becomes relevant for supercooling higher than 30 °C, as discussed above) [15].

Cox et al. [126] studied the effects of the interfacial layering by examining the nucleation rates, with changes induced by varying wettability via different morphologies. They concluded that interfacial layering of water above the surface could boost ice nucleation propensity, but only when the surface provided a potential energy landscape to the water molecules that favored ice-like bilayer structures to form. They also found ice nucleation to be inhibited at higher wettabilities.

MD studies of heterogeneous icing have additionally shed some light on the dynamic relationship between surface properties (e.g., lattice structure [122], charge [117], etc.) and interfacial region properties within the supercooled liquid (e.g., density peaks [115], hydrogen bonding [121], etc.) which favor ice nucleation. While these represent interesting discoveries, an overarching theory of ice nucleation has still not been developed, and papers still focus on testing one aspect of surface structure or morphology or chemistry at a time.

8.4.3.3 Understanding the Interfacial Region

Another challenging problem that is potentially tractable for MD simulations is that of the interfacial region characterization. Unique surface-induced molecular configurations in the interfacial region have also been shown to be key to minimizing ice adhesion; examples include: (a) a region of reduced density (up to ~1.5 Å thick for the SAM surfaces described earlier [127]) or density-depleted layer (DDL), can form near smooth hydrophobic surfaces [128], (b) an nm-thick Quasi-Liquid Layer (QLL) has been observed in simulations involving generic crystal surfaces, analogous to the lubricant layer on slippery low surface energy surfaces (but which unlike SLIPS does not deplete with time). Experiments have shown that the formation of DDL or QLL results in extremely low ice adhesion stress values, far lower than what surface-energy models predict [129]. Thus, designing novel icephobic surfaces could require understanding how to induce adhesion-minimising configurations within this nanoscale interfacial region. While the understanding of interfacial region structure has been increasingly advanced in recent years mostly through experimental studies as discussed in earlier sections, there remain numerous gaps. Given the length scale and the limitations of experimental studies, molecular simulations provide a useful alternative tool to get nanoscale insight into the structure and dynamics of this region, and probe into the formation of DDL and QLL.

In general, the results of the studies above show consistency between simulations and water models, reporting QLL thickness of at most 1 nm [130], which is about one order of magnitude below most experiments. Kling et al. [131] addressed the molecular structure and also the dynamics of the QLL at the ice surfaces as a function of temperature by MD simulations. Their results highlight substantial differences between 2D QLL and bulk water, stemming from reduced dimensionality and the template effect of the underlying bulk ice layers. Furthermore, they showed that a meaningful characterization of the ice surface originates from concurrently taking into account both local and midrange structural properties as well as dynamical properties. Thus, although MD simulations have looked at QLL forming at the ice-vapor interface, there are still open questions with regard to the structure and dynamics of the QLL in the ice-solid interface.

8.4.3.4 Ice Adhesion

The literature on ice adhesion using MD is limited to very few papers at this point, and is an area that we identify as poised for growth. Xiao et al. [132] were the first to investigate ice adhesion purely through molecular modeling, by applying horizontal and vertical shear tests to an ice block placed onto a silicon or graphene surface. They reported adhesion stress values between 250 and 450 MPa depending on the surface, which is two orders of magnitude higher than that observed in experiments. They attributed this overestimation to the high loading rate used in the simulations (6–7 orders of magnitude greater than experiment) which is necessary to get the ice detachment to occur within the timescale of MD simulations. They also investigated the role of a liquid layer at the interface, and found that it reduced adhesion. Metya et al. [133] studied ice adhesion to SLIPS by adding a lubricant layer between a graphitic surface and the ice above it, and applying a vertical shear force. They found that the adhesion stress depends both on the surface patterning and the lubricant thickness. Bao et al. [134] looked at ice adhesion on single-wall carbon nanotubes (curved surfaces) and found it to be up to 45% lower than simple (flat) graphene surfaces. They also reproduced the experimentally observed linear relationship between wettability (parameterized by carbon-oxygen interaction energy at the nanoscale) and the measured adhesion stress, which has since been independently reported [135]. Finally, a recent study used a surface inspired by "fish skin" to show that the effect of surface structure could influence the ice adhesion, by creating a pathway for cracks to propagate, enabling quicker ice rupture and detachment [136], which has conceptually been observed in the experiment. A common limitation in the listed studies is the high adhesion stress values reported (>100 MPa) due to the high loading imposed, and the small parameter space explored in each individual paper.

While the high loading is a problem that needs to be addressed, a novel way to rapidly search large parameter space may be to use advanced data science techniques to investigate high-dimensional data sets. A recent paper [137] attempts to combine molecular simulations with Machine Learning (ML), by training a popular supervised ML algorithm called *support vector machines* on a dataset of molecular simulation results, and then using the algorithm to predict ice adhesion (or to be more precise, probability of detachment for a given pulling force) for a range of surfaces for which the simulation outcome was not available to the algorithm. This approach showed encouraging results, indicating that ML algorithms may represent yet another way of getting around having to run many computationally expensive molecular simulations.

8.4.4 Summary

We have outlined in this section that icing is challenging to study using continuum simulations, and a high-resolution simulation method such as Molecular Dynamics (MD) is required. Following a comprehensive review of the MD literature, we have identified the following research gaps. First, nucleation studies are limited to testing one surface modification at a time, and a broader approach is needed. Second, the liquid layer at the solid-ice interface is of keen interest to experimentalists but is currently understudied using MD. Finally, designing new icephobic surfaces requires being able to predict and optimize adhesion stresses by modifying surface properties, to design surfaces for which no data exists. This is a complex, multi-factor optimization problem; a brute-force approach is not feasible as it would require too many simulations.

8.5 Conclusions and Perspective on Future Research

In this chapter, we have discussed how research has addressed the problem of icing, and specifically how surface properties can be tuned to prevent or control ice formation. Many advances have been done in the past ten years; nonetheless, many challenges and open questions are still open, both from a fundamental and a technological perspective.

In the next few years, bridging experiments and simulations will certainly help in improving our understanding of icing and providing new design rules for innovative surface solutions. Such a bridging is made possible because of recent progress on both sides: from an experimental perspective, micro/nano-fabrication capability as well as new characterization tools down to the nanoscale, and from the numerical side the increasing computational power of MD.

Nucleation mechanisms take place in nano-to-microsecond time scale and length scales of a few hundred to thousand molecules. These are difficult to be investigated in experiments. In principle, the scales of these mechanisms are ideal for Molecular Dynamics (MD) simulations. As a result, several computational studies provide valuable information about various factors affecting ice nucleation, such as surface wettability, interfacial structure, and interfacial dynamics. Using this information, experiments may be able to improve their design, achieving more reliable performance. Similarly, experimental results can lead to better targeted and useful simulations.

Accurately modeling ice adhesion requires both nanoscale information at the interface as well as macroscale details relating to the bulk ice; this requires a multiscale model that couples a high-resolution method for the interfacial region with a coarse-grained model for the bulk ice microstructure. Therefore, a continuum mechanics (CM) method (e.g., finite volumes) can be instead used to model the bulk ice microstructure which we identify as important but is beyond the capabilities of MD, and a hybrid MD-CM model prediction may be directly comparable to experimental data. Alternative approaches involve the use of machine learning based approaches [137], which is also likely to grow significantly in the future.

From the experimental standpoint, several techniques now enable the investigation of materials and their properties down to the micro- and nanoscale and will find more and more use in studying ice interaction with surfaces. To name a few examples: (i) X-ray tomography [138], (ii) environmental scanning electron microscopy [93, 139], (iii) atomic force microscopy combined with localized chemical characterization techniques such as Fourier Transform Infrared (FTIR) or Raman spectroscopy [18], (iv) atom probe tomography [140], and (v) nanoscale thermal metrology [141, 142]. Such experimental techniques, in combination with MD, will be helpful to investigate nanoscale phenomena, such as the quasi-liquid layer and its properties, helping to unravel the fundamental mechanism of ice formation on surfaces that are not yet completely understood.

In summary, this detailed investigation into the literature highlights not only the wide range of new developments, but also the vast potential for future growth in multiple identified areas. We conclude by stating that there's still plenty of room at the bottom [143] in the development of icephobic surfaces.

References

1. Poots G, Gent RW, Dart NP, Cansdale JT. Aircraft icing. Philos Trans R Soc London Ser A Math Phys Eng Sci 2000;358:2873–911. https://doi.org/10.1098/rsta.2000.0689.
2. Appiah-Kubi P, Martos B, Atuahene I, William S. U.S. Inflight Icing Accidents and Incidents, 2006 to 2010. 2013.
3. Farzaneh Masoud. Overview of Impact & Mitigation of Icing on Power Network Equipment Masoud. Https://WwwInmrCom 2019.
4. Dalili N, Edrisy A, Carriveau R. A review of surface engineering issues critical to wind turbine performance. Renew Sustain Energy Rev 2009;13:428–38. https://doi.org/10.1016/j.rser.2007.11.009.
5. Farzaneh M, Gauthier H, Castellana G, Englebrecht C, Elìasson ÁJ, Fikke S, et al. Coatings for Protecting Overhead Power in Winter Conditions Network Equipment. 2015.
6. Jones K, Mulherin N. An evaluation of the severity of the January 1998 ice storm in northern New England. Rep FEMA Reg 1 1998:33.
7. Henson W, Stewart R, Kochtubajda B. On the precipitation and related features of the 1998 Ice Storm in the Montréal area. Atmos Res 2007;83:36–54. https://doi.org/10.1016/j.atmosres.2006.03.006.
8. Gyakum JR, Roebber PJ. The 1998 ice storm-analysis of a planetary-scale event. Mon Weather Rev 2001;129:2983–97. https://doi.org/10.1175/1520-0493(2001)129<2983:TISAOA>2.0.CO;2.
9. Elíasson ÁJ. Wet-snow accumulation A study of two severe events in complex terrain in Iceland 2012.
10. André Leblond, Fikke SM, Wareing B, Chereshnyuk S, Elíasson ÁJ, Farzaneh M, et al. Guidelines for meteorological icing models, statistical methods and topographical effects. vol. 225. 2006.

11. Fikke S, Persson P-E, Wareing B, Chum J, Makkonen L, Ronsten G, et al. COST 727: Atmospheric Icing on Structures Measurements and data collection on icing: State of the Art. Meteo Schweiz 2007:115.
12. Huang X, Tepylo N, Pommier-Budinger V, Budinger M, Bonaccurso E, Villedieu P, et al. Ice-phobic surfaces that are wet. Appl Surf Sci 2011;6:5166–72. https://doi.org/10.1016/j.cis.2019.04.005.
13. Farzaneh M, Jakl F, Arabani MP, Elìasson AJ, Fikke SM, Gallego A, et al. Systems for prediction and monitoring of ice shedding, anti-icing and de-icing for power line conductors and ground wires. vol. 10. 2011. https://doi.org/10.1016/S1474-4422(11)70033-3.
14. Schutzius TM, Jung S, Maitra T, Eberle P, Antonini C, Stamatopoulos C, et al. Physics of Icing and Rational Design of Surfaces with Extraordinary Icephobicity. Langmuir 2015;31:4807–21. https://doi.org/10.1021/la502586a.
15. Amirfazli A, Antonini C. Fundamentals of Anti-Icing Surfaces. Non-wettable Surfaces Theory, Prep. Appl., Royal Society of Chemistry; 2016, p. 319–46. https://doi.org/10.1039/9781782623953-00319.
16. Stone H a. Ice-phobic surfaces that are wet. ACS Nano 2012;6:6536–40. https://doi.org/10.1021/nn303372q.
17. Irajizad P, Nazifi S, Ghasemi H. Icephobic surfaces: Definition and figures of merit. Adv Colloid Interface Sci 2019. https://doi.org/10.1016/j.cis.2019.04.005.
18. Huang X, Tepylo N, Pommier-Budinger V, Budinger M, Bonaccurso E, Villedieu P, et al. A survey of icephobic coatings and their potential use in a hybrid coating/active ice protection system for aerospace applications. Prog Aerosp Sci 2019. https://doi.org/10.1016/j.paerosci.2019.01.002.
19. Shen Y, Wu X, Tao J, Zhu C, Lai Y, Chen Z. Icephobic materials: Fundamentals, performance evaluation, and applications. Prog Mater Sci 2019;103:509–57. https://doi.org/10.1016/j.pmatsci.2019.03.004.
20. Volmer M, Weber A. Keimbildung in übersättigten Gebilden. Zeitschrift Für Phys Chemie 2017;119U:277–301. https://doi.org/10.1515/zpch-1926-11927.
21. Becker R, Döring W. Kinetische Behandlung der Keimbildung in übersättigten Dämpfen. Ann Phys 1935;416:719–52. https://doi.org/10.1002/andp.19354160806.
22. Turnbull D, Fisher JC. Rate of nucleation in condensed systems. J Chem Phys 1949;17:71–3. https://doi.org/10.1063/1.1747055.
23. Bai XM, Li M. Test of classical nucleation theory via molecular-dynamics simulation. J Chem Phys 2005;122:1–4. https://doi.org/10.1063/1.1931661.
24. Lamb D, Verlinde J. Physics and Chemistry of Clouds. Cambridge: Cambridge University Press; 2011. https://doi.org/10.1017/CBO9780511976377.
25. Ickes L, Welti A, Hoose C, Lohmanna U. Classical Nucleation Theory of homogeneous freezing of water: Thermodynamic and kinetic parameters. Phys Chem Chem Phys 2015;17:5514–37. https://doi.org/10.1039/c4cp04184d.
26. Sojoudi H, McKinley GH, Gleason KK. Linker-free grafting of fluorinated polymeric cross-linked network bilayers for durable reduction of ice adhesion. Mater Horizons 2015;2:91–9. https://doi.org/10.1039/c4mh00162a.
27. Guerin F, Laforte C, Farinas M-I, Perron J. Analytical model based on experimental data of centrifuge ice adhesion tests with different substrates. Cold Reg Sci Technol 2016;121:93–9. https://doi.org/10.1016/j.coldregions.2015.10.011.
28. Rønneberg S, He J, Zhang Z. The need for standards in low ice adhesion surface research: a critical review. J Adhes Sci Technol 2020;34:319–47. https://doi.org/10.1080/01694243.2019.1679523.
29. Jamil MI, Zhan X, Chen F, Cheng D, Zhang Q. Durable and Scalable Candle Soot Icephobic Coating with Nucleation and Fracture Mechanism. ACS Appl Mater Interfaces 2019;11:31532–42. https://doi.org/10.1021/acsami.9b09819.
30. Yang H, Ma C, Li K, Liu K, Loznik M, Teeuwen R, et al. Tuning Ice Nucleation with Supercharged Polypeptides. Adv Mater 2016;28:5008–12. https://doi.org/10.1002/adma.201600496.

31. Moriya T, Manabe K, Tenjimbayashi M, Suwabe K, Tsuchiya H, Matsubayashi T, et al. A superrepellent coating with dynamic fluorine chains for frosting suppression: effects of polarity, coalescence and ice nucleation free energy barrier. RSC Adv 2016;6:92197–205. https://doi.org/10.1039/C6RA18483A.
32. Bai G, Gao D, Liu Z, Zhou X, Wang J. Probing the critical nucleus size for ice formation with graphene oxide nanosheets. Nature 2019;576:437–41. https://doi.org/10.1038/s41586-019-1827-6.
33. Liu M, Hou Y, Li J, Tie L, Guo Z. Transparent slippery liquid-infused nanoparticulate coatings. Chem Eng J 2018;337:462–70. https://doi.org/10.1016/j.cej.2017.12.118.
34. Zheng L, Li Z, Bourdo S, Khedir RK, Asar MP, Ryerson CC, et al. Exceptional Superhydrophobicity and Low Velocity Impact Icephobicity of Acetone-Functionalized Carbon Nanotube Films. Langmuir 2011;27:9936.
35. Eberle P, Tiwari MK, Maitra T, Poulikakos D. Rational nanostructuring of surfaces for extraordinary icephobicity. Nanoscale 2014;6:4874–81. https://doi.org/10.1039/c3nr06644d.
36. Antonini C, Innocenti M, Horn T, Marengo M, Amirfazli A. Understanding the effect of superhydrophobic coatings on energy reduction in anti-icing systems. Cold Reg Sci Technol 2011;67:58–67. https://doi.org/10.1016/j.coldregions.2011.02.006.
37. Jung S, Dorrestijn M, Raps D, Das A, Megaridis CM, Poulikakos D. Are Superhydrophobic Surfaces Best for Icephobicity? Langmuir 2011;27:3059. https://doi.org/10.1021/la104762g.
38. Brassard JD, Laforte C, Guerin F, Blackburn C. Icephobicity: Definition and measurement regarding atmospheric icing. Adv Polym Sci 2019;284:123–43. https://doi.org/10.1007/12_2017_36.
39. Hejazi V, Sobolev K, Nosonovsky M. From superhydrophobicity to icephobicity: forces and interaction analysis. Sci Rep 2013;3:2194. https://doi.org/10.1038/srep02194.
40. Golovin K, Kobaku SPR, Lee DH, DiLoreto ET, Mabry JM, Tuteja A. Designing durable icephobic surfaces. Sci Adv 2016;2:e1501496. https://doi.org/10.1126/sciadv.1501496
41. Wang Y, Liu J, Li M, Wang Q, Chen Q. The icephobicity comparison of polysiloxane modified hydrophobic and superhydrophobic surfaces under condensing environments. Appl Surf Sci 2016;385:472–80. https://doi.org/10.1016/j.apsusc.2016.05.117.
42. Parent O, Ilinca A. Anti-icing and de-icing techniques for wind turbines: Critical review. Cold Reg Sci Technol 2011;65:88–96. https://doi.org/10.1016/j.coldregions.2010.01.005.
43. Sojoudi H, Wang M, Boscher ND, McKinley GH, Gleason KK. Durable and scalable icephobic surfaces: Similarities and distinctions from superhydrophobic surfaces. Soft Matter 2016;12:1938–63. https://doi.org/10.1039/c5sm02295a.
44. Ryzhkin IA, Petrenko VF. Physical Mechanisms Responsible for Ice Adhesion. J Phys Chem B 1997;101:6267–70. https://doi.org/10.1021/jp9632145.
45. Wilen LA, Wettlaufer JS, Elbaum M, Schick M. Dispersion-force effects in interfacial premelting of ice. Phys Rev B 1995;52:12426–33. https://doi.org/10.1103/PhysRevB.52.12426.
46. Menini R, Farzaneh M. Advanced icephobic coatings. J Adhes Sci Technol 2011;25:971–92. https://doi.org/10.1163/016942410X533372.
47. Somlo B, Gupta V. A hydrophobic self-assembled monolayer with improved adhesion to aluminum for deicing application. Mech Mater 2001;33:471–80. https://doi.org/10.1016/S0167-6636(01)00068-0.
48. Yang S, Xia Q, Zhu L, Xue J, Wang Q, Chen QM. Research on the icephobic properties of fluoropolymer-based materials. Appl Surf Sci 2011;257:4956–62. https://doi.org/10.1016/j.apsusc.2011.01.003.
49. Ghalmi Z, Farzaneh M. Experimental investigation to evaluate the effect of PTFE nanostructured roughness on ice adhesion strength. Cold Reg Sci Technol 2015;115:42–7. https://doi.org/10.1016/j.coldregions.2015.03.009.
50. Peng C, Xing S, Yuan Z, Xiao J, Wang C, Zeng J. Preparation and anti-icing of superhydrophobic PVDF coating on a wind turbine blade. Appl Surf Sci 2012;259:764–8. https://doi.org/10.1016/j.apsusc.2012.07.118.

51. Yamaguchi H, Kikuchi M, Kobayashi M, Ogawa H, Masunaga H, Sakata O, et al. Influence of molecular weight dispersity of poly{2-(perfluorooctyl)ethyl acrylate} brushes on their molecular aggregation states and wetting behavior. Macromolecules 2012;45:1509–16. https://doi.org/10.1021/ma202300r.
52. Yagüe JL, Gleason KK. Enhanced cross-linked density by annealing on fluorinated polymers synthesized via initiated chemical vapor deposition to prevent surface reconstruction. Macromolecules 2013;46:6548–54. https://doi.org/10.1021/ma4010633.
53. Valentini L, Bittolo Bon S, Hernández M, Lopez-Manchado MA, Pugno NM. Nitrile butadiene rubber composites reinforced with reduced graphene oxide and carbon nanotubes show superior mechanical, electrical and icephobic properties. Compos Sci Technol 2018;166:109–14. https://doi.org/10.1016/j.compscitech.2018.01.050.
54. He Z, Zhuo Y, He J, Zhang Z. Design and preparation of sandwich-like polydimethylsiloxane (PDMS) sponges with super-low ice adhesion. Soft Matter 2018;14:4846–51. https://doi.org/10.1039/c8sm00820e.
55. Yu D, Zhao Y, Li H, Qi H, Li B, Yuan X. Preparation and evaluation of hydrophobic surfaces of polyacrylate- polydimethylsiloxane copolymers for anti-icing. Prog Org Coatings 2013;76:1435–44. https://doi.org/10.1016/j.porgcoat.2013.05.036.
56. Irajizad P, Al-Bayati A, Eslami B, Shafquat T, Nazari M, Jafari P, et al. Stress-localized durable icephobic surfaces. Mater Horizons 2019;6:758–66. https://doi.org/10.1039/c8mh01291a.
57. Valentini L, Bittolo Bon S, Pugno NM, Hernandez Santana M, Lopez-Manchado MA, Giorgi G. Synergistic icephobic behaviour of swollen nitrile butadiene rubber graphene and/or carbon nanotube composites. Compos Part B Eng 2019;166:352–60. https://doi.org/10.1016/j.compositesb.2018.11.095.
58. Zhuo Y, Xiao S, Amirfazli A, He J, Zhang Z. Polysiloxane as icephobic materials – The past, present and the future. Chem Eng J 2021;405:127088. https://doi.org/10.1016/j.cej.2020.127088.
59. Lee JB, Dos Santos S, Antonini C. Water Touch-and-Bounce from a Soft Viscoelastic Substrate: Wetting, Dewetting, and Rebound on Bitumen. Langmuir 2016;32:8245–54. https://doi.org/10.1021/acs.langmuir.6b01796.
60. Chaudhury MK, Kim KH. Shear-induced adhesive failure of a rigid slab in contact with a thin confined film. Eur Phys J E 2007;23:175–83. https://doi.org/10.1140/epje/i2007-10171-x.
61. Wang C, Fuller T, Zhang W, Wynne KJ. Thickness dependence of ice removal stress for a polydimethylsiloxane nanocomposite: Sylgard 184. Langmuir 2014;30:12819–26. https://doi.org/10.1021/la5030444.
62. Liu Y, Ma L, Wang W, Kota AK, Hu H. An experimental study on soft PDMS materials for aircraft icing mitigation. Appl Surf Sci 2018;447:599–609. https://doi.org/10.1016/j.apsusc.2018.04.032.
63. Lv J, Yao X, Zheng Y, Wang J, Jiang L. Antiadhesion Organogel Materials: From Liquid to Solid. Adv Mater 2017;29:1–8. https://doi.org/10.1002/adma.201703032.
64. He Z, Xiao S, Gao H, He J, Zhang Z. Multiscale crack initiator promoted super-low ice adhesion surfaces. Soft Matter 2017;13:6562–8. https://doi.org/10.1039/c7sm01511a.
65. Golovin K, Kobaku SPR, Lee DH, DiLoreto ET, Mabry JM, Tuteja A. Designing durable icephobic surfaces. Sci Adv 2016;2. https://doi.org/10.1126/sciadv.1501496.
66. Wang Y, Yao X, Chen J, He Z, Liu J, Li Q, et al. Organogel as durable anti-icing coatings. Sci China Mater 2015;58:559–65. https://doi.org/10.1007/s40843-015-0069-7.
67. Wang Y, Yao X, Wu S, Li Q, Lv J, Wang J, et al. Bioinspired Solid Organogel Materials with a Regenerable Sacrificial Alkane Surface Layer. Adv Mater 2017;29:1–7. https://doi.org/10.1002/adma.201700865.
68. Rioboo R, Marengo M, Dall'Olio S, Voue M, De Coninck J. An innovative method to control the incipient flow boiling through grafted surfaces with chemical patterns. Langmuir 2009;25:6005–9. https://doi.org/10.1021/la900463b.
69. Boreyko JB, Hansen RR, Murphy KR, Nath S, Retterer ST, Collier CP. Controlling condensation and frost growth with chemical micropatterns. Sci Rep 2016;6:1–15. https://doi.org/10.1038/srep19131.

70. Nath S, Ahmadi SF, Boreyko JB. A Review of Condensation Frosting. Nanoscale Microscale Thermophys Eng 2017;21:81–101. https://doi.org/10.1080/15567265.2016.1256007.
71. Mangini D, Antonini C, Marengo M, Amirfazli A. Runback ice formation mechanism on hydrophilic and superhydrophobic surfaces. Cold Reg Sci Technol 2015;109:53–60. https://doi.org/10.1016/j.coldregions.2014.09.012.
72. Bar Dolev M, Braslavsky I, Davies PL. Ice-Binding Proteins and Their Function. Annu Rev Biochem 2016;85:515–42. https://doi.org/10.1146/annurev-biochem-060815-014546.
73. Davies PL. Ice-binding proteins: A remarkable diversity of structures for stopping and starting ice growth. Trends Biochem Sci 2014;39:548–55. https://doi.org/10.1016/j.tibs.2014.09.005.
74. Raymond JA, Devries AL. Adsorption inhibition as a mechanism of freezing resistance in polar fishes 1977;74:2589–93.
75. Gwak Y, Park J-I, Kim M, Kim HS, Kwon MJ, Oh SJ, et al. Creating Anti-icing Surfaces via the Direct Immobilization of Antifreeze Proteins on Aluminum. Sci Rep 2015;5:12019. https://doi.org/10.1038/srep12019.
76. Kasahara K, Waku T, Wilson PW, Tonooka T, Hagiwara Y. The inhibition of icing and frosting on glass surfaces by the coating of polyethylene glycol and polypeptide mimicking antifreeze protein. Biomolecules 2020;10:1–13. https://doi.org/10.3390/biom10020259.
77. Vance TDR, Bayer-Giraldi M, Davies PL, Mangiagalli M. Ice-binding proteins and the 'domain of unknown function' 3494 family. FEBS J 2019;286:855–73. https://doi.org/10.1111/febs.14764.
78. Duman JG. Animal ice-binding (antifreeze) proteins and glycolipids: An overview with emphasis on physiological function. J Exp Biol 2015;218:1846–55. https://doi.org/10.1242/jeb.116905.
79. Gibson MI. Slowing the growth of ice with synthetic macromolecules: Beyond antifreeze(glyco) proteins. Polym Chem 2010;1:1141–52. https://doi.org/10.1039/c0py00089b.
80. Kaleda A, Haleva L, Sarusi G, Pinsky T, Mangiagalli M, Bar Dolev M, et al. Saturn-Shaped Ice Burst Pattern and Fast Basal Binding of an Ice-Binding Protein from an Antarctic Bacterial Consortium. Langmuir 2019;35:7337–46. https://doi.org/10.1021/acs.langmuir.8b01914.
81. Drori R, Davies PL, Braslavsky I. When are antifreeze proteins in solution essential for ice growth inhibition? Langmuir 2015;31:5805–11. https://doi.org/10.1021/acs.langmuir.5b00345.
82. Mangiagalli M, Bar-Dolev M, Tedesco P, Natalello A, Kaleda A, Brocca S, et al. Cryo-protective effect of an ice-binding protein derived from Antarctic bacteria. FEBS J 2017;284:163–77. https://doi.org/10.1111/febs.13965.
83. Mangiagalli M, Sarusi G, Kaleda A, Bar Dolev M, Nardone V, Vena VF, et al. Structure of a bacterial ice binding protein with two faces of interaction with ice. FEBS J 2018;285:1653–66. https://doi.org/10.1111/febs.14434.
84. Mangiagalli M, Brocca S, Orlando M, Lotti M. The "cold revolution". Present and future applications of cold-active enzymes and ice-binding proteins. N Biotechnol 2020;55:5–11. https://doi.org/10.1016/j.nbt.2019.09.003.
85. Xu Q, Wilen LA, Jensen KE, Style RW, Dufresne ER. Viscoelastic and poroelastic relaxations of soft solid surfaces. ArXiv 2020;125:238002. https://doi.org/10.1103/physrevlett.125.238002.
86. Glover JD, McLaughlin CE, McFarland MK, Pham JT. Extracting uncrosslinked material from low modulus sylgard 184 and the effect on mechanical properties. J Polym Sci 2020;58:343–51. https://doi.org/10.1002/pol.20190032.
87. Peng J, Liu B, Gao SH, Zhu KY, Zhao YH, Li XH, et al. Enhanced anti-icing properties of branched PDMS coatings with self-regulated surface patterns. Sci China Technol Sci 2020;63:960–70. https://doi.org/10.1007/s11431-019-1482-x.
88. Guo H, Liu M, Xie C, Zhu Y, Sui X, Wen C, et al. A sunlight-responsive and robust anti-icing/deicing coating based on the amphiphilic materials. Chem Eng J 2020;402:126161. https://doi.org/10.1016/j.cej.2020.126161.

89. Li C, Li X, Tao C, Ren L, Zhao Y, Bai S, et al. Amphiphilic Antifogging/Anti-Icing Coatings Containing POSS-PDMAEMA-b-PSBMA. ACS Appl Mater Interfaces 2017;9:22959–69. https://doi.org/10.1021/acsami.7b05286.
90. Rahimi AR, Murphy M, Upadhyay V, Faiyaz K, Battocchi D, Webster DC. Amphiphilically modified self-stratified siloxane-glycidyl carbamate coatings for anti-icing applications. J Coatings Technol Res 2020. https://doi.org/10.1007/s11998-020-00402-8.
91. Ozbay S, Yuceel C, Erbil HY. Improved Icephobic Properties on Surfaces with a Hydrophilic Lubricating Liquid. ACS Appl Mater Interfaces 2015;7:22067–77. https://doi.org/10.1021/acsami.5b07265.
92. Subramanyam SB, Rykaczewski K, Varanasi KK. Ice adhesion on lubricant-impregnated textured surfaces. Langmuir 2013;29:13414–8. https://doi.org/10.1021/la402456c.
93. Rykaczewski K, Anand S, Subramanyam SB, Varanasi KK. Mechanism of frost formation on lubricant-impregnated surfaces. Langmuir 2013;29:5230–8. https://doi.org/10.1021/la400801s.
94. Kim P, Wong TS, Alvarenga J, Kreder MJ, Adorno-Martinez WE, Aizenberg J. Liquid-infused nanostructured surfaces with extreme anti-ice and anti-frost performance. ACS Nano 2012;6:6569–77. https://doi.org/10.1021/nn302310q.
95. Parker AR, Lawrence CR. Water capture by a desert beetle. Nature 2001;414:33–4.
96. Gao L, McCarthy TJ. Wetting 101°. Langmuir 2009;25:14105–15. https://doi.org/10.1021/la902206c.
97. Furmidge CGL. Studies at Phase Interfaces I. The Sliding of Liquiid Drops on Solid Surfaces and a Theory for Spray Retention. J Colloid Sci 1962;17:309--324. https://doi.org/10.1016/j.ultrasmedbio.2012.04.007.
98. Dotan A, Dodiuk H, Laforte C, Kenig S. The relationship between water wetting and ice adhesion. J Adhes Sci Technol 2009;23:1907–15. https://doi.org/10.1163/016942409X12510925843078.
99. Kulinich SA, Farzaneh M. Ice adhesion on super-hydrophobic surfaces. Appl Surf Sci 2009;255:8153–7. https://doi.org/10.1016/j.apsusc.2009.05.033.
100. Meuler AJ, Smith JD, Varanasi KK, Mabry JM, McKinley GH, Cohen RE. Relationships between Water Wettability and Ice Adhesion. ACS Appl Mater Interfaces 2010;11:3100. https://doi.org/10.1021/am1006035.
101. Nosonovsky M, Hejazi V. Why Superhydrophobic Surfaces Are Not Always Icephobic. ACS Nano 2012;6:8488–91. https://doi.org/10.1021/nn302138r.
102. Maitra T, Jung S, Giger ME, Kandrical V, Ruesch T, Poulikakos D. Superhydrophobicity vs. Ice Adhesion: The Quandary of Robust Icephobic Surface Design. Adv Mater Interfaces 2015;2. https://doi.org/10.1002/admi.201500330.
103. Ling EJY, Uong V, Renault-Crispo JS, Kietzig AM, Servio P. Reducing Ice Adhesion on Nonsmooth Metallic Surfaces: Wettability and Topography Effects. ACS Appl Mater Interfaces 2016;8:8789–800. https://doi.org/10.1021/acsami.6b00187.
104. Dhiman R, Chandra S. Freezing-induced splashing during impact of molten metal droplets with high Weber numbers. Int J Heat Mass Transf 2005;48:5625–38. https://doi.org/10.1016/j.ijheatmasstransfer.2005.05.044.
105. Schiaffino S, Sonin AA. Molten droplet deposition and solidification at low Weber numbers 1997;9. https://doi.org/10.1063/1.869434.
106. Maitra T, Antonini C, Tiwari MK, Mularczyk A, Imeri Z, Schoch P, et al. Supercooled water drops impacting superhydrophobic textures. Langmuir 2014;30:10855–61. https://doi.org/10.1021/la502675a.
107. Li H, Roisman I V., Tropea C. Influence of solidification on the impact of supercooled water drops onto cold surfaces. Exp Fluids 2015;56:133. https://doi.org/10.1007/s00348-015-1999-2.
108. Wright WB, Struk P, Bartkus T, Addy G. Recent Advances in the LEWICE Icing Model. SAE Tech Pap 2015;2015-June. https://doi.org/10.4271/2015-01-2094.
109. Bourgault Y, Boutanios Z, Habashi WG. Three-dimensional Eulerian approach to droplet impingement simulation using FENSAP-ICE, Part 1: Model, algorithm, and validation. J Aircr 2000;37:95–103. https://doi.org/10.2514/2.2566.

110. Son C, Kim T. Development of an icing simulation code for rotating wind turbines. J Wind Eng Ind Aerodyn 2020;203:104239. https://doi.org/10.1016/j.jweia.2020.104239.
111. Makkonen L, Laakso T, Marjaniemi M, Finstad KJ. Modelling and Prevention of Ice Accretion on Wind Turbines. Wind Eng 2002;25:3–21. https://doi.org/10.1260/0309524011495791.
112. Zhao Y, Guo Q, Lin T, Cheng P. A review of recent literature on icing phenomena: Transport mechanisms, their modulations and controls. Int J Heat Mass Transf 2020;159. https://doi.org/10.1016/j.ijheatmasstransfer.2020.120074.
113. Xue Y, Liu R, Li Z, Han D. A review for numerical simulation methods of ship–ice interaction. Ocean Eng 2020;215:107853. https://doi.org/10.1016/j.oceaneng.2020.107853.
114. Lupi L, Molinero V. Does hydrophilicity of carbon particles improve their ice nucleation ability? J Phys Chem A 2014;118:7330–7. https://doi.org/10.1021/jp4118375.
115. Cox SJ, Kathmann SM, Slater B, Michaelides A. Molecular simulations of heterogeneous ice nucleation . II . Peeling back the layers. J Chem Phys 2015;142. https://doi.org/10.1063/1.4919715.
116. Cox SJ, Kathmann SM, Purton JA, Gillan MJ, Michaelides A. Non-hexagonal ice at hexagonal surfaces: The role of lattice mismatch. Phys Chem Chem Phys 2012;14:7944–9. https://doi.org/10.1039/c2cp23438f.
117. Glatz B, Sarupria S. The surface charge distribution affects the ice nucleating efficiency of silver iodide. J Chem Phys 2016;145. https://doi.org/10.1063/1.4966018.
118. Pruppacher HR, James D Klett. Microphysics of Clouds and Precipitation. 1978. https://doi.org/10.1007/978-94-009-9905-3.
119. Pedevilla P, Cox SJ, Slater B, Michaelides A. Can Ice-Like Structures Form on Non-Ice-Like Substrates? The Example of the K-feldspar Microcline. J Phys Chem C 2016;120:6704–13. https://doi.org/10.1021/acs.jpcc.6b01155.
120. Bi Y, Cabriolu R, Li T. Heterogeneous ice nucleation controlled by the coupling of surface crystallinity and surface hydrophilicity. J Phys Chem C 2016;120:1507–14. https://doi.org/10.1021/acs.jpcc.5b09740.
121. Glatz B, Sarupria S. Heterogeneous Ice Nucleation: Interplay of Surface Properties and Their Impact on Water Orientations. Langmuir 2018;34:1190–8. https://doi.org/10.1021/acs.langmuir.7b02859.
122. Fitzner M, Sosso GC, Cox SJ, Michaelides A. The Many Faces of Heterogeneous Ice Nucleation: Interplay between Surface Morphology and Hydrophobicity. J Am Chem Soc 2015;137:13658–69. https://doi.org/10.1021/jacs.5b08748.
123. Lupi L, Hudait A, Molinero V. Heterogeneous nucleation of ice on carbon surfaces. J Am Chem Soc 2014;136:3156–64. https://doi.org/10.1021/ja411507a.
124. Li K, Xu S, Chen J, Zhang Q, Zhang Y, Cui D, et al. Viscosity of interfacial water regulates ice nucleation. Appl Phys Lett 2014;104. https://doi.org/10.1063/1.4868255.
125. Goertz MP, Houston JE, Zhu X-Y. Hydrophilicity and the viscosity of interfacial water. Langmuir 2007;23:5491–7. https://doi.org/10.1021/la062299q.
126. Cox SJ, Kathmann SM, Slater B, Michaelides A. Molecular simulations of heterogeneous ice nucleation. I. Controlling ice nucleation through surface hydrophilicity. J Chem Phys 2015;142. https://doi.org/10.1063/1.4919714.
127. Björneholm O, Hansen MH, Hodgson A, Liu LM, Limmer DT, Michaelides A, et al. Water at Interfaces. Chem Rev 2016;116:7698–726. https://doi.org/10.1021/acs.chemrev.6b00045.
128. Chattopadhyay S, Uysal A, Stripe B, Ha YG, Marks TJ, Karapetrova EA, et al. How water meets a very hydrophobic surface. Phys Rev Lett 2010;105:1–4. https://doi.org/10.1103/PhysRevLett.105.037803.
129. Chen D, Gelenter MD, Hong M, Cohen RE, McKinley GH. Icephobic surfaces induced by interfacial nonfrozen water. ACS Appl Mater Interfaces 2017;9:4202–14. https://doi.org/10.1021/acsami.6b13773.
130. Limmer DT, Chandler D. Premelting, fluctuations, and coarse-graining of water-ice interfaces. J Chem Phys 2014;141. https://doi.org/10.1063/1.4895399.
131. Kling T, Kling F, Donadio D. Structure and Dynamics of the Quasi-Liquid Layer at the Surface of Ice from Molecular Simulations. J Phys Chem C 2018;122:24780–7. https://doi.org/10.1021/acs.jpcc.8b07724.

132. Xiao S, He J, Zhang Z. Nanoscale deicing by molecular dynamics simulation. Nanoscale 2016;8:14625–32. https://doi.org/10.1039/c6nr02398c.
133. Metya AK, Singh JK. Ice adhesion mechanism on lubricant-impregnated surfaces using molecular dynamics simulations. Mol Simul 2019;45:394–402. https://doi.org/10.1080/08927022.2018.1513649.
134. Bao L, Huang Z, Priezjev N V., Chen S, Luo K, Hu H. A significant reduction of ice adhesion on nanostructured surfaces that consist of an array of single-walled carbon nanotubes: A molecular dynamics simulation study. Appl Surf Sci 2018;437:202–8. https://doi.org/10.1016/j.apsusc.2017.12.096.
135. Rønneberg S, Xiao S, He J, Zhang Z. Nanoscale correlations of ice adhesion strength and water contact angle. Coatings 2020;10:1–17. https://doi.org/10.3390/coatings10040379.
136. Xiao S, Skallerud BH, Wang F, Zhang Z, He J. Enabling sequential rupture for lowering atomistic ice adhesion. Nanoscale 2019;11:16262–9. https://doi.org/10.1039/c9nr00104b.
137. Ringdahl S, Xiao S, He J, Zhang Z. Machine Learning Based Prediction of Nanoscale Ice Adhesion on Rough Surfaces. Coatings 2020;11:33. https://doi.org/10.3390/coatings11010033.
138. Antonini C, Lee JBB, Maitra T, Irvine S, Derome D, Tiwari MKMKMKMK, et al. Unraveling wetting transition through surface textures with X-rays: Liquid meniscus penetration phenomena. Sci Rep 2014;4:4055. https://doi.org/10.1038/srep04055.
139. Enright R, Miljkovic N, Al-Obeidi A, Thompson C V, Wang EN. Condensation on superhydrophobic surfaces: the role of local energy barriers and structure length scale. Langmuir 2012;28:14424–32. https://doi.org/10.1021/la302599n.
140. El-Zoka AA, Kim SH, Deville S, Newman RC, Stephenson LT, Gault B. Enabling near-atomic–scale analysis of frozen water. Sci Adv 2020;6:1–12. https://doi.org/10.1126/sciadv.abd6324.
141. Caddeo C, Melis C, Ronchi A, Giannetti C, Ferrini G, Rurali R, et al. Thermal boundary resistance from transient nanocalorimetry: A multiscale modeling approach. Phys Rev B 2017;95:1–12. https://doi.org/10.1103/PhysRevB.95.085306.
142. Gandolfi M, Crut A, Medeghini F, Stoll T, Maioli P, Vallée F, et al. Ultrafast Thermo-Optical Dynamics of Plasmonic Nanoparticles. J Phys Chem C 2018;122:8655–66. https://doi.org/10.1021/acs.jpcc.8b01875.
143. Feynman R. There's Plenty of Room at the Bottom. Eng Sci 1960;23:22–36.

Chapter 9
A Mesoscale Modeling of Wetting: Theory and Numerical Simulations

Francesco Magaletti

Abstract This chapter illustrates a mesoscale, Diffuse Interface (DI), modeling of liquid-vapor systems close to solid surfaces. The model is built upon the Square Gradient Approximation (SGA) of the more general Density Functional Theory (DFT), to take into account the wetting properties of the solid surfaces and dynamic conditions. The formal derivation of the fluid-solid interaction potential is here reviewed, showing the connection with the Young-Laplace laws [40, 76] for sufficiently large bubbles/drops. The final section shows how the model can be numerically exploited to address the heterogeneous bubble nucleation process, including the description of thermal fluctuations into the model. Bubbles spontaneously appear during the simulation thanks to the fluid fluctuations, instead of being ad-hoc patched in the initial condition, allowing the statistical analysis of the nucleation process in terms of nucleation rate and spatial distribution. This opens the route for a multiscale simulation strategy, providing a continuum framework bridging the gap between the Molecular Dynamics (MD) simulations and the macroscopic Computational Fluid Dynamics (CFD).

Keywords Diffuse interface model · Wettability · Contact angle · Nucleation · Fluctuating hydrodynamics

9.1 Introduction

For any multiphase system of fundamental and technological relevance, wetting is a crucial phenomenon to be considered. Wetting refers to the behavior of a couple of immiscible fluids when in contact with a solid substrate. This can be found in a widespread number of industrial and natural areas such as paintings or printing [65], surface treatment for anti-frosting or for self-cleaning properties [24], penetration of liquids into porous materials [60], pesticides deposition in agricolture [4], boiling

F. Magaletti (✉)
Advanced Engineering Center, School of Computing Engineering and Mathematics, University of Brighton, Lewes Road, Brightong, BN2 4GJ, UK
e-mail: F.Magaletti@brighton.ac.uk

M. Marengo and J. De Coninck (eds.), *The Surface Wettability Effect on Phase Change*,
https://doi.org/10.1007/978-3-030-82992-6_9

[5], cavitation in confined systems [57], and many many others. In most of these applications the dimension of the droplets or bubbles involved largely exceed the range of molecular interaction. In these conditions, the classical description *à la* Young-Laplace [40, 76] in terms of macroscopic concept such as surface tension, γ_{LV}, and static contact angle, ϕ, seems rather adequate. The apparent scale separation can be, in fact, exploited to model the interface between the two fluids/phases as a mathematical surface of discontinuity (i.e. a sharp interface approximation). As a consequence, the Gibbs description of the interfacial properties in terms of interfacial excess quantities assigned to a formal dividing surface [29] is adopted. In this classical approach, the interface is a free boundary whose position and shape evolves in time. In the simplest equilibrium case two conditions must be fulfilled, namely: (i) the Laplace equation

$$\Delta p = p_V - p_L = \frac{2\gamma_{LV}}{R}, \tag{9.1}$$

where the case of a bubble with radius R and mean curvature $2/R$ is illustrated; (ii) the Young equation

$$\gamma_{LV}\cos(\phi) = \gamma_{SV} - \gamma_{SL}, \tag{9.2}$$

where γ_{SV} and γ_{SL} are the surface energies at the solid–vapor and solid–liquid interfaces, respectively.

Notwithstanding the assumptions of the classical capillarity model are physically sound, different phenomena are not well captured by this description. A single static contact angle, for example, is not able to explain the common observation of droplets suspended in equilibrium on an inclined, or even vertical, plate [14, 18]. The non-ideality of the substrate induced by surface roughness, even on a length scale much smaller with respect to the droplet size, produces a non-unique contact angle. The concept of contact angle hysteresis, with different advancing and receding angles, needs to be introduced for a satisfactory explanation. Additional parameters need to be introduced to properly capture the physical nature of the interface phenomena. When dealing with nanodroplets or nanobubbles, for example, the surface tension becomes a function of the drop/bubble size, and the celebrated Tolman length is used to tune this dependence [69]. The authors in [50] showed its crucial effect in order to properly capture bubble nucleation rates within the Classical Nucleation Theory framework.

The dynamic case is even trickier, with the moving contact line problem that runs into the well known theoretical paradox when a no-slip condition for the fluid velocity is expected [34]. Actually, a rolling motion of the two fluids near the contact line makes the no-slip assumption kinematically compatible with the motion of the contact line [13], however the velocity field needs to be multivalued at the contact line. This characteristic, together with the incompressibility and the Newtonian assumptions, produce a singularity on the stress and on the energy dissipation rate, making the no-slip condition dynamically incompatible with the motion of the contact line. The existence of a precursor film, i.e. a molecularly thin layer of fluid propagating ahead of the moving contact line, first introduced in the pioneering work by Hardy

[31], is able to resolve the paradox and the mathematical singularity. Its existence has been proven on very hydrophilic surfaces by scanning the molecular region through atomic force microscopy [80] and can be theoretically explained by introducing the concept of disjoining pressure [23]. A different strategy to avoid the stress singularity requires the local application of a Navier slip condition in the small contact region, whose width is on the order of magnitude of the slip length, [12, 33]. Unfortunately different slip models can reproduce the macroscopic behavior of the flow field by suitably tuning the model parameters, hindering the definite understanding of the problem. More recently, Molecular Dynamics (MD) simulations [53, 54, 68] shed new light on the nature of the physical phenomena occurring at the moving contact line: local slip and a relaxation dynamic for the stress at the contact region, related to a friction force at the contact line, are able to connect the continuum modeling perspective with the molecular observations at the nanoscale. However, how to associate a physical meaning to the values of the model parameters is still an open issue.

A very different perspective rises when realizing that the interface between the two fluids/phases cannot be treated as a sharp discontinuity. MD simulations confirm the existence of a small (few times the molecular size for temperature sufficiently below the critical [51]) interfacial region where the density field varies smoothly. The intrinsic density profile is actually sharper but is smoothed out by the capillary waves emerging due to thermal fluctuations at molecular scale [7]. The description in terms of the density field, generally called the Diffuse Interface (DI) approach, naturally captures all the peculiar features of the interfacial properties emerging at the nano scale. The size dependence of the surface tension, for example, is reproduced [16, 30], actually showing that the initial Tolman's approximation of a constant Tolman length is only valid for quite large critical nuclei. The finite thickness of the interface together with a diffusive mass transfer across the interface is also able to remove the singularity on the stress and on the dissipation rate in the problem of the moving contact line [59].

The DI approach is the focus of this review chapter and will be in depth analyzed in the following sections. In particular, Sect. 9.2 deals with the mathematical derivation of the mesoscale modeling, while in Sect. 9.3 numerical simulations of heterogeneous bubble nucleation are illustrated. Finally, Sect. 9.4 draws the concluding remarks and focuses on the future perspectives.

9.2 The Diffuse Interface Model for Multiphase Systems

As an alternative to the Sharp Interface philosophy, the interface between two immiscible fluids can be described as a thin layer where fluid properties change smoothy. Statistical mechanics [71] provides the link between surface tension and equilibrium density profile $\rho(z)$ of a planar liquid-vapour interface with normal in the z direction,

$$\gamma = \frac{k_B T \epsilon^2}{6m^2} \int_{-\infty}^{\infty} \left(\frac{d\rho}{dz} \right)^2 dz, \tag{9.3}$$

where k_B is the Boltzmann constant, m the molecular mass, T the temperature and ϵ the characteristic width of the interface. This expression associates the surface tension to the transition layer between the two different phases and implicitly calls the attention to a diffuse interface description of capillary phenomena.

The basic idea can be traced back to Van der Waals (1894—translated by Rowlinson in [74]) who suggested to stabilize the vapor-liquid interface by weakly non-local (gradient) terms in the free energy. The Van der Waals' Diffuse Interface model is introduced in terms of an Helmoholtz free energy functional

$$F[\rho, T] = \int_V \left(f_b(\rho, T) + \frac{\lambda}{2} \nabla\rho \cdot \nabla\rho \right) dV + \oint_{\partial V} f_w(\rho, T)\, dS, \tag{9.4}$$

where f_b is the free energy density (per unit volume) of the bulk fluid at mass density ρ and temperature T, and the square gradient term is the capillary contribution. As discussed in [1, 43] this corresponds to a gradient approximation of more general, non-local descriptions, like those exploited in Density Functional Theory [66]. As a consequence of the square gradient capillary contribution, at equilibrium, this model describes a smooth density profile transitioning between liquid and vapor on a typical scale ϵ, the thickness of the interface. Finally, the free energy contribution f_w takes into account the fluid-wall interactions describing the wetting properties of the surface.

In what follows, equilibrium conditions will be derived and manipulated to identify the relation between the equilibrium density profile and the surface tension. The isothermal case is considered here for the sake of simplicity, disregarding the temperature T in the following expressions. The interested reader is referred to [25] for the detailed derivation of the modelling in the most general case. For a closed system of given mass M_0, the constrained free energy functional (F_c) reads

$$F_c = F + l_1 \left(M_0 - \int_V \rho\, dV \right), \tag{9.5}$$

where l_1 is the Lagrange multiplier enforcing the mass constraint. By minimizing the constrained free energy through the application of the functional derivative,

$$\begin{aligned} \delta F_c[\rho] &= \delta \int_V \left(f_b(\rho) + \frac{\lambda}{2} \nabla\rho \cdot \nabla\rho - l_1 \rho \right) dV + \delta \oint_{\partial V} f_w(\rho)\, dS = \\ &= \int_V \left(\frac{\partial f_b}{\partial \rho} - \lambda \nabla^2 \rho - l_1 \right) \delta\rho\, dV + \oint_{\partial V} \left(\frac{\partial f_w}{\partial \rho} + \lambda \nabla\rho \cdot \hat{\boldsymbol{n}} \right) \delta\rho\, dS = 0, \end{aligned} \tag{9.6}$$

the Lagrange multiplier is identified as $l_1 = \mu_c^b - \lambda \nabla^2 \rho$ with $\mu_c^b = \partial f_b / \partial \rho$ the bulk chemical potential. It follows that, at equilibrium, the generalized chemical potential μ_c must be constant:

$$\mu_c = \mu_c^b - \lambda \nabla^2 \rho = const = \mu_c^{eq} \,. \tag{9.7}$$

The boundary term produces the additional requirement

$$\left(\lambda \nabla \rho \cdot \hat{\mathbf{n}} + \frac{\partial f_w}{\partial \rho} \right) \Bigg|_{\partial V} = 0 \,, \tag{9.8}$$

where $\hat{\mathbf{n}}$ is the outward normal, to be read as a non-linear boundary condition for the density accounting for the surface wettability.

In the case of a flat interface away from the boundaries, Eq. 9.7 rewrites

$$\mu_c = \mu_c^b(\rho) - \lambda \frac{d^2\rho}{ds^2} = \mu_{eq} \,, \tag{9.9}$$

with $\hat{\mathbf{s}}$ (i.e. $\rho = \rho(s)$) the normal direction to the interface. After multiplying by $d\rho/ds$ and integrating between the saturation vapor density, $\rho_\infty = \rho_V$, and a generic density value, ρ, one has

$$w_b(\rho) - w_b(\rho_V) = \frac{\lambda}{2} \left(\frac{d\rho}{ds} \right)^2 \,, \tag{9.10}$$

with $w_b = f_b - \mu_{eq}\rho$ the bulk Landau free energy density, namely the Grand Potential. Evaluating Eq. 9.10 far away from the interface, in the bulk liquid region where $\rho = \rho_L$, the equilibrium property $w_b(\rho_V) = w_b(\rho_L)$ is obtained. By definition, the surface tension is the excess grand potential

$$\gamma_{LV} = \int_{-\infty}^{s_i} (w[\rho] - w[\rho_V])\, ds + \int_{s_i}^{\infty} (w[\rho] - w[\rho_L])\, ds = \int_{-\infty}^{\infty} (w[\rho] - w[\rho_V])\, ds, \tag{9.11}$$

where s_i denotes the position of the Gibbs dividing surface, whose specific definition is irrelevant due to the equilibrium property $w[\rho_V] = w[\rho_L]$. The actual grand potential density w is defined, as usual, as the Legendre transform of the free energy and includes the capillary contribution, $w = f_b + \lambda/2 \nabla\rho \cdot \nabla\rho - \mu_c \rho$. At equilibrium the definition Eq. 9.11, combined with Eq. 9.10 and the condition $\mu_c = \mu_{eq}$, yields

$$\gamma_{LV} = \int_{-\infty}^{\infty} \left[f_b + \frac{1}{2}\lambda \left(\frac{d\rho}{ds} \right)^2 - \mu_{eq}\rho - w_b(\rho_V) \right] ds = \int_{-\infty}^{\infty} \left[w_b + \frac{1}{2}\lambda \left(\frac{d\rho}{ds} \right)^2 - w_b(\rho_V) \right] ds =$$
$$= \int_{-\infty}^{+\infty} \lambda \left(\frac{d\rho}{ds} \right)^2 ds = \int_{\rho_V}^{\rho_L} \sqrt{2\lambda\,(w_b(\rho) - w_b(\rho_V))}\, d\rho \,, \tag{9.12}$$

formally recovering the statistical mechanics result in Eq. 9.3.

The connection between the diffuse and sharp interface approaches is shown when considering the case of a spherical bubble in equilibrium with its liquid. Equation 9.7 is rewritten in spherical coordinate, where all the quantities are now function of the radial distance r only:

$$\mu_c^b - \lambda \frac{d^2\rho}{dr^2} - \lambda \frac{2}{r}\frac{d\rho}{dr} = \mu_{eq} \,. \tag{9.13}$$

By proceeding as for the planar case, multiplying by $d\rho/dr$ and integrating between $r = 0$, where $\rho = \rho_V$ and $d\rho/dr = 0$ because of the symmetry, and $r = \infty$, where $\rho = \rho_L$ and $d\rho/dr = 0$ because the uniform liquid phase is reached, the equilibrium reads

$$f_b(\rho_L) - f_b(\rho_V) - \int_0^\infty \lambda \frac{2}{\eta}\left(\frac{d\rho}{d\eta}\right)^2 d\eta = \mu_{eq}\left(\rho_L - \rho_V\right) . \tag{9.14}$$

In the above expression it can be recognized the pressure difference between the two phases by using the thermodynamic definition $p = -\partial(f_b/\rho)/\partial(1/\rho) = \rho\mu_c^b - f_b$,

$$\Delta p = p(\rho_V) - p(\rho_L) = \int_0^\infty \lambda \frac{2}{\eta}\left(\frac{d\rho}{d\eta}\right)^2 d\eta \,. \tag{9.15}$$

The right hand side can be approximated considering that the density variation will be important only close to the interface, at the bubble radius $r = R$. Furthermore, the spherical density equilibrium profile can be approximated by the profile of a planar interface, thus the remaining integral is approximatively equal to the surface tension defined in Eq. 9.12:

$$p(\rho_V) - p(\rho_L) = \int_0^\infty \lambda \frac{2}{\eta}\left(\frac{d\rho}{d\eta}\right)^2 d\eta \simeq \frac{2}{R}\int_0^\infty \lambda\left(\frac{d\rho}{d\eta}\right)^2 d\eta \simeq \frac{2\gamma_{LV}}{R} \,, \tag{9.16}$$

which recovers the Laplace equilibrium condition, Eq. 9.1.

The authors in [25] derived an explicit expression for the fluid-surface interaction term f_w, generalizing the result in [9] where two immiscible fluids of equal density was taken into account. They showed that f_w should follow a strict relation with the specific equation of state $f_b(\rho, T)$ used to describe the bulk properties of the fluid.

The starting point to derive the proper expression for f_w is the geometrical relation $\hat{\mathbf{s}} \cdot \hat{\mathbf{n}} = \cos\phi$, involving the angle between wall-normal and interface-normal directions (see Fig. 9.1). The equilibrium boundary condition, Eq. (9.8), is rewritten as

$$\frac{df_w}{d\rho} + \lambda \frac{d\rho}{ds}\cos\phi = 0, \tag{9.17}$$

using $\nabla\rho = d\rho/ds\,\hat{\mathbf{s}}$. Equation (9.10) is used to express the equilibrium interface-normal density variation, allowing to integrate the above equation as

$$f_w(\rho) = -\cos\phi \int_{\rho_V}^{\rho} \sqrt{2\lambda\left(w_b(\tilde{\rho}) - w_b(\rho_V)\right)}\, d\tilde{\rho} + f_w(\rho_V) \,, \tag{9.18}$$

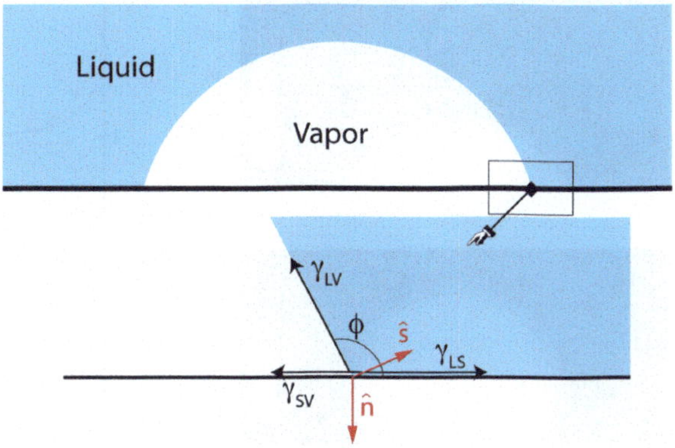

Fig. 9.1 Sketch of the capillary forces originating at the triple contact line. The vector normal to the wall is indicated with $\hat{\mathbf{n}}$, while $\hat{\mathbf{s}}$ indicates the vector normal to the interface, in the direction of the density gradient. The geometrical relation $\hat{\mathbf{s}} \cdot \hat{\mathbf{n}} = \cos\phi$ holds between the vectors and the contact angle ϕ. Reprinted with permission [25]

where $f_w(\rho_V) = \gamma_{SV}$ is the surface energy of a pure vapor. Similarly, for a pure liquid with density ρ_L, Eq. (9.18) provides

$$f_w(\rho_L) = \gamma_{SL} = -\cos\phi \int_{\rho_V}^{\rho_L} \sqrt{2\lambda\,(w_b\,(\tilde{\rho}) - w_b\,(\rho_V))}\, d\tilde{\rho} + \gamma_{SV} = -\gamma_{LV}\cos\phi + \gamma_{SV} \tag{9.19}$$

corresponding to the Young equation, Eq. 9.2, for the equilibrium contact angle. In the left panel of Fig. 9.2 the equilibrium configurations of a two-dimensional sessile vapor bubbles at different wetting surfaces (hydrophobic and hydrophilic) are shown. The measured contact angle corresponds to the prescribed angle ϕ, enforced through the boundary condition, Eq. 9.8, as expected.

As an additional feature of this mesoscale model, the fluid-solid interaction also describes density layering at the solid surface for non neutrally wettable surfaces ($\cos\phi \neq 0$). The right panel of Fig. 9.2 reports the wall-normal density profiles of a liquid in contact with a flat solid wall. In the hydrophilic case the liquid-solid interaction accumulates fluid at the wall, resulting in a local increase of density. Conversely, the lower affinity of hydrophobic walls produces a depletion region with layers of the order of a few nanometers. At nanoscale, layering is a common feature for liquids in contact with solid walls resulting from the interaction potential of the different molecules. Oscillations of the density field close to the wall is, in fact, observed through MD simulations [64] or X-ray scattering experiments [35, 77]. These oscillations are well captured by the more sophisticated, non-local, Density Functional Theory [20, 66, 67] while, due to the approximations introduced by the weakly non-local square gradient theory, the observed layering can be understood as a coarse-grained description of the actual phenomenology. A related approach, resulting in a similar monotonic density stratification at the wall, is described by introducing a disjoining pressure into the model, see e.g. [10], where a mean field theory is used to approximate the effect of the fluid-solid interaction potential.

When dealing with dynamic phenomena, the conservation laws for mass, momentum and total energy, together with the proper boundary conditions, rule the fluid dynamics of the system

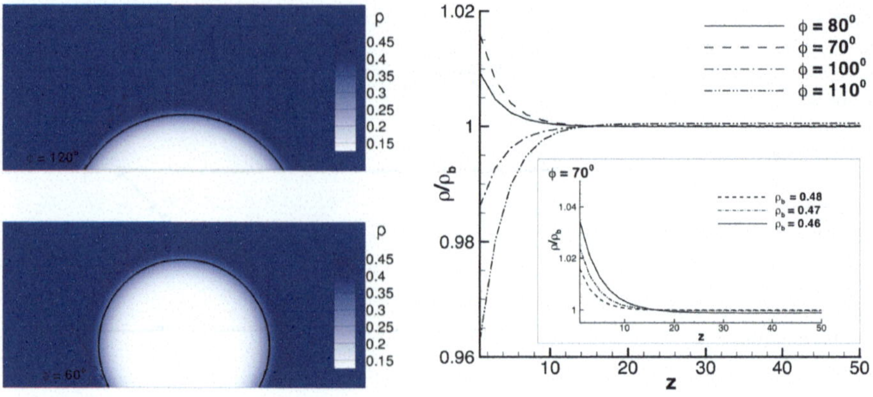

Fig. 9.2 Left panel: vapor-liquid contact angle for hydrophobic (top) and hydrophilic (bottom) solid surfaces. Right panel: Adsorption/depletion of the density field occurring in the wall normal direction, z, as a function of the contact angle. In the main plot the bulk liquid density is $\rho_b = 0.48$. All the quantities are properly rescaled with Lennard-Jones units, since an equation of state reproducing the thermodynamic properties of a LJ fluid [37] is used in this calculation. Hydrophobic walls, $\phi > 90°$, show a reduction of the density close to the wall; the opposite behavior is provided by hydrophilic walls. The inset shows that this layering effect is amplified when the degree of metastability is increased. Reprinted with permission [25]

$$\begin{aligned}
\frac{\partial \rho}{\partial t} + \nabla \cdot (\rho \mathbf{v}) &= 0,\\
\frac{\partial \rho \mathbf{v}}{\partial t} + \nabla \cdot (\rho \mathbf{v} \otimes \mathbf{v}) &= \nabla \cdot \boldsymbol{\tau},\\
\frac{\partial e}{\partial t} + \nabla \cdot (e \mathbf{v}) &= \nabla \cdot (\boldsymbol{\tau} \cdot \mathbf{v}) - \nabla \cdot \boldsymbol{q},
\end{aligned} \tag{9.20}$$

where e is the total energy density, sum of the internal, kinetic and capillary energies, and $\boldsymbol{\tau}, \boldsymbol{q}$ are the stress tensor and the heat flux, respectively. The classical procedure of non-equilibrium thermodynamic [15] allows to identify proper constitutive relations for these two terms also in the case of non-local free energy functional in Eq. 9.4, providing an extended version of the Navier-Stokes equations including capillary contributions

$$\boldsymbol{\tau} = -p\boldsymbol{I} + \mu\left(\nabla\mathbf{v} + \nabla\mathbf{v}^T\right) - \frac{2}{3}\mu\nabla\cdot\mathbf{v}\boldsymbol{I} + \lambda\left(\frac{1}{2}|\nabla\rho|^2 + \rho\nabla^2\rho\right)\boldsymbol{I} - \lambda\nabla\rho\otimes\nabla\rho, \tag{9.21}$$

$$\mathbf{q} = -k\nabla T + \lambda\rho\nabla\cdot\mathbf{v}\nabla\rho, \tag{9.22}$$

where μ and k are the dynamic viscosity and the thermal conductivity of the fluid. The interested reader is referred to [36, 45] for the detailed derivation.

With respect to the classical Navier-Stokes equations, where the velocity or the stress and the temperature or the heat flux are enforced at the boundaries, an additional boundary condition is required due to the higher differential order introduced by the capillary contributions in the stress tensor. From a physical standpoint the system requires the wettability properties to be specified at boundaries, hence Eq. 9.8 is enforced with the definition in Eq. 9.18

$$\frac{\partial \rho}{\partial n} = \cos\phi \sqrt{\frac{2\left(w_b(\rho) - w_b(\rho_V)\right)}{\lambda}} . \tag{9.23}$$

9.3 Application to the Simulation of Heterogeneous Bubble Nucleation in Pool Boiling Conditions

The DI approach has been exploited to numerically address the detailed dynamics of cavitation bubbles [44, 47, 48] and to study bubble evaporation [41, 42, 62]. This model provides a deterministic description of the two-phase fluid system, which is a good approximation at the larger hydrodynamic scales. However, at the length scale of the interface, thermal fluctuations become relevant requiring a stochastic description of the system [8, 56, 63]. Stochasticity is even more relevant when dealing with nucleation, i.e. the very first inception of the phase change phenomena. Liquids, in fact, can resist in superheated (temperature above the boiling point) or tensile (pressure below the saturation pressure) conditions without the formation of any vapor bubble for a long time, since nucleation is an activated process: an energy barrier needs to be surmounted to trigger the phase change process. Thermal fluctuations are the responsible for such activation, randomly providing large enough regions with low density acting as nucleation sites.

In order to properly study the detailed characteristics of nucleate boiling, for example, the spontaneous formation of bubbles is crucial since the spatial and time distribution of the nucleation sites is, in fact, one of the unknown of the problem. As a consequence, a deterministic approach where the vapor nucleii are initially patched might miss the essential features of the nucleation process and affect the results. In order to properly capture the stochastic nature of bubble formation, the DI model has been recently coupled with the Fluctuating Hydrodynamics (FH) theory [11, 27, 61]. Originally proposed by Landau and Lifshits (L&L) [39] and later developed in several works such as [19, 22, 55, 58], the FH theory introduces thermal fluctuations in the continuum framework described by the Navier–Stokes equations for simple fluids. L&L's idea was to treat the thermodynamic fluxes, namely, stress tensor and energy flux, as stochastic processes whose statistical properties can be inferred by enforcing the fluctuation-dissipation balance (FDB). The original L&L theory deals with linear fluctuations around equilibrium states but the theory has been generalized to linear fluctuations around non-equilibrium steady states [70], and also to the unsteady nonlinear case [19, 81]. More recently FH has been extended to capillary fluids [61] and applied to study the spinodal decomposition induced by thermal noise [11], the interface fluctuations near the contact line [3], the nucleation of cavitation bubbles [25–28] and the onset of nucleate boiling [46].

The main aspects of the Fluctuating Diffuse Interface (FDI) model are here summarized. The stochastic evolution of the two–phase system is described by the conservation laws in Eq. 9.20 supplemented with the stochastic forcing terms in the momentum and energy equations:

$$\begin{aligned} \frac{\partial \rho}{\partial t} + \nabla \cdot (\rho \mathbf{v}) &= 0, \\ \frac{\partial \rho \mathbf{v}}{\partial t} + \nabla \cdot (\rho \mathbf{v} \otimes \mathbf{v}) &= \nabla \cdot \boldsymbol{\tau} + \nabla \cdot \tilde{\boldsymbol{\tau}}, \\ \frac{\partial e}{\partial t} + \nabla \cdot (\mathbf{v} e) &= \nabla \cdot \left[\left(\boldsymbol{\tau} + \tilde{\boldsymbol{\tau}} \right) \cdot \mathbf{v} \right] - \nabla \cdot \left(\mathbf{q} + \tilde{\boldsymbol{q}} \right), \end{aligned} \tag{9.24}$$

where $\tilde{\boldsymbol{\tau}}$ is the stochastic stress tensor, expressed in terms of a delta-correlated white noise symmetric tensor $\boldsymbol{\eta}(\mathbf{x}, t)$, whose components are

$$\tilde{\tau}_{ij} = \sqrt{2 k_B T \mu} \left[\eta_{ij} - \delta_{ij} \frac{1}{3} \eta_{kk} \right], \tag{9.25}$$

with δ_{ij} the Kronecker delta and with

$$\begin{aligned} \langle \eta_{ij}(\mathbf{x}, t) \rangle &= 0, \\ \langle \eta_{ij}(\mathbf{x}, t) \eta_{i'j'}(\mathbf{x}', t') \rangle &= \left(\delta_{ii'} \delta_{jj'} + \delta_{ji'} \delta_{ij'} \right) \delta(\mathbf{x} - \mathbf{x}') \delta(t - t'). \end{aligned}$$

Similarly $\tilde{\mathbf{q}}$ is the stochastic heat flux, expressed in terms of a different delta-correlated white noise $\boldsymbol{\zeta}(\mathbf{x}, t)$, with components

$$\tilde{q}_i = \sqrt{2 k_B T^2 k} \zeta_i, \tag{9.26}$$

and whose statistical properties follow

$$\begin{aligned} \langle \zeta_i(\mathbf{x}, t) \rangle &= 0, \\ \langle \zeta_i(\mathbf{x}, t) \zeta_{i'}(\mathbf{x}', t') \rangle &= \delta_{ii'} \delta(\mathbf{x} - \mathbf{x}') \delta(t - t'). \end{aligned}$$

The numerical simulation of heterogeneous bubble nucleation has been performed in [25, 46] by solving the set of stochastic PDEs in Eqs. 9.24, following [2, 17]. A pseudo-random number generation algorithm [49] provides uniformly distributed random values that are processed with the Box-Muller transformation [6] to obtain a sequence of gaussian-distributed numbers to be used as the noise components η_{ij} and ζ_i in each spatial computational point and at each time instant. The complete simulation is constituted by several runs, each seeded with different gaussian noise, consisting in a set of statistically independent samples of the non-stationary stochastic process leading to bubble formation. A statistical analysis is then performed on the results extracted in each run, to provide the nucleation sites distribution and the nucleation rate.

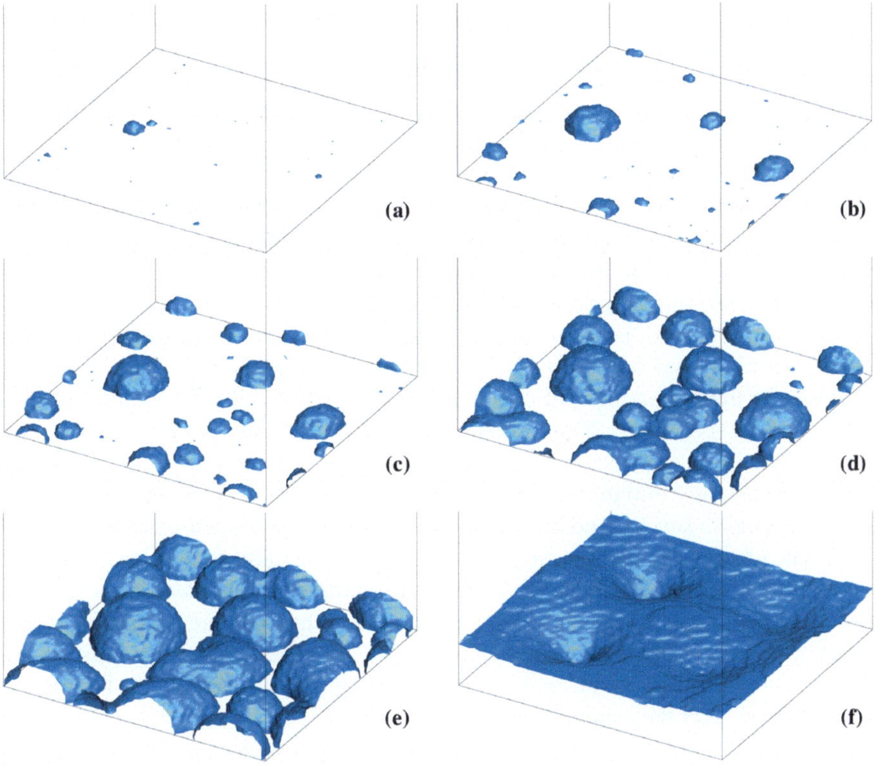

Fig. 9.3 Vapor bubble evolution during nucleate boiling with constant heat flux enforced on the bottom boundary. The figures refer to one sample randomly extracted from the set of different simulations in the same condition. Reprinted with permission [46]

Some snapshots extracted from a numerical experiment of boiling are reported in Fig. 9.3, where the liquid is initially at saturation conditions and the bottom boundary is heated with a constant heat flux q_w. Few small vapor embryos appear in the snapshot (a) but only the larger ones are able to grow and initiate the phase transition process, snapshots (b, c). At a later stage, snapshots (d, e), some of the growing bubbles experience apparent coalescence events, very rapidly producing large vapor regions. In the meanwhile new bubbles are produced due to the continuously applied heat flux. Finally, snapshot (f), a vapor layer covers the heated surface.

More advanced information can be extracted by evaluating the relative position of the bubbles. At variance with the Classical Nucleation Theory expectation, where the nucleation event is considered rare enough to be approximated as an isolated process leading to a uniform spatial distribution of bubbles, the mesoscale simulations show regions of clustered nucleated bubbles alternated to regions without nucleation events. This suggests a different nucleation mechanism, in line with the findings in [73], consisting in an *activated instability* process where the vapor bubbles surmounting the free energy barrier initiate a local instability that facilitates new nucleations in the neighboring liquid. The radial distribution function (rdf) shown in the left panel of Fig. 9.4 quantitatively measures this non-uniform spatial distribution of bubbles. The rdf at both times shows a pick ($rdf > 1$) at small distances, suggesting that the vapor embryos tend to appear close to each other with a higher probability. The pick is more evident at later times ($t = 17{,}000$ in the figure), when some of the embryos are larger than the critical bubble dimension. The statistical information provided by the rdf is visually confirmed in the two configurations reported in the insets, where the top view of the system clearly shows that the bubbles tend to cluster in small portion of the hot surface or close to bigger bubbles.

The non-uniform rdf is also indirectly related to a high probability of coalescence between the bubbles because highly clustered embryos would increase the chances of collisions. In the right panel of Fig. 9.4 the transition probability of a vapor embryo with radius R is reported. This probability refers to the chance of a vapor region with a given size to grow beyond the critical dimension (surmounting the free energy barrier) and actually initiate the phase transition. As expected, small embryos have a low probability to survive and most of them, in fact, are re-condensed into the liquid phase. The effective transition state is determined as the radius associated with 50% probability of transition. The plot evidences the importance of coalescence events on the evaporation success rate, showing that almost two thirds of the nucleated bubbles merged with other bubbles during the process. Coalescence, in fact, extends the life of a vapor region and enhances the chance to experience a strong rare fluctuation that locally triggers the evaporation.

9.4 Conclusions

The DI model provides an effective mesoscopic description of liquid-vapor systems in a continuum framework, capturing the intense—but smooth—density variation in the narrow interfacial region observed at the molecular level. The interfacial properties, like the surface tension and the equilibrium contact angle, are shown to be linked to the equilibrium density profile. Moreover, the model is able to recover the celebrated Young-Laplace relations for macroscopic bubbles and drops, hence covering a widespread range of length scales and bridging the gap between a molecular and a macroscopic description of the two-phase system. A dedicated effort to assess the validity of the continuum assumption when dealing with ultrafast (ps-ns)/nanoscale (nm) phenomena is however still required. Comparisons of continuum theories with MD results [21, 30, 52, 54, 79] would allow to determine when atomistic effects start playing a significant role or some hypotheses underneath continuum simulations break down.

From a computational standpoint, the presented DI model is characterised by a strongly reduced computational cost when compared with all-atom simulations. A rough estimate shows that a single fluctuating DI simulation of vapor bubble nucleation in a cubic domain with 350 nm length, and on a time scale of 10 ns,

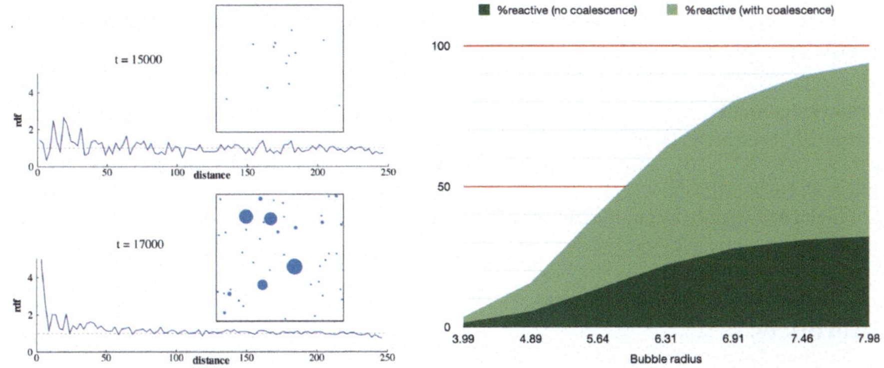

Fig. 9.4 Left: Radial distribution function of the bubble relative position at two different time instants. The solid blue curve is the rdf calculated from the simulation data while the dashed grey line at $rdf = 1$ represents the case of homogeneously distributed bubbles as a reference. The two insets represent a top view of the bubble configurations, with the dots size reported with the correct proportion between bubble dimension and wall size. Right: Transition probability as a function of the embryo radius. Among the reactive trajectories, those bubbles that do not experience any coalescence event during their dynamics are reported in dark green in the area chart. The percentage of those who merged with other bubbles at least once, is reported in light green. The effective critical radius, $R_c = 5.9$, is associated with a transition probability of 50%. Reprinted with permission [46]

only costs 10 k core hours. A similar MD simulation, would require 500 million particles and would cost approximatively 500 k core hours. On the opposite side of the length scale spectrum, the DI model becomes unaffordable on systems larger than few microns, even if implemented with advanced Adaptive Mesh Refinement techniques. Different approaches are then required for the simulation of macroscopic two-phase flows, such as the DI model with artificially enlarged interface like the phase-field Cahn-Hilliard model [1], or the Volume Of Fluid (VOF) technique [32, 75].

Given the intrinsic multiscale nature of the wetting phenomena, two strategies can be envisaged to properly address this complex process: (i) MD and DI parametric simulations can be performed to extract effective closure relations and ad-hoc boundary conditions to be imposed to the macroscale fluid dynamics simulations; (ii) multiscale simulation approaches can be conceived by dividing the domain in areas of competence, with the macroscopic fields evolved with two-phase macroscale models, like VOF, while in the triple line regions—where the two phases come in contact with the wall—increasingly detailed models, like DI or MD, could address the more complex interactions. None of the two approaches is however error-free. On one hand, problem-specific conditions occurring during the macroscopic evolution of the two-phase flow could fall outside the range of validity of the empiric laws extracted with the parametric simulations. On the other hand, specific strategies need to be implemented to treat the exchange of information between the two models in the overlap regions, potentially affecting the results. This subject is recently receiving an exponentially increasing interest but still requires huge developments to be effectively exploited to address real-life conditions [38, 72, 78].

Acknowledgements A large part of this work has received funding from the European Union's Horizon 2020 research and innovation programme under the Marie Skłodowska-Curie grant agreement No [836693]. The author also acknowledges the CINECA award under the Iscra project (project id: IscrB_HET-NUCL), and the PRACE-DECI16 (project id: WETONB), for the availability of high performance computing resources and support.

References

1. Anderson, D., McFadden, G., & Wheeler, A. (1998). Diffuse-interface methods in fluid mechanics. *Annual Review of Fluid Mechanics*, *30*, 139–165.
2. Balboa, F., Bell, J. B., Delgado-Buscalioni, R., Donev, A., Fai, T. G., Griffith, B. E., & Peskin, C. S. (2012). Staggered schemes for fluctuating hydrodynamics. *Multiscale Modeling & Simulation*, *10*, 1369–1408.
3. Belardinelli, D., Sbragaglia, M., Gross, M., & Andreotti, B. (2016). Thermal fluctuations of an interface near a contact line. *Physical Review E*, *94*, 052803.
4. Bergeron, V., Bonn, D., Martin, J. Y., & Vovelle, L. (2000). Controlling droplet deposition with polymer additives. *Nature*, *405*, 772–775.
5. Bourdon, B., Rioboo, R., Marengo, M., Gosselin, E., & De Coninck, J. (2012). Influence of the wettability on the boiling onset. *Langmuir*, *28*, 1618–1624.
6. Box, G., Muller, M. E. et al. (1958). A note on the generation of random normal deviates. *The Annals of Mathematical Statistics*, *29*, 610–611.
7. Braga, C., Smith, E. R., Nold, A., Sibley, D. N., & Kalliadasis, S. (2018). The pressure tensor across a liquid-vapour interface. *The Journal of chemical physics*, *149*, 044705.
8. Buff, F., Lovett, R., & Stillinger Jr, F. (1965). Interfacial density profile for fluids in the critical region. *Physical Review Letters*, *15*, 621.
9. Cahn, J. W. (1977). Critical point wetting. *The Journal of Chemical Physics*, *66*, 3667–3672.
10. Carey, V., & Wemhoff, A. (2005). Thermodynamic analysis of near-wall effects on phase stability and homogeneous nucleation during rapid surface heating. *International journal of heat and mass transfer*, *48*, 5431–5445.
11. Chaudhri, A., Bell, J. B., Garcia, A. L., & Donev, A. (2014). Modeling multiphase flow using fluctuating hydrodynamics. *Physical Review E*, *90*, 033014.
12. Cox, R. (1986). The dynamics of the spreading of liquids on a solid surface. part 1. viscous flow. *Journal of Fluid Mechanics*, *168*, 169–194.
13. Davis, S. H. et al. (1974). On the motion of a fluid-fluid interface along a solid surface. *Journal of Fluid Mechanics*, *65*, 71–95.
14. De Gennes, P.-G., Brochard-Wyart, F., & Quéré, D. (2013). *Capillarity and wetting phenomena: drops, bubbles, pearls, waves*. Springer Science & Business Media.
15. De Groot, S. R., & Mazur, P. (2013). *Non-equilibrium thermodynamics*. Courier Dover Publications.
16. Dell'Isola, F., Gouin, H., & Rotoli, G. (1996). Nucleation of spherical shell-like interfaces by second gradient theory: Numerical simulations. *European Journal of Mechanics, B/Fluids*, *15*, 545–568.
17. Delong, S., Griffith, B. E., Vanden-Eijnden, E., & Donev, A. (2013). Temporal integrators for fluctuating hydrodynamics. *Physical Review E*, *87*, 033302.
18. Dussan, E. (1979). On the spreading of liquids on solid surfaces: static and dynamic contact lines. *Annual Review of Fluid Mechanics*, *11*, 371–400.
19. Español, P. (1998). Stochastic differential equations for non-linear hydrodynamics. *Physica A: Statistical Mechanics and its Applications*, *248*, 77–96.

20. Evans, R., Stewart, M. C., & Wilding, N. B. (2017). Drying and wetting transitions of a lennard-jones fluid: Simulations and density functional theory. *The Journal of Chemical Physics*, *147*, 044701.
21. Fernandez-Toledano, J.-C., Blake, T., Lambert, P., & De Coninck, J. (2017). On the cohesion of fluids and their adhesion to solids: Young's equation at the atomic scale. *Advances in colloid and interface science*, *245*, 102–107.
22. Fox, R. F., & Uhlenbeck, G. E. (1970). Contributions to non-equilibrium thermodynamics. i. theory of hydrodynamical fluctuations. *Physics of Fluids (1958-1988)*, *13*, 1893–1902.
23. Frumkin, A. (1938). Phenomena of wetting and adhesion of bubbles. i. *Zh. Fiz. Khim*, *12*, 337–45.
24. Fürstner, R., Barthlott, W., Neinhuis, C., & Walzel, P. (2005). Wetting and self-cleaning properties of artificial superhydrophobic surfaces. *Langmuir*, *21*, 956–961.
25. Gallo, M., Magaletti, F., & Casciola, C. M. (). Heterogeneous bubble nucleation dynamics. *Journal of Fluid Mechanics*, *906*.
26. Gallo, M., Magaletti, F., & Casciola, C. M. (2018a). Fluctuating hydrodynamics as a tool to investigate nucleation of cavitation bubbles. *Multiphase Flow: Theory and Applications*, (p. 347).
27. Gallo, M., Magaletti, F., & Casciola, C. M. (2018b). Thermally activated vapor bubble nucleation: The landau-lifshitz–van der waals approach. *Phys. Rev. Fluids*, *3*, 053604. https://link.aps.org/doi/10.1103/PhysRevFluids.3.053604. https://doi.org/10.1103/PhysRevFluids.3.053604.
28. Gallo, M., Magaletti, F., Cocco, D., & Casciola, C. M. (2020). Nucleation and growth dynamics of vapour bubbles. *Journal of Fluid Mechanics*, *883*.
29. Gibbs, J. W. (1906). *The scientific papers of J. Willard Gibbs* volume 1. Longmans, Green and Company.
30. Gránásy, L. (1998). Semiempirical van der waals/cahn–hilliard theory: size dependence of the tolman length. *The Journal of chemical physics*, *109*, 9660–9663.
31. Hardy, W. B. (1919). Iii. the spreading of fluids on glass. *The London, Edinburgh, and Dublin Philosophical Magazine and Journal of Science*, *38*, 49–55.
32. Hirt, C. W., & Nichols, B. D. (1981). Volume of fluid (vof) method for the dynamics of free boundaries. *Journal of computational physics*, *39*, 201–225.
33. Hocking, L. (1976). A moving fluid interface on a rough surface. *Journal of Fluid Mechanics*, *76*, 801–817.
34. Huh, C., & Scriven, L. (1971). Hydrodynamic model of steady movement of a solid/liquid/fluid contact line. *Journal of colloid and interface science*, *35*, 85–101.
35. Huisman, W. J., Peters, J. F., Zwanenburg, M. J., de Vries, S. A., Derry, T. E., Abernathy, D., & van der Veen, J. F. (1997). Layering of a liquid metal in contact with a hard wall. *Nature*, *390*, 379–381.
36. Jamet, D., Lebaigue, O., Coutris, N., & Delhaye, J. (2001). The second gradient method for the direct numerical simulation of liquid–vapor flows with phase change. *Journal of Computational Physics*, *169*, 624–651.
37. Johnson, J. K., Zollweg, J. A., & Gubbins, K. E. (1993). The lennard-jones equation of state revisited. *Molecular Physics*, *78*, 591–618.
38. Koumoutsakos, P. (2005). Multiscale flow simulations using particles. *Annu. Rev. Fluid Mech.*, *37*, 457–487.
39. Landau, L. D., & Lifshits, E. M. (1959). *Fluid mechanics, by LD Landau and EM Lifshitz* volume 11. Pergamon Press Oxford, UK.
40. Laplace, P.-S. (1807). Theory of capillary attraction. *Supplements to the 10th book of Celestial Mechanics*.
41. Laurila, T., Carlson, A., Do-Quang, M., Ala-Nissila, T., & Amberg, G. (2012). Thermohydrodynamics of boiling in a van der waals fluid. *Physical Review E*, *85*, 026320.
42. Liu, J., Landis, C. M., Gomez, H., & Hughes, T. J. (2015). Liquid–vapor phase transition: Thermomechanical theory, entropy stable numerical formulation, and boiling simulations. *Computer Methods in Applied Mechanics and Engineering*, *297*, 476–553.

43. Lutsko, J. F. (2011). Density functional theory of inhomogeneous liquids. iv. squared-gradient approximation and classical nucleation theory. *The Journal of chemical physics*, *134*, 164501.
44. Magaletti, F., Gallo, M., Marino, L., & Casciola, C. M. (2015a). Dynamics of a vapor nanobubble collapsing near a solid boundary. In *Journal of Physics: Conference Series* (p. 012012). IOP Publishing volume 656.
45. Magaletti, F., Gallo, M., Marino, L., & Casciola, C. M. (2016). Shock-induced collapse of a vapor nanobubble near solid boundaries. *International Journal of Multiphase Flow*, *84*, 34–45.
46. Magaletti, F., Georgoulas, A., & Marengo, M. (2020). Unraveling low nucleation temperatures in pool boiling through fluctuating hydrodynamics simulations. *International Journal of Multiphase Flow*, (p. 103356).
47. Magaletti, F., Marino, L., & Casciola, C. (2015b). Shock wave formation in the collapse of a vapor nanobubble. *Physical Review Letters*, *114*, 064501.
48. Magaletti, F., Marino, L., & Casciola, C. M. (2015c). Diffuse interface modeling of a radial vapor bubble collapse. In *Journal of Physics: Conference Series* (p. 012028). IOP Publishing volume 656.
49. Matsumoto, M., & Nishimura, T. (1998). Mersenne twister: a 623-dimensionally equidistributed uniform pseudo-random number generator. *ACM Transactions on Modeling and Computer Simulation (TOMACS)*, *8*, 3–30.
50. Menzl, G., Gonzalez, M. A., Geiger, P., Caupin, F., Abascal, J. L., Valeriani, C., & Dellago, C. (2016). Molecular mechanism for cavitation in water under tension. *Proceedings of the National Academy of Sciences*, *113*, 13582–13587.
51. Nijmeijer, M., Bakker, A., Bruin, C., & Sikkenk, J. (1988). A molecular dynamics simulation of the lennard-jones liquid–vapor interface. *The Journal of chemical physics*, *89*, 3789–3792.
52. Pourali, M., Meloni, S., Magaletti, F., Maghari, A., Casciola, C. M., & Ciccotti, G. (2014). Relaxation of a steep density gradient in a simple fluid: Comparison between atomistic and continuum modeling. *The Journal of Chemical Physics*, *141*, 154107.
53. Qian, T., Wang, X.-P., & Sheng, P. (2003). Molecular scale contact line hydrodynamics of immiscible flows. *Physical Review E*, *68*, 016306.
54. Ren, W., & E, W. (2007). Boundary conditions for the moving contact line problem. *Physics of fluids*, *19*, 022101.
55. Rubı, J., & Mazur, P. (2000). Nonequilibrium thermodynamics and hydrodynamic fluctuations. *Physica A: Statistical Mechanics and its Applications*, *276*, 477–488.
56. Sanyal, M., Sinha, S., Huang, K., & Ocko, B. (1991). X-ray-scattering study of capillary-wave fluctuations at a liquid surface. *Physical review letters*, *66*, 628.
57. Scognamiglio, C., Magaletti, F., Izmaylov, Y., Gallo, M., Casciola, C. M., & Noblin, X. (2018). The detailed acoustic signature of a micro-confined cavitation bubble. *Soft matter*.
58. Sengers, J. V., & de Zárate, J. M. O. (2007). Thermal fluctuations in non-equilibrium thermodynamics. *Journal of Non-Equilibrium Thermodynamics*, *32*, 319–329.
59. Seppecher, P. (1996). Moving contact lines in the cahn-hilliard theory. *International journal of engineering science*, *34*, 977–992.
60. Shahidzadeh-Bonn, N., Tournié, A., Bichon, S., Vié, P., Rodts, S., Faure, P., Bertrand, F., & Azouni, A. (2004). Effect of wetting on the dynamics of drainage in porous media. *Transport in porous media*, *56*, 209–224.
61. Shang, B. Z., Voulgarakis, N. K., & Chu, J.-W. (2011). Fluctuating hydrodynamics for multiscale simulation of inhomogeneous fluids: Mapping all-atom molecular dynamics to capillary waves. *The Journal of chemical physics*, *135*, 044111.
62. Shen, B., Yamada, M., Hidaka, S., Liu, J., Shiomi, J., Amberg, G., Do-Quang, M., Kohno, M., Takahashi, K., & Takata, Y. (2017). Early onset of nucleate boiling on gas-covered biphilic surfaces. *Scientific reports*, *7*, 2036.
63. Sides, S. W., Grest, G. S., & Lacasse, M.-D. (1999). Capillary waves at liquid-vapor interfaces: A molecular dynamics simulation. *Physical Review E*, *60*, 6708.
64. Sikkenk, J., Indekeu, J., Van Leeuwen, J., & Vossnack, E. (1987). Molecular-dynamics simulation of wetting and drying at solid-fluid interfaces. *Physical review letters*, *59*, 98.

65. Soltman, D., Smith, B., Kang, H., Morris, S., & Subramanian, V. (2010). Methodology for inkjet printing of partially wetting films. *Langmuir*, *26*, 15686–15693.
66. Tarazona, P., & Evans, R. (1984). A simple density functional theory for inhomogeneous liquids: Wetting by gas at a solid-liquid interface. *Molecular Physics*, *52*, 847–857.
67. Tarazona, P., Marconi, U. M. B., & Evans, R. (1987). Phase equilibria of fluid interfaces and confined fluids: non-local versus local density functionals. *Molecular Physics*, *60*, 573–595.
68. Thompson, P. A., & Robbins, M. O. (1989). Simulations of contact-line motion: slip and the dynamic contact angle. *Physical review letters*, *63*, 766.
69. Tolman, R. C. (1949). The effect of droplet size on surface tension. *The journal of chemical physics*, *17*, 333–337.
70. Tremblay, A.-M., Arai, M., & Siggia, E. (1981). Fluctuations about simple nonequilibrium steady states. *Physical Review A*, *23*, 1451.
71. Triezenberg, D., & Zwanzig, R. (1972). Fluctuation theory of surface tension. *Physical Review Letters*, *28*, 1183–1185.
72. Tryggvason, G., Dabiri, S., Aboulhasanzadeh, B., & Lu, J. (2013). Multiscale considerations in direct numerical simulations of multiphase flows. *Physics of Fluids*, *25*, 031302.
73. Uline, M. J., & Corti, D. S. (2007). Activated instability of homogeneous bubble nucleation and growth. *Physical review letters*, *99*, 076102.
74. Van der Waals, J. (1979). The thermodynamic theory of capillarity under the hypothesis of a continuous variation of density. *Journal of Statistical Physics*, *20*, 200–244.
75. Welch, S. W., & Wilson, J. (2000). A volume of fluid based method for fluid flows with phase change. *Journal of computational physics*, *160*, 662–682.
76. Young, T. (1805). Iii. an essay on the cohesion of fluids. *Philosophical transactions of the royal society of London*, (pp. 65–87).
77. Yu, C.-J., Richter, A., Datta, A., Durbin, M., & Dutta, P. (2000). Molecular layering in a liquid on a solid substrate: an x-ray reflectivity study. *Physica B: Condensed Matter*, *283*, 27–31.
78. Zhang, J., Borg, M. K., & Reese, J. M. (2017). Multiscale simulation of dynamic wetting. *International Journal of Heat and Mass Transfer*, *115*, 886–896.
79. Zhang, Y., Sprittles, J. E., & Lockerby, D. A. (2020). Nanoscale thin-film flows with thermal fluctuations and slip. *Physical Review E*, *102*, 053105.
80. Zhao, Y. (2014). Moving contact line problem: Advances and perspectives. *Theoretical and Applied Mechanics Letters*, *4*, 034002.
81. Zubarev, D., & Morozov, V. (1983). Statistical mechanics of nonlinear hydrodynamic fluctuations. *Physica A: Statistical Mechanics and its Applications*, *120*, 411–467.

Chapter 10
Molecular Dynamics Simulations for the Design of Engineering Processes

Juan Carlos Fernández-Toledano

Molecular dynamics (MD) is the computational technique used to describe the solution of the classical Newton equation of motion for a set of many particles with the general objective of studying the dynamical properties of the system. The development of MD runs in parallel with the evolution of the modern computational tools since a first period between the late 50's to the 70's [1–6] where the foundations of this new technique were established and showing that MD could be used to model real systems with predictions compatible with experimental results. A fast explosion increment of popularity appeared in the 80's that continue until the present time where MD has become a standard tool in many research fields from biotechnology, development of new materials, chemical reactions, solid physics, and of course, the analysis of the dynamic of a liquid in contact with a solid plate (wetting dynamics).

The key point to understand the popularity of MD simulations can be found in the few inputs that we need to introduce to study a particular system. We need to consider the individual blocks that define the system (atoms or molecules) and how these blocks interact among them (interaction potentials). Then, we just solve the Newton equations to access the trajectory of the different particles and analyze them using the framework of statistical mechanics to construct observables that can be compared with experimental measurements or theoretical predictions. Therefore, once we correctly set up an MD simulation, we have something very similar to an experimental system where we measure the different properties (surface tension, viscosity, contact angles, properties of the interfaces, etc.) or even discover new physical phenomena. Moreover, we can change with great precision the different elements of the system to have access to the microscopic mechanism behind the macroscopic properties which is something completely impossible in a real experiment.

J. C. Fernández-Toledano (✉)
Laboratory of Surface and Interfacial Physics (LPSI), Department of Physics, University of Mons, Mons, Belgium
e-mail: carlos.toledano@umons.ac.be

M. Marengo and J. De Coninck (eds.), *The Surface Wettability Effect on Phase Change*,
https://doi.org/10.1007/978-3-030-82992-6_10

However, if we think of the huge amount of atoms that a real system contains, we can visualize the main limitation of the MD simulations: we have a strong restriction in the size and in the time scale accessible from the simulation. Typically, we can establish the frontier in the order of one micrometer in size and one microsecond in time, and then, MD is restringed to model phenomena at the nanoscale. Nevertheless, the advances in nanotechnology since the last two decades have propitiated the development of new experimental techniques able to gather very precise measurements of different systems at the nanoscale, and therefore, we start to have the possibility to compare directly MD simulations with experimental measurements. Also, the development of multiscale simulations could provide a very powerful tool to model realistic macroscopic systems with great precision where MD techniques are used to model the different interfaces in combination with CFD computations everywhere else [7–11]. Then, the MD part of the multiscale simulation automatically provides, for example, the velocity dependence of the contact angles, the slip length, the liquid structure in the proximities of the plate, etc. With this information, the CFD technique solves the Navier–Stokes equation for a macroscopic system overcoming the time and size restrictions of the standard MD simulations.

This chapter is organized as follows. First, we introduce a brief description of molecular dynamics as well as the different techniques used to measure different properties of the system. Then, we present two examples of how MD simulations can be useful in some engineering applications. In the first example, we use MD to find the possible microscopic mechanism of the so-called hydrodynamic assist, i.e. the possibility of increasing the coating speed beyond its expected limit by modifying the flow in a highly confined liquid. This is an example of how MD can be used to 'discover' new physics. In the second case, we use MD simulation to check the validity of an approximated theoretical equation for the force balance of a contact line pinned in a heterogeneity.

10.1 MD Description

10.1.1 Newton Equation and Integration Scheme

As we have introduced above, MD simulations basically consist in solving the Newton equations for each one of the atoms i that constitutes the simulation

$$m_i \frac{\partial \vec{r}_i}{\partial t} = \vec{F}_i \tag{10.1}$$

Here, m_i and $\vec{r}_i$ are the mass and the position of each atom i of the system, t is the time and $\vec{F}_i$ is the total force exerted on the atom i as a consequence of the interaction with other atoms and/or the presence of an external force field. Then, the evolution of the trajectory of each one of the atoms can be obtained by solving the Eq. 10.1.

The most used technique to solve this equation is the well-known Verlet algorithm [4] which can be obtained from the Taylor expansion of the atom position $\vec{r}_i$ around the time t. As a result, the atom location of an atom i at time $t + \Delta t$ can be computed from the position of this particle at two earlier time steps, $t - \Delta t$ and t, where Δt is the time step used to integrate the Newton equation.

$$\vec{r}_i(t + \Delta t) = 2\vec{r}_i(t) - \vec{r}_i(t - \Delta t) + \vec{F}_i(t)\Delta t^2/m_i \tag{10.2}$$

This integrator has the advantage of being reasonably fast from a computational point of view and it has great stability with time. In general, the most time-consuming part of MD simulations is the computation of the interaction between the different atoms of the system, and therefore, this integrator is used in combination with a neighbor list [12] that can reduce the computation time in one order of magnitude. This list contains the index of the atoms that are close enough to have a non-negligible interaction, i.e. that have a distance among them lower that some cut-off distance r_{cut} and then, we can exclude from the calculation all atoms that have almost no contribution to the force acting over one particular atom. Other equivalent integration schemes provide identical trajectories than the Verlet algorithm as the Leap Frog algorithm [13] or the velocity-corrected Verlet algorithm [14] which provides a better estimation of the atom velocities with more cost in computational time. In general, the standard velocity-Verlet algorithm is adequate for most MD simulations although for some applications it could be convenient to use integration algorithms that use high-order derivatives of the particle position [12, 13]. However, these algorithms should be used with precaution due to the non-time reversibility and storage and time consumption cost associate with them in comparison with the standard Verlet algorithm.

The methods to solve Newton's equation described above correspond to a situation where the number of atoms N, the volume V, and the energy E remains constant during the system evolution, i.e. to the so-called microcanonical ensemble where the system temperature T can change during the computation. However, in the majority of the practical situations, we want to study a system where T remains constant and the energy E is free to evolve. In this case, we use the canonical ensemble where NVT remains constant during the simulation. To model the canonical ensemble, we need to introduce an additional element on the simulation to control the behavior of the temperature: the thermostat. In an MD simulation, the temperature of a system composed N identical atoms of mass m is defined from the average of the variance of the atom velocity $\langle v^2 \rangle = \sum_i^N v_i^2$ as

$$k_B T = \frac{1}{3} m \langle v^2 \rangle \tag{10.3}$$

Here, k_B is the Boltzmann constant. In some cases, we only need to reach the equilibrium state of a set of atoms at a given temperature T_0. For example, when we focus on the analysis of the equilibrium properties of the system. In other cases, we are interested in the evolution of the system from equilibrium when we modify some parameters. Then, the simplest method used to find the equilibrium state at some particular temperature consists of the rescaling of the atom velocities by a factor $\beta = \sqrt{T_0/T_i}$ during the equilibration of the system where T_i and T_0 are the instantaneous and the target temperature, respectively. Then, we artificially add or remove energy from the system until we reach a stationary regime characterized by the fluctuation of the total energy around a constant value that corresponds to the equilibrium state. This method corresponds to the so-called velocity scaling thermostat [15]. Once we have reached the equilibrium state for the target temperature, we can remove the velocity scaling thermostat and, if the system is in a real equilibrium, the instantaneous temperature T_i will fluctuates around the desired value, $\langle T_i \rangle = T_0$. Of course, the trajectories of the atoms during the equilibration process are non-physical, but this a practical method to artificially drive a system to equilibrium.

When we are interested in the study of more realistic atoms displacement in a non-equilibrium simulation, as the analysis of a drop spreading over a substrate, a more sophisticated method is necessary to maintain a constant temperature without affecting the physics of the problem. In the case of the modeling of the wetting phenomena, we always have liquids in contact with solid plates that we can use in practice as a thermostat. If we let the solid atoms to vibrate and to exchange momentum with the liquid, we can apply the velocity scaling thermostat only to the solid phase, and then, the collision between liquid and solid atoms will induce a sufficient heat transfer to maintain the liquid temperature around the target value T_0. This method mimics the energy transfer existent in a real experiment where the liquid maintains its temperature during a dynamic process due to the interactions with its environment. Also, it allows us to model realistic systems where the liquid temperature can vary due to a physical process or even the existence of temperature gradients.

Another interesting thermostat widely used that also produces realistic trajectories of the atoms is the well-known Nose–Hoover [16, 17] thermostat that is, in general, the recommended method when we want to impose the temperature fluctuation around a fix constant value T_0. Although this thermostat allows the temperature fluctuation around the selected temperature, it avoids the presence of temperature gradients that can have physical meaning in different wetting phenomena.

Other ensembles could be also interesting for many different applications. For example, the NPT ensemble where the pressure P is kept constant but the volume is free to vary, or the μVT ensemble where the number of atoms can change during the simulation meanwhile the chemical potential remains constant [13]. However, in the different systems presented in this chapter, we will restring the NVE and NVT ensembles with a velocity Verlet algorithm to integrate the movement equations with an integration time step of $\Delta t = 5$ ps.

10.1.2 Interaction Potentials

One of the key elements to perform an MD simulation of a particular system is the description of the interactions acting over each one of the particles inside the system.

We can decompose the interactions in four groups: an external force field V_{ext}, the pairwise interaction V_{pair}, the bond interaction V_{bond} and the three-body interactions $V_{3\text{-}body}$. Then, the total energy of the system E can be expressed as

$$E = V_{ext} + V_{pair} + V_{bond} + V_{3\text{-}body} \tag{10.4}$$

- V_{ext} corresponds to an external field (as gravity or electromagnetic for example) acting over a set of particles on the system.
- V_{pair} defines the interaction between two pairs of atoms i and j located at $\vec{r}_i$ and $\vec{r}_j$, and its magnitude only depends on the distance between them $r_{ij} = |\vec{r}_i - \vec{r}_j|$ and, in general, it is the most relevant contribution interaction in an MD simulation. There is a large number of possible pair interactions we can introduce depending on the system we want to model. For example, neutral non-polar atoms are typically model with a Lennard-Jones potential that includes the contribution of a medium-range Van der Waals interaction plus a short-range repulsion to avoid the atom overlapping. Due to its simplicity, this is a well studied potential able to describe a phase transition between solid/liquid/gas [18] and widely used to simulated the wetting phenomena. Charged or polar molecules can be described with coulombic potentials that decay to zero at much longer distances having more cost from a computational point of view. Some particular combinations of different coulombic potentials are used to construct realistic water models like the classical SPC, SPC/E, and TIP3P potentials [19–21]. For more realistic interactions, it is possible to use the large database of empirical tabulated potentials that are widely used to model realistic biological processes [22].
- V_{bond} defines the details of the interaction of the atoms inside a molecule. The simpler bond interaction is the harmonic potential between atoms inside a molecule, the finite extensible nonlinear elastic potential (FENE) [23] typically used to model polymers, or dihedral potentials [24] used to define preferred bond angles between the atoms inside a molecule.
- $V_{3\text{-}body}$ is a three-body interaction potential used to obtain the precise behavior of a system under phase transitions [25] or as an effective potential to perform coarse-grained simulations as the Stillinger–Weber potential for silicium [26] also used to model water with remarkably precision but with a fraction of the computational cost compared with the classic electrostatic models [27].

10.1.2.1 Simplistic Liquid and Solid Model for Wetting MD Simulations

To model the wetting process, we need basically to put a liquid in contact with a solid substrate. Also, we should be able to modify the affinity between the liquid and the solid to study a range of different wettabilities. Also, we should allow the exchange of energy between the liquid and the solid and therefore, we can use the substrate as a thermostat for the liquid avoiding the introduction of thermostats that can hide some physical phenomena. Moreover, it could be interesting to change the inner properties of the liquid as its viscosity or surface tension. A widely used model that contains all these ingredients is a simple Lennard Jones liquid formed by linear chains of N atoms in contact with a solid where the atoms can vibrate to exchange momentum with the liquid. By tuning the length of the chains (the number of atoms per molecule), we can modify the viscosity, the surface tension, and control the evaporation of the liquid. The interaction between the liquid and the solid and then, the solid/liquid affinity, can be also modeled with a Lennard Jones interaction. The classical expression for the Lennard-Jones potential is

$$V_{LJ}(r_{ij}) = 4\epsilon\left(\left(\frac{a}{r_{ij}}\right)^{12} - \left(\frac{a}{r_{ij}}\right)^{6}\right) \tag{10.5}$$

Here, ϵ is the depth of the interaction potential between, and a is the distance at which the potential is zero that defines the effective diameter of the atoms. The balance between the attractive and repulsive interactions generate a minimum in the potential located at $r_{ij}^m = 2^{1/6}a$. Typically, the standard cut-off distance for the Lennard-Jones potential used to define the neighbor list is set to $r_c = 2.5a$. In the different examples provided along with this chapter, for both solid and liquid atoms, we set as the effective diameter $a = 0.35$ nm and mass $m_0 = 12$ g/mol that are typical values for a carbon atom. The depth of the Lennard-Jones potential is set to $\epsilon = C_{AB}\epsilon_0$, where the non-dimensional coupling parameter C_{AB} enables us to adjust the relative affinities between the different types of atoms: solid–solid (SS), liquid–liquid (LL), and solid–liquid (SL). To reduce the number of parameters in the simulation, we can set C_{AB} the value 1.0 for both the LL and SS interactions ($C_{LL} = C_{SS} = 1.0$), but for the SL interaction C_{SL}, can be tunned to explore a range of liquid–solid affinities. We set the temperature of the system as $T = 33$ K and the depth of the potential to be $\epsilon_0 = k_B T = 4.56 \times 10^{-22}$ J.

The liquid molecules considered are 8-atom linear chains where the adjacent atoms are linked by a FENE bond potential

$$V_{bond}(r_{ij}) = -0.5\alpha R_0^2 \ln\left(1 - \frac{r_{ij}^2}{R_0^2}\right) \tag{10.6}$$

where r_{ij} is the distance between the adjacent atoms, $\alpha = \epsilon_0/\text{Å}^2 = 0.0456$ J/m^2 is the strength of the potential, $R_0 = 1.4a$ is the maximum bond length.

The solid is constructed as an arrangement of atoms in a rectangular square-planar lattice and having a lattice parameter of $2^{1/6}a$ corresponding to the equilibrium distance defined by the Lennard-Jones potential. The solid atoms are allowed to vibrate thermally around their initial positions according to a harmonic potential:

$$V_h(r) = 1000 k_B T \frac{|\vec{r} - \vec{r}_0|^2}{a^2} \tag{10.7}$$

where $\vec{r}_0$ is the initial position of the solid atoms.

Initially, we apply a velocity scaling thermostat to all atoms of the system until the system reaches the target temperature T_0 and then, we remove this thermostat from the liquid atoms before measuring the different magnitudes. In the different examples provided, the momentum exchange between the liquid and the solid atoms is sufficient to maintain constant the liquid temperature.

10.1.3 Liquid Characterization

Once we have defined the interaction potential between the different atoms of the system. all the macroscopic properties as density, surface tension, diffusion coefficient, or viscosity are not input parameters but magnitudes that emerge as the result of these interactions. We need then to find strategies to measure these magnitudes as in a real experiment. Here, we present the different methods used to characterize the main properties of an MD liquid.

10.1.3.1 Pressure

Irving and Kirkwood [28] derived an exact expression for the calculation of the pressure tensor in any point of the space by using the microscopic velocities and forces acting over each atom of the system

$$P_{\alpha\beta} = \frac{1}{V}\left(\sum_{i}^{N} r_{i\alpha} f_{i\beta} + \sum_{i}^{N} m_i v_{i\alpha} v_{i\beta}\right) \tag{10.8}$$

Here, α and β are the component (x, y or z) of the tensor, V is the volume of the considered region, N is the number of atoms inside this region. Finally, $v_{i\alpha}$ and $f_{i\alpha}$ are the α component of the i particle velocity and force, respectively.

10.1.3.2 Surface Tension

To compute the surface tension in MD simulations we use its mechanical interpretation in terms of the pressure tensor: in the proximities of a flat liquid/vapor interface which normal is oriented in the z direction, there is an anisotropy between the normal $P_N = P_{zz}$ and the tangential $P_T = 0.5(P_{xx} + P_{yy})$ component of the pressure tensor with respect the interface. P_N is constant across a flat interface and its value is equal to the liquid pressure in the bulk P_0 but P_T evolves from P_0 in the bulk to a different value at the interface [29]. Then, the surface tension is simply a measurement of this anisotropy [12]

$$\sigma_{lv} = \int (P_N(z) - P_T(z))dz \tag{10.9}$$

Here, the integral is performed along the whole liquid–vapor interface. To measure the surface tension of our liquid, we create a plane liquid slab of 48,840 atoms surrounded by a vacuum phase, with the surfaces normal to the z axis and periodical boundary conditions in the x and y directions in a simulation box with dimensions $L_x = 23.2$ nm, $L_y = 12.25$ nm and $L_z = 17.8$ nm. Then, we compute the pressure tensor from Eq. 10.8 inside bins of thickness dz = 0.5 nm parallel to the liquid/vacuum interface and we apply the Eq. 10.9 to obtain a final value of $\sigma_{lv} = 2.84 \pm 0.56$ mN/m at $T = 33$ K. In the simulation, the surface tension strongly depends on the temperature of the system. A temperature increment will lead to a reduction of the surface tension as in a real system [30]. Also, the emergence of temperature gradient will produce gradients of surface tension that will induce a Marangoni effect in the system [31].

10.1.3.3 Diffusion Coefficient and Viscosity

The diffusion is the characterization of the motion of the particles in absence of flow and it can be expressed in terms of the macroscopic diffusion transport coefficient D_0. In MD simulations, D_0 can easily be obtained from the mean-squared distance over which the particles have moved in a time interval Δt, i.e. the mean squared displacement $\langle \Delta r^2(\Delta t) \rangle \equiv \langle |\vec{r}_i(t_0 + \Delta t) - \vec{r}_i(t_0)|^2 \rangle_{t_0}$. At long times, the mean squared displacement of a set of particles at equilibrium depends linearly with the diffusion coefficient [13]

$$\langle \Delta r^2(\Delta t) \rangle = 6 D_0 \Delta t \tag{10.10}$$

The formal connection between the liquid viscosity η_L and the microscopic properties of the liquid atoms is a Green–Kubo equation that links η_L with the integral of the auto-correlation function of the off-diagonal element of the pressure tensor of the liquid at equilibrium [13].

$$\eta_L = \frac{V}{3k_B T} \int \langle P_{xy}(0) P_{xy}(t) + P_{xz}(0) P_{xz}(t) + P_{yz}(0) P_{yz}(t) \rangle dt \tag{10.11}$$

Where V is the system volume. The values of the computed diffusion coefficient and liquid viscosity for the considered Lennard-Jones liquid at $T = 33$ K are $D_0 = 12.1 \pm 0.6$ nm^2/ns and $\eta_L = 0.270 \pm 0.012$ mPa.s. In the case of a system composed by spherical particles with short-range interactions (as the Lennard Jones potential), we can approach the relation between the diffusion coefficient and liquid viscosity through the Stokes–Einstein equation [12]

$$D_0 = \frac{k_B T}{3\pi \eta_L a} \tag{10.12}$$

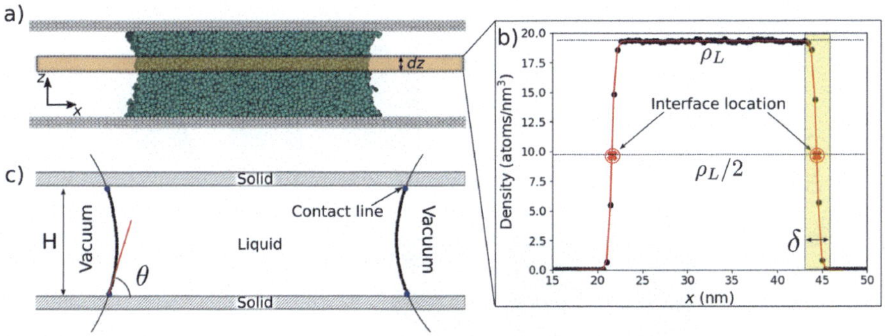

Fig. 10.1 **a** Atom projection in the x-z plane with an example of the bin used to locate the profile. **b** Density profile along x with the interface located where the density decays a 50% of the bulk value. **c** Circle fitting of the meniscus location to compute the contact angle θ and the contact line location

10.1.3.4 Liquid Density, Interfaces and Contact Angle

One of the strengths of the MD simulations applied to study wetting in comparison with other computational techniques is the spontaneous formation of the different interfaces of the system without the introduction of any additional element in the computation.

Consider our simple Lennard-Jones liquid confined along the z axis between two solid plates with periodical boundary conditions in the y direction. At $T = 33$ K, we have a phase separation between a liquid and a gas (or vacuum) phase if the simulation box is large enough and then, two free liquid–vacuum interfaces spontaneously appear along the x direction as shown in Fig. 10.1a. We can compute the liquid–vacuum profile from the liquid atom positions by projecting all the liquid atoms onto a single x–z plane. Then, we subdivide this plane into rectangular bins of thickness dz along the z direction, and we compute the density profile along x inside of these bins. These profiles have a constant value at the center corresponding to the value of the liquid density in the bulk $\rho_L = 19.30 \pm 0.06$ atoms/nm^3 (384.6 ± 1.2 kg/m^3). In the proximities of the interface, the density rapidly decays to zero (or the gas density) as represented in Fig. 10.1b. This decay defines the location of the interface and it can be fitted with a sigmoidal function. The location of the interface for each bin is set to be where the density decays a 50 % concerning the bulk value (the equimolar surface). If we repeat this process for each bin we reconstruct the full liquid–vacuum interface as shown in Fig. 10.1c. Also, the fitting parameters for the sigmoidal function provide the thickness of the interface associated to each bin being $\delta = 1.55 \pm 0.05$ nm at the center of the meniscus and $\delta = 2.18 \pm 0.05$ nm at the contact line.

Once we have computed the liquid–vapor profile, we can use the same techniques used in a real experiment to determine the value of the contact angles. Due to the absence of gravity, the menisci have constant curvature when averaged over time;

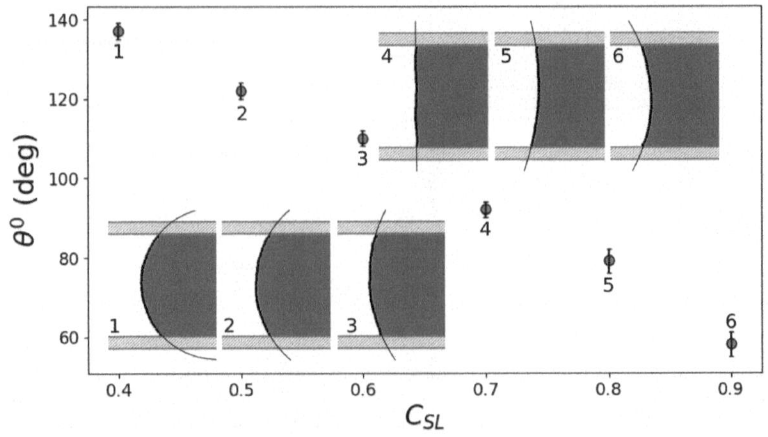

Fig. 10.2 Equilibrium contact angles versus the adimensional coupling parameter

hence, the contact angle and contact line location may be found by fitting an arc of a circle to the meniscus profile and measuring its tangent at the solid as sketched in Fig. 10.1c. If we fix the properties of the liquid, the value of the contact angle will only be affected by the affinity between the liquid and the solid that we can modify by tuning the coupling parameter C_{SL} for the Lennard-Jones potential as shown in Fig. 10.2.

10.1.3.5 Flow Diagrams

The different velocity profiles showed in this chapter are obtained from the computation of the full flow diagram of the corresponding simulation. To compute the flow profiles in the liquid bridge geometry, we project the liquid atoms onto the x–z plane, which we divided into a squared grid of side 0.5 nm. Then, we associate with each bin the average velocity of the atoms inside along a given time interval. The flow inside the 3D drops is measured similarly, except that the bins are arranged in concentric cylindrical shells around the z-axis that crosses the mass center of the drop.

10.2 Hydrodynamic Assist in Forced Wetting

When a liquid drop moves along a substrate, experiments show that the contact angles evolve from the equilibrium value θ^0 to a dynamic value θ being the magnitude of this change depends on the contact-line velocity v_{cl}: the contact angle increases in the direction of movement (advance angle) and it decreases in the opposite direction (receding angle). The finding of a precise relation between θ and v_{cl} is the core of the wetting dynamics problem that has been a source of controversy for at least 50 years [32–36] and it is not fully resolved yet. In the last two decades, MD simulations have become a valuable tool to study the problem of the wetting dynamics to elucidate the relation between u_{cl} and θ as well as to understand the behavior of the microscopic contact angle at the molecular level [37–47]. The importance to have a better understanding of the wetting dynamic lies in the huge number of technological applications that involve this problem.

Two main models have been developed to model the dependence of the contact angles with the contact line velocity. The first one is the macroscopic hydrodynamic theory which considers that the main resistance to the contact line motion comes from the viscous dissipation of the liquid due to the bending of the liquid–vapor interface at a scale below that at which the angle is measured, but larger than the length of the interface [48–50]. This interface bending is then responsible for the deviation of the apparent angle θ_{app} at the macroscale from the microscopic angle. In the hydrodynamic approach, the local microscopic contact angle is often considered to be unaffected by the displacement of the contact line and therefore keeps its equilibrium value θ^0. However, a large number of MD simulations have shown clearly that the microscopic contact angle can be affected by the velocity of the contact line withouth the interface blending postulated by the classic hydrodynamic theory. Therefore, another theoretical approach is typically used to explain the relation between θ and v_{cl} observed in MD simulations. This second model is the molecular kinetic theory (MKT) which considers that the displacement of the contact line is the result of the collective motion of the liquid atoms affected by the out-of-balance capillary force $\sigma_{lv}(\cos\theta - \cos\theta^0)$ where θ is the value of the microscopic dynamic contact angle. Then, the MKT considers that θ is truly velocity-dependent and its variation is due to a local friction force acting at the contact line. Here, the friction is caused by the interaction of the liquid molecules with the potential energy landscape of the solid surface [51–53]. The relation proposed between the contact line velocity and the microscopic contact angle is

$$u_{cl} = 2\kappa^0\lambda \sinh\left(\frac{\lambda^2\sigma_{lv}(\cos\theta - \cos\theta^0)}{2k_B T}\right) \tag{10.13}$$

Here, κ^0 and λ are the characteristic frequency and length of the thermal displacement of the liquid atoms in the proximities of the contact line which in the MKT framework must be viewed as a three-phase zone (TPZ), i.e. a finite region in the space where the solid/liquid, the solid/gas and the liquid/gas interfaces met. In the region of low contact line velocities, we can substitute the hyperbolic sinus by its argument, and then, we obtain a linear relationship between the driven force and the contact line velocity

$$\zeta v_{cl} = \sigma_{lv}(\cos\theta^0 - \cos\theta) \tag{10.14}$$

where $\zeta = k_B T/(\kappa^0\lambda^3)$ corresponds with the coefficient for the contact line friction.

One particular example of a wetting dynamic is the forced wetting case that refers to any method by which the contact line is driven across the substrate at a constant velocity v_{cl}. Forced wetting appears in many practical situations as liquid coating [54], rolling drops on an incline [55, 56] or pushed liquid bridges in a tube [57] among many other examples. For sufficient large v_{cl}, the advance angle approaches the limit value of 180°. This set a maximum value for the contact line velocity v_{cl}^{max} which is critical, for example, in liquid coating where the air is entrained for $v_{cl} > v_{cl}^{max}$ [58]. The precise mechanism behind air entrainment is not fully understood, although it is known that it can be postponed to higher velocities by modifying the flow with some kind of confinement as it has been observed in curtain coating experiments which defines what it has been called *hydrodynamic assist* [59, 60]. MD simulation of forced wetting has been a powerful tool to understand this phenomenon [44] that opens the door to optimize forced wetting processes by increases the contact line velocity over its expected limit.

To understand the role of strong confinement on the dynamic of wetting, we perform MD simulations of a non-confined system, as the spontaneous drop spreading case, and we also model a highly confined geometry as a liquid bridge inside a microchannel. Then, we compare the relation between θ and v_{cl} in both cases.

10.2.1 Spontaneous Spreading

When a liquid drop is put in contact with a substrate, the out-of-balance capillary force $\sigma_{lv}(\cos\theta^0 - \cos\theta)$ drives the contact line towards its equilibrium configuration. In this classic example of the wetting dynamic, the contact angle θ decreases from an initial value of 180° until it reach the equilibrium value θ^0. During this process, the contact radius $r_c(t)$ between the liquid and the solid also evolves between 0 to the final contact radius. The change ratio of the contact radius, i.e. the contact line velocity $v_{cl} = \dot{r}_c$, is intimately related with the variation of the dynamic angle, $\theta(t) \propto v_{cl}(t)$.

We study the spontaneous spreading dynamic by creating a Lennard Jones spherical liquid drop of 27,000 8-atom molecules with a radius of 13.9 nm after the equilibration. Then, we put the drop in contact with a plate composed of 74,748 atoms, we set the liquid–solid affinity defined by the coupling C_{SL} and we let the drop spontaneously spread. Typical snapshots of the spreading process are shown in Fig. 10.3a for a system with $\theta^0 = 60.5 \pm 1.8°$. The measurement of the contact angle and the location of the contact line is obtained by using the methods described in Sect. 10.1.3.4. Figure 10.3b illustrates the evolution of the contact radius from an initial zero value to the equilibrium contact radius. From these curves, we can compute the contact line velocity by fitting a ratio of polynomials function and computing its analytical derivative. The fitting function used is represented with dashed lines in Fig. 10.3b having in all cases a coefficient of determination $r^2 \geq 0.97$. Figure 10.3c shows the evolution of the contact angle for all the solid–liquid affinities considered. In all

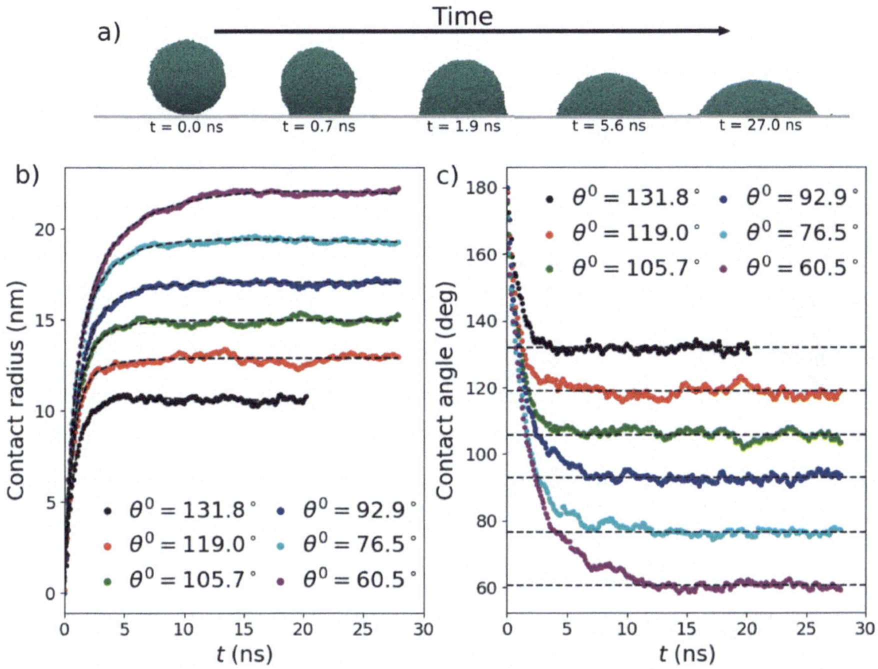

Fig. 10.3 **a** Snapshots for the $\theta^0 = 60.5 \pm 1.8°$ simulation at different time steps. **b** Radius of contact and **c** dynamic contact angle evolution. The dashed lines represent the fitting used to compute the contact line velocity and the equilibrium contact angle, respectively

cases, the contact angle has an initial value of 180° and it decreases with time until it fluctuates around the equilibrium value represented with the dashed lines.

From the computed dynamic contact angles and contact line velocities, we can rationalize the results in terms of the MKT model presented in Eq. 10.13. Figure 10.4a illustrate the out-of-balance capillary force $\sigma_{lv}(\cos\theta^0 - \cos\theta)$ versus the contact line velocity v_{cl}. In all cases, we observe a linear dependence between both magnitudes in the proximities of equilibrium where v_{cl} is small. Then, we can estimate the contact line friction coefficient ζ from the slope of the linear fitting in this region. The resulting contact line friction is shown in Fig. 10.4b and, as expected, the contact line friction increases with the affinity between the liquid of the solid: as the attraction between the liquid molecules and the solid is increased, the displacement of the liquid in contact with the plate is slower which result in a higher friction.

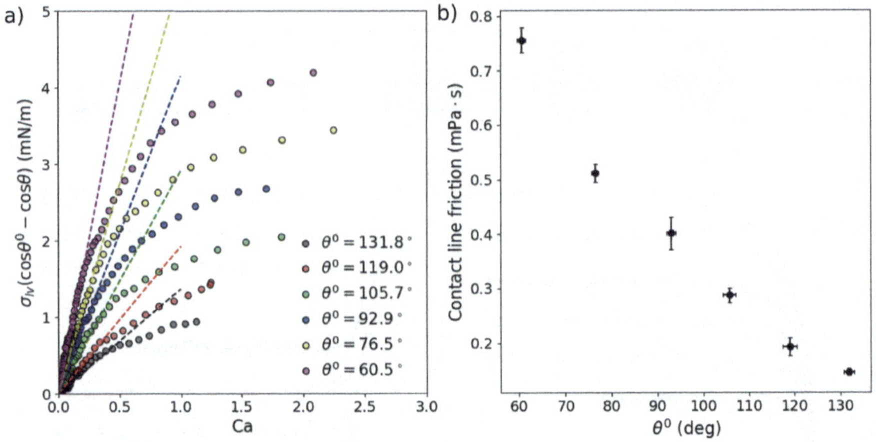

Fig. 10.4 **a** Driven force $\sigma_{lv}(\cos\theta^0 - \cos\theta)$ versus the contact line velocity for the different solid–liquid affinities. The lines represent the linear fitting in the low-velocity region. **b** $(\theta)^3 - (\theta^0)^3$ versus the capillary number. The lines represent the linear fittings

10.2.2 Forced Wetting

We now consider an example of forced wetting in a strongly confined geometry constructed with a Lennard-Jones liquid bridge of 6105 8-atoms linear chains trapped along the z axis between two plates of 16,275 atoms each. We introduce the flow in the liquid by moving the plates along the x axis at constant velocity v_p but in opposite directions. Then, the location of the mass center of the liquid remains stationary and the contact line velocity in the frame of reference of the plate is $v_{cl} = -v_p$. For a range of solid–liquid affinities modeled with the coupling parameter C_{SL}, we study different plate velocities and we measure the advance and receding angles computed with the methods presented in Sect. 10.1.3.4. An example of the variation on the liquid–vapor interface for $\theta^0 = 95.6 \pm 1.7°$ at different velocities of the plates are shown in the snapshots of Fig. 10.5a.

Once we have measured the dynamic angles, we can represent the driven force versus the contact line velocity as we did in the spreading case. As shown in Fig. 10.5b, we observe a linear relationship between the driven force and the contact line velocity for all liquid–solid affinities. Then, we can use the linear version of the MKT from Eq. 10.14 to fits the data and extract the contact line friction from the slope. The resulting contact line friction and its comparison with the spontaneous spreading friction are shown in Fig. 10.5c. As it has been pointed out firstly by Blake et al. [44] for a similar MD simulation, the measured values of the friction for the forced wetting case are significantly smaller than the friction computed in spontaneous spreading. This reduction of the contact-line friction observed in force wetting concerning the spontaneous case means a weaker relation between the contact line velocity and the dynamic contact angles, i.e. in the forced wetting simulations it is

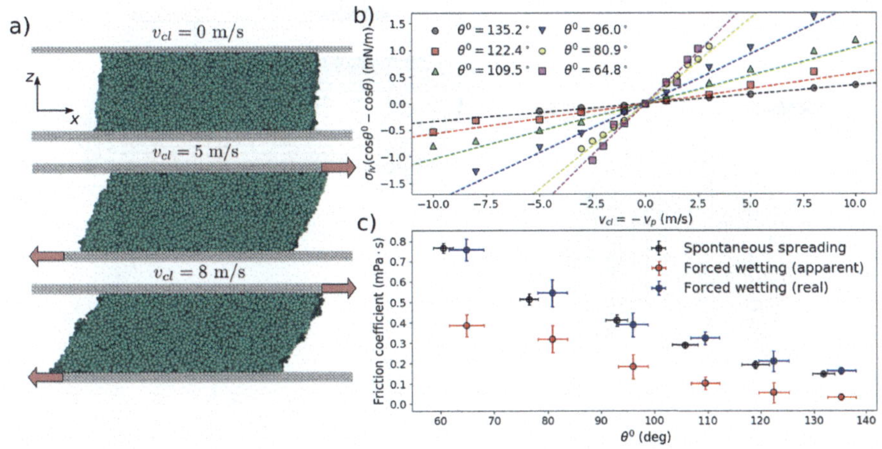

Fig. 10.5 **a** Snapshots for the liquid bridge with $\theta^0 = (95.6 \pm 1.7)°$ at different contact line velocities. **b** Driven force versus contact line velocity. **c** Comparison of the contact-line friction coefficients measured for spontaneous spreading and forced wetting simulations

necessary to introduce higher contact line velocities in comparison with spontaneous spreading to reach the same dynamic angle. In both spontaneous and forced wetting simulations, the interactions and the solid structure are identical but the simulations reveal significantly different flows. Figure 10.6a illustrates the flow diagram during spontaneous spreading at some time. The tangential flow of the liquid with respect to the plate is negligible everywhere except in the proximities of the contact line where the velocity of the atoms matches with the contact line velocity, i.e. there is no slip. The situation is different for the forced wetting case where the movement of the two plates generates a steady rotational flow as shown in Fig. 10.6b. Due to the strong confinement, the tangential velocity of the liquid atoms close to the surface is lower than the plate velocity. In Fig. 10.6c we represent the tangential velocity profile along z at the center of the liquid bridge for $\theta^0 = 95.6 \pm 1.7°$ and the plates moving at $v_p = 7$ m/s. We observe the linear profile characteristic for a Couette flow but the value of the velocity extrapolated at the intersection with the plates is significantly smaller which reveals the presence of slip.

Blake et al. [44] proposed that the presence of the slip velocity v_{slip} introduce a strong shear stress along the whole solid–liquid interface and then, also at the contact line. Therefore, the shear stress due to the rotational flow introduce a new force contribution that need to be added to the capillary force $\sigma_{lv}(\cos\theta^0 - \cos\theta)$. The extended MKT theory including slip is [44]

$$v_{cl} = 2\kappa^0\lambda \sinh\left[\frac{\lambda^2}{k_B T}\left(\sigma_{lv}(\cos\theta^0 - \cos\theta) + \zeta\, v_{slip}\right)\right] \qquad (10.15)$$

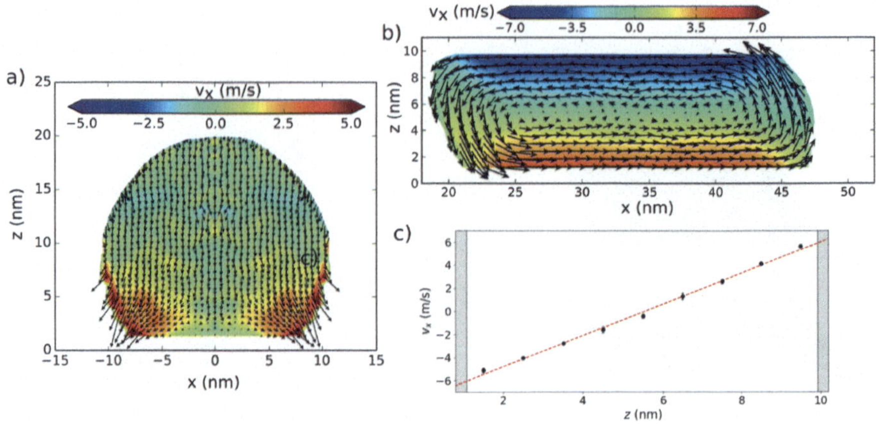

Fig. 10.6 Flow diagrams for **a** spontaneous spreading with $\theta^0 = 106°$ and **b** forced wetting with $\theta^0 = 121°$ with $v_p = 10$ m/s. **c** Profile of v_x versus z at the center of the bridge computed from the **c** flow diagram

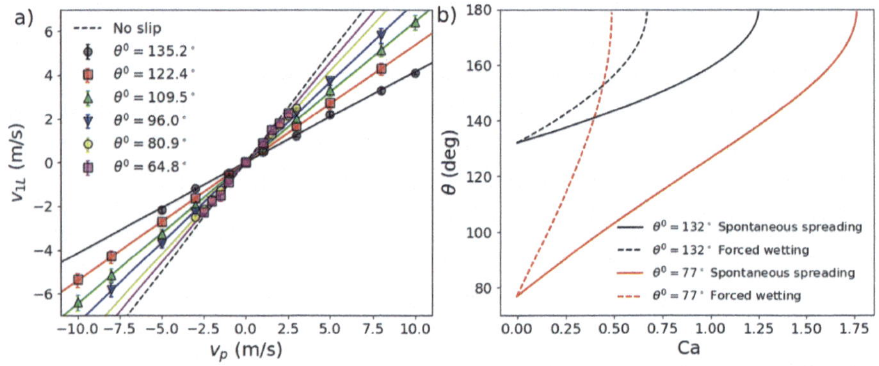

Fig. 10.7 **a** Velocity of the first layer of liquid in contact with the plate v_{1L} versus the velocity of the plate v_p. **b** Comparison of the velocity dependence of the contact angle with the capillary number Ca

Then, the forced wetting contact-line friction showed in Fig. 10.4c corresponds indeed to apparent friction containing the effect of the slip. To check this, we can measure the slip velocities by subtracting to the plate velocity v_p the velocity of the first layer of liquid in contact with the plate v_{1L} that can be obtained from the flow profiles. In all cases, there is a linear relation between v_{1L} and v_p as shown in Fig. 10.7a. Then, we use Eq. 10.15 to determine the real contact-line friction which is very similar to the values computed for spontaneous spreading where the slip is absent. Therefore, the discrepancies in the wetting dynamic observed for the spontaneous and forced wetting cases can be satisfactory explained due to the effect of slip.

In conclusion, the manipulation of the flow can be used to promote slip and then to increase the contact line velocity beyond its initial limit as it has been experimentally observed and defined as hydrodynamical assist [59, 60]. We can visualize this phenomenon by solving Eq. 10.15 with the values for the contact line friction obtained for spontaneous spreading and considering $\lambda = 0.43$ nm as found for an identical plate structure [43]. Figure 10.7b illustrate the different wetting dynamic computed for spontaneous spreading and forced wetting for two different wettabilities where we observe that the effect of the flow can double the maximum contact line speed compare with spontaneous spreading. Due to the reduction of the slip, the differences between forced and spontaneous wetting become less pronounced as the solid–liquid affinity increases.

10.3 Drop Pinning on an Incline

The contact line pinning is another example of an important wetting phenomenon in different technological areas. The microscopic roughness and heterogeneities on the surfaces have a macroscopic effect on the resistance of the contact line to move until the contact angle reaches a critical value. The advancing θ_a and the receding θ_r contact angles define the threshold required by the contact line to advance or recede, respectively. The difference between these angles determines the hysteresis $H = \theta_a - \theta_r$. Depending on the surface treatment and the liquid involved, the hysteresis typically lays between few degrees to several tens of degrees [61]. The hysteresis effect on the pinning can be understood as a resistance force to movement $F_H = \sigma_L(\cos\theta_a - \cos\theta_r)$ as a consequence of the unbalance Young's force [35]. Thermodynamically, this pinning force corresponds to local differences on the work of adhesion between the liquid and the plate.

In some applications, we want to minimize the contact line pinning to promote the displacement of a liquid over a surface. For example, low hysteresis surfaces are required to manufacture superhydrophobic coatings [62] or surfaces with guidelines to have precise control of the liquid displacement [63]. In that case, F_H will be small, and then, a small force will be enough to induce the drop to roll. However, in other cases, we want to take advantage of the contact line pinning to perform some operations as the ordered deposition of nanoparticles immersed in a liquid through the contact line [64]. In this case, we require precise control of the pinning–depinning process that could be achieved by introducing a surface patterning or chemical heterogeneities on the solid phase.

One standard experimental method used to characterize this hysteresis consists of the deposit of a drop on a surface which is progressively tilted by an angle α. When α is small, the shape of the droplet is deformed but remains pinned on the substrate. However, when α is larger than the critical tilting angle α_c, the droplet starts to roll. Furmidge proposed the following relation between this critical angle and the hysteresis force [65]

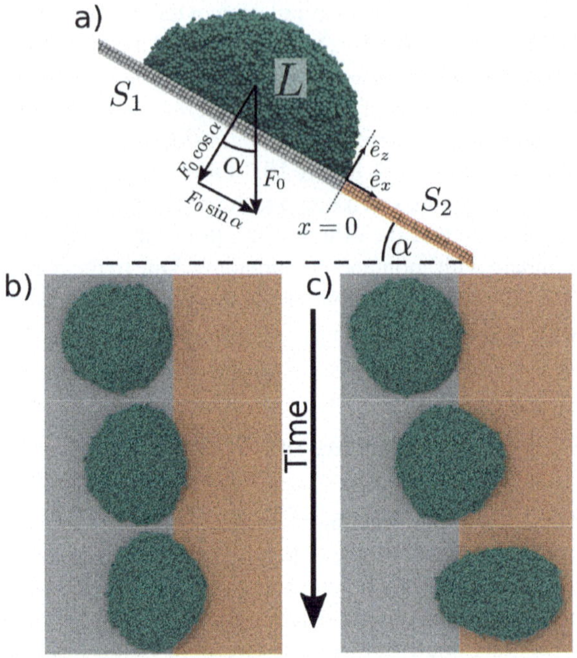

Fig. 10.8 **a** Sketch of a liquid drop in an incline at the proximities of the S_1–S_2 junction. **b** Top view snapshot of the simulation at different times with pinning for $F_0 = 99.62$ pN and **c** without pinning for $F_0 = 132.82$ pN

$$mg \sin\alpha_c = \frac{k}{2} w F_H = \frac{k}{2} w \sigma_L (\cos\theta_r - \cos\theta_a) \tag{10.16}$$

where m is the mass of the liquid droplet, g the gravity constant, w the width of the drop in the direction perpendicular to inclination. The parameter k is the so-called retention force factor which is typically consider as close to 2 [65].

When the contact line is pinned, the drop remains stationary and then, the net force acting over the drop must be equal to zero. If the surface is tilted along x, the force balance in that direction is [66]

$$mg \sin\alpha + \gamma \oint_{CL} dl\hat{n} \cdot \hat{e}_x \cos\theta = 0 \tag{10.17}$$

The first term is the gravity contribution, and the second is the total force along the contact line projected in x.

The force balance showed in Eq. 10.17 is essential to determine the conditions to have contact-line pinning. However, it is experimentally challenging to measure the distribution of the local contact angles along the contact line when there is an external force acting over the drop. One strategy to avoid this problem consists in the study of the ideal problem of a drop pinned in a single 'defect' created as the junction between two ideal substrates S_1 and S_2 with different wettabilities. Then, it may be possible to find an approximation for Eq. 10.17 using magnitudes that can be measured easily in the lab. Figure 10.8a. shows a sketch of this simplified geometry with the junction located at $x = 0$. Here, the upper half-plane $x < 0$ is relatively hydrophilic with an equilibrium angle θ_1 while the lower one $x > 0$ is hydrophobic and characterized by the equilibrium angles θ_2.

De Coninck et al. [66] studied the simplified geometry described above and proposed the existence of three possible scenarios: (a) a part of the contact line just follows the junction, (b) there is a small part in the central region that penetrates the hydrophobic, but still a part of the CL remains in the junction and (c) part of the contact line crosses completely to the hydrophobic. For the case (c) they proposed a simplification of Eq. 10.17 as

$$mg\sin\alpha + \gamma L(\cos\theta_2 - \cos\theta_1) = 0 \tag{10.18}$$

where L is the length of the intersection of the contact line. That is an equation equivalent to the Furmidge formula of Eq. 10.16 with k being exactly equal to 2 as reported in experimental studies and where L plays the role of w. Also, all the magnitudes of this equation are experimentally accessible.

Unfortunately, it is very complicated to check experimentally the validity of the Eq. 10.18 and compare it with the exact result from Eq. 10.17 due to the difficulty to measure the local contact angle distribution along the contact line. However, we can use MD simulations to model this problem and have access to all the required magnitudes to perform this comparison.

10.3.1 MD and Pinning

During the last decade, MD simulations applied to the contact-line pinning have been used to improve our understanding of the pinning force [67], the stick-slip of the contact line during evaporation [68–70], the pinning of drops with dispersed nanoparticles [71] and the pinning–depinning process on surfaces with chemical heterogeneities [72] or flexible nanopillars [73]. Here, we present as an example of how MD can be used as a real experiment to validate a theoretical approach like the one presented above.

We consider initially a spherical drop of a Lennard-Jones liquid with 5000 8-atoms molecules and a solid of 66,102 atoms. The plane $x = 0$ splits the plate it two half-planes S_1 and S_2 with different wettabilities. Se set the affinities between S_1-L and S_2-L to have an equilibrium contact angle of $\theta_1 = 70°$ and $\theta_2 = 135°$, respectively. Then, we put the drop in S_1 close to the junction S_1/S_2 and we let to spread and reach equilibrium. Once the equilibrium is achieved, we add an external force to model the gravity as $\mathbf{f_0} = f_0 \sin\alpha \hat{e}_x + f_0 \cos\alpha \hat{e}_z$ acting on every liquid atom. Therefore, the total force acting on the drop is $\mathbf{F_0} = N\mathbf{f_0}$ where N is the total number of liquid atoms. We fix the tilting angle $\alpha = 30°$ and we tune just the force $F_0 = |\mathbf{F_0}|$. For small values of F_0, the drop will reach a stationary state where the contact line remains pinned in the junction as illustrated in Fig. 10.8b for $F_0 = 99.62$ pN. When the force is larger than a critical value F_0^c the drop overpasses the junction and rolls over S_2 as shown in Fig. 10.8c for $F_0 = 132.82$ pN.

To compute the contact line location and the distribution of the local contact angle, we first need to determine the location of the liquid–vacuum (L–V) interface as the region where the density decays a 50% with respect to the value in the bulk and averaged over 50 instantaneous configurations separated by 5×10^{-3} ns. Then, the contact line is located as the intersection between the L–V interface and the plate. Some examples of the averaged contact line pinned on the junction are shown in Fig. 10.9a–d for different values of the external force F_0. From the intersection of the contact line with the junction, we can measure the length L. We define x_m as the value of x for which we have the maximum thickness in y.

For the computation of the local contact angles, we first calculate the normal direction to the contact line in each one of the points of this contact line, and then, we translate this normal line along with the L–V interface at a different height. This provides the profile associated with each point of the contact line. Then, we fit as usual a circular arch to the profile and compute the corresponding angle. Figure 10.9e represent an example of the distribution of the contact angles along y for $F_0 = 79.69$ pN. For $x < x_m$ the value of the contact angle is constant and equal to the equilibrium contact angle of the hydrophilic solid S_1, $\theta_{min} = \theta_1$. For $x > x_m$ there is a variation of the local contact angle from the equilibrium value in S_1, θ_1 to a maximum value θ_{max}. In Fig. 10.10a we represent θ_{min} and θ_{max} versus the external force F_0, where we observe that the value of the angle for $x < x_m$ is always equal to $\theta_{min} = \theta_1$. However, θ_{max} increases with F_0 and for some critical force F_0^c, $\theta_{max} = \theta_2$, and the contact line depins and therefore, the drop can escape from the junction and rolls along S_2.

Once we have measured the local contact angles, we can compute numerically the integral of the second term from Eq. 10.17 and in Fig. 10.10b we plot it versus the expected value $F_0 \sin\alpha$. Both contributions of the forces are in almost perfect agreement and then, the balance between them results in a total net force equal to zero. We can now compare this result with the approximate Eq. 10.18 computed with the measured values of L, θ_{min} and θ_{max} showed also in Fig. 10.10b. in this case, the approximated model works remarkably well for the larges forces where the contact line is pinned and able to cross the junction (as in Fig. 10.9c, d) corresponding with the case (c) described above and then, where the simplified Eq. 10.18 is valid [66]. For lower forces, a large part of the contact line is trapped at the junction as in Fig. 10.9a, o slightly penetrate on the hydrophobic solid as in Fig. 10.9a which corresponds with the cases (a) and (b) described in [66] where the full Eq. 10.17 should be used to describe the force balance.

In this example, we have shown how MD is an interesting tool not only to test the validity of a theoretical model as the approximate Eq. 10.18 but also to characterize the deviation observed when the system properties move out of the range of the application required by the theory as illustrated in Fig. 10.10b. Once we have checked Eq. 10.18, it is possible to use it to estimate the maximum contact radius r_c that a drop can have to be trapped in a junction between two solids of different wettabilities by approximating $L \approx 2r_c$. Therefore, this equation can be useful, for example, to

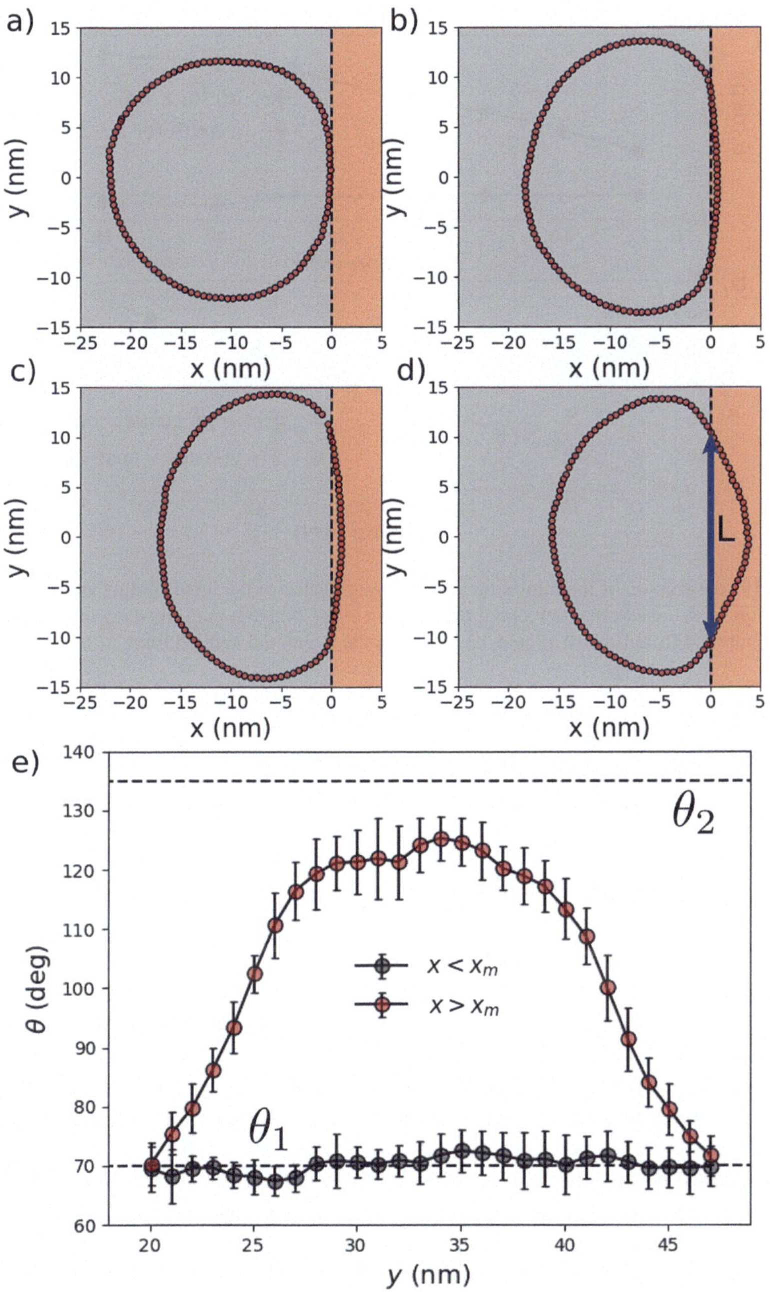

Fig. 10.9 Averaged contact line location for **a** $F_0 = 6.64$ pN, **b** $F_0 = 66.41$ pN, **c** $F_0 = 79.69$ pN and **d** $F_0 = 99.62$ pN. **e** Local contact angle versus y for $x < x_m$ and $x > x_m$ with $F_0 = 79.69$ pN

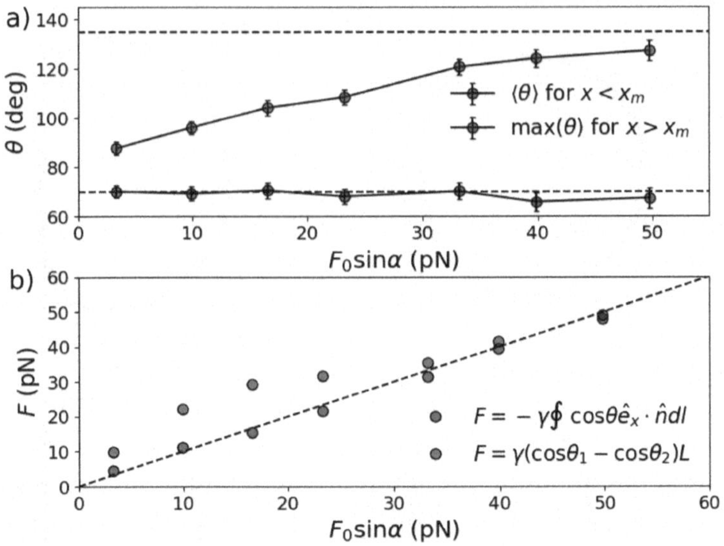

Fig. 10.10 **a** Variation of the minimum and maximum value of the local contact angle around the contact line with the external tangential force $F_0 \sin\alpha$. **b** Comparison of the tangential contact line forces computed from Eqs. 10.17 and 10.18 versus the tangential applied force

design a surface combining two different wettabilities to avoid the presence of drops larger than some particular size for a given tilting angle α.

10.4 Concluding Remarks

In this chapter, we have introduced the basis of the molecular dynamic technique, and we have given a flavor of its interest in engineering applications, providing two particular examples. On the one hand, we have seen how MD can be used to understand at the molecular level an experimental observation as the hydrodynamic assist that helps us to develop theoretical models. In this case, we have seen how the presence of slip between the liquid and the substrate modifies the wetting dynamics that could explain why it is possible experimentally to increase the coating velocity beyond its expected limit in very confined geometries as in the curtain coating method. On the other hand, we showed an example of MD simulations applied to check the validity of a proposed theoretical model and how it behaves out of its range of validity. Here, we have performed MD simulations to test the validity of an approximated force balance equation for a drop pinned on a junction between two substrates with different wettabilities.

In summary, MD technique can be used to model any physical phenomena although, as we point out in the introduction of this chapter, it has strong restrictions in the length and time scale accessible with this technique. Nevertheless, the improvement in the combined MD/CFD methods could circumvent these limitations, and being a new revolution in our understanding of the physical phenomena behind many applications that will help us to improve the current technology and develop new ones unimaginable nowadays.

References

1. B. Alder and T. Wainwright, " Phase transition for a hard sphere system," J. Chem. Phys. **27**, 1208 (1957) https://doi.org/10.1063/1.1743957
2. A. Rahman, "Correlations in the motion of atoms in liquid argon," Phys. Rev. **136**, A405 (1964) https://doi.org/10.1103/PhysRev.136.A405
3. J. Orban and A. Bellemans, "Velocity-inversion and irreversibility in a dilute gas of hard disks," Physics Letters A **24**, 620–621 (1967) https://doi.org/10.1016/0375-9601(67)90651-2
4. L. Verlet, "Computer "experiments" on classical fluids. I. Thermodynamical properties of Lennard-Jones molecules," Phys. Rev. **159**, 98 (1967) https://doi.org/10.1103/PhysRev.159.98
5. G. D. Harp and B. J. Berne, "Time-correlation functions, memory functions, and molecular dynamics," https://doi.org/10.1103/PhysRevA.2.975 Phys. Rev. A **2**, 1970 (1970)
6. A. Rahman and F. H. Stillinger, "Molecular dynamics study of liquid water," https://doi.org/10.1063/1.1676585 J. Chem. Phys. **55**, 3336 (1970)
7. J. Zhang, M. K. Borg, and J. M. Reese, "Multiscale simulation of dynamic wetting," Int. J. Heat Mass Transf. **115**, 886 – 896 (2017) https://doi.org/10.1016/j.ijheatmasstransfer.2017.07.034
8. H. Minaki and S. Li, "Multiscale modeling and simulation of dynamic wetting," Comput. Methods in Appl. Mech. Eng. **273**, 273 – 302 (2014) https://doi.org/10.1016/j.cma.2014.01.030
9. S. Gim, H.-K. Lim, and H. Kim, "Multiscale simulation method for quantitative prediction of surface wettability at the atomistic level" J. Phys. Chem. Lett. **9**, 1750–1758 (2018) https://doi.org/10.1021/acs.jpclett.8b00466
10. Q. Yuan and Y.-P. Zhao, "Multiscale dynamic wetting of a droplet on a lyophilic pillar-arrayed surface," https://doi.org/10.1017/jfm.2012.539 J. Fluid Mech. **716**, 171–188 (2013)
11. R. Zhang, S. Farokhirad, T. Lee, and J. Koplik, "Multiscale liquid drop impact on wettable and textured surfaces," Phys. Fluids **26**, 082003 (2014) https://doi.org/10.1063/1.4892083
12. D. Frenkel and B. Smit, *Understanding Molecular Simulation* (Academic Press, San Diego), (2002) https://doi.org/10.1016/B978-0-12-267351-1.X5000-7
13. M. P. Allen and D. J. Tildesley, *Computer simulation of liquids* (Clarendon Press, New York, 1989)
14. W. C. Swope, H. C. Andersen, P. H. Berens, and K. R. Wilson, " A computer simulation method for the calculation of equilibrium constants for the formation of physical clusters of molecules: Application to small water clusters," https://doi.org/10.1063/1.442716 J. Chem. Phys. **76**, 637 (1982)
15. K. Hermansson, G. C. Lie, and E. Clementi, "On velocity scaling in molecular dynamics simulations," https://doi.org/10.1002/jcc.540090303 J. Comput. Chem **9**, 200–203 (1988)
16. S. Nose, "A unified formulation of the constant temperature molecular dynamics methods," https://doi.org/10.1063/1.447334 J. Chem. Phys. **81**, 511–519 (1984)
17. W. G. Hoover and B. L. Holian, "Kinetic moments method for the canonical ensemble distribution," Phys. Lett. A **211**, 253–257 (1996) https://doi.org/10.1016/0375-9601(95)00973-6
18. B. Smit, "Phase diagrams of lennard–jones fluids," J. Chem. Phys. **96**, 8639–8640 (1992) https://doi.org/10.1063/1.462271

19. H. J. C. Berendsen, J. P. M. Postma, W. F. van Gunsteren, and J. Hermans, "Wetting of low energy surfaces," in *I*ntermolecular Forces, edited by B. Pullma (Marcel Dekker, Reidel, Dordrecht, 1981) p. 331
20. H. J. C. Berendsen, J. R. Grigera, and T. P. Straatsma, "The missing term in effective pair potentials," J. Phys. Chem. **91**1, 6269–6271 (1987) https://doi.org/10.1021/j100308a038
21. W. L. Jorgensen, J. Chandrasekhar, J. D. Madura, R. W. Impey, and M. L. Klein, "Comparison of simple potential functions for simulating liquid water," J. Chem. Phys. **79**, 926–935 (1983) https://doi.org/10.1063/1.445869
22. D. Wolff and W. Rudd, "Tabulated potentials in molecular dynamics simulations," Comput. Phys. Commun **120**, 20 – 32 (1999) https://doi.org/10.1016/S0010-4655(99)00217-9
23. R. B. Bird, C. F. Curtiss, R. C. Armstrong, and O. Hassager, *D*ynamics of polymeric liquids, vol. 2: Kinetic Theory (Wiley, New York, 1987)
24. A. D. MacKerell, D. Bashford, M. Bellott, R. L. Dunbrack, J. D. Evanseck, M. J. Field, S. Fischer, J. Gao, H. Guo, S. Ha, D. Joseph-McCarthy, L. Kuchnir, K. Kuczera, F. T. Lau, C. Mattos, S. Michnick, T. Ngo, D. T. Nguyen, B. Prodhom, W. E. Reiher, B. Roux, M. Schlenkrich, J. C. Smith, R. Stote, J. Straub, M. Watanabe, J. Wiórkiewicz-Kuczera, D. Yin, and M. Karplus, "All-atom empirical potential for molecular modeling and dynamics studies of proteins," J. Phys. Chem. B **102**, 3586–616 (1998) https://doi.org/10.1021/jp973084f
25. S. R. Pruitt, H. Nakata, T. Nagata, M. Mayes, Y. Alexeev, G. Fletcher, D. G. Fedorov, K. Kitaura, and M. S. Gordon, "Importance of three-body interactions in molecular dynamics simulations of water demonstrated with the fragment molecular orbital method," https://doi.org/10.1021/acs.jctc.5b01208 J. Chem. Theory Comput. **12**, 1423–1435 (2016)
26. F. H. Stillinger and T. A. Weber, "Computer simulation of local order in condensed phases of silicon," Phys. Rev. B **31**, pages 5262 (1985) https://doi.org/10.1103/PhysRevB.31.5262
27. V. Molinero and E. B. Moore, "Water modeled as an intermediate element between carbon and silicon," J Phys Chem B. **113**, 4008–16 (2009) https://doi.org/10.1021/jp805227c
28. J. H. Irving and J. G. Kirkwood, "The statistical mechanics theory of transport processes. iv. the equations of hydrodynamics," J. Chem. Phys. **1**8, 817 (1950) https://doi.org/10.1063/1.1747782
29. G. Navascués and M. V. Berry, "The statistical mechanics of wetting," Mol. Phys. **34**, 649–664 (1977) https://doi.org/10.1080/00268977700102021
30. K. S. Silmore, M. P. Howard, and A. Z. Panagiotopoulos, "Vapour-liquid phase equilibrium and surface tension of fully flexible Lennard–Jones chains," Mol. Phys. **115**, 320–327 (2017) https://doi.org/10.1080/00268976.2016.1262075
31. G. Fuchs and H. C. Kuhlmann, "Molecular dynamics simulation of Marangoni flow in a cavity," International Conference on Nanochannels, Microchannels, and Minichannels , 871–877 (2007) https://doi.org/10.1115/ICNMM2007-30137
32. E. B. Dussan V., "On the spreading of liquid on solid surfaces: static and dynamic contact lines," (1979)
33. P. G. De Gennes, "Wetting: Statics and dynamics," Reviews of Modern Physics **57**, 827–863 (1985) https://doi.org/10.1103/RevModPhys.57.827
34. T. D. Blake, "The physics of moving wetting lines," https://doi.org/10.1016/j.jcis.2006.03.051 Journal of Colloid and Interface Science **299**, 1–13 (2006)
35. D. Bonn, J. Eggers, J. Indekeu, and J. Meunier, "Wetting and spreading," Reviews of Modern Physics **81**, 739–805 (2009) https://doi.org/10.1103/RevModPhys.81.739
36. J. H. Snoeijer and B. Andreotti, "Moving contact lines: Scales, regimes, and dynamical transitions," Annual Review of Fluid Mechanics **45**, 260–292 (2013) https://doi.org/10.1146/annurev-fluid-011212-140734
37. J. Koplik, J. R. Banavar, and J. F. Willemsen, "Molecular dynamics of fluid flow at solid surfaces," https://doi.org/10.1063/1.857376 Physics of Fluids A: Fluid Dynamics **1**, 781–794 (1989)
38. P. A. Thompson and M. O. Robbins, "Simulations of contact-line motion: Slip and the dynamic contact angle," Physical Review Letters **63**, pages 766–769 (1989) https://doi.org/10.1103/PhysRevLett.63.766

39. M. J. de Ruijter, T. D. Blake, and J. De Coninck, "Dynamic wetting studied by molecular modeling simulations of droplet spreading," https://doi.org/10.1021/la990171l Langmuir **15**, 7836–7847 (1999)
40. D. R. Heine, G. S. Grest, and E. B. Webb, "Spreading dynamics of polymer nanodroplets," Physical Review E **68**, 061603 (2003) https://doi.org/10.1103/PhysRevE.68.061603
41. D. Seveno, G. Ogonowski, and J. De Coninck, "Liquid coating of moving fiber at the nanoscale," Langmuir **20**, 8385–8390 (2004) https://doi.org/10.1021/LA049574Y
42. J. De Coninck and T. Blake, "Wetting and Molecular Dynamics Simulations of Simple Liquids," https://doi.org/10.1146/annurev.matsci.38.060407.130339 Annual Review of Materials Research **38**, 1–22 (2008)
43. E. Bertrand, T. D. Blake, J. D. Coninck, and J. De Coninck, "Influence of solid-liquid interactions on dynamic wetting: a molecular dynamics study," https://doi.org/10.1088/0953-8984/21/46/464124 Journal of Physics: Condensed Matter **21**, 464124 (2009)
44. T. D. Blake, J.-C. C. Fernandez-Toledano, G. Doyen, and J. De Coninck, "Forced wetting and hydrodynamic assist," https://doi.org/10.1063/1.4934703 Physics of Fluids **27**, 112101 (2015)
45. A. V. Lukyanov and A. E. Likhtman, "Dynamic Contact Angle at the Nanoscale: A Unified View," https://doi.org/10.1021/acsnano.6b01630 ACS Nano **10**, 6045–6053 (2016)
46. J. C. Fernandez-Toledano, T. D. Blake, P. Lambert, and J. De Coninck, "On the cohesion of fluids and their adhesion to solids: Young's equation at the atomic scale," https://doi.org/10.1016/j.cis.2017.03.006 Advances in Colloid and Interface Science **245**, 102–107 (2017)
47. J. C. Fernandez-Toledano, T. D. Blake, P. Lambert, and J. De Coninck, "Moving contact lines and Langevin formalism," J. Colloid Interface Sci. **562**, 287–292 (2020a)
48. R. J. Hansen and T. Y. Toong, "Interface behavior as one fluid completely displaces another from a small-diameter tube," https://doi.org/10.1016/0021-9797(71)90014-2 Journal of Colloid And Interface Science **36**, 410–413 (1971)
49. O. V. Voinov, "Hydrodynamics of wetting," https://doi.org/10.1007/BF01012963 Fluid Dynamics **11**, 714–721 (1976)
50. R. G. Cox, "The dynamics of the spreading of liquids on a solid surface. Part 1. Viscous flow," https://doi.org/10.1017/S0022112086000332 Journal of Fluid Mechanics **168**, 169 (1986)
51. T. D. Blake and J. M. Haynes, "Kinetics of liquid/liquid displacement," J. Colloid Interface Sci. **30**, 421–423 (1969)
52. T. Blake, "Dynamic Contact Angles and Wetting Kinetics," in *Wettability* (Ed. Marcel Dekker, 1993) pp. 252–309
53. T. D. Blake and J. De Coninck, "The influence of solid-liquid interactions on dynamic wetting," https://doi.org/10.1016/S0001-8686(01)00073-2 Advances in colloid and interface science **96**, 21–36 (2002)
54. P. Gao, L. Li, J. Feng, H. Ding, and X.-Y. Lu, "Film deposition and transition on a partially wetting plate in dip coating," https://doi.org/10.1017/jfm.2016.64 J. Fluid Mech. **791**, 358–383 (2016)
55. H.-Y. Kim, H. Lee, and B. Kang, "Sliding of liquid drops down an inclined solid surface," https://doi.org/10.1006/jcis.2001.8156 J. Colloid Interface Sci. **247**, 372–380 (2002)
56. J. Snoeijer, E. Rio, N. Le Grand, and L. Limat, "Self-similar flow and contact line geometry at the rear of cornered drops," https://doi.org/10.1063/1.1946607 Phys. Fluids **17**, 72101 (2005)
57. A. Barajas and R. Panton, "The effects of contact angle on two-phase flow in capillary tubes," https://doi.org/10.1016/0301-9322(93)90007-H Int. J. Multiphase Flow **19**, pages 337–346 (1993)
58. T. Blake and K. Ruschak, "A maximum speed of wetting," https://doi.org/10.1038/282489a0 Nature **282**, 489 –491 (1979)
59. T. D. Blake, A. Clarke, and K. J. Ruschak, "Hydrodynamic assist of dynamic wetting," https://doi.org/10.1002/aic.690400205 AIChE J. **40**, 229–242 (1994)
60. E. Vandre, M. S. Carvalho, and S. Kumar, "Delaying the onset of dynamic wetting failure through meniscus confinement," https://doi.org/10.1002/aic.690400205 J. Fluid Mech. **707**, 496–520 (2012)

61. J. R. E. Johnson and R. Dettre, "Wetting of low energy surfaces," in *Wettability, edited by J. C.Berg* (Marcel Dekker, New York, 1993) pp. 1–74
62. Y. F. Huang, C. Huang, Y. L. Zhong, and S. P. Yi, "Preparing superhydrophobic surfaces with very low contact angle hysteresis," https://doi.org/10.1179/1743294412Y.0000000087 Surface Engineering **29**, 633–636 (2013)
63. T. Dong and T. J. McCarthy, "Superhydrophobic, low-hysteresis patterning chemistry for water-drop manipulation," https://doi.org/10.1021/acsami.7b15739 ACS Appl. Mater. Interfaces **47**, 41126–41130 (2017)
64. V. Flauraud, M. Mastrangeli, G. Bernasconi, J. Butet, D. T. L. Alexander, E. Shahrabi, O. J. F. Martin, and J. Brugger, "Nanoscale topographical control of capillary assembly of nanoparticles," https://doi.org/10.1038/nnano.2016.179 Nature Nanotech **12**, 73–80 (2017)
65. C. G. L. Furmidge, "Studies at phase interfaces," https://doi.org/10.1016/0095-8522(62)90011-9 J. Colloid Sci. **17**, 309–324 (1962)
66. J. De Coninck, J.-C. Fernandez-Toledano, F. Dunlop, and T. E. Huillet, "Pinning of a drop by a junction on an incline," https://doi.org/10.1103/PhysRevE.96.042804 Phys. Rev. E **96**, 042804 (2017)
67. H. Kusudo, T. Omori, and Y. Yamaguchi, "Extraction of the equilibrium pinning force on a contact line exerted from a wettability boundary of a solid surface through the connection between mechanical and thermodynamic routes," https://doi.org/10.1063/1.5124014 Chem. Phys. **151**, 154501 (2019)
68. F.-C. Wang and H.-A. Wu, "Pinning and depinning mechanism of the contact line during evaporation of nano-droplets sessile on textured surfaces," https://doi.org/10.1039/C3SM50530H Soft Matter **9**, 5703–5709 (2013)
69. J. Zhang, F. Müller-Plathe, and F. Leroy, "Pinning of the contact line during evaporation on heterogeneous surfaces: Slowdown or temporary immobilization? insights from a nanoscale study," https://doi.org/10.1021/acs.langmuir.5b01097 Langmuir **31**, 7544–7552 (2015)
70. J. Zhang, H. Huang, and X.-Y. Lu, "Pinning-depinning mechanism of the contact line during evaporation of nanodroplets on heated heterogeneous surfaces: A molecular dynamics simulation," https://doi.org/10.1021/acs.langmuir.9b00796 Langmuir **35**, 6356–6366 (2019)
71. B. Shi and E. B. Webb, "Self-pinning of a nanosuspension droplet: Molecular dynamics simulations," https://doi.org/10.1103/PhysRevE.94.012614 Phys. Rev. E **94**, 012614 (2016)
72. J.-C. Fernandez-Toledano, C. Rigaut, M. Mastrangeli, and J. De Coninck, "Controlling the pinning time of a receding contact line under forced wetting conditions," https://doi.org/10.1016/j.jcis.2020.01.054 J. Colloid Interface Sci. **565**, 449 – 457 (2020b)
73. F. Wang and H. Wu, "Molecular origin of contact line stick-slip motion during droplet evaporation," Sci. Rep. **5**, 17521 (2015) https://doi.org/10.1038/srep17521

Chapter 11
Multi-scale Multiphase Flow Gas–Liquid–Solid Interfacial Equation Based on Thermodynamic and Mathematical Approach

Yukihiro Yonemoto and Tomoaki Kunugi

11.1 Introduction

There exist many interfacial phenomena related to liquid, gas and solid phases around our daily lives. In the engineering field, the efficiency of industrial devices and the quality of products which are related to applications of fuel injection, spray-cooling/coating, heat exchanger, chemical reactor and flotation [1–5] are affected by interfacial phenomena. If the interfacial phenomena can be completely predicted by numerical approach, the economical cost is extremely reduced and research and development for industrial products will be promoted.

At the interfaces between gas, solid and liquid, many complicated phenomena such as evaporation, condensation, wetting, and electrokinetic phenomena are observed [6–8]. These phenomena may be concerned with thermodynamics, heat and mass transfer and hydrodynamics. However, in the numerical simulation of multiphase flow, the treatment of the interfacial interaction is very difficult because there are a lot of physical factors as shown in Fig. 11.1. For example, in the interaction between the vortex and the bubble, viscosity, density and surface tension of liquid would be important factors that are related to the breakup of a bubble. In the interaction between the bubbles, mass transfer and Coulomb force in addition to hydrodynamic force and liquid properties may be related to bubble coalescence where a liquid film between bubbles would play an important role [9–12]. The wetting phenomenon is a problem characterized by interaction among gas, liquid and solid phases. The wettability of liquid on the solid surface is quite complex because of the fluid motion from gas and liquid phases, and the solid surface structure influence the phenomenon [13–15]. In addition, such kind of factors is characterized by a wide range of temporal and

Y. Yonemoto (✉)
Kumamoto University, Kumamoto, Japan
e-mail: yonemoto@mech.kumamoto-u.ac.jp

T. Kunugi
Zhejiang University, Hangzhou, P. R. China

M. Marengo and J. De Coninck (eds.), *The Surface Wettability Effect on Phase Change*,
https://doi.org/10.1007/978-3-030-82992-6_11

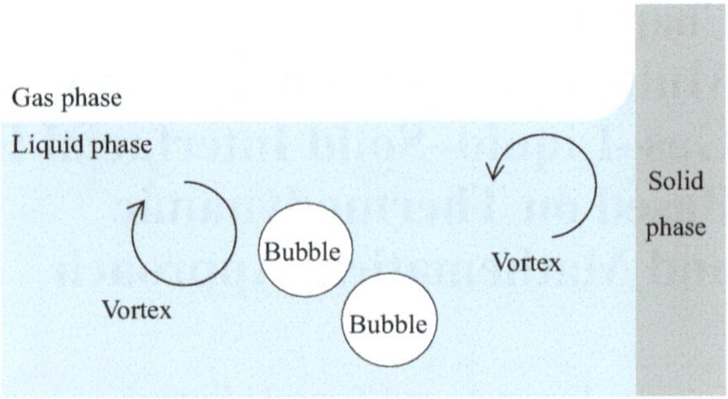

Fig. 11.1 Schematic of interfacial interactions among liquid, gas and solid phases

spatial scales. At the present time, there are no numerical models that can universally consider all interfacial problems characterized by gas, liquid and solid phases although each phenomenon is investigated in detail from experimental, numerical and theoretical point of views [16–23]. Therefore, the development of a numerical model to simulate gas–liquid–solid interfacial phenomena that can consider various temporal and spatial scales is very important.

Although there may be many unresolved problems for the interfacial phenomena, at least, bubble coalescence that is gas–liquid interfacial problem, and wetting phenomena that is three-phase problems would contain important fundamental physics for understanding more complicated phenomena such as phase change on the solid surface. Therefore, this chapter mainly focuses on those interfacial problems.

11.2 Gas–Liquid Interfacial Model

11.2.1 Interfacial Interaction on the Basis of Conventional Approach

With respect to bubble coalescence, there are many studies of experimental, theoretical and numerical approaches, and some ideas are discussed for understanding the phenomena. In the experimental approaches, the effects of specific ion and electrolyte solution on the interaction are investigated [24–31]. From the researches, the importance of electrostatic interaction between gas–liquid interfaces is revealed. In the theoretical approaches, bubble coalescence is mainly discussed on the basis of the lubrication theory where a thin liquid film between two bubbles or two interfaces is considered [9–12]. This theory is a very popular approach for the interfacial interaction. In this approach, the liquid drains from the liquid film, and bubbles finally coalesce. The drainage of the liquid and the merging of two interfaces means the film rupture. The mechanism of the film rupture is actually very complicated. In the bridging-dewetting mechanism [32–35], it is considered that solid particles immersed in the liquid film induce a film rupture. In this model, the rupture time of the liquid film depends on the wettability of the solid particles. However, the theoretical approaches on the basis of the liquid film concept always premise the existence of the liquid film. If the liquid film can be realized in multiphase flow simulation, the theory can be applied as a coalescence model. However, in an actual situation, bubble interaction is treated as a statistical problem, and is mainly considered by the frequency/efficiency of coalescence between bubbles [36, 37]. If the scale of the system becomes small, the statistical approach is no longer available in the multiphase flow simulation.

In the numerical simulation where the scale of the system is not so large, it is important to consider the geometrical change in the gas–liquid interface. Here, we show some popular interfacial models.

The first one is the Continuum Surface Force (CSF) model proposed by Brackbill et al. [38]. In the CSF model, the discontinuous interface is interpreted as the interface which has thickness h as shown in Fig. 11.2. The thickness h enables to define the continuous function. Therefore, it is possible to integrate the physical values over the interface. Finally, the surface tension is estimated mathematically by taking the limitation ($h \rightarrow 0$). By this limitation, surface tension is incorporated into Navier–Stokes (NS) equation as a body force.

The concept of the CSF model is based on the Eulerian method. On the other hand, the front tracking model is based on not only Eulerian but also the Lagrangian method [39]. In this model, the interface has no thickness. The interface is represented by some markers as shown in Fig. 11.3. Therefore, the interface consists of line-segment in two dimensions or polygon elements in three dimensions by using markers. These markers enable to make the tangential vector by which the surface tension is estimated on the basis of differential geometry. Finally, this surface tension is combined with the delta function in order to make the surface tension work at the interface, and incorporated into NS equation.

However, the two models mentioned above cannot treat the interfacial interaction. Specifically, in the CSF model, bubble coalescence is unphysical. In the front tracking model, the interfacial grid points are added or deducted artificially when the

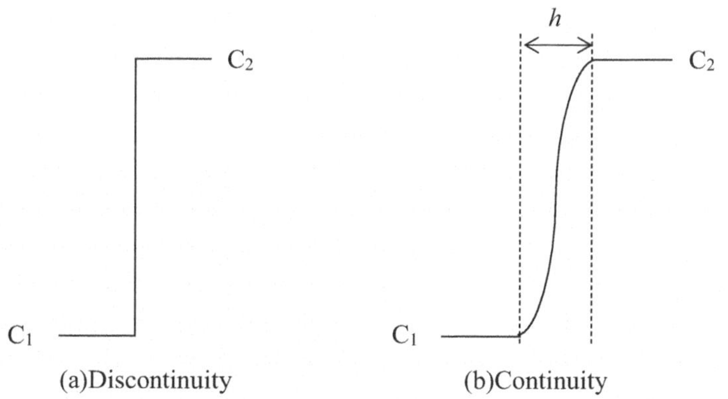

Fig. 11.2 Schematic of the interface in CSF (Continuum Surface Force) model

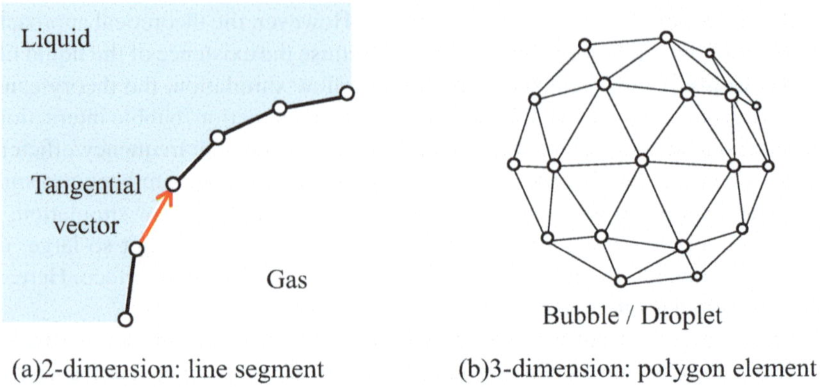

Fig. 11.3 Schematic of the interface in front tracking model

interface interacts with each other. This is because these models are mainly developed on the basis of mechanical energy balance where the interface is assumed as the mathematical interface with zero thickness. Therefore, interfacial physics cannot be considered on the interface. In order to treat the interfacial phenomena that are mainly characterized by thermodynamics, an interfacial model that does not depend on a geometrical approach will be needed.

Recently, the diffuse interface method is a popular approach to simulate multiphase flow [40–44]. This method is based on the thermodynamic concept, which is called the phase-field theory or van der Waals theory [45, 46]. In the phase-field theory, surface tension is thought as excess energy due to the presence of the interface as shown in Fig. 11.4.

It is free energy on the interface. Then, the free energy consists of homogeneous term and gradient term as follows:

Fig. 11.4 Schematic of a diffuse interface

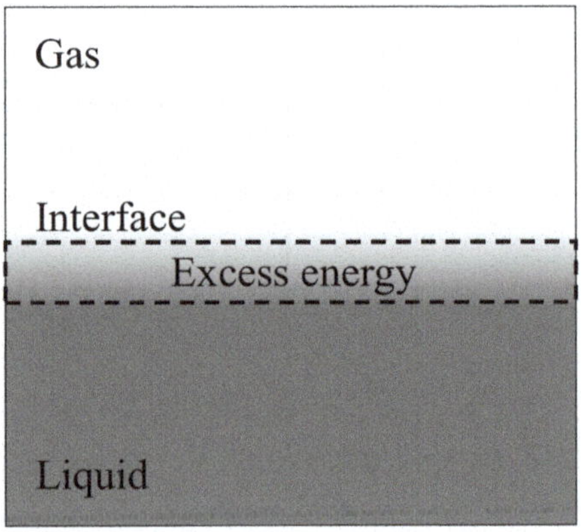

$$f(\phi, \nabla\phi) = f_0(\phi) + \frac{1}{2}k|\nabla\phi|^2, \tag{11.1}$$

where k, n, and ϕ are the gradient energy parameter [J m^{-1}], normal direction against the interface [m], and the order parameter [−]. By the gradient term, the interface has a thickness. The gradient term is concerned with surface tension. The surface tension γ [N/m] is defined by the following relation:

$$\gamma = \int_{-\infty}^{+\infty} k\left(\frac{\partial\phi}{\partial n}\right)^2 dn. \tag{11.2}$$

Although the detailed explanation of the numerical procedure is omitted here, surface tension force in the NS equation is evaluated by the gradient of chemical potential which is obtained by the free energy equation. Then, the interface is transported by convective Cahn–Hilliard equation, which is basically reaction–diffusion equation.

$$\frac{\partial\phi}{\partial t} + \boldsymbol{u}\cdot\nabla\phi = \nabla\cdot M\nabla\mu, \tag{11.3}$$

where ϕ is taken by concentration of material in the original phase field theory, and M and μ represent the mobility [mol^2 s kg m^{-3}] and the chemical potential [J mol^{-1}], respectively. The diffuse interface method is widely applied to two-phase flow simulation and wetting behaviour. Therefore, this method may have a potential to consider the interfacial interaction. However, in the diffuse interface method, the interface is transported by the Cahn–Hilliard equation where the mobility is used. The mobility is basically an ion scale parameter. Thus, the interface should not be transported by the ion scale which means the vector $\boldsymbol{u}$ in Eq. (11.3) is not the fluid convection scale. Therefore, the diffuse interface method cannot discriminate temporal and spatial scales of phenomena. However, the phase-field theory is a promising theory to be able to consider the interfacial phenomena that are related to the thermodynamics. The development of an interfacial model on the basis of this kind of theory is very important in order to describe various physical phenomena with the consideration of temporal and spatial scales precisely.

11.2.2 Development of Interfacial Model for Interfacial Interaction

There are many complicated interfacial phenomena that are characterized by multi-scale physics with various temporal and spatial scales. In this section, the possibility of the multiscale concept for multiphase flow phenomena is shown as one methodology [47].

11.2.2.1 Multi-scale Concept and Modeling

In the modeling of the gas–liquid interface, an effect of electrostatic potential due to contamination such as specific ion on the interfacial interaction is considered because the existence of the specific ion cannot be ignored from the experimental fact [24–29]. Then, the free energy equation is reevaluated from a viewpoint of the molecular scale. Concretely speaking, in the derivation of the free energy, the concepts of the Lattice gas model and continuum limit are applied. Figure 11.5 shows a concept of a multi-scale interfacial model. The phase-field theory assumes that the interface has a finite thickness and is characterized by the free energy. This means that one can treat various interfacial phenomena related to thermodynamics. Thus, the discussion of the interfacial modeling is made on the basis of the consideration for the three assumptions with the images in Fig. 11.5. The first one is that the interface has a finite thickness like a fluid membrane. This assumption means a mesoscopic interface is considered between micro and macroscopic interfaces. The second one is that the interface consists of a few hundred molecules. Then, the last assumption is that the continuum approximation holds in this interface. By these assumptions, the physical values can be defined at the interface.

In the present model, molecular motion is expressed by the lattice gas model as shown in Fig. 11.6. In the lattice gas model, random molecular motion is considered

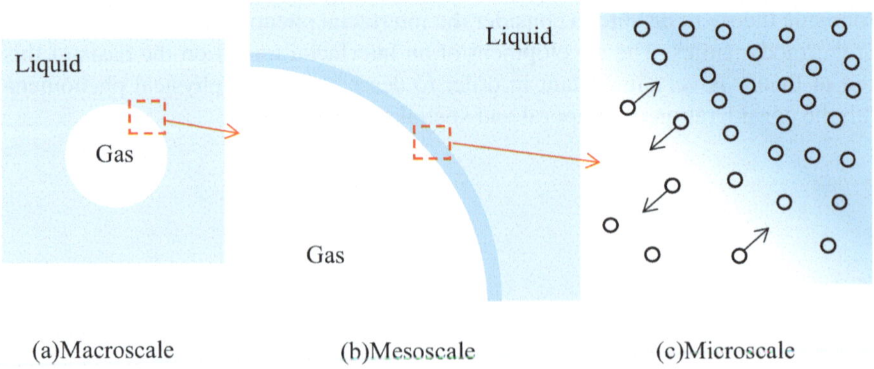

Fig. 11.5 Multi-scale concept for interface

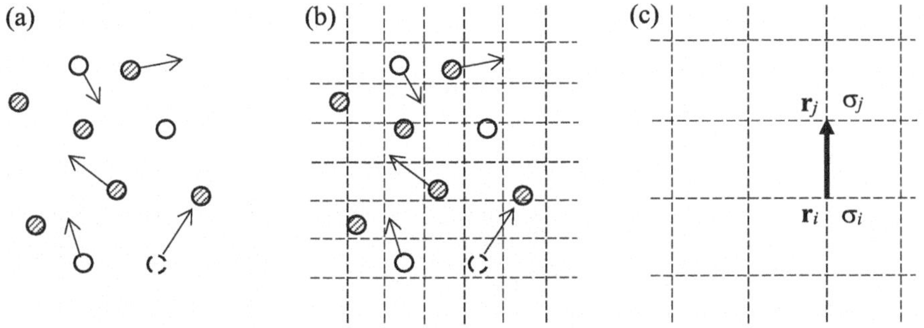

Fig. 11.6 Schematic diagram of lattice gas model, **a** Random molecular motion, **b** Molecular motion: considered in square or cubic lattice with width l_0 [m], **c** Existence of molecule on lattices: $\sigma_i = 1$ (Molecular B) and $\sigma_i = 0$ (Molecular A)

in square or cubic lattice with width l_0. Then, the variables σ [−] that represent the existence of the molecules are defined at the lattice points. For example, if the variable takes unity, molecule B exists at the lattice point. If the variable takes zero, molecule A exists at the point. On the basis of this concept, the Hamiltonian H of the molecule is derived as follows [48]:

$$H = -\frac{1}{2}\sum_{ij} Y_{ij}, \tag{11.4}$$

and,

$$Y_{ij} = W_{ij}^{AA}(1-\sigma_i)(1-\sigma_j) + W_{ij}^{BB}\sigma_i\sigma_j + W_{ij}^{AB}\left\{\sigma_i(1-\sigma_j) + \sigma_j(1-\sigma_i)\right\}. \tag{11.5}$$

Here, each W_{ij} represents the intermolecular potential. Then, from some experimental evidence such as specific ions, we think that the contamination at the interface will contribute to the bubble interaction, and it is assumed that the contamination would induce a deviation of charge at the interface as shown in Fig. 11.7. By the deviation of charge, an electric double layer would form around the interface. Therefore, the electric interaction energy is added to the Hamiltonian as follows:

$$H = -\frac{1}{2}\sum_{ij} Y_{ij} - \frac{1}{2}\sum_{i}^{N} z_i e V(\boldsymbol{r}_i)\sigma_i. \tag{11.6}$$

Here, it is assumed that the contamination is uniformly adsorbed to the interface and the adsorption process is ignored.

The Hamiltonian is microscopic interaction energy between lattices. Therefore, in order to evaluate the Hamiltonian as a continuum value, the thermodynamic averaging procedure is taken using the partition function of the molecule. Concretely speaking,

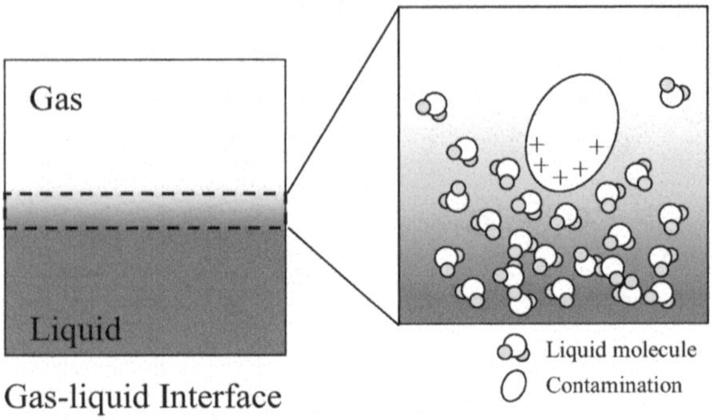

Fig. 11.7 Electric double layer at the interface

the existence of the molecule is statistically averaged at the lattice point by the following relation:

$$\langle\sigma_i\rangle_0 = \int_\Gamma \sigma_i \exp(-\frac{H_0}{k_B T})\mathrm{d}\overline{\Gamma}. \quad (11.7)$$

In Eq. (11.7), H_0, k_B, T and d $\overline{\Gamma}$ represent Hamiltonian, Boltzmann constant, temperature and infinitesimal volume in phase space $\overline{\Gamma}$, respectively. By this averaging procedure, the molecule condition where the existence is evaluated by $\sigma_i = 0$ or 1 can be interpreted as the local continuum condition. Then, the continuum limit ($l_0 \to 0$) is taken by considering infinitesimal lattice width l_0 in order to treat larger scale. For example,

$$(\phi_i - \phi_j)\boldsymbol{n} = \left(l_0 \frac{\phi_i - \phi_j}{l_0}\right)\boldsymbol{n} \to_{l_0\to 0} l_0 \nabla\phi, \quad (11.8)$$

where $\phi_i \to \phi$, $\phi_j \to \phi$, and $\boldsymbol{n}$ is the unit vector between lattice (i) and (j). Through the above discussion, the free energy equation is finally derived.

$$F = \int_V \left[f_0(\psi) + \frac{d}{2}(\nabla\psi)^2\right]dV. \quad (11.9)$$

Here,

$$f_0(\psi) = -\frac{a}{2}\psi^2 + \frac{b}{4}\psi^4 + cz_i eV\psi, \quad (11.10)$$

$$a = \frac{U - 4k_B T}{l_0^3}, \quad (11.11)$$

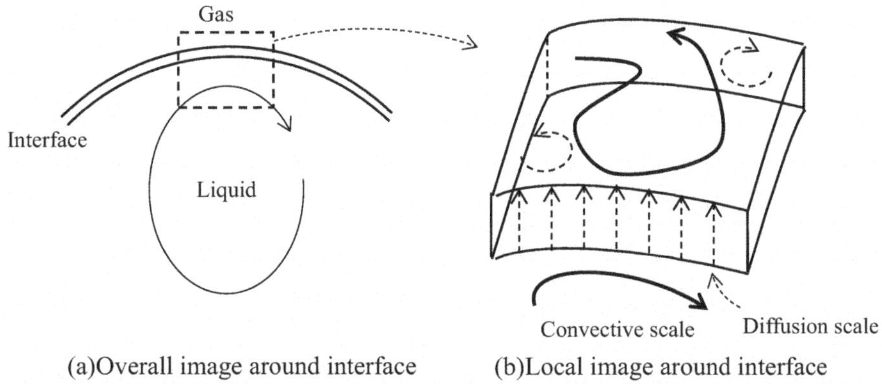

Fig. 11.8 Conceptual image for the distinction of temporal and spatial scales: fluid motion and mass transfer

$$b = \frac{16k_B T}{3l_0^3}, \tag{11.12}$$

$$c = \frac{1}{2l_0^3}, \tag{11.13}$$

$$d = \frac{U}{2l_0}. \tag{11.14}$$

In this equation, the electric potential due to the contamination is included in the third term in Eq. (11.10). In Eqs. (11.11) and (11.14), U [J] is the intermolecular potential.

As a next step, the free energy equation is incorporated into the conventional NS equation. The important point is to discriminate the temporal and spatial scales of phenomena as mentioned above. For example, consider an interface as shown in Fig. 11.8.

There is dynamic fluid flow around the interface as shown in Fig. 11.8a. Then, there is mass transfer through the interface (Fig. 11.8b). The characteristic lengths may be different from each other. On the basis of this concept, the Chapman–Enskog expansion is applied to temporal and spatial operators of D/Dt and ∇ using a small parameter ε.

$$\frac{\mathrm{D}}{\mathrm{D}t} = \frac{\mathrm{D}}{\mathrm{D}t^{(0)}} + \varepsilon\frac{\mathrm{D}}{\mathrm{D}t^{(1)}} + \varepsilon^2\frac{\mathrm{D}}{\mathrm{D}t^{(2)}} + \cdots, \tag{11.15}$$

$$\nabla = \nabla^{(0)} + \varepsilon\nabla^{(1)} + \varepsilon^2\nabla^{(2)} + \cdots. \tag{11.16}$$

Here, the superscript (k) ($k = 0, 1, 2, \ldots$) represents the scale of the phenomena. Then, the parameter ε consists of characteristic lengths of interface and fluid motion as follows.

$$\varepsilon = \frac{\delta}{L}. \tag{11.17}$$

In this equation, δ[m] and L [m] represent the characteristic lengths of the interface and that of the vertical fluid flow, respectively. By substitution of Eqs. (11.15) and (11.16) into conventional NS equation, the following relation is derived:

$$\rho \frac{D\boldsymbol{u}}{Dt^{(0)}} + \varepsilon\rho \frac{D\boldsymbol{u}'}{Dt^{(1)}} = -\nabla^{(0)} \cdot \boldsymbol{T} - \varepsilon\nabla^{(1)} \cdot \boldsymbol{T}' + \rho\boldsymbol{g}. \tag{11.18}$$

Here, Eqs. (11.15) and (11.16) are considered until $O(\varepsilon)$, where the physical values in the mesoscopic scale are assumed to be represented by velocity $\boldsymbol{u}'$ and stress tensor $\boldsymbol{T}'$ that correspond to the physical quantities at the interface. Moreover, we assume that the stress tensor is represented by the Maxwell relation. Thus, thermodynamic pressure is defined by the following relation:

$$\boldsymbol{P}_{\text{therm}} = -\left(\frac{\partial F}{\partial V}\right)_T \boldsymbol{I}. \tag{11.19}$$

Finally, by considering that the thickness of the interface is significantly small compared to a minimum scale of an eddy such as the Kolmogorov scale, and by performing a simple tensor analysis [47], the multi-scale multiphase flow equation is derived as follows:

$$\rho \frac{D\boldsymbol{u}}{Dt^{(0)}} + \varepsilon\rho \frac{D\boldsymbol{u}'}{Dt^{(1)}} = -\nabla^{(0)} \cdot \boldsymbol{T} + \varepsilon\nabla^{(1)} \cdot (f_0(\psi)\boldsymbol{I}) - \varepsilon d\nabla^{(1)}\psi\left(\nabla^{(1)} \cdot \nabla^{(1)}\psi\right) + \rho\boldsymbol{g}. \tag{11.20}$$

11.2.2.2 Model for Interfacial Interaction

The multi-scale multiphase flow equation is mainly solved by combining two interfacial interaction models. One is the contamination at the interface, which is related to the electrostatic potential. The other is a mass transfer between interfaces. The importance of the mass transfer is indicated by previous experimental observation of microbubble flow [49]. In the experiment, microbubble coalescence was observed and the results indicated that the mass transfer was related to the bubble coalescence. Figure 11.9 shows an example of the air microbubble coalescence in water flow. The sampling rate is 7,140 fps and the void fraction is 0.2%. These figures show a time evolution of microbubble flow. In this experiment, two microbubbles in contact with each other as shown in Fig. 11.9a–c. The diameter of one bubble is 94 μm, and the other one is 49 μm before coalescence. In Fig. 11.9d, two microbubbles coalesce with each other where a yielded bubble diameter is ca. 98 μm. It is found that one bubble

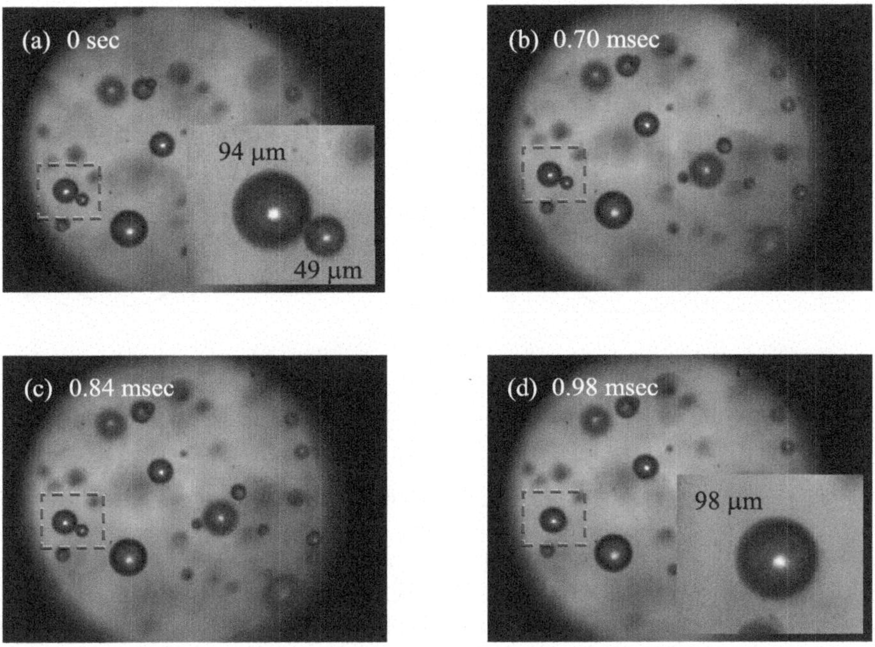

Fig. 11.9 Image for microbubble coalescence at Re = 4440 with void fraction 0.2 [%] where electrical conductivity is 4.7 μS/cm

is in contact with the other one until the coalescence of two bubbles. Other than this case, some patterns for the bubble interaction were observed such as sliding of one bubble along the surface of another one, and immediate coalescence after the contact of two bubbles. However, it was very difficult to predict these behaviors with the film thinning theory, and the result indicated that microbubble coalescence may concern with other complicated phenomena such as mass transfer [49]. According to Henry's law, the amount of a gas dissolved in a liquid is proportional to the gas pressure at the bubble surface in equilibrium when the system is at a constant temperature. In a real system, it is difficult to make perfectly purified water where there is no contamination. Therefore, a relationship between the dissolved gas and the contamination in the liquid would be an important factor in the interfacial interaction. Therefore, one possible model for the coalescence model can be developed on the basis of the mass diffusion and the nucleation theory [50], although there may be more possible mechanisms for the bubble coalescence.

In the coalescence model, the diffusion equation is solved on the basis of some assumptions.

$$\frac{\mathrm{D}C}{\mathrm{D}t} = D_{\mathrm{diff}} \nabla \cdot \nabla C. \tag{11.21}$$

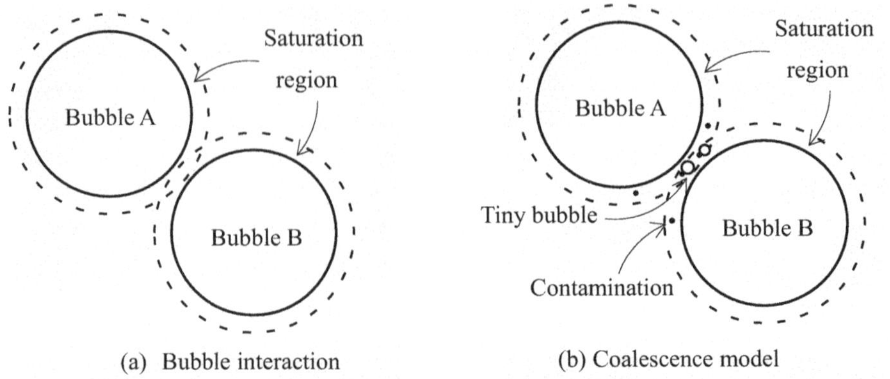

Fig. 11.10 Schematic of bubble interaction and nucleation model for bubble coalescence

In this equation, C and D_{diff} represent the concentration [mol m^{-3}] and diffusion coefficient [m^2 s^{-1}], respectively. As shown in Fig. 11.10a, there is a saturation region around a bubble due to the bubble's own pressure, and the overlapping region arises when the two bubbles interact with each other. Thus, it is assumed that the mass diffusion may arise due to a change in the free energy related to the electric potential in the overlapping region, then the dissolved gas reaches a steady state instantaneously when two bubbles mutually interact. In addition, it is assumed that there is contamination in the liquid, and the contamination induces tiny bubbles in the overlapping region as shown in Fig. 11.10b. These tiny bubbles induce bubble coalescence. This concept is modeled on the basis of the nucleation theory as follows:

$$r_{\text{crit}} = \frac{2\sigma \nu_0}{k_B T\left(\ln \frac{P_{gA}}{P_{gB}} + \beta_0 \ln \Omega_{\text{dust}}\right)}. \tag{11.22}$$

In this equation, Ω_{dust}, β_0 and r_{crit} represent the electric conductivity [μS cm^{-1}], arbitrary parameter [−] and critical radius [m] of the inception for a tiny bubble. The dissolved gas is simulated by diffusion equation, and r_{crit} is compared with an averaged radius r_{ave} that is calculated from the amount of the diffused gas in the overlapping region of the liquid.

11.2.2.3 Numerical Result for Bubble–Bubble Interaction

In the developed interfacial model, the multi-scale multiphase flow equation has two scales for macroscopic scale denoted by the superscript of 0 and mesoscopic scale denoted by that of 1. In the strict sense, two different scales must be considered in the simulation. But in the present test case, mesoscopic scale is only considered for the sake of simplicity. Thus, the following equations are mainly used with the equation of continuity for the simulation [47, 50].

Fig. 11.11 Schematic of the numerical domain for microbubble interaction

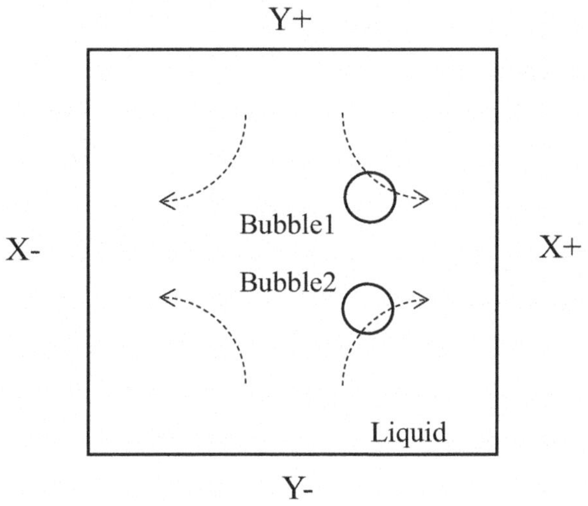

$$\rho\frac{\mathrm{D}\boldsymbol{U}}{\mathrm{D}t} = -\nabla^{(1)}\cdot\boldsymbol{T} + -\varepsilon d\nabla^{(1)}\psi\left(\nabla^{(1)}\cdot\nabla^{(1)}\psi\right) + \rho\boldsymbol{g}, \tag{11.23}$$

where $\boldsymbol{T}$ is a stress tensor expressed as $\boldsymbol{T} = P\boldsymbol{I}\text{-}\tau$ where P is the mechanical pressure and τ is shear stress. The volume tracking method used for capturing the gas–liquid interface is based on the Multi-interface Advection and Reconstruction Solver (MARS) method [51]. This volume tracking method is similar to the Piece-wise Linear Interface Construction (PLIC) [52] algorithm and is on the basis of the VOF method [53].

Figure 11.11 shows a numerical condition of the test case for bubble interaction. Two-dimensional simulation is mainly performed. As the boundary condition, X+ and X− are pressure free. Then, the fluid flows from Y+ in the downward direction and Y− in the upward direction with a constant velocity that eventually yields a flow denoted by the dashed arrows. By this condition, two bubbles can interact with each other. Three cases are mainly simulated. In case 1, both the electric potential and the nucleation theory are not considered. In case 2, only the electric potential is considered. In case 3, both models are considered. The gravity is neglected.

Figure 11.12 shows the numerical results of the bubble interactions. In case 1, two bubbles simply coalesce with each other when they interact. This case may correspond to the simulation with the use of the CSF model. In case 2, the electric potential is only considered in the interfacial interaction. Two bubbles do not coalesce due to the repulsive force between the interfaces. In case 3 where the electric potential and the nucleation model are considered, two bubbles flow with keeping a liquid film between them, and finally coalesce with each other. The timing of the bubble coalescence between case 1 and case 3 is different. From these results, it is found that our interfacial model can treat interfacial interaction such as coalescence and repulsion.

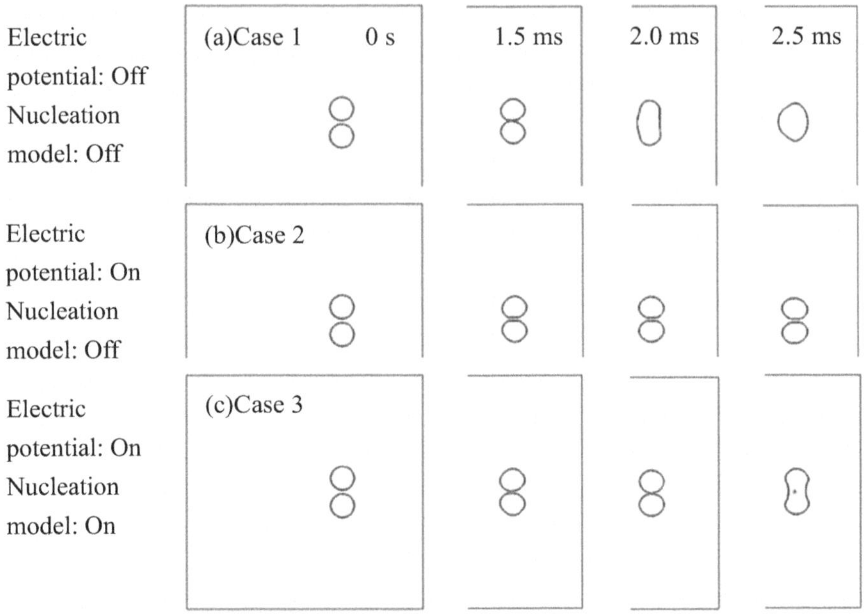

Fig. 11.12 Numerical results for microbubble interaction. **a** Case 1: Electric potential $V = 0$ [V], **b** Case 2: $V = -1.0 \times 10^{-5}$ [V], **c** Case3: $V = -1.0 \times 10^{-5}$ [V] and $\Omega_{dust} = 103.146$ [μS cm^{-1}]

One example of the results for the three-dimensional numerical simulation that was performed for the confirmation of a more realistic situation is shown in Fig. 11.13 [54]. In this case, the model of the interfacial electrostatic potential was only considered. The thin liquid film was observed between two microbubble interfaces as shown in Fig. 11.13a. Then, the pressure in the liquid film is lower than the bulk pressure around the microbubbles as shown in Fig. 11.13b. This result indicates the important

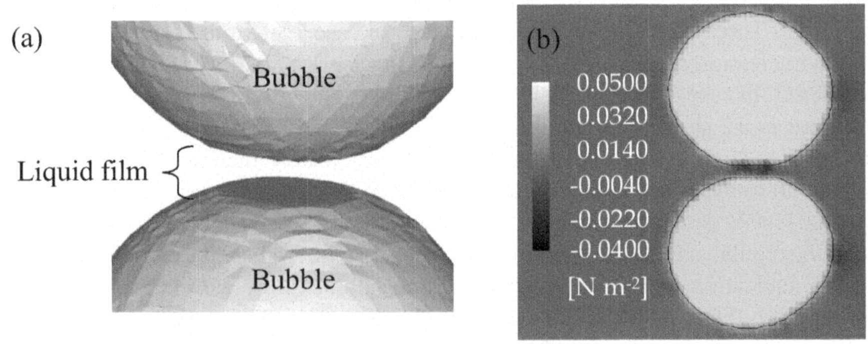

Fig. 11.13 Microbubble interaction in three-dimensional case ($V = -1.0 \times 10^{-5}$ [V]): **a** Liquid film in three dimensions, **b** Pressure distribution in the liquid film

knowledge that the bulk liquid flows into the thin liquid film to satisfy the continuity and maintain a certain thickness of the liquid film. This means that the general drainage explanation based on the lubrication theory [9–12], which is based only on hydrodynamics, cannot explain the premise of the existence of a liquid film between two microbubbles. Thus, the present numerical results could prove that the contamination at the interface (electrostatic potential) is very important with respect to the bubble interaction.

11.2.2.4 Multi-scale Model for Gas–Liquid Interface Through Micro to Macro

The modeling discussed in the previous section is mainly focused on the mesoscopic scale. In this section, a relationship between a macroscopic interface and a multi-scale multiphase flow equation is discussed [55].

The jump condition is a kind of boundary condition at the interface. There are two types of jump conditions. One is called the primary jump condition, and the other is called the secondary jump condition [56, 57]. The primary condition includes mass, momentum, total energy, and entropy evolution jump condition. The secondary condition is about entropy jump condition. In the present study, we focus on the momentum jump condition and call it the conventional jump condition.

The conventional jump condition is mainly derived using mathematical concepts such as stokes theorem, divergence theorem of Gauss, and differential geometry. The test volume shown in Fig. 11.14a is used in the concepts. Thus, the conventional jump condition is obtained by volume integral and surface integration as shown in Fig. 11.14b. In the derivation, it is thought that the interface has no thickness and surface tension is the force acting on a line-segment on the surface.

$$\dot{m}_g \boldsymbol{u}_g + \dot{m}_l \boldsymbol{u}_l - \left\{(-P_g)\boldsymbol{n}_g + \boldsymbol{\tau}_g \cdot \boldsymbol{n}_g\right\} - \{(-P_l)\boldsymbol{n}_l + \boldsymbol{\tau}_l \cdot \boldsymbol{n}_l\} - 2H\sigma \boldsymbol{n}_g + \frac{d\sigma}{ds}\boldsymbol{t} = 0, \tag{11.24}$$

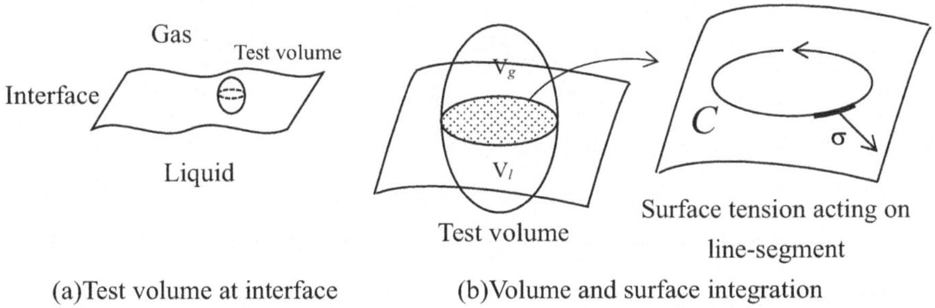

Fig. 11.14 Mathematical concept for the treatment of interfacial jump condition

where the first and second terms in the left-hand side of Eq. (11.24) represent the evaporation term, and $\dot{m}_k(k = g, l)$ [kg m^{-2} s^{-1}] is related to mass transfer through the interface. σ [N m^{-1}], P_k [N m^{-2}] and s [m] are the surface tension coefficient, pressure, and coordinate along the interface, respectively. The mean curvature is expressed by H ($H = (\kappa_1 + \kappa_2)/2$, where κ_1 and κ_2 [m^{-1}] are the principal curvatures). $\boldsymbol{n}_k$, $\boldsymbol{t}$ and τ_k are the unit normal, unit tangential vectors, and shear stress, respectively. The jump condition at the interface is characterized by the curvature related to the shape of the interface, which means that the interface is a mathematical interface with zero thickness. Thus, the jump condition is considered to be a macroscopic interfacial equation, and this form is used as a reference for macroscopic interface in the present study.

The jump condition can be also derived from our interfacial model discussed in the above section (Eq. 11.20). We call it thermodynamic jump condition. Before the derivation of the jump condition, two conditions are considered. One is the interface which has a thickness. This condition means that excess free energy exists in the interface, and the surface tension coefficient is estimated by integrating an excess free energy over the interface as expressed in Eq. (11.2). This concept is defined in van der Waals theory and this formula is actually limited to the flat surface. However, the formula of free energy is used in the derivation. Then, another condition is that the jump condition is based on the interfacial coordinate. Based on two conditions, we model the interface as shown in Fig. 11.15.

From the first condition, the interface is divided into some elemental interfaces as shown in Fig. 11.15a. From the second condition, the interfacial coordinates on the elemental interfaces are set as shown in Fig. 11.15b. Here, we focus on the local interface, thus the bending of the interface is small. The base vectors are assumed to be orthogonal relation. The mathematical operators become as follows:

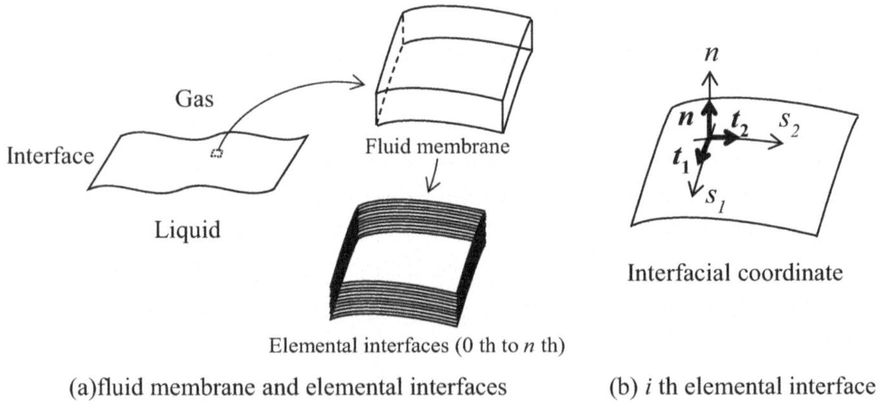

Fig. 11.15 Concept of fluid membrane consisting elemental interfaces, and interfacial coordinate on an elemental interface

$$\nabla = \boldsymbol{t}_1 \frac{\partial}{\partial s_1} + \boldsymbol{t}_2 \frac{\partial}{\partial s_2} + \boldsymbol{n} \frac{\partial}{\partial n}, \tag{11.25}$$

$$\nabla \cdot \nabla = (\kappa_1 + \kappa_2) \frac{\partial}{\partial n} + \frac{\partial^2}{\partial s_1^2} + \frac{\partial^2}{\partial s_2^2} + \frac{\partial^2}{\partial n^2}. \tag{11.26}$$

Based on this concept of the interface, the multi-scale multiphase flow equation is transformed and integrated over the interface. Then, the thermodynamic jump condition is finally derived as follows:

$$\begin{aligned} 0 = & \underbrace{\dot{m}_{\mathrm{G}}^{(0)} u_{\mathrm{G}}^{(0)} \boldsymbol{n}_{\mathrm{G}} + \dot{m}_{\mathrm{L}}^{(0)} u_{\mathrm{L}}^{(0)} \boldsymbol{n}_{\mathrm{L}}}_{(a)} + \underbrace{\varepsilon \left(\dot{m}_{\mathrm{G}}^{(1)} u_{\mathrm{G}}^{(1)} \boldsymbol{n}_{\mathrm{G}} + \dot{m}_{\mathrm{L}}^{(1)} u_{\mathrm{L}}^{(1)} \boldsymbol{n}_{\mathrm{L}} \right)}_{(b)} \\ & \underbrace{-[(-P_{\mathrm{G}} - \varepsilon \pi_{\mathrm{G}}) \boldsymbol{n}_{\mathrm{G}} + \boldsymbol{n}_{\mathrm{G}} \cdot \boldsymbol{\tau}_{\mathrm{G}}]}_{(c)} \underbrace{-[(-P_{\mathrm{L}} - \varepsilon \pi_{\mathrm{L}}) \boldsymbol{n}_{\mathrm{L}} + \boldsymbol{n}_{\mathrm{L}} \cdot \boldsymbol{\tau}_{\mathrm{L}}]}_{(d)} \\ & \underbrace{-\boldsymbol{n}_{\mathrm{G}} \varepsilon (\kappa_1 + \kappa_2) \sigma}_{(e)} \underbrace{-\varepsilon \left(\boldsymbol{t}_1 \frac{1}{2} \frac{\partial \sigma}{\partial s_1} + \boldsymbol{t}_2 \frac{1}{2} \frac{\partial \sigma}{\partial s_2} \right)}_{(f)} + additonal\ terms. \end{aligned} \tag{11.27}$$

For the sake of simplicity, detailed expression of the additional terms is omitted. In Eq. (11.27), π_i ($=cz_i eV\psi$) represents the contamination that is related to the electric potential.

11.3 Gas–Liquid–Solid Interfacial Model

In this section, a situation where a droplet adheres to a solid substrate is discussed because this situation would be the simplest case of wetting phenomenon. Then, the multi-scale interfacial model is tried to apply to the wettability problem.

11.3.1 Existing Models for Wetting Phenomena

Considered a situation where a droplet adheres to a solid surface as shown in Fig. 11.16a, and reaches an equilibrium state, the Young equation holds at the contact line. However, if an external force affects the droplet like as an inclination of the solid surface and fluid flow around the droplet, this situation becomes a complex one. The contact line moves and the dynamic situation occurs.

In the numerical simulation, the wetting problem is treated as the boundary condition. Figure 11.16b shows a cross-sectional view of a droplet on a solid surface. More simple approach is the use of the static constant contact angle θ_s, and then the normal vector $\boldsymbol{n}$ of the liquid surface near the solid is calculated by the following relation:

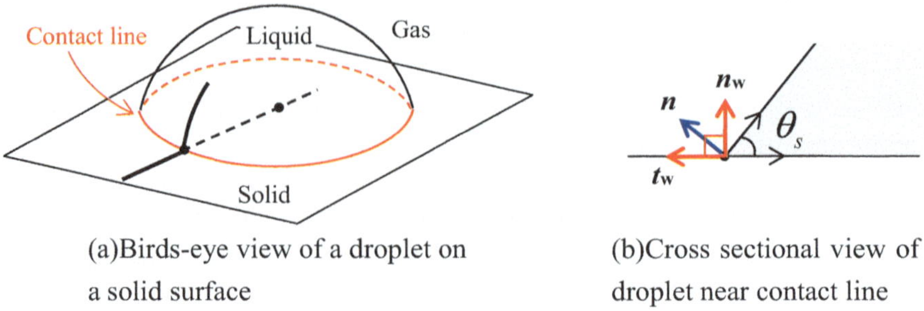

Fig. 11.16 Droplet on a solid surface

$$\boldsymbol{n} = \boldsymbol{n}_w \cos\theta_s + \boldsymbol{t}_w \sin\theta_s, \tag{11.28}$$

where $\boldsymbol{n}_\mathrm{w}$ and $\boldsymbol{t}_\mathrm{w}$ are the unit vectors normal and tangential to the wall, respectively. Alternatively, there is a model that uses the relationship between the dynamic contact angle and the velocity of the contact line. For this approach, there are some models such as the hydrodynamic model [58, 59], Tanner's law [60], and molecular kinetic theory [61]. In this kind of approach, the dynamic contact angle is determined from the contact line velocity [62]. In the recent numerical simulation, the solid–liquid wall condition is considered on the basis of Navier slip condition (Eq. 11.29) to avoid stress singularity problem at the contact line [63, 64].

$$u_w = \beta \frac{\partial u}{\partial y}. \tag{11.29}$$

The β is a slip coefficient and has a microscopic length scale. Although the slip coefficient is thought to be related to fluid viscosity, surface roughness, wettability, etc. [65], this coefficient is actually determined from experiments due to the difficulty in theoretical prediction. In the numerical simulation, the moving contact line problem is solved by associating Eq. (11.29) with viscous shear stress on the solid surface and unbalanced Young stress that is related to the contact angle [66].

In the diffuse interface method, some boundary conditions to consider the wetting problem are proposed. For example, the static contact angle is used to determine the gradient of the free surface near the contact line by the following relation [67]. The methodology is similar to the concept in Eq. (11.28).

$$\boldsymbol{n}_w \cdot \nabla\phi = -|\nabla\phi| \cos\theta_s. \tag{11.30}$$

In this equation, $\boldsymbol{n}_\mathrm{w}$ and ϕ are the unit vector normal to the solid surface and the order parameter that identifies the liquid and gas phases, respectively. The other model is called the non-equilibrium wetting condition as follows [68].

$$-\mu_f \delta \frac{\partial \phi}{\partial t} = \alpha \boldsymbol{n}_w \cdot \nabla \phi - \sigma \cos \theta_s g'(\phi). \tag{11.31}$$

In this equation, δ and α represent the interfacial thickness and a phase field parameter that are related to surface tension and the interfacial thickness. μ_f is the friction factor at the contact line. In this model, the contact line velocity is controlled by the friction factor. This factor is determined from experimental evidence.

However, the main concept of the above-mentioned models relies on direct or indirect usage of the experimentally measured contact angle and is just trying to determine the instantaneous gradient of liquid surface geometrically. If the contact line motion stops, the gradient of the liquid surface will go back to an initial value. Therefore, these approaches will not be able to treat wetting behaviors where contact angle hysteresis occurs, and deformed solid surface due to adhesion without special additional conditions. The contact angle is not physical value, and the contact angle should be secondarily determined from some kind of balance like as a force or energy.

11.3.2 Multi-scale Model for Gas–Liquid–Solid Interface

In this section, we try to consider other possibilities in order to treat the wetting problem that includes solid physical properties. Specifically, the momentum interfacial jump condition has been shown in the previous section, and the interfacial jump condition is applied to the three-phase contact line problem in order to obtain some hints for the modeling [69].

At first, we briefly explain how we apply the interfacial jump condition to the three-phase contact line. The momentum interfacial jump condition is described by Eq. (11.24). Then, it is assumed that the jump condition holds at each interface where a point on the contact line is considered as the origin. Thus, the base vectors are defined on each interface as shown in Fig. 11.17. In this figure, point P is the origin. The base vectors on liquid–gas and solid–liquid interfaces can be expressed by that of the solid–gas interface. For example, the base vector on the gas–liquid interface is associated with that of the solid–gas interface by the rotation of the vector as follows.

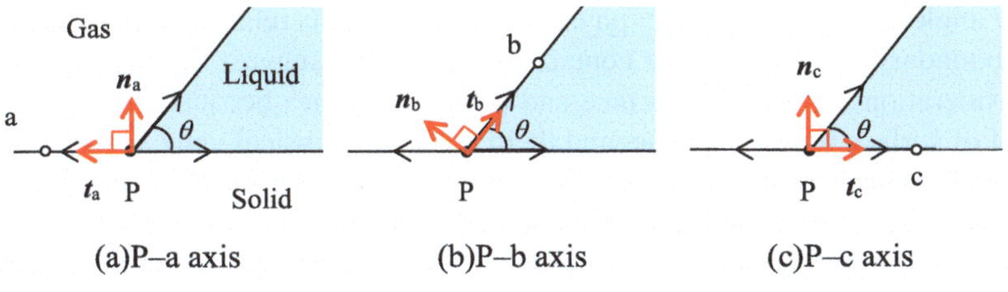

Fig. 11.17 Definition of base vectors on each interface

$$\begin{cases} \boldsymbol{t}_{\mathrm{b}} = -\cos\theta \boldsymbol{t}_{\mathrm{a}} + \sin\theta \boldsymbol{n}_{\mathrm{a}} \\ \boldsymbol{n}_{\mathrm{b}} = \sin\theta \boldsymbol{t}_{\mathrm{a}} + \cos\theta \boldsymbol{n}_{\mathrm{a}} \end{cases}. \tag{11.32}$$

Then, the base vector on solid–liquid interface is an opposite direction against the solid–gas interface. Therefore, the opposite sign is taken as follows:

$$\begin{cases} \boldsymbol{t}_{\mathrm{a}} = -\boldsymbol{t}_{\mathrm{c}} \\ \boldsymbol{n}_{\mathrm{a}} = \boldsymbol{n}_{\mathrm{c}} \end{cases}. \tag{11.33}$$

A force balance at the contact line is considered by integrating the jump condition within an infinitesimal area.

$$\begin{aligned} \boldsymbol{F} &= \int \boldsymbol{f} \mathrm{d}A \\ &= \int (\boldsymbol{f}_a + \boldsymbol{f}_b + \boldsymbol{f}_c) \mathrm{d}A \\ &= \boldsymbol{F}_a + \boldsymbol{F}_b + \boldsymbol{F}_c. \end{aligned} \tag{11.34}$$

By the substitution of the base vectors for Eqs. (11.31) and (11.32) into Eq. (11.33), the force balance equation is finally derived as follows:

$$\begin{aligned} \boldsymbol{F} = \boldsymbol{n}_{\mathrm{G_a}} \iint_A & [\,(m_{\mathrm{G_b}} u_{\mathrm{G_b}} - m_{\mathrm{L_b}} u_{\mathrm{L_b}}) \cos\theta + (P_{\mathrm{G_a}} - P_{\mathrm{S_a}}) + (P_{\mathrm{G_b}} - P_{\mathrm{L_b}}) \cos\theta + (P_{\mathrm{L_c}} - P_{\mathrm{S_c}}) \\ & - (2H_{\mathrm{a}}\sigma_{\mathrm{a}} + 2H_{\mathrm{b}}\sigma_{\mathrm{b}} \cos\theta + 2H_{\mathrm{c}}\sigma_{\mathrm{c}})\,] \mathrm{d}s_{\mathrm{a}} \mathrm{d}w - \boldsymbol{n}_{\mathrm{G_a}} \int \sigma_{\mathrm{b}} \sin\theta \mathrm{d}w \\ & + \boldsymbol{n}_{\mathrm{G_a}} \iint_A [(\boldsymbol{\tau}_{\mathrm{S_a}} - \boldsymbol{\tau}_{\mathrm{G_a}}) + (\boldsymbol{\tau}_{\mathrm{L_b}} - \boldsymbol{\tau}_{\mathrm{G_b}}) \cos\theta + (\boldsymbol{\tau}_{\mathrm{S_c}} - \boldsymbol{\tau}_{\mathrm{L_c}})] \mathrm{d}A \\ & + \boldsymbol{t}_a \iint_A [(m_{\mathrm{G_b}} u_{\mathrm{G_b}} - m_{\mathrm{L_b}} u_{\mathrm{L_b}}) \sin\theta + (P_{\mathrm{G_b}} - P_{\mathrm{L_b}}) \sin\theta - 2H_{\mathrm{b}}\sigma_{\mathrm{b}} \sin\theta] \mathrm{d}A + \\ & \boldsymbol{t}_{\mathrm{G_a}} \int (-\sigma_{\mathrm{a}} + \sigma_{\mathrm{b}} \cos\theta + \sigma_{\mathrm{c}}) \mathrm{d}w + \iint_A (\boldsymbol{\tau}_{\mathrm{L_b}} - \boldsymbol{\tau}_{\mathrm{G_b}}) \cdot \boldsymbol{t}_a \sin\theta \mathrm{d}A \end{aligned} \tag{11.35}$$

In this derivation, mass transfer through the solid interface is ignored. The normal and tangential components of surface tension at the contact line appear. It is found that the contact angle is affected by many physical quantities. This relation may indicate a kind of boundary condition at the contact line, and may suggest the importance of the consideration of the solid surface and phase properties because the normal component of surface tension appears and there are some physical values related to the solid phase. However, the interfacial jump condition is for the fluid interface. Therefore, a further discussion for a solid phase will be needed.

The following relation is an equilibrium relation of solid where deformation and inertia are assumed to be small:

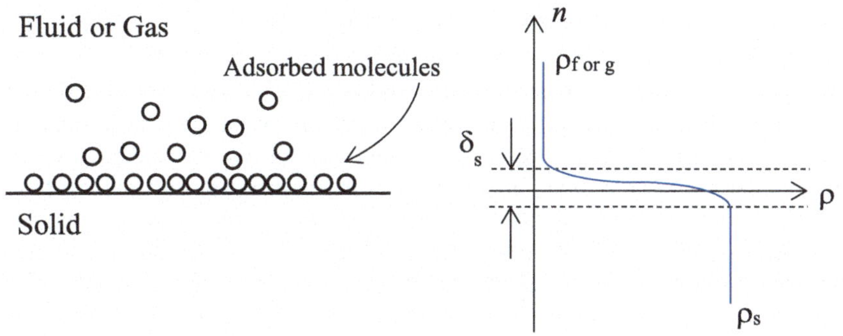

Fig. 11.18 Schematic of fluid or gas molecules on solid surface

$$0 \approx \nabla \cdot \boldsymbol{\sigma} + \rho_s \boldsymbol{b} \tag{11.36}$$

σ, ρ_s and $\boldsymbol{b}$ are stress component [N m^{-2}], density of solid [kg m^{-3}], and force per unit mass [N kg^{-1}], respectively. Here, the solid surface is normally considered as a sharp interface. However, in an actual situation, adsorbed fluid or gas molecules will exist on the solid surface as shown in Fig. 11.18. In addition, the molecules adjacent to the solid surface would behave as a solid [70]. This means that some kind of a thin transition region would exist at the solid interface. Here, the free energy density of the solid surface can be defined at the solid interface if it can be assumed that the transition region is smaller than that of the liquid–gas interface. By the assumptions, the surface energy would be defined by the following expressions.

$$\gamma_{\mathrm{sg}} = d_{\mathrm{sg}} \int_{-\infty}^{+\infty} \left(\frac{\partial \psi_{\mathrm{sg}}}{\partial n} \right)^2 dn, \tag{11.37}$$

$$\gamma_{\mathrm{sl}} = d_{\mathrm{sl}} \int_{-\infty}^{+\infty} \left(\frac{\partial \psi_{\mathrm{sl}}}{\partial n} \right)^2 dn. \tag{11.38}$$

Based on the previous assumption of the solid interface, the multi-scale expansion discussed in the Sect. 11.2.2.1 is applied to Eq. (11.36).

$$0 \approx \nabla^{(0)} \cdot \boldsymbol{\sigma} + \varepsilon_s \nabla^{(1)} \cdot \boldsymbol{\sigma}^{surf} + \rho_s \boldsymbol{b}. \tag{11.39}$$

The second term on the right-hand side of this equation is assumed to be related to the solid surface property, and would be an important term for the interaction between solid and liquid at the interface. It is assumed that this term would affect the fluid motion as an external force, and the new term is added in the multi-scale multiphase flow equation like the following relation:

$$\rho \frac{\mathrm{D}\boldsymbol{u}}{\mathrm{D}t^{(0)}} + \varepsilon \rho \frac{\mathrm{D}\boldsymbol{u}'}{\mathrm{D}t^{(1)}} = -\nabla^{(0)} \cdot \boldsymbol{T} + \varepsilon \nabla^{(1)} \cdot (f_0(\psi)\boldsymbol{I})$$

$$- \varepsilon d \nabla^{(1)} \psi \left(\nabla^{(1)} \cdot \nabla^{(1)} \psi \right) - \varepsilon_s \nabla_t^{(1)} \cdot \boldsymbol{\sigma}^{surf} + \rho \boldsymbol{g}. \quad (11.40)$$

In Eq. (11.40), the fourth term on the right-hand side is newly added. If the surface stress can be modeled by a similar treatment in our gas–liquid interfacial model, this term may work as a Marangoni-like force because the free energy would change across the contact line. However, in order to complete this model, further consideration of a relationship between surface stress and the free energy equation will be needed because the physics of the solid surface contains plastic and elastic contribution, and the geometrical effect of surface roughness on the wettability.

11.4 Summary

In this chapter, a multi-scale multiphase flow equation for the gas–liquid interface is shown for an example of multi-scale modeling. Through the modeling, the bubble coalescence is simulated as a test case on the basis of the derived equation. Then, the multi-scale multiphase flow equation for the solid surface problem is discussed by referring to the idea for the application of interfacial jump condition (sharp interface model) to the droplet wettability, and for multi-scale expansion. If the thermodynamic treatment for the interface (diffuse interface model) can be applied to the solid surface, the present modeling indicates that the newly derived term would work as a Marangoni-like force due to the variation in the free energy across the contact line. If the present idea is valid, it indicates that characterization of the solid surface will be needed because the surface stress must be considered, which is related to the surface structure and physical properties of plasticity and elasticity. In order to evaluate the validity of the present model, further modification and consideration would be needed. However, the present work indicates the importance that not only fluid dynamic but also solid property and physics must be considered for understanding the wetting phenomena, and the temporal and spatial scales should be discriminated.

References

1. N. K. Fukumasu, J. I. Yanagihara, G. C. K. Filho, "Influence of fuel infection frequency on droplet dispersion in ethanol pulsed spray flames", Applied Thermal Eng., 81, pp. 198–209, 2015.
2. J. Kim, "Spray cooling heat transfer: The state of the art", Int. J. Heat and Fluid Flow, 28, pp. 753–767, 2007.
3. W.-L. Cheng, W.-W. Zhang, H. Chen, L. Hu, "Spray cooling and flash evaporation cooling: The current development and application", Renew. Sustain. Energy Rev., 55, pp. 614–628, 2016.
4. F. Pironti, D. Mizrahi, A. Acosta, D. G.-Mendizabal, "Liquid-solid wetting factor in trickle-bed reactors: its determination by a physical method", Chem. Eng. Sci., 54, pp. 3793–3800, 1999.
5. C. Owusu, D. Fornasiero, J. A.-Mensah, M. Zanin, "Influence of pulp aeration on the flotation of chalcopyrite with xanthate in chalcopyrite/pyrite mixtures", Int. J. Mineral Proc., 134, pp. 50–57, 2015.

6. D. Zang, S. Tarafdar, Y. Y. Tarasevich, M. D. Choudhury, T. Dutta, "Evaporation of a droplet: from physics to applications", Phys. Rep., 804, p. 1056, 2019.
7. E. Y. Gatapova, A. A. Semenov, D. V. Zaitsev, O. A. Kabov, "Evaporation of a sessile water drop on a heated surface with controlled wettability", Colloid. Surf. A: Physicochem. Eng. Asp., 441, pp. 776–785, 2014.
8. A. V. Delgado, F. G.-Caballero, R. J. Hunter, L. K. Koopal, J. Lyklema, "Measurement and interpretation of electrokinetic phenomena", J. Colloid Int. Sci., 309, pp. 194–224, 2007.
9. A. K. Chesters, G. Hofman, "Bubble coalescence in pure liquids", Applied Sci. Res., 38 pp. 353–361, 1982.
10. M. J. Prince, H. W. Blanch, "Bubble coalescence and break-up in air-sparged bubble columns", AIChE J., 36(10), pp. 1485–1499, 1990.
11. I. U. Vakarelski, R. Manica, X. Tang, S. J. O'Shea, G. W. Stevens, F. Grieser, R. R. Dagastine, D. Y. C. Chan, "Dynamic interactions between microbubbles in water", PNAS, 107(25), pp. 11177–11182, 2010.
12. J. Solsvik, S. Nystrand, B. Gawel, G. Øye, H. A. Jakobsen, "Theoretical and experimental investigation of film drainage time", Energy Procedia, 64, pp. 33–42, 2015.
13. D. Bonn, J. Eggers, J. Indekeu, J. Meunier, E. Rolley, Wetting and spreading, Rev. Mod. Phys., 81, 739, 2009.
14. J. De Coninck, M. J. de Ruijter, M. Voue, Dynamics of wetting, Curr. Opin. Colloid and Int. Sci., 6(1), 49-53, 2001.
15. J. H. Snoeijer, B. Andreotti, Moving contact lines: scales, regimes, and dynamical transitions, Annu. Rev. Fluid Mech., 45, 269-292, 2013.
16. R. J. Hansen, T. Y. Toong, "Dynamic contact angle and its relationship to forces of hydrodynamic origin", J. Colloid Int. Sci., 37(1), pp. 196–207, 1971.
17. R. Sedev, "The molecular-kinetic approach to wetting dynamics: achievements and limitations", Adv. Colloid Int. Sci., 222, pp. 661–669, 2015.
18. J. Zawala, D. Kosior, K. Malysa, "Formation and influence of the dynamic adsorption layer on kinetics of the rising bubble collisions with solution/gas and solution/solid interfaces", Adv. Colloid Int. Sci., 222, pp. 765–778, 2015.
19. M. Firouzi, T. Howes, A. V. Nguyen, "A quantitative review of the transition salt concentration for inhibiting bubble coalescence", Adv. Colloid Int. Sci., 222, pp. 305–318, 2015.
20. L. Wang, "Modeling of bubble coalescence in saline water in the presence of flotation frothers", Int. J. Mineral Proc., 134, pp. 41–49, 2015.
21. J. Zawala, A. Wiertel, A. Niecikowska, K. Malysa, "Influence of external vibrations on bubble coalescence time at water and oil surfaces – Experiments and modelling", Colloid Surf. A: Physicochemical and Eng. Asp., 519, pp. 137–145, 2017.
22. S. Li, B. Y. Ni, "Simulation on the interaction between multiple bubbles and free surface with viscous effects", Eng. Anal. Bound. Elem., 68, pp. 63–74, 2016.
23. C. W. Extrand, S. I. Moon, "Experimental measurement of forces and energies associated with capillary rise in a vertical tube", J. Colloid Int. Sci., 407, pp. 488–492, 2013.
24. V. S. J. Craig, "Bubble coalescence and specific-ion effects", Curr. Opin. Colloid Interface Sci., 9, pp. 178–184, 2004.
25. C. L. Henry, V. S. J. Craig, "Ion-specific influence of electrolytes on bubble coalescence in nonaqueous solvents", Langmuir, 24, pp. 7979–7985, 2008.
26. R. R. Lessard, S. A. Zieminski, "Bubble coalescence and gas transfer in aqueous electrolytic soolutions", Ind. Eng. Chem. Fundam., 10(2), pp. 260–269, 1971.
27. S Marčelja, "Short-range forces in surface and bubble interaction", Curr. Opin. Colloid Interface Sci., 9, pp. 165–167, 2004.
28. Y. H. Tsang, Y.-H. Koh, D. L. Koch, "Bubble-size dependence of the critical electrolyte concentration for inhibition of coalescence", J. Colloid Interface Sci., 275, pp. 290–297, 2004.
29. Jr. C. P. Ribeiro, D. Mewes, "The influence of electrolytes on gas hold-up and regime transition in bubble columns", Chem. Eng. Sci., 62, pp. 4501–4509, 2007.
30. S Marčelja, "Selective coalescence of bubbles in simple electrolytes", J. Phys. Chem. B, 110, pp. 13062–13067, 2006.

31. D. Langevin, "Bubble coalescence in pure liquids and in surfactant solutions", 20, pp. 92–97, 2015.
32. P. R. Garrett, "Preliminary considerations concerning the stability of a liquid heterogeneity in a plane parallel liquid film", J. Colloid Interface Sci., 76(2), pp. 587–590, 1980.
33. G. C. Frye, J. C. Berg, "Antifoam action by solid particles", J. Colloid Interface Sci., 127(1), pp. 222–238, 1989.
34. R. Aveyard, B. D. Beake, J. H. Clint, "Wettability of spherical particles at liquid surfaces", J. Chem. Soc., Faraday Trans., 92(21), pp. 4271–4277, 1996.
35. N. D. Denkov, "Mechanisms of foam destruction by oil-based antifoams", Langmuir, 20, pp. 9463–9505, 2004.
36. M. Colombo, M. Fairweather, "RANS simulation of bubble coalescence and break-up in bubbly two-phase flows", Chem. Eng. Sci., 146, pp. 207–225, 2016.
37. T. Cong, X. Zhang, "Numerical study of bubble coalescence and breakup in the reactor fuel channel with a vaned grid", Energies, 11, 256, 2018.
38. J. U. Brackbill, D. B. Kothe, C. Zemach, "Continuum method for modeling surface tension", J. Comput. Phys., 100, pp. 335–354, 1992.
39. G. Tryggvason, B. Bunner, A. Esmaeeli, D. Juric, N. Al-Rawahi, W. Tauber, J. Han, S. Nas, Y. –J. Jan, "A front-tracking method for the computations of multiphase flow", J. Comput. Phys., 169, pp. 708–759, 2001.
40. D. M. Anderson, G. B. McFadden, "Diffuse-interface methods in fluid mechanics", Annu. Rev. Fluid Mech., 30, pp. 139–165, 1998.
41. H. Ding, P. D. M. Spelt, C. Shu, "Diffuse interface model for incompressible two-phase flows with large density ratios", J. Comput. Phys., 226, pp. 2078–2095, 2007.
42. D. Jacqmin, "Calculation of two-phase Navier-Stokes flows using phase-field modeling", J. Comput. Phys., 155, pp. 96–127, 1999.
43. D. Jacqmin, "Contact-line dynamics of a diffuse fluid interface", J. Fluid Mech., 402, pp. 57–88, 2000.
44. D. Jamet, O. Lebaigue, N. Coutris, J. M. Delhaye, "The second gradient method for the direct numerical simulation of liquid-vapor flows with phase change", J. Comput. Phys., 169, pp. 624–651, 2001.
45. J. W. Cahn, J. E. Hilliard, "Free energy of a nonuniform system, I. Interfacial energy", J. Chem. Phys., 28(2), pp. 258–267, 1958.
46. J. S. Rowlinson, B. Widom, Molecular theory of capillarity, Oxford University Press, 1984.
47. Y. Yonemoto, T. Kunugi, "Multi-scale modeling of the gas-liquid interface based on mathematical and thermodynamic approaches", Open Transport Phenom. J., 2, pp. 69–79, 2010.
48. S. A. Safran, Statistical thermodynamics of surfaces, interfaces, and membranes, ADDISON-WESLEY PUBLISHING COMPANY, 1994.
49. Y. Yonemoto, Y. Yanagisawa, H. Kawara, T. Kunugi, "Coalescence of microbubble", J. JSEM, 8(1), pp. 38–44, 2008.
50. Y. Yonemoto, T. Kunugi, "Modelling of microbubble coalescence", WIT Trans. Eng. Sci., 79, pp. 273–284, 2013.
51. T. Kunugi, "MARS for multiphase calculation", Comput. Fluid Dynamics, 9, 563, 2001.
52. D. L. Youngs, Numerical methods for fluid dynamics, Academic Press, 1982.
53. C. W. Hirt, B. D. Nichols, "Volume of fluid method for the dynamics of free boundaries", J. Comput. Phys., 39, pp. 201–225, 1981.
54. T. Kunugi, Y. Yonemoto, "Simulation of Microbubble Coalescence and Repulsion", In Micro- and Nanobubbles: Fundamentals and Applications, pp. 26–34, Edited by H. Tsuge, Pan Stanford Publishing Pte. Ltd., 2014.
55. Y. Yonemoto, T. Kunugi, "Macroscopic gas-liquid interfacial equation based on thermodynamic and mathematical approaches". In Mass transfer – Advanced Aspects, Edited by Hironori Nakajima, ISBN: 978–953–307–636–2, InTech, 2011.
56. J. M. Delhaye, "Jump conditions and entropy surfaces in two-phase systems. Local instant formulation", Int. J. Multiphase Flow, 1, pp. 395–409, 1974.

57. L. E. Scriven, "Dynamics of a fluid interface: equation of motion for Newtonian surface fluids", Chem. Eng. Sci., 12, pp. 98–108, 1960.
58. O. V. Voinov, "Hydrodynamics of wetting", Fluid Dynamics, 11, pp. 714–721, 1976.
59. R. G. Cox, "The dynamics of the spreading of liquids on a solid surface. Part 1. Viscous flow", J. Fluid Mech., 168, pp. 169–194, 1986.
60. L. H. Tanner, "The spreading of silicone oil drops on horizontal surfaces", J. Phys. D: Appl. Phys., 12, pp. 1473–1484, 1979.
61. T. D. Blake, "The physics of moving wetting lines", J. Colloid Interface Sci., 299, pp. 1–13, 2006.
62. I. Malgarinos, N. Nikolopoulos, M. Marengo, C. Antonini, M. Gavaises, "VOF simulation of the contact angle dynamics during the drop spreading: Standard models and a new wetting force model", Adv. Colloid Interface Sci., 212, pp. 1–20, 2014.
63. S. Ganesan, S. Rajasekaran, L. Tobiska, "Numerical modeling of the non-isothermal liquid droplet impact on a hot solid substrate", Int. J. Heat and Mass Transfer, 78, pp. 670–687, 2014.
64. D. R. Noble, A. Kucala, M. J. Martinez, "A conformal decomposition finite element method for dynamic wetting applications", Proc. of ASME 2017 Fluids Engineering Division Summer Meeting, FEDSM2017–69378.
65. R. S. Voronov, D. V. Papavassiliou, L. L. Lee, "Review of fluid slip surperhydrophobic surfaces and its dependence on the contact angle", Ind. Eng. Chem. Res., 47, pp. 2455–2477, 2008.
66. X. Shang, Z. Lue, E. Y. Gatapova, L. A. Kabov, B. Bai, "GNBC-based front-tracking method for the three-dimensional simulation of droplet motion on a solid surface", Comput. Fluids, 172, pp. 181–195, 2018.
67. H. Ding, P. D. M. Spelt, "Onset of motion of a three-dimensional droplet on a wall in shear flow at moderate Reynolds numbers", J. Fluid Mech., 599(25), pp. 341–362, 2008.
68. A. Carlson, Capillarity and dynamic wetting, PhD thesis, KTH Royal Institute of Technology, 2012.
69. Y. Yonemoto, T. Kunugi, "Macroscopic wettability based on an interfacial jump condition", Phys. Rev. E, 81, 056310, 2010.
70. J.-J.Shu, J. B. M. Teo, W. K. Chan, "Fluid velocity slip and temperature jump at a solid surface", Applied Mech. Rev., 69(2), 020801, 2017.

Chapter 12
Vapor Nucleation in Metastable Liquids: the Continuum Description

Mirko Gallo, Francesco Magaletti, Dario Abbondanza, and Carlo Massimo Casciola

Abstract Liquid–vapor phase change is of importance across a wide spectrum of fundamental and applied disciplines. The phenomenology of vapor formation is very complex due to the large range of spatio-temporal scales involved. Here, the microscopic features of vapor embryos nucleation coexist with the macroscopic bubble dynamics. This multiscale nature of the phenomenon makes nucleation challenging both from a theoretical and experimental point of view. In this work, we aim to retrace the state of art of continuum description of liquid–vapor phase change, starting from the classical nucleation theory, which provides a basic description of the phenomenon up to the phase field description and fluctuating hydrodynamics theory.

Keywords Liquid–vapor phase transition · Nucleation · Diffuse interface models · Fluctuating hydrodynamics

---

The original version of this chapter was revised: This chapter added newly. The updated information for this chapter is available at https://doi.org/10.1007/978-3-030-82992-6_13.

M. Gallo (✉) · D. Abbondanza · C. M. Casciola
Department of Mechanical and Aerospace Engineering-DIMA, Sapienza Universitá di Roma, 00184 Rome, Italy
e-mail: mirko.gallo@uniroma1.it

D. Abbondanza
e-mail: dario.abbondanza@uniroma1.it

C. M. Casciola
e-mail: carlomassimo.casciola@uniroma.it

F. Magaletti
Advanced Engineering Centre, School of Architecture, Technology and Engineering, University of Brighton, Lewes Road, Brighton BN2 4GJ, UK
e-mail: F.Magaletti2@brighton.ac.uk

M. Marengo and J. De Coninck (eds.), *The Surface Wettability Effect on Phase Change*,
https://doi.org/10.1007/978-3-030-82992-6_12

12.1 Introduction

Everyday experience shows that a liquid transforms into a vapor when the pressure is decreased below a certain threshold—a process called cavitation—or the temperature is raised sufficiently—boiling. This is certainly one of the most common examples of a phase transition, and it is noteworthy that, despite our familiarity with it, the process is as yet not fully understood.

Phase transitions are complex multiscale problems, whose incipit is the nucleation of embryos of the new phase in the mother phase. Nuclei of the new phase spontaneously appear in the liquid due to thermal fluctuations. This observation underlines the role of microscopic dynamics for nucleation, whose origin rests at the atomistic level. Despite its microscopic origin, the characteristic time scales of nucleation may be orders of magnitude larger than the molecular characteristic time. Moreover, once started, the phenomenon becomes extremely fast, sometimes explosive, in reaching macroscopic scales. All these properties are characteristic features of the activated processes, where a free energy barrier keeps the system in the metastable mother phase and the transition to the new phase occurs due to rare fluctuations—hence the initial slowness—able to drive the system over the barrier to let it abruptly fall in a stable free energy minimum. The barrier height, hence the frequency of barrier crossings, depends on the thermodynamics conditions, i.e., temperature and pressure and on the level of confinement (e.g., volume) of the liquid [1]. Typical aspect of the liquid–vapor phase transition is that the fluid inertia may play a significant role in the overall dynamics, especially for macroscopically large systems.

When vapor nuclei locally appear, the liquid accelerates and its pressure decreases. Inside the bubbles, the pressure is as low as the vapor pressure (for water at ambient temperature the vapor pressure is about 2.3 kPa) hence they can survive, and even grow, as long as they remain in low-pressure regions. However, when the flow transports them in a region with higher pressure, they suddenly become unstable and collapse.

Collapse is particularly relevant for cavitation, where pressure changes—decrease and increase—are induced by large scale, inertia-dominated fluid motions. Overall, vapor bubble implosion encompasses a range of complex phenomena, including large bubble deformation and topological changes, shockwave emission and propagation through the liquid, phase transition to and from supercritical conditions, and intense pressure and temperature peaks on the order of dozens GPa and 1000 K, respectively.

The detailed experimental assessment of bubble nucleation dynamics is conceivably difficult. Quantitative measurements of, e.g., local pressures and temperatures, bubble sizes at nucleation and their successive dynamics are hampered by the small space-time scales [2–5] (ns/μs in terms of time and nm/μs in terms of space) and by the randomness of the individual nucleation processes whose location cannot be predicted in advance.

On the other hand, developing a fully consistent and comprehensive theory remains a big theoretical challenge. The current understanding of nucleation phenomena is mainly built upon simplified, quasi-static descriptions like classical nucleation theory (CNT) [6]. CNT applies macroscopic thermodynamics to microscopic scales. Spatially uniform embryos characterized by the thermodynamic properties of the stable phase (the vapor) are assumed to be separated from the uniform, unstable liquid by a sharp interface across which the relevant thermodynamic properties change abruptly. Although these assumptions produce qualitatively good results under many circumstances, they are unsatisfactory from the strictly theoretical point of view and can be shown to be unable to predict crucial features like, e.g., vanishing free energy barriers at spinodal conditions. More fundamental approaches, like the density functional theory (DFT) [7], can overcome some of these limitations. Still, they do not cope with coupling the phase transition with macroscopic dynamics.

Molecular dynamics (MD) simulations [8] and, more appropriately, the MD-based advanced sampling techniques for rare events [9] represent powerful tools to investigate thermally activated processes. However, concerning the former, the computational cost limits the application to small systems (less than few tens of nanometers) and to very short times, preventing the study of hydrodynamic interactions, while the latter, still limited to extremely small systems and targeted at the extraction of the free energy profile along with the transition, can only deal with quasi-equilibrium processes.

In fact, non-equilibrium and spatiotemporal variations are crucial to understand cavitation and boiling in realistic conditions and require approaches able to bridge the gap between vapor embryo nucleation and macroscopic motion.

Recently, a diffuse interface model able to encompass bubble nucleation [10–13] has been developed in the framework of continuum fluid mechanics. The model embeds thermal fluctuations in the context of the so-called Fluctuating Hydrodynamics and has been exploited to address cavitation in its entirety, starting from the bubble inception to the macroscopic, inertia-dominated bubble motion. The model, consisting of the Navier–Stokes system expressing mass, momentum and energy conservation, includes Van der Waals capillarity and is augmented with stochastic terms tailored to reproduce the Einstein-Boltzmann probability distribution for the fields. The resulting set of stochastic partial differential equations provides a thermodynamically consistent framework for the numerical simulation of liquid/vapor phase transition under general conditions, describing the entangled phenomena occurring during phase transitions, from nucleation to highly non-linear dynamics, including extreme events such as bubble collapse and shock wave emission.

The purpose of this paper is to provide first the basic background on the liquid–vapor phase transition, to successively retrace the derivation of the fluctuating hydrodynamics model and demonstrate its potential with applications to cavitation, leaving aside boiling which is discussed in a different chapter.

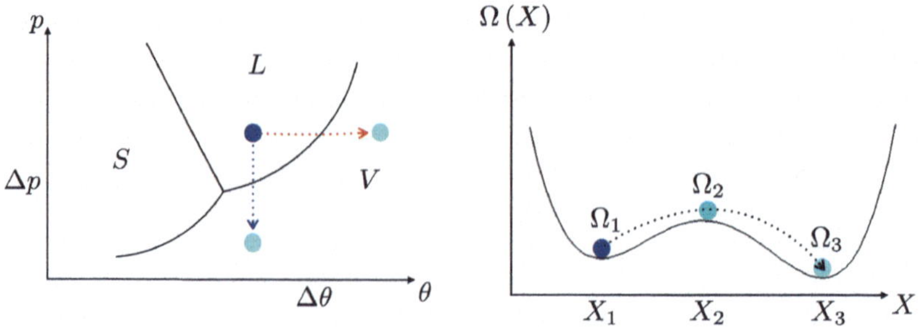

Fig. 12.1 Left panel: the sketch qualitatively describes the state of matter as a function of pressure and temperature. The *S-zone* represent the solid state, the *L-zone* the liquid one and the *V-zone* the vapor phase. Right panel: the sketch qualitatively describes the energetic configurations of a thermodynamic system as a function of a generic reaction coordinate X. The states 1, 2, 3 represent the metastable, the critical and the stable configurations, respectively

12.2 Nucleation and Phase Transition

12.2.1 Metastability

The aim of this section is to review the classical aspects of the liquid–vapor phase transition. With reference to the phase diagram of Fig. 12.1, starting from the liquid state (blue circle in the left panel of the figure), boiling and cavitation are represented by the red and the blue paths, respectively. In this oversimplified view, phase transition occurs when crossing the saturation line that demarcates the respective existence regions of liquid and vapor, respectively.

Actually, additional ingredients are needed to achieve a more realistic description of the liquid–vapor phase transition [14]. It is a known fact indeed that, after reaching the boiling temperature or the vapor pressure, the liquid can remain longly trapped in a metastable state near saturation, in overheated or stretched conditions. The origin of the metastability can be traced back to the existence of a finite free energy barrier between the liquid and the vapor state that renders the liquid/vapor phase change an activated process.

The height of the energy barrier is related to the overheating/stretching levels in the liquid and may range from infinity at saturation to zero in spinodal conditions. Depending on the barrier height, the system gets trapped in a metastable state for longer or shorter times.

In order to clarify the context, it is useful to remind the nature of the free energy that is used in describing the transition. Given that the system is in a prescribed equilibrium state, say at fixed volume V, temperature θ and number of particles N, system microstates are statistically distributed according to the appropriate distribution function (pdf), the canonical pdf, $\exp(-H/(k_B\theta))/Z$, in the specific case, where k_B is the Boltzmann constant, H the Hamiltonian of the system and Z the partition function that normalizes the pdf. From basic statistical mechanics, the thermodynamic potential associated with the conditions prescribed in the example is the Helmholtz free energy, $F(V, \theta, N) = -k_B\theta \ln Z$. In these conditions, let $p(V_b)$ be the probability to observe a vapor bubble with volume V_b, for an agreed upon definition of bubble volume. The quantity $\Omega(V_b) = -k_B\theta \ln[p(v)]$ is called the Landau free energy associated with the collective variable V_b (the name collective variable, sometimes called progress variable or reaction coordinate in the chemistry literature, comes from realizing that such variable selects the collection of microstates—a collective state—consistent with observing a bubble of volume V_B).

In general, given an equilibrium system and a generic reaction coordinate X, the (Landau) free energy landscape $\Omega(X)$ illustrated on the right panel of Fig. 12.1 describes the progress of the transition between (collective) states of high probability (low free energy). In the example sketched in the figure, the free energy profile features two local minima, the absolute minimum at X_3 and a second relative minimum at X_1. Both of them identify a collective state with a high probability of occurrence. That is a way to say that, once they randomly appear, they have a good chance to last for long, since any nearby state has a lesser probability. The two minima are separated by a relative maximum at X_2, called the transition state. The absolute minimum is the thermodynamically stable (collective) state that can eventually be reached through a spontaneous evolution, if the system was allowed sufficient time to complete the transition. The relative minimum X_1 is a metastable state. In general terms, this means that the system is locally at equilibrium, implying that the state will eventually be recovered for perturbations that displace the state by not too much. In the present context, the perturbations that are alluded to are the spontaneous density fluctuations that are induced by the thermal agitation at molecular level.

In order to bring the system to the transition state, the free energy must increase by the amount $\Delta\Omega^* = \Omega_2 - \Omega_1$, with $\Omega_{1/2} = \Omega(X_{1/2})$, called the transition free energy barrier. In presence of a barrier, the transition is said to be a thermally activated event, since the fluctuations displacing the system from the local equilibrium state have thermal origin. In this case, the life time τ of the metastable state is related to the energy barrier as $\tau \propto \exp(\Delta\Omega^*/k_B\theta)$, suggesting the definition of metastability as a stability limited over time, see Sect. 12.2.2 for additional details.

Metastable states are omni-present in nature, since the simple act of observing a state implies it has to persist sufficiently long. For many of them, the transition time can be quite long. At ambient conditions, the diamond phase of carbon is a metastable state while the stable form being graphite. On statistical grounds, the diamond-graphite phase transition is sooner or later expected to occur. However, the lifetime of the metastable diamond is remarkably long, order of million years [15, 16]. Metastability is also observed in supercooled water [17, 18], (e.g freezing rain, icing aircraft) or in emulsions and colloids [19, 20], and in mechanical systems, for instance in avalanches [21] and more in general with *sandpile-like* systems [22].

The liquid–vapor phase transition involves the transition from a metastable state corresponding to liquid to a stable state corresponding to vapor. Figure 12.2 shows the phase diagram of a Lennard-Jones fluid [23] with the *binodal* and the *spinodal* lines reported in the inset as blu and azure lines. In the $\rho - \theta$ plane, the binodal (or coexistence) line is identified as the set of points having same temperature, chemical

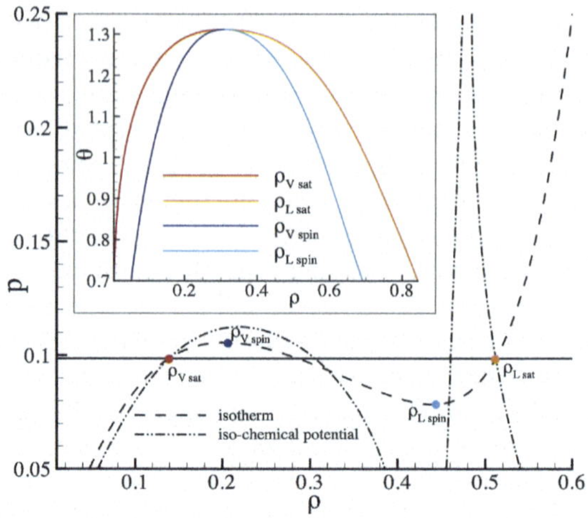

Fig. 12.2 Phase diagram for the Lennard-Jones EoS [23]. In the main plot, the isotherm $\theta = 1.25$ and the iso-chemical potential $\mu = \mu_{sat}$ with the saturation value are reported with dashed and dash-dotted lines, respectively. The saturation densities are identified as the two points with equal temperature, chemical potential and pressure; the red circle represents the vapor saturation point and the orange circle the liquid one. The other two circles, blue and light blue, represent the spinodal points, vapor and liquid, respectively, identified on the isotherm where $\partial p/\partial \rho = 0$. In the inset, the loci of all the saturation and spinodal points at different temperatures are reported in the $\rho - \theta$ plane

potential and pressure. The spinodal lines (red and orange lines) are identified along isotherms where $\partial p/\partial \rho = 0$. Metastable liquids are represented by a point placed between binodal and spinodal limits (the orange and azure line), and it is separated from their stable states (homogeneous vapor phase), by an energy barrier that must be surmounted to bring the system in the new, more stable phase. Starting from an ideally homogeneous liquid phase, thermal fluctuations induce the formation of vapor nuclei. After the nuclei reach a critical size, they start expanding surrounded by their mother phase, in a complex non-equilibrium process, leading the system to decompose in two different phases. This stage is the "incipit" of the phase transition and it is called *nucleation*.

Depending on thermodynamic conditions, the mean time required for the occurrence of a fluctuation able to produce a supercritical embryo can be very long, for this reason, nucleation and consequently phase change can be seen as a *rare event*.

12.2.2 Random Walker in a Bistable Potential and the Kramers Theory

A simple model able to qualitatively reproduce the phenomenology of phase transitions in metastable liquids consists of a random walker with position **X** wandering over a bistable Landau free energy landscape $\Omega(\mathbf{X})$, where $\mathbf{X} \in \mathcal{S}$ with $\mathcal{S}$ the state space. For definiteness, it will be assumed that the state space is the entire d-dimensional space, $\mathcal{S} = \mathbb{R}^d$, with the free energy growing unbounded at large distance, $\Omega(\mathbf{X}) \to \infty$ as $|\mathbf{X}| \to \infty$. In this setting, Kramers theory [24] provides the *mean first passage time* $\langle\tau\rangle$ for the diffusion across a barrier of the random walker trapped in the neighborhood of a free energy minimum (a metastable basin).

The random walker is assumed to obey the Langevin dynamics

$$\frac{\mathrm{d}\mathbf{X}}{\mathrm{d}t} = \boldsymbol{\mu}(\mathbf{X}) + (2\mathbf{D})^{1/2}\boldsymbol{\xi}(t)\,, \tag{12.1}$$

where $\boldsymbol{\xi}(t)$ is a zero-mean, delta-correlated Gaussian process with correlation matrix $\langle\boldsymbol{\xi}(t)(t) \otimes \boldsymbol{\xi}(t)(t^{'})\rangle = \mathbf{I}\delta(t - t^{'})$ where $\otimes$ is the tensor product, $\mathbf{D} = D\mathbf{I}$ the diffusion tensor, here assumed diagonal as appropriate for the following discussion, and $\mathbf{I}$ is the identity matrix. The drift $\boldsymbol{\mu}$ is taken to be proportional through the friction coefficient α to the negative free energy gradient,

$$\boldsymbol{\mu} = -\alpha\frac{\partial\Omega}{\partial\mathbf{X}}\,. \tag{12.2}$$

According to the dynamics described by the above Langevin equation, the probability $P(\mathbf{X}, t)$ of finding the random walker at position $\mathbf{X}$ at time t obeys the Fokker–Planck equation

$$\frac{\partial P(\mathbf{X}, t)}{\partial t} = -\mathcal{F}P(\mathbf{X}, t)\,, \tag{12.3}$$

with

$$\mathcal{F} = \frac{\partial}{\partial\mathbf{X}} \cdot \boldsymbol{\mu}(\mathbf{X}) - \frac{\partial}{\partial\mathbf{X}} \otimes \frac{\partial}{\partial\mathbf{X}} : \mathbf{D} \tag{12.4}$$

the Fokker–Planck operator (in the above operator notation, the $\mathbf{X}$-derivatives are understood to act on whatever they multiply from the left). The Fokker–Planck equation needs to be supplemented with appropriate boundary, $P \to 0$ as $|\mathbf{X}| \to \infty$, and initial conditions, $P(\mathbf{X}, 0) = P_0(\mathbf{X})$.

It is worth stressing that, for $\alpha = \alpha_0$ and $D = D_0$ constant, the equilibrium pdf

$$P_{eq}(\mathbf{X}) = \frac{e^{-\beta\Omega(\mathbf{X})}}{\int_{\mathcal{S}} e^{-\beta\Omega(\mathbf{X})} d\mathbf{X}}\,, \tag{12.5}$$

where, as usual, $\beta = 1/(k_B\theta)$, satisfies the time independent version of Eq. (12.3), provided the relation

$$\frac{D_0\beta}{\alpha_0} = 1 \tag{12.6}$$

between friction constant and diffusion coefficient is obeyed. This is an instance of fluctuation-dissipation theorem stating that the random walker diffusion is intimately intertwined with the dissipative process that relaxes the system toward the free energy minima through friction. Also worth noting is that the equilibrium solution (12.5) is consistent with the statistical interpretation of the Landau's free energy profile $\Omega(\mathbf{X})$ illustrated in Sect. 12.2.1, see also Sect. 12.2.3 for more detail.

We are now ready to discuss Kramer's theory on the mean first passage time. Let us denote $\mathcal{B} \subset \mathcal{S}$ a metastable locus in $\mathcal{S}$. The probability $P(\mathbf{X}, t|\mathbf{Y}, t_0)$ of the transition from the state $\mathbf{Y}$ at the time t_0 to the state $\mathbf{X}$ at the time t obeys the very same Fokker–Planck equation introduced above

$$\frac{\partial P\left(\mathbf{X}, t|\mathbf{Y}, t_0\right)}{\partial t} = -\mathcal{F} P\left(\mathbf{X}, t|\mathbf{Y}, t_0\right), \tag{12.7}$$

now with initial conditions

$$P\left(\mathbf{X}, t_0|\mathbf{Y}, t_0\right) = \delta\left(\mathbf{X} - \mathbf{Y}\right) \qquad \mathbf{X} \in \mathcal{B},$$

which assign to the walker zero probability of moving away from the initial position $\mathbf{Y}$ in no time. As appropriate to describe barrier crossing, the boundary data consist in absorbing conditions at the metastable basin boundary,

$$P\left(\mathbf{X}, t|\mathbf{Y}, t_0\right) = 0 \qquad \mathbf{X} \in \partial\mathcal{B},$$

since the walker is assumed to start somewhere in the metastable domain and to *disappear* as it gets to the edge of the metastable basin.

The probability that the trajectory $\mathbf{X}(t, \mathbf{Y}, t_0)$ emanating from $\mathbf{Y}$ at time t_0 is contained in the set $\mathcal{B}$ can be estimated by introducing the pdf $\pi\left(\tau|\mathbf{Y}\right)$ of the first passage time $\tau(\mathbf{Y})$, defined as the time the walker needs to reach the absorbing boundary from the initial position $\mathbf{Y}$. In fact the probability that the trajectory up to time t is contained in the metastable domain,

$$\Pi\left(t|\mathbf{Y}, t_0\right) = \int_{\mathcal{B}} P\left(\mathbf{X}, t|\mathbf{Y}, t_0\right) \mathrm{d}\mathbf{X}, \tag{12.8}$$

is given by the probability $P\left(\tau(\mathbf{Y}) > t\right)$ that the time $\tau(\mathbf{Y})$ to reach $\partial\mathcal{B}$ starting from $\mathbf{Y}$ is in the future of the current time t

$$\Pi\left(t|\mathbf{Y}, t_0\right) = P\left(\tau(\mathbf{Y}) > t\right) = \int_t^{+\infty} \pi\left(\tau|\mathbf{Y}\right) \mathrm{d}\tau\,. \tag{12.9}$$

Thus, the mean value of τ is

$$\langle\tau\left(\mathbf{Y}\right)\rangle = \int_0^{+\infty} \tau\, \pi\left(\tau|\mathbf{Y}\right) \mathrm{d}\tau = -\int_0^{+\infty} \tau\, \frac{\partial \Pi\left(\tau|\mathbf{Y}, 0\right)}{\partial \tau} \mathrm{d}\tau\,, \tag{12.10}$$

that, after integrating by parts, leads to

$$\langle \tau (\mathbf{Y}) \rangle = \int_0^{+\infty} \mathrm{d}\tau \int_{\mathcal{B}} P(\mathbf{X}, \tau | \mathbf{Y}, 0)\, \mathrm{d}\mathbf{X} . \tag{12.11}$$

In order to obtain the explicit expression of the mean first passage time from Eq. (12.10), it is instrumental to introduce the transition probability distribution, $P(\mathbf{Z}, q|\mathbf{X}, t)$, of the states $\mathbf{X}$ at time t from which the walker will reach the target $\mathbf{Z}$ at a future time $q > t$. Such probability distribution evolves in time according to the Backward Kolmogorov Equation

$$\frac{\partial P(\mathbf{Z}, q|\mathbf{X}, t)}{\partial t} - \mathcal{F}^{\dagger} P(\mathbf{Z}, q|\mathbf{X}, t) = 0 , \tag{12.12}$$

with the adjoint Fokker–Planck operator defined as

$$\mathcal{F}^{\dagger} = -\mu(\mathbf{X}) \cdot \frac{\partial}{\partial \mathbf{X}} - \mathbf{D} : \frac{\partial}{\partial \mathbf{X}} \otimes \frac{\partial}{\partial \mathbf{X}} . \tag{12.13}$$

By applying the adjoint operator $\mathcal{F}^{\dagger}$ to Eq. (12.11), on account of Eq. (12.13), one has

$$\mathcal{F}^{\dagger} \langle \tau (\mathbf{Y}) \rangle = -\int_0^{+\infty} \mathrm{d}\tau \int_{\mathcal{B}} \frac{\partial P(\mathbf{X}, \tau | \mathbf{Y}, 0)}{\partial \tau} \mathrm{d}\mathbf{X} = \int_0^{+\infty} \pi(\tau|\mathbf{Y})\, \mathrm{d}\tau = 1 , \tag{12.14}$$

where time homogeneity enjoyed by $P(\mathbf{X}, t|\mathbf{Y}, t_0)$, implying $P(\mathbf{X}, t|\mathbf{Y}, t_0) = P(\mathbf{X}, t - t_0|\mathbf{Y}, 0)$, is exploited. In explicit form, Eq. (12.14) reads

$$-\mu(\mathbf{X}) \cdot \frac{\partial \langle \tau (\mathbf{X}) \rangle}{\partial \mathbf{X}} - \mathbf{D} : \left(\frac{\partial}{\partial \mathbf{X}} \otimes \frac{\partial}{\partial \mathbf{X}} \right) \langle \tau (\mathbf{X}) \rangle = 1 , \tag{12.15}$$

providing a partial differential equation for the mean first passage time. The appropriate boundary condition for this elliptic equation is $\langle \tau (\mathbf{X}) \rangle = 0$ on $\partial\mathcal{B}$. For future reference, it is worth stressing that in deriving the equation for the mean first passage time, the assumption of constant friction and diffusion coefficient does not play any role, i.e., Eq. (12.15) holds with full generality.

The above general theory can be specialized for a one-dimensional problem characterized by a bistable potential $\Omega(X)$ (see right panel of Fig. 12.1). $\Omega(X)$ possesses two minima at X_1 and X_3 (metastable and stable state, respectively) separated by the transition state X_2 where a local maximum is achieved. Furthermore, $\Omega(X)$ is assumed to rapidly grow away from the region of interest that includes the two, stable and metastable, basins (coercive function). As in the general setting, the aim is to determine the mean time required to reach the stable basin around X_2 starting from a neighborhood of X_1. The system is assumed to be governed by the one dimensional version of Eq. (12.1)

$$\frac{dX}{dt} = -\alpha_0 \frac{d\Omega}{dX} + \sqrt{2D_0}\xi(t)\,, \tag{12.16}$$

where the explicit expression of the drift μ in terms of free energy gradient with (constant) friction α_0 and diffusion coefficient D_0 has been used. The mean first passage time $\langle\tau\rangle$ than follows by solving the one-dimensional version of Eq. (12.15),

$$\alpha_0 \frac{d\Omega}{dX}\frac{d}{dX}\langle\tau(X)\rangle - D_0 \frac{d^2}{dX^2}\langle\tau(X)\rangle = 1\,. \tag{12.17}$$

By multiplying both sides of Eq. (12.17) by the integrating factor $\exp(-\beta\Omega)/D$, after taking the fluctuation dissipation balance (12.6) into account, the equation is rearranged as

$$\frac{d}{dX}\left(\exp(-\beta\Omega)\frac{d\langle\tau(X)\rangle}{dX}\right) = -\frac{\exp(-\beta\Omega)}{D}\,. \tag{12.18}$$

Equation (12.18) can be integrated twice, first on the set $(-\infty, X)$ and then on the set (X, X_3), providing

$$\langle\tau(X)\rangle = \int_X^{X_3} \exp(\beta\Omega(x))\,dx \int_{-\infty}^{x} \frac{\exp(-\beta\Omega(y))}{D}dy\,, \tag{12.19}$$

where X_3 is assumed to be an absorbing boundary ($\langle\tau(X_3)\rangle = 0$). The above double integral can be substantially simplified, since significant contributions to the outer integral only come from the neighborhood of X_2, where $\exp(\beta\Omega)$ is large and, equivalently, the inner integral is built up from the neighborhood of X_1, where $\exp(-\beta\Omega)$ is dominant. Hence, selecting the initial state in the metastable basin, the mean first passage time is explicitly obtained as

$$\langle\tau\rangle \simeq \int_{\cup} \exp\left(-\frac{\Omega(X)}{k_B\theta}\right)dX \int_{\cap}\frac{1}{D}\exp\left(\frac{\Omega(X)}{k_B\theta}\right)dX\,, \tag{12.20}$$

where the symbol $\cup$ represents the neighborhood of X_1 (the metastable basin) and $\cap$ the neighborhood of the transition state X_2. The above integrals can be evaluated by using a saddle point argument. With reference to the right panel of Fig. 12.1, let $\Omega_{\cup/\cap}(\mathbf{X}) = \Omega_{1/2} \pm 1/2\Omega''_{1/2}(\mathbf{X} - \mathbf{X}_{1/2})^2$ be the second order Taylor expansion of the free energy in the neighborhood of the two extremas, minimum at $\mathbf{X}_1$ and maximum at $\mathbf{X}_2$. The two integrals in Eq. (12.20) are then approximated as

$$\begin{aligned}\langle\tau\rangle \simeq \exp\left(\frac{\Omega_2-\Omega_1}{k_B\theta}\right)\int_{-\infty}^{\infty}\exp\left[-\frac{1/2\Omega''_1(\mathbf{X}-\mathbf{X}_1)^2}{k_B\theta}\right]dX \\ \frac{1}{D_0}\int_{-\infty}^{\infty}\exp\left[-\frac{1/2\Omega''_2(\mathbf{X}-\mathbf{X}_2)^2}{k_B\theta}\right]dX\,,\end{aligned} \tag{12.21}$$

where for future reference, it is assumed that the diffusivity depends on position and varies slowly in the neighborhood of the maximum. The Gaussian integrals are readily evaluated, to yield

$$\langle \tau \rangle \simeq \frac{\exp(\beta \Delta \Omega_*)}{D(\mathbf{X}_2)} \sqrt{\frac{2\pi k_B \theta}{\Omega_1''}} \sqrt{\frac{2\pi k_B \theta}{\Omega_2''}} \,. \tag{12.22}$$

12.2.3 A Digression on Landau's Free Energy

In order to frame CNT within the general context of statistical mechanics and allow comparison with alternative approaches, it may be useful to consider a toy model aimed at clarifying the connection between system free energy and the Landau free energy of the collective variable(s) selected to describe the phase transition.

Assume a cylinder of volume V is split into two parts, part 1, with volume $v_1 = v$ and part 2, with volume $v_2 = V - v$, by a freely movable, massless piston. The two resulting subsystems are taken to contain $n_1 = n$ and $n_2 = N - n$ identical particles. With the temperature kept constant by contact with a heat bath and a constant total volume, the pdf of the microstate is described by the canonical distribution. The peculiarity here is that the volume v is free to fluctuate, constituting an internal variable for the overall system. The internal variable, together with momenta and positions of the particles in the two subsystems, collectively denoted Γ_1 and Γ_2, respectively, needs to be specified to determine the microstate of the system. Hence, the canonical probability distribution is

$$\mathrm{p}(\Gamma_1, \Gamma_2, v; \theta, V, N, n) = \frac{1}{Z} e^{-\left[H_1(\Gamma_1, v) + H_2(\Gamma_2, V - v)\right]/(k_B \theta)} \tag{12.23}$$

where the Hamiltonian functions H_1 and H_2 explicitly depend on subsystem volume and couple together only through the constraint on the total volume. As usual, the partition function

$$Z(\theta, V, N, n) = \int_0^V Z_1(v) Z_2(V - v) dv \,, \tag{12.24}$$

where, as a peculiarity of the present system, an integral over the internal variable is involved, provides the (Helmholtz) free energy

$$F(\theta, V, N, n) = -k_B \theta \ln[Z] = -k_B \theta \ln \left[\int_0^V Z_1(v) Z_2(V - v) dv \right] \,. \tag{12.25}$$

The pdf of the internal variable is obtained as the marginal pdf of the microstate,

$$\mathrm{p}(v;\theta,V,N,n)=\int d\Gamma_1 d\Gamma_2 p(\Gamma_1,\Gamma_2,v)=\frac{Z_1(v)Z_2(V-v)}{Z}, \tag{12.26}$$

while the microstate pdf conditioned to partition volume follows by Bayes theorem

$$\mathrm{p}(\Gamma_1,\Gamma_2|v;\theta,V,N,n)=\frac{\mathrm{p}(\Gamma_1,\Gamma_2,v)}{\mathrm{p}(v)}=\frac{e^{-H_1(\Gamma_1,v)/(k_B\theta)}}{Z_1}\frac{e^{-H_2(\Gamma_2,V-v)/(k_B\theta)}}{Z_2}. \tag{12.27}$$

If, for some reasons, one is interested in a collective state characterized by the partition volume, v is the appropriate collective variable and the corresponding Landau free energy is

$$\Omega(v;\theta,V,N,n)=-k_B\theta\ln \mathrm{p}(v)=-k_B\theta\ln\left[\frac{Z_1(v)Z_2(V-v)}{Z}\right]. \tag{12.28}$$

From their definitions, it is clear that system free energy $F(\theta,V,N,n)$ and Landau free energy $\Omega(v;\theta,V,N,N)$ are entirely different quantities, in general. However, in this case, the two are strictly related, making sometimes difficult to keep the two concepts distinct. To make similarities and differences as clear as possible, it is useful to introduce the subsystems (Helmholtz) free energies, $F_{1/2}(\theta,v_{1/2},n_{1/2})=-k_B\theta\ln Z_{1/2}$. The Landau free energy Ω can be then rewritten as

$$\Omega(v;\theta,V,N,n)=F_1(v)+F_2(V-v)-F \tag{12.29}$$

while the overall system free energy becomes

$$F(\theta,V,N,n)=-k_B\theta\ln\left[\int_0^V e^{-[F_1(v)+F_2(V-v)]/(k_B\theta)}dv\right]. \tag{12.30}$$

By definition, the most probable value of v, v_0, is obtained by minimizing Ω. As expected, minimization yields $p_1(v)=p_2(V-v)$, where $p_{1/2}=-\partial F_{1/2}/\partial v_{1/2}$ ($v_1=v$, $v_2=V-v$) are the subsystems pressures (the symbol used for pressures, p, should not be confused with that for pdf's, p).

Let us now see how the two free energies are related. For macroscopic systems, the integral in the last expression for the free energy F can be estimated by a saddle point argument as

$$\int_0^V e^{-[F_1(v)+F_2(V-v)]/(k_B\theta)}dv\simeq$$
$$e^{-(F_{10}+F_{20})/(k_B\theta)}\int_0^V e^{-\left[-(\partial p_1/\partial v_1+\partial p_2/\partial v_2)(v-v_0)^2\right]/(2k_B\theta)}dv,$$

where, for large systems, the last integral is $[2\pi k_B\theta/(-(\partial p_1/\partial v_1+\partial p_2/\partial v_2))]^{1/2}$, and the subscript 0 denotes quantities at the maximally probable state $v=v_0$. Hence, the system free energy, in the thermodynamic limit, is

$$F \simeq F_{10}(v_0) + F_{20}(V - v_0)\,, \tag{12.31}$$

where the neglected term is order $1/V$. By noticing that, by subtracting F, which is independent of v, from the expression for $\Omega(v)$, the Landau free energy can be rewritten as

$$\Delta\Omega(v) = F_1(v) + F_2(V - v)\,. \tag{12.32}$$

For this elementary example, $F = \min_v \Omega(v)$, a property that generally holds, with the proviso of macroscopic sizes, whenever the system can be split into parts characterized by an internal variable—the partition volume, in the present case—which is an appropriate collective variable to parameterize the relevant collections of microstates.

The above analysis is easily extended to different conditions. An example is provided by a slight generalization of the system just discussed, where N particles of one species are dissolved in a solute, and the movable piston is replaced by a semipermeable membrane, giving rise to osmotic effects.

As it will be discussed in the next subsection, particular interest is reserved, in the context of vapor nucleation, to grand canonical systems, constrained on total volume, temperature and chemical potential. Here, the role of the movable piston is played by the interface with the number of particles in the two subsystems, bubble and surrounding liquid, together with the subsystems' volume free to vary. In this case, however, the additional element represented by the surface tension needs to be included to account for the real nature of the interface across which the thermodynamic densities change in a finite though narrow interfacial layer.

The notion of an energy associated with the interface is strictly connected with the notion of Gibbs dividing surface. The dividing surface is the ideal (sharp) surface that separates the nominal regions occupied by the two phases under the assumption that such phases were homogeneous up to the dividing interface across which their properties experience a jump. The usual way to define the position of the dividing surface is by requiring that the mass of each phase is exactly expressed by the volume integral of the respective uniform density. This implies that no mass excess/defect needs to be attached to the interface. On the other hand, the effect of fluid stratification at the interface does not allow to conserve at the same time the (free) energy. For this reason, a suitable surface energy is introduced to account for the excess (free) energy.

12.2.4 Classical Nucleation Theory

Classical nucleation theory (CNT) [6, 25, 26] provides the fundamental understanding of bubble nucleation, both for homogenous (bubble nucleating in the bulk) and heterogenous conditions (bubble forming on extraneous nucleating agents). The simplest example of heterogenous nucleation consists in a vapor bubble forming on a flat solid surface with specified contact angle ϕ.

The (unbounded) system consists of the half space above a planar solid wall everywhere occupied by the metastable liquid except that for a spherical cup that forms a vapor bubble laying on the flat solid wall. In the context of CNT, liquid and vapor are considered homogenous up to the interface. Chemical potential μ and temperature θ are prescribed and the relevant free energy density per unit volume (density is more appropriate to discuss unbounded system) is the grand potential density.

Following the considerations illustrated in the previous subsection, the Landau free energy for a bubble of volume V_b can be expressed in terms of the grand potential of liquid and vapor, using the bubble radius as collective variable to parameterize the free energy landscape,

$$\begin{aligned}\Delta\Omega\,(R,\phi) = \int_{V_b}(\omega_V - \omega_L)dV &= -\Delta p V_V\,(R,\phi) + \gamma_{LV} A_{LV}\,(R,\phi) + \\ &+ (\gamma_{SV} - \gamma_{LS})\,A_{SV}\,(R,\phi)\,,\end{aligned} \tag{12.33}$$

where the diverging contribution of the unbound liquid phase has been subtracted and $R = [3/(4\pi)V_B]^{1/3}$ is the radius of the spherical cup. The free energy depends on the vapor-liquid pressure difference $\Delta p = p_V - p_L$, and on the surface energies $\gamma_{LV}, \gamma_{SV}, \gamma_{LS}$ associated with the liquid–vapor, solid–vapor and liquid–solid interfaces. In the above expression, the area of each interface is uniquely determined by simple trigonometric expressions involving bubble volume (alternatively, spherical cap radius) and contact angle.

The Young equilibrium contact angle $\phi = \cos^{-1}(\gamma_{LS} - \gamma_{SV})/\gamma_{LV})$, where the angle is measured from the vapor-solid interface (i.e., $\phi > \pi/2$ corresponds to hydrophilic chemistry), allows for re-expressing the relevant geometric quantities as $A_{SV} = \pi R^2 \sin^2\phi$, $A_{LV} = 2\pi R^2(1 - \cos\phi)$, $V_b(R,\phi) = V_b(R,\pi)\Psi(\phi)$, where $\Psi(\phi) = 1/4(1 - \cos\phi)^2(2 + \cos\phi)$ is a purely geometrical factor given by the ratio of the spherical cup volume and volume of the whole sphere with same radius. Thus, starting from a homogeneous metastable liquid and denoting by $\Delta\Omega_{hom} = -\Delta p V_V(R,\pi) + \gamma_{LV} A_{LV}(R,\pi)$ the free energy of a spherical bubble of radius R in the bulk liquid, the corresponding free energy of a spherical cup at the wall is

$$\Delta\Omega\,(R,\phi) = \Delta\Omega_{hom}\,(R)\,\Psi\,(\phi)\,. \tag{12.34}$$

The free energy attains a maximum, the critical state, at the critical radius R^*,

$$R^* = \frac{2\gamma_{LV}}{\Delta p}\,, \tag{12.35}$$

and the free energy barrier in nucleating the bubble from the metastable liquid is

$$\Delta\Omega^* = \Delta\Omega\left(R^*, \phi\right) = \Delta\Omega^*_{hom}\Psi\left(\phi\right) = \frac{16}{3}\pi\frac{\gamma^3_{LV}}{\Delta p^2}\Psi\left(\phi\right). \tag{12.36}$$

Since the dependence on the contact angle in Eq. (12.34) factors out, the critical radius is the same both for heterogeneous and homogenous nucleation. The barrier $\Delta\Omega^*$ do instead depend on the angle, and it is lower for heterogeneous nucleation than it is for the homogeneous case, $\Delta\Omega_{hom}$ ($\Psi(\phi) \leq 1$). For obvious geometrical reasons, the critical volume $V^* = 4/3\pi R^{*3}\Psi(\phi)$ is also smaller for the heterogeneous case. For the especially simple case of a neutrally wetting liquid $\phi = \pi/2$, the barrier toward homogeneous nucleation is twice as large as the barrier for nucleating over a perfectly flat wall. It follows that, in general, the probability of observing a bubble nucleating on the solid surface is significantly larger than it is in the bulk.

Once the Landau's free energy profile is available, Eq. (12.33), the Kramer's theory illustrated in Sect. 12.2.2 can be used to estimate the mean time needed to nucleate a bubble, Eq. (12.20), provided a Langevin equation assumed to describe the bubble evolution is available. The underlying assumption is that the bubble radius/volume executes a random walk over the free energy landscape, under the effect of thermal fluctuations that make the bubble size change randomly in time.

A clever proposal by [27] for nucleation in the bulk liquid (homogenous nucleation) assumes that the validity of the macroscopic Rayleigh–Plesset equation for the bubble dynamics extends down to the dimension of a nucleating bubble. The overdamped form of the equation, as appropriate to small scales where inertia is negligible, is used, and the deterministic evolution is made stochastic by adding a (zero average, delta correlated in time, Gaussian) noise of proper amplitude. The equation can be written both in terms of bubble radius R and bubble volume $v = 4\pi/3R^3$,

$$\dot{v} = -\frac{3v}{4\eta}\left[2\gamma\left(\frac{4\pi}{3v}\right)^{1/3} - \Delta p\right] + \sqrt{2D_v}\xi\,, \qquad (a)$$

$$\dot{R} = -\frac{1}{4\eta}\left[2\gamma - \Delta pR\right] + \sqrt{2D_R}\xi\,, \qquad (b) \tag{12.37}$$

where D_v and D_R are two different, though related, diffusivities and η is the liquid viscosity. Noise strength, related to the diffusion coefficient, should then be determined by suitable extension of the fluctuation-dissipation theorem.

The free energy, Eq. (12.33), specialized for bulk nucleation ($\gamma_{SL} = \gamma_{SV} = 0$), see the left panel in Fig. 12.3,

$$\Omega = \gamma_{LV}4\pi R^2 - \Delta p\frac{4\pi}{3}R^3 = \gamma_{LV}4\pi\left(\frac{3v}{4\pi}\right)^{2/3} - \Delta pv\,, \tag{12.38}$$

allows to rewrite the respective right-hand side of each of the above two equations as

$$\dot{v} = -\alpha_v \frac{d\Omega}{dv} + \sqrt{2D_v}\xi\,, \qquad (a)$$
$$\dot{R} = -\alpha_R \frac{d\Omega}{dR} + \sqrt{2D_R}\xi\,, \qquad (b) \tag{12.39}$$

where the friction coefficients are $\alpha_v(v) = 3v/(4\mu)$ and $\alpha_R(R) = 1/(16\pi\eta R)$, respectively. In these stochastic equations, the friction coefficient depends on the state variable.

The fluctuation dissipation theorem as introduced in Sect. 12.2.2 cannot be directly used here, due to the dependency of the friction coefficient on the stochastic variable and needs then to be restated. Focusing on the equation for the bubble radius, for definiteness, it is instrumental to rewrite the above Langevin equations in dimensionless form, rescaling with the barrier height $\Delta\Omega^* = 16\pi/3(\gamma^3/\Delta p^4)$ and using the dimensionless radius $r = R/R^*$. After rescaling time with $T = (R^*)^2/(\alpha_R^*\Delta\Omega^*)$, with $\alpha_R^* = 1/(16\pi\eta R^*)$ the friction coefficient at the transition state, normalizing the diffusion coefficient by $D_R^* = \alpha_R^*\Delta\Omega^*$ and realizing that the noise term ξ should be rescaled by $1/\sqrt{T^*}$, Eq. (12.39.b) becomes

$$\frac{dr}{d\tau} = -\frac{1}{r}\frac{d}{dr}\left(3r^2 - 2r^3\right) + \sqrt{2\tilde{D}_R(r)}\hat{\xi}(\tau)\,, \tag{12.40}$$

where all the dimensional parameters have been absorbed in the dimensionless diffusion coefficient and $\langle\hat{\xi}(\tau_1)\hat{\xi}(\tau_2)\rangle = \delta(\tau_1 - \tau_2)$. Reverting to dimensional quantities, the diffusion coefficient reads $D_R(R) = \alpha_R^*\Delta\Omega^*\tilde{D}_R(R/R^*)$.

Based on standard statistical mechanics arguments, the solution of the associated Fokker–Planck equations should have the form $\mathrm{p}_{eq}(R) = C\exp(-\beta\Omega(R))$ which, using dimensionless variables, reads $\mathrm{p}_{eq}(r) = R^*C\exp(-\beta\Delta\Omega^*\tilde{\omega}(r))$, with $\omega(r) = 3r^2 - 2r^3$.

The associated Fokker–Planck equation is

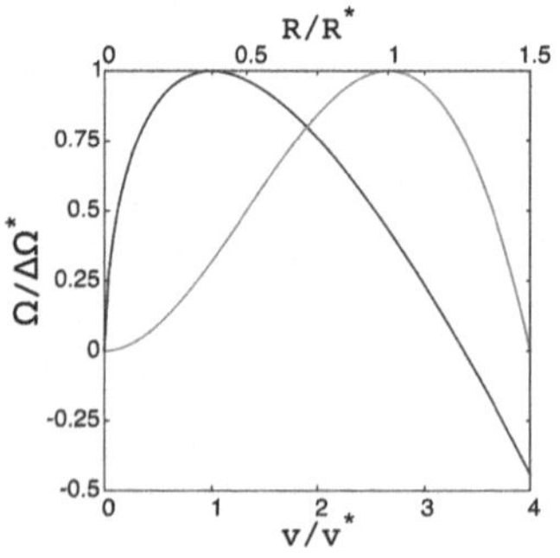

Fig. 12.3 Landau free energy profile for homogeneous nucleation in the CNT context as a function of bubble volume (purple curve) and bubble radius (green curve). The free energy is normalized by the barrier height while volume and radius are normalized by the corresponding critical values

$$\frac{d}{dr}\left[\tilde{\alpha}_R(r)\frac{d\omega}{dr}\mathrm{p}_{eq}(r)\right]+\frac{d^2}{dr^2}\left[\tilde{D}_R(r)\mathrm{p}_{eq}(r)\right]=0\,, \tag{12.41}$$

where $\tilde{\alpha}_R = \alpha_R/\alpha_R^*$ and the dimensionless diffusion coefficient is assumed to depend on the state variable.

The first move to obtain the fluctuation dissipation theorem in the generalized form needed here consists in requiring that the postulated Boltzmann form for the equilibrium pdf should satisfy Eq. (12.41). This procedure leads to a differential equation for the diffusion coefficient,

$$\frac{d^2\tilde{D}_R}{dr^2}-2B\omega'\frac{d\tilde{D}_R}{dr}+B\left[B\left(\omega'\right)^2-\omega''\right]\tilde{D}_R+\left[\left(\tilde{\alpha}_R\omega'\right)'-\tilde{\alpha}_R\left(\omega'\right)^2\right]=0\,, \tag{12.42}$$

where a prime denotes differentiation with respect to r. The only parameter entering the above equation is the dimensionless barrier height $B = \beta\Delta\Omega_*$.

After noting that, at the critical point, Eq. (12.42) reduces to

$$\frac{d^2\tilde{D}_R}{dr^2}(r^*)-\left[\tilde{\alpha}(r^*)-B\tilde{D}_R(r^*)\right]\omega''(r^*)=0\,, \tag{12.43}$$

where the free energy profile curvature at the saddle is $\omega''_* = -6$, with $\tilde{\alpha}* = 1$ by definition, one concludes that requiring $\tilde{D}_R^* = 1/B$ is tantamount to forcing an inflection point saddle for the diffusion coefficient at the transition state (compare with Eq. (12.6)). This prescription is the second move to obtain the fluctuation dissipation theorem and implies that $\tilde{D}_R \simeq 1 + \tilde{D}'_R(r_*)(r - r_*) + \mathcal{O}(r - r_*)^3$.

The solution of the initial value problem for Eq. (12.42) with data assigned at the critical point is

$$\tilde{D}_R(r)=\left[D_R^*+D_R'^*\left(r-r_*\right)\right]e^{-B\left(\omega_*-\omega\right)}-e^{B\omega}\int_{r_*}^{r}\alpha\omega' e^{-B\omega}dr\,, \tag{12.44}$$

which is a family of solutions depending on the slope, $\tilde{D}_R'^*$, of the diffusion coefficient at the critical point. The freedom of assigning the slope can be exploited to prevent the diffusion coefficient from growing negative. A good choice is $\tilde{D}_R'^* = 0$. This is the third and last move to obtain the (generalized) fluctuation dissipation theorem. The diffusion coefficient is shown in the left panel of Fig. 12.4.

Before discussing the escape rate, which will lead to the sought-for expression for the bubble nucleation rate, it is instrumental to consider the integral

$$\tilde{I}_\cap=\int_\cap\frac{e^{B\omega(r)}}{\tilde{D}_R(r)}dr\sim e^B\sqrt{\frac{2\pi}{B|\omega''_*|}}\,, \tag{12.45}$$

where the integration is across the saddle point (no need to specify the actual integration domain in the asymptotic limit of large activation barriers, refer to Sect. 12.2.2

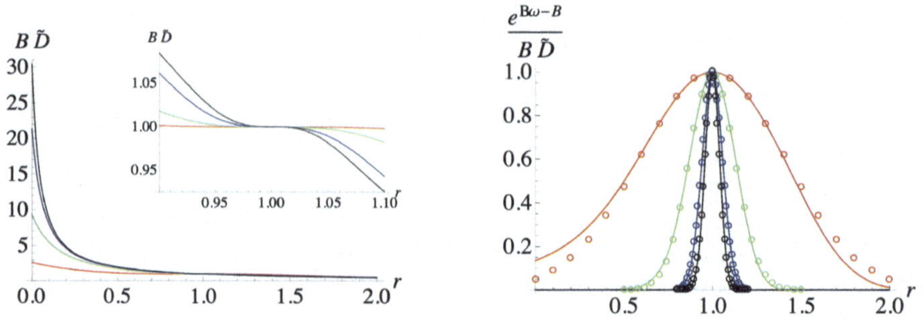

Fig. 12.4 Left: dimensionless diffusion coefficient (premultiplied by the dimensionless barrier height B) versus bubble radius for the stochastic dynamics of the bubble radius at increasing dimensionless barrier: $B = 1$ (red), $B = 10$, (green), $B = 50$, (blue), $B = 100$, (black). The slope at the critical point is $\tilde{D}_R'^{*} = 0$ and $\tilde{D}_R > 0$ for $r \geq 0$. At the transition point $\tilde{D}_R^* = 1/B, d\tilde{D}_R^*/dr = 0$ and $d^2\tilde{D}_R^*/dr^2 = 0$. Right: $\exp[B(\omega - \omega_*)]/(B\tilde{D}_R)$ (solid lines, same color coding) and saddle point approximation $\exp[1/2B\omega_*''(r - r_*)^2]/(BD_R^*)$ (symbols)

to find the analogous in the context of classical Kramer's theory). The approximation is based on the well-known saddle point argument reviewed in Sect. 12.2.2 and is illustrated in the right panel of Fig. 12.4 where the integrand $e^{B\omega(r)}/\tilde{D}_R(r)$ is shown to be already well approximated by $e^B e^{-1/2B|\omega_*''|(r-r_*)^2}/\tilde{D}_R^*$, with accuracy increasing with barrier height.

Using Eqs. (12.45) and (12.20), the escape rate from the potential well is

$$\kappa = \frac{1}{\langle\tau\rangle} = \frac{1}{I_\cup I_\cap} \simeq \sqrt{\frac{|\Delta\Omega_*''|}{2\pi k_B\theta}} \frac{D_* e^{-\Delta\Omega_*/(k_B\theta)}}{\int_\cup e^{-\Delta\Omega(R)/(k_B\theta)} dR}, \tag{12.46}$$

where dimensional expressions are now used. On account of the statistical interpretation of Landau free energy (Sect. 12.2.3), the probability of observing the value R for the bubble radius is $\mathrm{p}(R) \propto \exp[-\beta\Delta\Omega(R)]$, which shows that the last term on the right-hand side is the probability

$$\hat{\mathrm{p}}(R_*) = \exp[-\beta\Delta\Omega(R_*)]/\int_0^{R^*} \exp[-\beta\Delta\Omega(R)]dR$$

of observing a critical bubble normalized to the total probability of having a sub-critical bubble (the free energy well $\cup$ consists of the interval $0 \leq R \leq R_*$). Once the escape rate is known, the nucleation rate (number of nucleated bubbles per unit time and unit volume) is obtained by multiplying the escape rate of each walker by the number of walkers per unit volume, i.e., the number of nucleation sites per unit volume, N_s/V,

$$J_R = \frac{N_s \kappa}{V} \simeq \sqrt{\frac{|\Delta\Omega''(R_*)|}{2\pi k_B \theta}} \frac{D_R^* N_s \hat{\mathrm{p}}(R_*)}{V}, \tag{12.47}$$

where $N_s \hat{\mathrm{p}}(R^*)$ is the expected number of nucleating bubble in the system, $\Delta\Omega''(R_*) = -6\Delta\Omega_*/R_*^2$ and $D_R^* = \Delta\Omega_*/(16\pi\eta R_* \beta)$.

The above reasoning can be also applied when the collective variable is the bubble volume, Eq. (12.38.b), instead of the bubble radius. In this case, the nucleation rate reads

$$J_v \simeq \sqrt{\frac{|\Delta\Omega''(v_*)|}{2\pi k_B \theta}} \frac{D_v^* N_s \hat{\mathrm{p}}(v_*)}{V}, \tag{12.48}$$

where now $\Delta\Omega''_* = -2\Delta\Omega_*/(3v_*^2)$, $D_v^* = 3v_*\Delta\Omega_*/(4\eta\beta)$ and

$$\hat{\mathrm{p}}(v_*) = \exp(-\beta\Delta\Omega_*) / \int_0^{v^*} \exp[-\beta\Delta\Omega(v)]dv\,.$$

Clearly, the two expressions for the nucleation rate show the same dependency on the relevant variables but differ by a numerical prefactor,

$$\frac{J_v}{J_R} = \frac{\int_0^1 \exp(-\beta\Delta\Omega)\rho^2 d\rho}{\int_0^1 \exp(-\beta\Delta\Omega) d\rho}\,.$$

Concerning heterogeneous nucleation, the nucleation rates on a planar surface can be obtained, on a heuristic ground, from the above expressions by introducing the contact angle-dependent geometrical factor $\Psi(\phi)$ in the expressions for the free energy profile, $\Delta\Omega\,(R,\phi) = \Delta\Omega_{hom}\,(R)\,\Psi\,(\phi)$, Eq. (12.34). It should be mentioned, however, that, from a theoretical point of view, the validity of the Rayleigh–Plesset like dynamics implied by Eq. (12.38) cannot be rigorously justified in this case.

As a final comment related to CNT, all terms in the expression for the nucleation rate can be obtained from the theory except from the number of nucleation sites [27] for which no closed-form theory is available and resort to simulation/experimentation is needed.

12.2.5 Beyond Classical Nucleation Theory

CNT is invaluable in providing a clear basic understanding of nucleation. However, the very assumptions that render it handy and conceptually clear give rise to several drawbacks. Among the most notable ones, the prediction on the barrier height can be significantly inaccurate as a consequence of the sharp-interface assumption, which depicts the interface as a zero thickness surface separating uniform adjoining states. In fact the critical nuclei are most often so small that the vapor distribution inside the nucleus is quite far from being uniform and the density, as an example, cannot even reach the expected bulk value. As a consequence, the nucleation barriers are incorrectly evaluated, especially close to spinodal conditions, where CNT does not reproduce the vanishing of the barrier.

More advanced theories like density functional theory (DFT) [7, 28] and CNT extensions [29] can provide more precise estimates of the barriers and can correct certain mispredictions of CNT. At any rate, also these more complete descriptions still deal with the nucleation of a single bubble and cannot predict the number of active nucleation sites needed to estimate the nucleation rate.

In this context, molecular dynamics simulations [8, 30] deserve a separate discussion. They offer the possibility to simulate bubble nucleation from basic principles, provided the system is sufficiently small and the barrier is not high [27]. In fact, Eq. (12.46) shows that the time to be awaited before nucleation occurs may be long and grows exponentially with the energy barrier, ruling out, in general, the possibility of a brute force approach. These limitations fostered the development of rare event methods and their application to nucleation [13, 31–35].

These comments stress the importance of developing mesoscale models dedicated to the study of nucleation in its entirety, starting from the inception phase up to the fully non-linear hydrodynamic motion. Along this line, a promising approach is the phase field model (Van der Waals), which, in stationary conditions, recovers the DFT description with a squared-gradient approximation of the free energy [36] with the big advantage of including the coupling with hydrodynamics [37–39]. In its original form, the model is deterministic. Adding thermal fluctuation in the spirit of Fluctuating Hydrodynamics [40, 41] leads to a system of stochastic partial differential equation (spde's) able to model vapor bubble nucleation in both homogeneous [10, 11] and heterogeneous conditions [12, 42]. The model provides good accuracy at a cheap computational cost and allows simulations of macroscopic systems. In fact, the typical size of a simulation on a small computational cluster ($200 \times 200 \times 200\,\mathrm{nm}^3$) is comparable with the largest brute force MD simulations [8] on a tier-0 machine. Moreover, the simulation can reach time scales on the order of microseconds, as opposed to the nanosecond extension of MD approaches. This enormous difference in simulation time and system size allows to address the simultaneous nucleation of several vapor bubbles and to follow their expansion and coalescence dynamics coupled with the macroscopic velocity field. These hydrodynamic effects—not captured by conventional techniques—greatly influence the nucleation dynamics, specially in closed systems [10], where "*bubble crowding*" strongly affects the nucleation rate.

12.3 Phase Field Description of Fluids

12.3.1 Thermodynamic of Non-homogeneous Systems

The aim of this section is to describe the thermodynamic equilibrium of a two-phase system. For the sake of simplicity, we will focus on a closed system with fixed temperature and volume (Canonical Ensemble), so that the Helmholtz free energy is the proper thermodynamic potential. Van der Waals was the first to recognize that a description based only on the local values of temperature and phase indicators was insufficient to describe the structure of the fluid layer separating the phases. The idea was introduced that a non-local term, depending on the phase field spatial gradients, needs to be added to the free energy to describe the smooth transition zones separating the phases (see [28, 36, 43] for a detailed derivation in the context of DFT).

The typical expression of the (Helmholtz) free-energy functional F in terms of temperature, θ, and the phase indicator, Φ, fields takes the form

$$F[\Phi,\theta] = \int_V f_b(\Phi,\theta) + f_s(\nabla\Phi, ..., \nabla\Phi \otimes ... \otimes \nabla\Phi)\,\mathrm{d}V + \int_{\partial V} f_w(\Phi,\theta)\,\mathrm{d}S\,, \tag{12.49}$$

where f_b is the bulk free-energy contribution, f_w accounts for the solid–fluid interactions, and f_s is the interface contribution depending of the gradients of the phase indicator Φ. At fixed temperature $\theta = \theta_0$, the equilibrium condition is reached when the first variation of the functional (12.49) with respect to the phase indicator Φ is zero, leading to the Euler-Lagrange equation

$$\frac{\delta F}{\delta \Phi} = \frac{\partial f_b}{\partial \Phi} + \sum_{k=1}^{N} (-1)^k \nabla^{(k)} : \frac{\partial f_s}{\partial\left(\nabla^{(k)}\Phi\right)} = 0\,, \tag{12.50}$$

where the superscript k on the differential operator ∇ denotes the rank k tensor operator defined as the n-fold tensor product of ∇ with itself. Equation (12.50) is a partial differential equation for phase indicator field Φ with boundary conditions

$$\frac{\partial f_w}{\partial \Phi} + g(\nabla\Phi, ..., \nabla\Phi \otimes ... \otimes \nabla\Phi, \mathbf{n}) = 0\,, \tag{12.51}$$

with $\mathbf{n}$ the outward unit normal to the domain boundary. The function g arises from the boundary terms when integrating by part, due to the solid–fluid interactions described by the boundary term f_w.

12.3.2 The Van der Waals Theory

In the last part of nineteenth century, scientists were starting to recognize that the separation surface between two thermodynamic phases could have a finite thickness. Van der Waals, based on phenomenological assumptions, proposed a gradient theory that led him to predict the interface thickness of a fluid near the critical point. In the framework of a general phase field theory, Van der Waals's approach assumes the density field as the relevant phase indicator, and the density gradient square norm as a surface contribution basically localized at the liquid–vapor interface where the density gradient is large (see. Eq. (12.52) below). The model is extremely powerful both for steady and unsteady conditions, providing a robust description of interfacial flows that naturally accounts for topology modification of the regions occupied by the two phases and the phase change between them [38]. For illustrative purposes, the solid–liquid free energy contribution is initially neglected to be reconsidered in the forthcoming sections.

For a closed system, with a given mass M_0, the constrained Helmholtz free energy of a two phase flow in the Van der Waals square gradient approximation [44–46] is:

$$F[\rho,\theta] = \int_V f_b(\rho,\theta) + \frac{\lambda}{2}\nabla\rho\cdot\nabla\rho \,\mathrm{d}V + \oint_{\partial V} f_w(\rho,\theta)\,\mathrm{d}S \qquad (12.52)$$
$$+ l\left(M_0 - \int_V \rho \,\mathrm{d}V\right),$$

with l a Lagrange multiplier and $f_b(\rho,\theta)$ the Helmholtz free energy density per unit volume of the homogeneous fluid at temperature θ and mass density ρ. The coefficient $\lambda(\rho,\theta)$, in general a function of the thermodynamic state, embodies all the information on the interfacial properties of the liquid–vapor system (i.e., surface tension and interface thickness). At given temperature, equilibrium is characterized by the minimum of the free energy functional (12.52) with respect to variations of the density distribution ρ. The relevant Euler-Lagrange equation is

$$\mu_c^b - \nabla\cdot(\lambda\nabla\rho) - l = 0\,, \qquad (12.53)$$

where $\mu_c^b = \partial f_b/\partial\rho|_\theta$ is the classical chemical potential, and the Lagrange multiplier is identified as the chemical potential of a uniform phase in equilibrium with the system. The equation defines a generalized chemical potential $\mu_c = \mu_c^b - \nabla\cdot(\lambda\nabla\rho)$ that must be constant at equilibrium. The consequence of this equilibrium condition is better illustrated for planar liquid–vapor interface separating the two bulk, homogenous phases (liquid and vapor, respectively). The only inhomogeneity direction is denoted by ν and λ is assumed to be constant. Determining the equilibrium density distribution amounts to finding a solution of

$$\mu_c = \mu_c^b(\rho) - \lambda\frac{\mathrm{d}^2\rho}{\mathrm{d}\nu^2} = \mu_{eq}\,. \qquad (12.54)$$

The boundary conditions for this second-order ordinary differential equation are obtained by evaluating the generalized chemical potential in the two bulk phases far away from the interface, namely in the bulk liquid and bulk vapor where $d\rho/d\nu = 0$. It follows $\mu_{eq} = \mu_c^b(\rho_V) = \mu_c^b(\rho_L)$.

The solution of Eq. (12.54) is readily obtained by multiplying through by $d\rho/ds$ and integrating between $\rho_\infty = \rho_V$ and ρ,

$$w_b(\rho) - w_b(\rho_V) = \frac{\lambda}{2}\left(\frac{d\rho}{d\nu}\right)^2 , \tag{12.55}$$

where $w_b(\rho) = f_b(\rho) - \mu_{eq}\rho$. Equation (12.55) shows that w_b has the same value in both the bulk phases, where the spatial derivative of mass density vanishes: $w_b(\rho_L) = w_b(\rho_V)$.

The grand potential, defined as the Legendre transform of the free energy,

$$\Omega = F - \int_V \rho \frac{\delta F}{\delta \rho} dV = \int_V w dV , \tag{12.56}$$

has the density (*actual* grand potential density)

$$w[\rho] = f - \mu_c \rho = f_b + \frac{\lambda}{2}\left(\frac{d\rho}{d\nu}\right)^2 - \left(\mu_c^b - \lambda\frac{d^2\rho}{d\nu^2}\right)\rho , \tag{12.57}$$

implying that, in the bulk, $w = w_b$, i.e., w_b is the *bulk* grand potential density.

Given the form of $w_b(\rho)$, the solution of Eq. (12.55) provides the (implicit expression for the) equilibrium density profile $\rho(\nu)$,

$$\nu = \sqrt{\frac{\lambda}{2}} \int_{\rho_V}^{\rho} \frac{d\rho}{\sqrt{w_b(\rho) - w_b(\rho_V)}} , \tag{12.58}$$

where the integration constant is fixed by assuming $\nu = 0$ as the location of the interface. The equilibrium profile (12.58) describes two bulk liquid and vapor domains separated by a thin layer. The layer thickness can be estimated as

$$\epsilon = \frac{\rho_L - \rho_V}{d\rho/d\nu|_{max}} . \tag{12.59}$$

The equilibrium condition, Eq. (12.55), provides the interface thickness in terms of the *bulk* grand potential density $w_b(\rho)$ and of the parameter λ,

$$\epsilon = (\rho_L - \rho_V)\sqrt{\frac{\lambda}{2\left[w_b(\bar{\rho}) - w_b(\rho_V)\right]}} , \tag{12.60}$$

without explicitly addressing the density profile. $\bar{\rho}$ is the density corresponding to the maximum of $\mathrm{d}\rho/\mathrm{d}\nu$, achieved where $\mathrm{d}w_b/\mathrm{d}\rho = 0$, in Eq. (12.55).

The surface tension can be defined as the excess (*actual*) grand potential density,

$$\gamma = \int_{-\infty}^{S_i} (w[\rho] - w[\rho_V])\, ds + \int_{S_i}^{+\infty} (w[\rho] - w[\rho_L])\, \mathrm{d}\nu = \\ = \int_{-\infty}^{+\infty} (w[\rho] - w[\rho_V])\, \mathrm{d}\nu, \tag{12.61}$$

where S_i is the position of the Gibbs dividing surface, whose precise value is ininfluential since $w[\rho_V] = w[\rho_L]$ (we stress that, e.g., $w[\rho_V]$ should be interpreted as the functional (12.57) evaluated at the constant density field ρ_V). Given the definition of $w[\rho]$, Eq. (12.57), and exploiting the equilibrium condition for the chemical potential, Eq. (12.54), it follows that

$$\gamma = \int_{-\infty}^{+\infty} \left[f_b + \frac{1}{2}\lambda \left(\frac{\mathrm{d}\rho}{\mathrm{d}\nu}\right)^2 - \mu_{eq}\rho - w_b(\rho_V) \right] \mathrm{d}\nu = \\ = \int_{-\infty}^{+\infty} \left[w_b + \frac{1}{2}\lambda \left(\frac{\mathrm{d}\rho}{\mathrm{d}\nu}\right)^2 - w_b(\rho_V) \right] \mathrm{d}\nu\,. \tag{12.62}$$

Using Eq. (12.55) one finds

$$\gamma = \int_{-\infty}^{+\infty} \lambda \left(\frac{\mathrm{d}\rho}{\mathrm{d}\nu}\right)^2 \mathrm{d}\nu = \int_{\rho_V}^{\rho_L} \sqrt{2\lambda\, (w_b(\rho) - w_b(\rho_V))}\, d\rho\,, \tag{12.63}$$

where the second expression can be evaluated with no a priori knowledge of the equilibrium density profile. It worths stressing that, as for the interface thickness, the surface tension only depends on the form of the *bulk* grand potential density $w_b(\rho)$ in the density range between the two equilibrium values, $[\rho_V; \rho_L]$, and on the parameter λ. In [11] the authors report a comparison between the diffuse interface prediction of surface tension and molecular dynamics simulations for a Lennard-Jones fluid. A constant value of λ was assumed to reproduce the simulation data showing how the Van der Waals model correctly captures the temperature dependence of the surface tension.

Let us now focus on the boundary term arising when minimizing Eq. (12.52)

$$\left(\lambda \nabla\rho \cdot \hat{\mathbf{n}} + \frac{\partial f_w}{\partial \rho} \right)\Bigg|_{\partial V} = 0\,, \tag{12.64}$$

where $\hat{\mathbf{n}}$ is the outward normal, to be read as a (non-linear) boundary condition for the density. The free energy contribution f_w arises from the fluid-wall interactions and accounts for the wetting properties of the surface. The analytic form of f_w is derived by the authors in [12], leading to

$$f_w(\rho) = \cos\phi \int_{\rho_V}^{\rho} \sqrt{2\lambda\left(w_b(\tilde{\rho}) - w_b(\rho_V)\right)}\, \mathrm{d}\tilde{\rho} + f_w(\rho_V)\ . \tag{12.65}$$

Considering that the surface energy of the vapor/liquid in contact with the solid is $f_w(\rho_{V/L}) = \gamma_{SV/L)}$, using Eq. 12.63, leads to the classical Young–Laplace equation

$$f_w(\rho_L) = \gamma_{LS} = \gamma_{LV}\cos\phi + \gamma_{SV}\ . \tag{12.66}$$

12.3.3 The String Method

It is worth noticing that, when either the liquid or the vapor is stable, the minimization of the free energy functional leads to a homogeneous phase. When the liquid or the vapor is metastable, three solutions at constant chemical potential are found instead. Two of them correspond to the homogeneous vapor and liquid, respectively. The third one is a two-phase solution with a critical nucleus of the stable phase surrounded by the metastable phase (vapor/liquid in the case of bubble/droplet, respectively).

Focusing on bubbles, the two-phase solution $\rho_{crit}(\mathbf{x})$ where the critical bubble is surrounded by the metastable liquid at $\rho = \rho_L^{met}, \theta = \bar{\theta}$ and $\mu_c(\rho_L^{met}, \bar{\theta}) = \mu^{met}$ is central to the present discussion. In a mathematical sense, the critical bubble is an unstable solution of Eq. (12.53), which requires specialized numerical techniques. Among other possibilities, the string method [47] is especially suited, in particular when the minimum free energy path (MFEP) joining the metastable (e.g., the liquid) to the stable fluid (e.g., the vapor ensuing from cavitation) is also sought for. The stable state will correspond to the presence of an equilibrium bubble enclosed by the liquid (at fixed volume, mass and temperature, the stable state is indeed in a vapor bubble surrounded by the liquid phase). The MFEP can be visualized as the continuous sequence of density configurations, $\rho(\alpha) = \rho(\mathbf{x}, \alpha)$, with α a parameter, the system assumes when transitioning from the metastable to the stable state. In other words, the MFEP is a curve in the space of density fields connecting the metastable to the stable density field. It satisfies

$$\left(\frac{\delta F}{\delta \rho}\right)^{\perp} [\rho(\mathbf{x}, \alpha)] = 0 \tag{12.67}$$

with F the free energy functional of Eq. (12.52). The symbol $\perp$ denotes the component of the chemical potential orthogonal to the path,

$$\left(\frac{\delta F}{\delta \rho}\right)^{\perp} [\rho(\mathbf{x}, \alpha)] = \left(\frac{\delta F}{\delta \rho}\right) [\rho(\mathbf{x}, \alpha)] - \frac{\displaystyle\int_V \frac{\delta F}{\delta \rho}[\rho(\mathbf{x}, \alpha)]\frac{d\rho}{d\alpha} dV}{\displaystyle\int_V \frac{d\rho}{d\alpha}\frac{d\rho}{d\alpha} dV}\frac{d\rho}{d\alpha}\ . \tag{12.68}$$

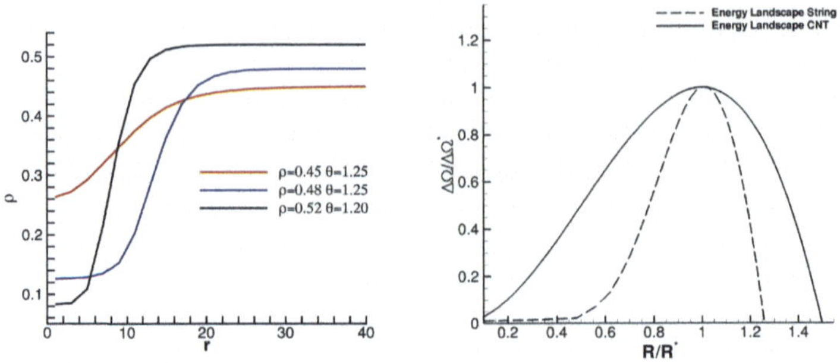

Fig. 12.5 Left panel: critical density profiles as evaluated by the string method [11]. Right panel: energy landscape comparison between CNT and string method [13]

The discrete form of the path, called the *string*, consists of a finite number of configurations $\rho(\mathbf{x}, \alpha_k)$, $k = 1, N_s$, called *images*. The string method numerically approximates the path with a discrete set consisting of N_s configurations $\{\rho^k(\mathbf{x})\}$. The head of the string ($k = 1$) is initialized as a uniform density field corresponding to the uniform metastable liquid $\rho(\mathbf{x}) = \rho_L^{met}$; the tail ($k = N_s$) is initialized as a guessed tanh-density profile adjoining the liquid and the vapor density to approximate a vapor bubble. All the intermediate images on the initial configuration of the string are obtained by interpolation of these two density fields with respect to the arclength.

The details of the algorithm used for relaxing the string to its final configuration satisfying Eq. (12.68) are illustrated in [13]. There the string images $\rho^k(\mathbf{x})$ are evolved over a pseudo-time τ according to an Allen-Cahn dynamics

$$\frac{\partial \rho^k}{\partial \tau} = \mu_c^{met} - \mu_c \, . \qquad (12.69)$$

It is worthwhile noting that the transition path geometry depends, in general, on the relaxation dynamics used to evolve the string. In an over-damped regime, steepest descent relaxation is used as a reference theory. In this context, the string converges to the MFEP connecting the local minimum to the saddle point that under the above assumption is also the most probable transition path [48]. However critical cluster as well as the energy barriers do not depend on the relaxation dynamics.

In the left panel of Fig. 12.5, critical density profiles as evaluated by the string method are reported for different thermodynamic states of the metastable liquid. The data refer to an equation of state reproducing a Lennard-Jones fluid [23]. The critical bubbles are characterized by a thick interface, with the bulk vapor density slowly achieved inside the nucleus and the liquid density recovered at a radial distance order of twice the critical radius. This is a crucial difference with CNT where the density profile changes sharply across a zero thickness interface. As shown in the right panel of Fig. 12.5, the free energy landscape is significantly different. As a consequence, crucial predictions, concerning, e.g., the nucleation rate, show considerable discrepancies [13].

12.4 Thermal Fluctuations in the Continuum Settings

12.4.1 Thermal Fluctuations of Capillary Fluids

The aim of the present section is to retrace the theory of fluctuating hydrodynamics, extending the Landau–Lifshitz approach to diffuse interface thermodynamics. According to the Einstein theory, the static correlation functions of a thermodynamic system in equilibrium can be evaluated from the entropy deviation ΔS from its equilibrium value S_0 [49]. For single component systems, ΔS can be expressed as a functional of fluctuating fields of mass density, $\delta\rho(\mathbf{x}, t)$, velocity, $\delta\mathbf{v}(\mathbf{x}, t)$, and temperature, $\delta\theta(\mathbf{x}, t)$

$$\Delta S = S - S_0 = \Delta S\left[\delta\rho, \delta\mathbf{v}, \delta\theta\right] = \int_V \left[s\left(\mathbf{x}, t\right) - s_0\right] dV, \tag{12.70}$$

where the integration is over the system volume V, $s\left(\mathbf{x}, t\right)$ is the entropy density per unit volume, and s_0 is its equilibrium value, (*i.e., S_0 is the entropy maximum).* For closed and isolates systems the entropy is constrained to a given total mass (M_0) and energy (E_0). Hence, the functional to be maximized at equilibrium is the constrained entropy ΔS_c expressed as

$$\Delta S_c = \Delta S + k_1\left(M_0 - \int_V \rho\, dV\right) + k_2\left(E_0 - \int_V e\, dV\right), \tag{12.71}$$

where k_1 and k_2 are two Lagrange multipliers.

For the sake of simplicity, we will focus here on a bulk system, the extension to fluids confined by hydrophobic/hydrophilic walls being described in our recent work [12]. The entropy S as a functional of the density and temperature fields is the functional derivative of the free energy with respect to temperature

$$S\left[\rho, \theta\right] = \int_V -\frac{\delta F}{\delta\theta}\, dV = \int_V -\frac{\partial f_b}{\partial\theta}\, dV = \int_V s_b(\rho, \theta)\, dV, \tag{12.72}$$

where the second equality holds if λ does not depend on temperature, and the last identity follows by the classical definition of the bulk entropy density s_b. Thus the constrained functional in Eq. (12.71) reads

$$\Delta S_c = \Delta S_b + k_1\left(M_0 - \int_V \rho\, dV\right) + k_2\left(E_0 - U - \int_V \frac{1}{2}\rho\mathbf{v}\cdot\mathbf{v}\, dV\right), \tag{12.73}$$

where the internal energy functional U is defined in terms of free energy as

$$U = F - \int dV \frac{\delta F}{\delta\theta}\theta = \int_V dV\left(u_b\left(\rho, \theta\right) + \frac{1}{2}\lambda\nabla\rho\cdot\nabla\rho\right), \tag{12.74}$$

with $u_b = f_b - \theta \partial f_b / \partial \theta$ the bulk internal energy density. The two Lagrange multipliers k_1 and k_2 are found by imposing that the first variation of the functional in Eq. (12.73) evaluated in the equilibrium state, must be zero:

$$\delta \Delta S_c \left[\rho_0, 0, \theta_0\right] = 0, \tag{12.75}$$

The above equation leads to $k_1 = -\mu_{c0}/\theta_0$, $k_2 = 1/\theta_0$, where μ_{c0} is the equilibrium chemical potential. For small fluctuations, the entropy functional can be expanded in a Taylor series around the equilibrium state with respect to the variables $\mathbf{U} = (\rho, \nabla\rho, \ldots, \nabla^n \rho, \theta, \mathbf{v})^T$ as follows

$$\begin{aligned} \Delta S_c &= \int_V \Delta s_c(\rho, \nabla\rho, \ldots, \nabla^n\rho, \theta, \mathbf{v})\, dV = \\ &= \int_V dV \left[\sum_i \left.\frac{\partial \Delta s_c}{\partial U_i}\right|_0 \delta U_i + \frac{1}{2} \sum_{i,j} \left.\frac{\partial^2 \Delta s_c}{\partial U_i \partial U_j}\right|_0 \delta U_i \delta U_j + \ldots \right] \end{aligned} \tag{12.76}$$

All terms appearing in the right-hand side of Eq. (12.76) can be rearranged in terms of suitable thermodynamic coefficients and of the fluctuations of density, temperature and velocity, e.g.,

$$\begin{aligned} \mathrm{d}s_b &= \frac{1}{\theta}\mathrm{d}u_b - \frac{\mu_b}{\theta}\mathrm{d}\rho\,, \\ \mathrm{d}u_b &= \rho c_v \mathrm{d}\theta + \left(\mu_b + \theta \left.\frac{\partial s_b}{\partial \rho}\right|_\theta\right)\mathrm{d}\rho\,, \\ \mathrm{d}\mu_b &= \frac{c_T^2}{\rho}\mathrm{d}\rho + \left(\frac{1}{\rho}\left.\frac{\partial p}{\partial \theta}\right|_\rho - \frac{s_b}{\rho}\right)\mathrm{d}\theta\,, \end{aligned} \tag{12.77}$$

where c_v is the specific heat at constant specific volume, c_T the isothermal speed of sound, p the pressure.

Assuming that the fluid is very close to equilibrium and the fluctuations are small with respect to the mean value, the entropy functional can be approximated by a quadratic form in the fluctuating fields,

$$\Delta S_c \simeq -\frac{1}{2}\int_V dV \left[\frac{c_{T0}^2}{\theta_0 \rho_0}\delta\rho^2 - \frac{\lambda}{\theta_0}\delta\rho\left(\nabla^2 \delta\rho\right) + \frac{\rho_0}{\theta_0}\delta\mathbf{v}\cdot\delta\mathbf{v} + \frac{\rho_0 c_{v0}}{\theta_0^2}\delta\theta^2\right]. \tag{12.78}$$

For future reference, it is worth expressing the above integral as

$$\Delta S_c \simeq -\frac{1}{2}\int_V \int_V dV_{\mathbf{x}} dV_{\tilde{\mathbf{x}}} \left\{ \delta\mathbf{v}(\mathbf{x}) \frac{\rho_0}{\theta_0} \delta\left(\mathbf{x}-\tilde{\mathbf{x}}\right) \cdot \delta\mathbf{v}(\tilde{\mathbf{x}}) + \right.$$
$$+ \delta\rho(\mathbf{x}) \left[\frac{c_{T0}^2}{\theta_0 \rho_0} \delta\left(\mathbf{x}-\tilde{\mathbf{x}}\right) - \frac{\lambda}{\theta_0} \nabla_{\mathbf{x}}^2 \delta\left(\mathbf{x}-\tilde{\mathbf{x}}\right) \right] \delta\rho(\tilde{\mathbf{x}}) +$$
$$\left. + \delta\theta(\mathbf{x}) \frac{\rho_0 c_{v0}}{\theta_0^2} \delta\left(\mathbf{x}-\tilde{\mathbf{x}}\right) \delta\theta(\tilde{\mathbf{x}}) \right\} , \tag{12.79}$$

where integration by parts is used twice to move the Laplacian ∇^2 from the density to the Dirac delta function. Equation (12.79) can be rewritten in operator form as

$$\Delta S_c = -\frac{1}{2}\int_V \mathbf{\Delta}^\dagger \mathcal{H}\, \mathbf{\Delta}\, dV , \tag{12.80}$$

where $\mathbf{\Delta} = (\delta\rho, \delta\mathbf{v}, \delta\theta)$ is the vector of the the fluctuating fields and $\mathcal{H}$ is a diagonal, positive definite matrix operator

$$\Gamma(\mathbf{x}) = \left(\mathcal{H}\mathbf{\Delta}\right)(\mathbf{x}) = \int_V \mathbf{H}(\mathbf{x},\tilde{\mathbf{x}})\mathbf{\Delta}(\tilde{\mathbf{x}}) dV_{\tilde{\mathbf{x}}} = \int_V \hat{\mathbf{H}}(\mathbf{x})\delta(\mathbf{x}-\tilde{\mathbf{x}})\mathbf{\Delta}(\tilde{\mathbf{x}}) dV_{\tilde{\mathbf{x}}} , \tag{12.81}$$

where

$$\hat{\mathbf{H}}(\mathbf{x}) = \begin{pmatrix} \hat{H}_{\delta\rho\delta\rho} & 0 & 0 \\ 0 & \mathbf{I}\hat{H}_{\delta\mathbf{v}\delta\mathbf{v}} & 0 \\ 0 & 0 & \hat{H}_{\delta\theta\delta\theta} \end{pmatrix} = \begin{pmatrix} \frac{c_{T0}^2}{\theta_0\rho_0} - \frac{\lambda}{\theta_0}\nabla_{\mathbf{x}}^2 & 0 & 0 \\ 0 & \frac{\rho_0}{\theta_0}\mathbf{I} & 0 \\ 0 & 0 & \frac{\rho_0 c_{v0}}{\theta_0^2} \end{pmatrix} \tag{12.82}$$

involves differential operators and $\mathbf{I}$ is the 3×3 identity matrix. Note that, indeed the Laplace operator $-\nabla^2$ appearing in the first line, which is, in general, non-negative, is strictly positive under the constraint of mass conservation since the mean spatial density fluctuation is identically zero.

Under these assumptions, the more general probability distribution functional for the fluctuating fields $\mathbf{\Delta}$ [50]

$$P_{eq}\left[\mathbf{\Delta}\right] = \frac{1}{Z}\exp\left(\frac{\Delta S_c}{k_B}\right) , \tag{12.83}$$

can be rewritten by using the second-order approximation, Eq. (12.80),

$$P_{eq}\left[\Delta\right] = \frac{1}{Z}\exp\left(-\frac{1}{2k_B}\int_V \Delta^\dagger \mathcal{H}\, \Delta\, dV\right) . \tag{12.84}$$

Since $\hat{\mathbf{H}}$ is diagonal, this expression can be factorized

$$P_{eq}[\mathbf{\Delta}] = P_{\delta\rho}[\delta\rho] P_{\delta\mathbf{v}}[\delta\mathbf{v}] P_{\delta\theta}[\delta\theta] , \tag{12.85}$$

with

$$P_{\delta\rho} = \frac{1}{Z_{\delta\rho}} \exp\left[-\frac{1}{2\mathrm{k}_B} \int\int d\mathbf{x}d\mathbf{x}' \delta\rho(\mathbf{x}) H_{\delta\rho\delta\rho} \delta(\mathbf{x}-\mathbf{x}') \delta\rho(\mathbf{x}') \right], \quad (12.86)$$

$$P_{\delta\mathbf{v}} = \frac{1}{Z_{\delta\mathbf{v}}} \exp\left[-\frac{1}{2\mathrm{k}_B} \int\int d\mathbf{x}d\mathbf{x}' \delta\mathbf{v}^T(\mathbf{x}) H_{\delta\mathbf{v}\delta\mathbf{v}} \delta(\mathbf{x}-\mathbf{x}') \delta\mathbf{v}(\mathbf{x}') \right], \quad (12.87)$$

$$P_{\delta\theta} = \frac{1}{Z_{\delta\theta}} \exp\left[-\frac{1}{2\mathrm{k}_B} \int\int d\mathbf{x}d\mathbf{x}' \delta\theta(\mathbf{x}) H_{\delta\theta\delta\theta} \delta(\mathbf{x}-\mathbf{x}') \delta\theta(\mathbf{x}') \right]. \quad (12.88)$$

The normalization constant Z follows as

$$Z = \int D\delta\rho D\delta\mathbf{v} D\delta\theta \exp\left(-\frac{1}{2k_B} \int_V \mathbf{\Delta}^\dagger \mathcal{H} \mathbf{\Delta} \, dV \right) = Z_{\delta\rho} Z_{\delta\mathbf{v}} Z_{\delta\theta} . \quad (12.89)$$

The correlation function

$$\mathbf{C}_{\mathbf{\Delta}}(\mathbf{x}) = \langle \mathbf{\Delta} \otimes \mathbf{\Delta}^\dagger \rangle = \quad (12.90)$$
$$= \frac{1}{Z} \int D\delta\rho D\delta\mathbf{v} D\delta\theta \, \mathbf{\Delta} \otimes \mathbf{\Delta}^\dagger \exp\left(\frac{1}{k_B} \int_V \Delta s_c(\delta\rho, \delta\mathbf{v}, \delta\theta) \, dV \right)$$

can now be evaluated in closed form by integrating Gaussian path integrals. To this end, it is helpful to resort to the characteristic functional [51] of the pdf which, for a generic process $X(\mathbf{x})$, is

$$\Phi[\chi] = \int DX \; P[X] \exp\left(\int \chi(\mathbf{x}) X(\mathbf{x}) \, dV_{\mathbf{x}} \right) . \quad (12.91)$$

For a Gaussian process governed by the pdf

$$P[X] = \frac{1}{Z} \exp\left(\int -\frac{1}{2} X(\mathbf{x}) A(\mathbf{x}, \tilde{\mathbf{x}}) X(\tilde{\mathbf{x}}) dV_{\mathbf{x}} dV_{\tilde{\mathbf{x}}} \right) \quad (12.92)$$

the characteristic functional reduces to

$$\Phi[\chi] = \int DX \exp\left[-\frac{1}{2} \int\int d\hat{\mathbf{x}} d\tilde{\mathbf{x}} X(\hat{\mathbf{x}}) \, A(\hat{\mathbf{x}}, \tilde{\mathbf{x}}) \, X(\tilde{\mathbf{x}}) + \int \chi(\hat{\mathbf{x}}) \, X(\hat{\mathbf{x}}) \, d\hat{\mathbf{x}} \right], \quad (12.93)$$

and is easily evaluated by completing the square (see Appendix A) as

$$\Phi[\chi] = \Phi[0] \exp\left(\frac{1}{2} \int\int d\hat{\mathbf{x}} d\tilde{\mathbf{x}} \, \chi(\hat{\mathbf{x}}) \, G(\hat{\mathbf{x}}, \tilde{\mathbf{x}}) \, \chi(\tilde{\mathbf{x}}) \right), \quad (12.94)$$

where

$$G(\hat{\mathbf{x}}, \tilde{\mathbf{x}}) = A^{-1}(\hat{\mathbf{x}}, \tilde{\mathbf{x}}) \quad (12.95)$$

(we stress that A is the kernel of an operator $\mathcal{A}$, such that A^{-1} should be understood as the kernel of the inverse $\mathcal{A}^{-1}$). The two-point correlation can be written in terms of the characteristic functional as

$$C_{XX}\left(\hat{\mathbf{x}}, \tilde{\mathbf{x}}\right) = \langle X\left(\hat{\mathbf{x}}\right) X\left(\tilde{\mathbf{x}}\right)\rangle = \\ = \left(\frac{1}{\Phi[0]}\frac{\delta}{\delta\chi\left(\hat{\mathbf{x}}\right)}\frac{\delta}{\delta\chi\left(\tilde{\mathbf{x}}\right)}\Phi[\chi]\right)_{\chi=0} = G\left(\hat{\mathbf{x}}, \tilde{\mathbf{x}}\right) . \tag{12.96}$$

In the present case, Eq. (12.84), the kernel of the operator A is given by

$$A\left(\hat{\mathbf{x}}, \tilde{\mathbf{x}}\right) = \frac{1}{\mathrm{k}_B}\hat{\mathbf{H}}\delta\left(\hat{\mathbf{x}} - \tilde{\mathbf{x}}\right) , \tag{12.97}$$

implying the equation

$$\int \mathbf{A}(\mathbf{x}, \mathbf{x}'')\mathbf{G}\left(\mathbf{x}'', \mathbf{x}'\right) dV_{\mathbf{x}''} = \\ \frac{1}{\mathrm{k}_B}\int \hat{\mathbf{H}}\delta(\mathbf{x} - \mathbf{x}'')\mathbf{G}\left(\mathbf{x}'', \mathbf{x}'\right) dV_{\mathbf{x}''} = \hat{\mathbf{U}}\delta(\mathbf{x} - \mathbf{x}') , \tag{12.98}$$

which, written in terms of operators, corresponds to the equation $\mathcal{A}\mathcal{A}^{-1} = \mathcal{U}$, with $\mathcal{U}$ the identity operator on the space of fluctuations. $\hat{\mathbf{U}}$ is the identity matrix acting on the five-dimensional tangent space at a given position $\mathbf{x}$, $\mathbf{\Delta}(\mathbf{x}) = (\delta\rho(\mathbf{x}), \delta\mathbf{v}(\mathbf{x}), \delta\theta(\mathbf{x}))$. In particular, since the matrix $\hat{\mathbf{H}}$ is diagonal, the $\delta\rho\,\delta\rho$ component of the above equation is

$$\int A_{\delta\rho\delta\rho}(\mathbf{x}, \mathbf{x}'')G_{\delta\rho\delta\rho}\left(\mathbf{x}'', \mathbf{x}'\right) dV_{\mathbf{x}''} = \\ \mathrm{k}_B^{-1}\int \left[\left(\frac{c_T^2}{\theta_0\rho_0} - \frac{\lambda}{\theta_0}\nabla_{\mathbf{x}}^2\right)\delta(\mathbf{x} - \mathbf{x}'')\right] G_{\delta\rho\delta\rho}\left(\mathbf{x}'', \mathbf{x}'\right) dV_{\mathbf{x}''} = \delta(\mathbf{x} - \mathbf{x}'). \tag{12.99}$$

After integration by parts, Eq. (12.99) reads

$$\frac{c_T^2}{\theta_0\rho_0\mathrm{k}_B}G_{\delta\rho\delta\rho}\left(\hat{\mathbf{x}}, \tilde{\mathbf{x}}\right) - \frac{\lambda}{\theta_0\mathrm{k}_B}\nabla_{\hat{\mathbf{x}}}^2 G_{\delta\rho\delta\rho}\left(\hat{\mathbf{x}}, \tilde{\mathbf{x}}\right) = \delta\left(\hat{\mathbf{x}} - \tilde{\mathbf{x}}\right) \tag{12.100}$$

After Fourier transformation, the equation becomes

$$\hat{G}\left(\mathbf{k}\right) + \frac{\lambda\rho_0}{c_T^2}\mathbf{k}\cdot\mathbf{k}\,\hat{G}\left(\mathbf{k}\right) = \frac{\rho_0\mathrm{k}_B\theta_0}{c_T^2} , \tag{12.101}$$

allowing to express the solution (the Green's function for the Helmholtz equation) as

$$G_{\delta\rho\delta\rho}\left(\hat{\mathbf{x}}, \tilde{\mathbf{x}}\right) = \int d\mathbf{k} \frac{\mathrm{k}_B \rho_0 \theta_0}{c_T^2 + \rho_0 \lambda \, \mathbf{k} \cdot \mathbf{k}} e^{i\mathbf{k}\cdot(\hat{\mathbf{x}}-\tilde{\mathbf{x}})} . \tag{12.102}$$

Explicitly performing the inverse Fourier transform yields

$$G_{\delta\rho\delta\rho}\left(\hat{\mathbf{x}}, \tilde{\mathbf{x}}\right) = C_{\delta\rho\delta\rho}\left(\hat{\mathbf{x}}, \tilde{\mathbf{x}}\right) = \frac{\mathrm{k}_B \theta_0}{4\pi \lambda \left|\hat{\mathbf{x}} - \tilde{\mathbf{x}}\right|} \exp\left(- \left|\hat{\mathbf{x}} - \tilde{\mathbf{x}}\right| \sqrt{\frac{c_T^2}{\rho_0 \lambda}} \right) \tag{12.103}$$

where we have recognized that $\mathbf{G} = \mathbf{C}_{\mathbf{\Delta}}$, Eq. (12.96). The same procedure can be used to reconstruct the entire correlation tensor $\mathbf{C}_\Delta = \langle \mathbf{\Delta} \otimes \mathbf{\Delta}^\dagger \rangle$, Eqs. (12.104–12.107):

$$\mathbf{C}_\Delta = \begin{pmatrix} C_{\delta\rho\delta\rho} & 0 & 0 \\ 0 & \mathbf{C}_{\delta\mathbf{v}\delta\mathbf{v}} & 0 \\ 0 & 0 & C_{\delta\theta\delta\theta} \end{pmatrix} , \tag{12.104}$$

with

$$C_{\delta\rho\delta\rho} = \frac{\mathrm{k}_B \theta_0}{4\pi \lambda \left|\hat{\mathbf{x}} - \tilde{\mathbf{x}}\right|} \exp\left(- \left|\hat{\mathbf{x}} - \tilde{\mathbf{x}}\right| \sqrt{\frac{c_T^2}{\rho_0 \lambda}} \right) , \tag{12.105}$$

$$\mathbf{C}_{\delta\mathbf{v}\delta\mathbf{v}} = \frac{k_B \theta_0}{\rho_0} \mathbf{I} \delta\left(\hat{\mathbf{x}} - \tilde{\mathbf{x}}\right) , \tag{12.106}$$

$$C_{\delta\theta\delta\theta} = \frac{k_B \theta_0^2}{\rho_0 c_v} \delta\left(\hat{\mathbf{x}} - \tilde{\mathbf{x}}\right) . \tag{12.107}$$

It can be concluded that, in the Gaussian approximation, the equilibrium correlations for velocity and temperature are short-ranged (delta-correlated in space, actually) and the cross-correlation of the fluctuating fields are zero.

12.4.2 *Fluctuation Dissipation Balance for the Capillary Navier–Stokes Equations*

The time evolution of the two-phase, vapor–liquid, system obeys mass, momentum and energy conservation. Standard procedures of non-equilibrium thermodynamics [52] allow to write the hydrodynamic equations endowed with capillarity (see [46, 53] for the detailed derivation) in which stress tensor and energy flux are modified *wrt* the usual Navier–Stokes equations. The stress tensor reads

$$\Sigma = \left[-p + \frac{\lambda}{2} |\nabla \rho|^2 + \lambda \rho \nabla \cdot (\lambda \nabla \rho) \right] \mathbf{I} - \lambda \nabla \rho \otimes \nabla \rho + \mu \left[(\nabla \mathbf{u} + \nabla \mathbf{u}^T) - \frac{2}{3} \nabla \cdot \mathbf{u} \mathbf{I} \right] , \tag{12.108}$$

with $p = -\rho^2 \partial(f_b/\rho)/\partial\rho = f_b - \mu_c^b \rho$ the pressure and μ the dynamic viscosity. The energy flux turns out to be

$$\mathbf{q} = \lambda\rho\nabla\rho\nabla\cdot\mathbf{u} - k\nabla\theta\,, \tag{12.109}$$

with k the thermal conductivity.

At mesoscopic scale, the above equations need to be augmented with thermal fluctuations. In this case, the stochastic evolution of the system is described by the conservation laws of mass, momentum and energy,

$$\begin{aligned}
\frac{\partial\rho}{\partial t} + \nabla\cdot(\rho\mathbf{u}) &= 0\,,\\
\frac{\partial\rho\mathbf{u}}{\partial t} + \nabla\cdot(\rho\mathbf{u}\otimes\mathbf{u}) &= \nabla\cdot\Sigma + \nabla\cdot\delta\Sigma\,,\\
\frac{\partial E}{\partial t} + \nabla\cdot(\mathbf{u}E) &= \nabla\cdot(\Sigma\cdot\mathbf{u} - \mathbf{q}) + \nabla\cdot(\delta\Sigma\cdot\mathbf{u} - \delta q)\,,
\end{aligned} \tag{12.110}$$

where $\mathbf{u}$ is the fluid velocity, E is the total energy density, $E = \mathcal{U} + 1/2\rho|\mathbf{u}|^2 + 1/2\lambda|\nabla\rho|^2$, with $\mathcal{U}$ the internal energy density. In the momentum and energy equations, Σ and $\mathbf{q}$ are the deterministic components of stress tensor and energy flux, respectively, Eqs. (12.108, 12.109). while the terms with the prefix δ are the stochastic parts. They are required to satisfy the Fluctuation-Dissipation–Balance (FDB) in order to recover the equilibrium correlations, Eq. (12.104).

The system of conservation laws, Eq. (12.110), can be rewritten in a compact notation

$$\frac{\partial\mathbf{U}}{\partial t} = \mathbf{N}[\mathbf{U}] + \mathbf{f}\,, \tag{12.111}$$

where the components of the vector $\mathbf{U}$ are the conserved fields $\mathbf{U} = \{\rho, \rho\mathbf{u}, E\}$, $\mathbf{N}$ is the deterministic non-linear operator, and $\mathbf{f}$ the stochastic forcing. The stochastic force $\mathbf{f}(x,t)$ is represented by a Gaussian process as

$$\langle\mathbf{f}(\tilde{x},t)\otimes\mathbf{f}^\dagger(\hat{x},t')\rangle = \mathbf{Q}(\tilde{x},\hat{x})\delta(t-t')\,, \tag{12.112}$$

where delta correlation in time is assumed. The stochastic term $\mathbf{f}$ is related to a standard Wiener process $\mathbf{W}dt = d\mathbf{B}$

$$\mathbf{f} = \mathbf{K}\mathbf{W}\,, \tag{12.113}$$

where $\mathbf{W} = \left\{W_\rho, \mathbf{W}_\mathbf{u}, W_\theta\right\}^T$, with $\mathbf{W}_\mathbf{u} = \left(W_{u_x}, W_{u_y}, W_{u_z}\right)^T$, is a Gaussian delta correlated process whose correlation reads

$$\langle\mathbf{W}(\tilde{y},t)\otimes\mathbf{W}(\hat{y},t')\rangle = \mathbf{I}\delta(\tilde{y}-\hat{y})\delta(t-t')\,, \tag{12.114}$$

with $\mathbf{I}$ a (5×5) identity matrix in the space of $\mathbf{W}$. The linear operator $\mathbf{K}$ reads

$$\mathbf{K} = \begin{pmatrix} 0 & 0 & 0 \\ 0 & \frac{\sigma_u}{\rho_0} & 0 \\ 0 & 0 & -\frac{\sigma_\theta}{\rho_0 c_v} \end{pmatrix} \tag{12.115}$$

with $\sigma_{\mathbf{u}}$ and σ_θ operators to be determined by enforcing the FDB.

The system of equations (12.111) can be linearized around the mean solution $\{\rho_0, \mathbf{0}, \theta_0\}$,

$$\partial_t \boldsymbol{\Delta} = \mathbf{L}\,\boldsymbol{\Delta} + \mathbf{f}\,, \tag{12.116}$$

where $\mathbf{L}$ is the linearized Navier–Stokes operator with capillarity

$$\mathbf{L} = \begin{pmatrix} 0 & -\rho_0 \nabla\cdot & 0 \\ -\frac{c_T^2}{\rho_0}\nabla + \lambda\nabla\nabla^2 & \frac{\mu}{\rho_0}\left(\nabla^2 + \frac{1}{3}\nabla\nabla\cdot\right) & -\frac{1}{\rho_0}\partial_\theta p \nabla \\ 0 & -\frac{\theta_0}{\rho_0 c_v}\partial_\theta p \nabla\cdot & \frac{k}{\rho_0 c_v}\nabla^2 \end{pmatrix}, \tag{12.117}$$

and $\boldsymbol{\Delta} = \{\delta\rho, \delta\mathbf{u}, \delta\theta\}$ the vector of fields fluctuations. Here the energy equation has been rewritten in terms of temperature to make it simpler to manage. Such a procedure provides a set of stochastic partial differential equations, whose equilibrium (statistically stationary) solution is a set of Gaussian fields.

The solution of Eq. (12.116) is formally expressed as [54]

$$\boldsymbol{\Delta}(x, t) = \int_0^t e^{\mathbf{L}(t-s)}\mathbf{f}(s)\,ds + e^{\mathbf{L}t}\boldsymbol{\Delta}_0\,, \tag{12.118}$$

where the last term that keeps memory of the initial conditions vanishes for large times (L is a dissipative operator). Hence, the equilibrium correlation reads

$$\langle \boldsymbol{\Delta}(\tilde{x}, t) \otimes \boldsymbol{\Delta}^\dagger(\hat{x}, t)\rangle = \int_0^t e^{\mathbf{L}(t-s)}\mathbf{Q}\,e^{\mathbf{L}^\dagger(t-s)}ds\,, \tag{12.119}$$

with $\mathbf{Q}$ as introduced in Eq. (12.112). The above integral can be solved by using a Hermitian non-singular operator $\mathbf{E}^{-1}$ such that $\mathbf{Q}$ factorizes as

$$\mathbf{Q} = -\mathbf{L}\mathbf{E}^{-1} - \mathbf{E}^{-1}\mathbf{L}^\dagger\,. \tag{12.120}$$

Using Eq. (12.120), the integrand in Eq. (12.119) is understood as the derivative with respect to the delay time s of $e^{\mathbf{L}(t-s)}\mathbf{E}^{-1}\,e^{\mathbf{L}^\dagger(t-s)}$. Thus Eq. (12.119) leads to

$$\lim_{t\to\infty}\langle \boldsymbol{\Delta} \otimes \boldsymbol{\Delta}^\dagger\rangle = \mathbf{E}^{-1} = \mathbf{C}_{\boldsymbol{\Delta}}\,, \tag{12.121}$$

i.e., the operator $\mathbf{E}^{-1}$ is the correlation matrix $\mathbf{C}_{\boldsymbol{\Delta}}$, Eq. (12.104).

Given the expression for $\mathbf{Q}$, Eq. (12.120), and the identity $\mathbf{E}^{-1} = \mathbf{C}_{\boldsymbol{\Delta}}$ it follows

$$\mathbf{Q} = -\left(\mathbf{L}\mathbf{C}_{\boldsymbol{\Delta}} + \mathbf{C}_{\boldsymbol{\Delta}}\mathbf{L}^\dagger\right) = \left(\mathbf{M} + \mathbf{M}^\dagger\right) = 2\mathrm{k}_B\mathbf{O}\,, \tag{12.122}$$

where $\mathbf{M} = -\mathbf{L}\mathbf{C}_\Delta$ and $\mathbf{O}$ can be recognized as the Onsager tensor. Relationship (12.122) is the FDB for the Capillary Navier–Stokes equations, which ultimately determines the stochastic noise.

Finally, the operators σ_u and σ_θ are identified by substituting Eq. (12.113) in Eq. (12.112) and enforcing the FDB, Eq. (12.122)

$$\mathbf{Q}(\tilde{x}, \hat{x})\delta(t - t') = \mathbf{K}\langle \mathbf{W}\mathbf{W}^\dagger \rangle \mathbf{K}^\dagger = 2\mathrm{k}_B \mathbf{O}\delta(t - t') \, , \qquad (12.123)$$

hence

$$\mathbf{K}\mathbf{K}^\dagger = 2\mathrm{k}_B \mathbf{O} = -\left(\mathbf{L}\mathbf{C}_\Delta + \mathbf{C}_\Delta \mathbf{L}^\dagger\right) = \mathbf{M} + \mathbf{M}^\dagger \, . \qquad (12.124)$$

The matrix $\mathbf{M}$ can be obtained by using the expression for $\mathbf{L}$ in Eq. (12.117), and the correlation matrix $\mathbf{C}_\Delta$ from Eq. (12.104), leading to

$$\mathbf{M} = \begin{pmatrix} 0 & m_{12} & 0 \\ m_{21} & \mathbf{m}_{22} & m_{23} \\ 0 & m_{32} & m_{33} \end{pmatrix} , \qquad (12.125)$$

whose entries are:

$$m_{12} = m_{21} = \mathrm{k}_B \theta_0 \nabla \delta \left(\mathbf{x} - \hat{\mathbf{x}}\right) , \qquad (12.126)$$

$$m_{23} = m_{32} = \frac{\mathrm{k}_B \theta_0^2}{\rho_0^2 c_v} \partial_\theta p \nabla \delta \left(\mathbf{x} - \hat{\mathbf{x}}\right) , \qquad (12.127)$$

$$\mathbf{m}_{22} = -\frac{\mu_0 \mathrm{k}_B \theta_0}{\rho_0^2} \left(\mathbf{I}\nabla^2 + \frac{1}{3}\nabla \otimes \nabla\right) \delta \left(\mathbf{x} - \hat{\mathbf{x}}\right) , \qquad (12.128)$$

$$m_{33} = -\frac{\mathrm{k}_B \theta_0^2 k}{\rho_0^2 c_v^2} \nabla^2 \delta \left(\mathbf{x} - \hat{\mathbf{x}}\right) . \qquad (12.129)$$

The sum of $\mathbf{M}$ with its hermitian conjugate $\mathbf{M}^\dagger$ provides

$$\mathbf{M} + \mathbf{M}^\dagger = \mathbf{K}\mathbf{K}^\dagger = \begin{pmatrix} 0 & 0 & 0 \\ 0 & 2\,\mathbf{m}_{22} & 0 \\ 0 & 0 & 2\,m_{33} \end{pmatrix} . \qquad (12.130)$$

The system of equations (12.130) has to be satisfied component-wise

$$\sigma_\theta \sigma_\theta^\dagger = -2\mathrm{k}_B \theta_0^2 k \nabla^2 \delta \left(\hat{\mathbf{x}} - \tilde{\mathbf{x}}\right) , \qquad (12.131)$$

$$\sigma_{\mathbf{u}} \otimes \sigma_{\mathbf{u}}^\dagger = -2\mu_0 \mathrm{k}_B \theta_0 \left(\mathbf{I}\nabla^2 + \frac{1}{3}\nabla \otimes \nabla\right) \delta \left(\hat{\mathbf{x}} - \tilde{\mathbf{x}}\right) , \qquad (12.132)$$

providing an explicit expression for the stochastic fluxes

$$\delta\Sigma = \sqrt{2\mu_0 \mathrm{k}_B\theta_0}\tilde{\mathbf{W}}_{\mathbf{u}} - \frac{1}{3}\sqrt{2\mu \mathrm{k}_B\theta}\,\mathrm{Tr}\left(\tilde{\mathbf{W}}_{\mathbf{u}}\right)\mathbf{I}, \tag{12.133}$$

$$\delta q = \sqrt{2k\mathrm{k}_B\theta_0^2}\mathbf{W}_\theta\,. \tag{12.134}$$

In Eq. (12.133), $\tilde{\mathbf{W}}_{\mathbf{u}} = \left(\mathbf{W}_{\mathbf{u}} + (\mathbf{W}_{\mathbf{u}})^T\right)/\sqrt{2}$ is a stochastic symmetric tensor field, and $\mathbf{W}_\theta$ is a stochastic vector, with the following statistical properties

$$\left\langle W^u{}_{\alpha\beta}(\hat{x},\hat{t})W^u{}_{\gamma\delta}(\tilde{x},\tilde{t})\right\rangle = \delta_{\alpha\gamma}\delta_{\beta\delta}\delta(\hat{x}-\tilde{x})\delta(\hat{t}-\tilde{t}), \tag{12.135}$$

$$\left\langle W^\theta{}_{\alpha}(\hat{x},\hat{t})W^\theta{}_{\beta}(\tilde{x},\tilde{t})\right\rangle = \delta_{\alpha\beta}\delta(\hat{x}-\tilde{x})\delta(\hat{t}-\tilde{t})\,. \tag{12.136}$$

It is straightforward to show that expressions (12.133), (12.134) are consistent with the equations (12.131), (12.132)

$$\langle\sigma_\theta W_\theta W_\theta^\dagger \sigma_\theta^\dagger\rangle = \langle\nabla_{\hat{\mathbf{x}}}\cdot\delta q\left(\hat{\mathbf{x}},t\right)\nabla_{\tilde{\mathbf{x}}}\cdot\delta q\left(\tilde{\mathbf{x}},t\right)\rangle = -2\mathrm{k}_B\theta_0^2 k\nabla^2\delta\left(\hat{\mathbf{x}}-\tilde{\mathbf{x}}\right), \tag{12.137}$$

$$\begin{aligned}\langle\sigma_u\mathbf{W}_{\mathbf{u}}\otimes\mathbf{W}_{\mathbf{u}}{}^\dagger\sigma_u{}^\dagger\rangle &= \langle\nabla_{\hat{\mathbf{x}}}\cdot\delta\Sigma\left(\hat{\mathbf{x}},t\right)\otimes\nabla_{\tilde{\mathbf{x}}}\cdot\delta\Sigma\left(\tilde{\mathbf{x}},t\right)\rangle\\ &= -2\mu_0\mathrm{k}_B\theta_0\left(\mathbf{I}\nabla^2 + \frac{1}{3}\nabla\otimes\nabla\right)\delta\left(\hat{\mathbf{x}}-\tilde{\mathbf{x}}\right).\end{aligned} \tag{12.138}$$

Moreover, by means of the FDB, the covariance of the stochastic process reads

$$\left\langle\delta\Sigma(\hat{x},\hat{t})\otimes\delta\Sigma^\dagger(\tilde{x},\tilde{t})\right\rangle = \mathbf{Q}^\Sigma\delta(\hat{x}-\tilde{x})\delta(\hat{t}-\tilde{t}), \tag{12.139}$$

with

$$\mathbf{Q}^\Sigma{}_{\alpha\beta\upsilon\eta} = 2\mathrm{k}_B\theta\mu\left(\delta_{\alpha\upsilon}\delta_{\beta\eta} + \delta_{\alpha\eta}\delta_{\beta\upsilon} - \frac{2}{3}\delta_{\alpha\beta}\delta_{\upsilon\eta}\right), \tag{12.140}$$

and

$$\left\langle\delta q(\hat{x},\hat{t})\otimes\delta q^\dagger(\tilde{x},\tilde{t})\right\rangle = \mathbf{Q}^{\mathbf{q}}\delta(\hat{x}-\tilde{x})\delta(\hat{t}-\tilde{t}), \tag{12.141}$$

with

$$\mathbf{Q}^{\mathbf{q}}{}_{\alpha\beta} = 2\mathrm{k}_B\theta^2 k\delta_{\alpha\beta}\,. \tag{12.142}$$

12.4.3 Bubble Nucleation Experiments Through the Capillary Landau–Lifshitz-Navier–Stokes Equations

In this section, we aim at presenting thermally activated vapor bubble nucleation simulations by using the capillary Landau–Lifshitz-Navier–Stokes equations derived in the previous section. Numerical experiments are conducted both for homogeneous and heterogeneous nucleation, for different levels of metastability. In particular, bubble nucleation is investigated in a metastable liquid enclosed in a box with periodic boundary conditions in all the directions for the homogeneous case, and two flat solid surfaces in z direction and periodicity in $x - y$ to address heterogeneous nucleation. In the latter case also the role of wall wettability is investigated by imposing the boundary condition for the density field as exposed in Sect. 12.3 (see Eq. 12.64), providing the nucleation rate as a function of the contact angle ϕ. The fluid is characterized by an equation of state that recovers the properties of a Lennard-Jones fluid [23].

The following reference quantities $\sigma = 3.4 \times 10^{-10}$ m as length, $\epsilon = 1.65 \times 10^{-21}$ J as energy, $m = 6.63 \times 10^{-26}$ kg as mass, $U_r = (\epsilon/m)^{1/2}$ as velocity, $T_r = \sigma/U_r$ as time, $\theta_r = \epsilon/k_B$ as temperature, $\mu_r = \sqrt{m\epsilon}/\sigma^2$ as shear viscosity, $c_{vr} = mk_B$ as specific heat at constant volume and $k_r = \mu_r c_{vr}$ as thermal conductivity are adopted. So that the dimensionless fields are defined as $\rho^* = \rho/\rho_r, \theta^* = \theta/\theta_r, \mathbf{u}^* = \mathbf{u}/U_r$. The system volume has been discretized on a equi-spaced grid ($\Delta x = \Delta y = \Delta z = 10$).

Several metastable conditions have been investigated, and here, we report in detail the results of different simulations at initial temperature $\theta^* = \theta^*_{eq} = 1.25$ and $\theta^* = \theta^*_{eq} = 1.20$. The proposed mesoscale approach allows for exploring very large system (in comparison with molecular dynamics simulations) and for a very long time, addressing the complete nucleation dynamics: starting from embryos formation up to macroscopic motions. Only ten runs for each simulation have been carried out in order to perform statistical averages of the results since the macroscopical observables, like the nucleation rates, have demonstrated to be statistically robust. For a detailed explanation of bubble identification procedure, we refer to [11, 12].

A few snapshots of the system evolution in the different metastable conditions are shown in the left panels of Figs. 12.6 and 12.7, representing the nucleation dynamics in homogeneous and heterogeneous conditions, respectively. In both cases, starting from a homogeneous liquid phase, the hydrodynamic fluctuations lead the system to spontaneously decompose in two different phases. The vapor nuclei starts forming with a complex shape, far from a spherical one, as observed in other works [8]. After reaching the critical size, they start expanding up to a stable equilibrium state. This new thermodynamic state is characterized by the presence of several stable vapor bubbles in equilibrium with the surrounding liquid. The number and the dimension of the bubbles in the latter stage are strictly connected with the initial metastable condition, as evident from the left of panels of Figs. 12.6 and 12.7.

In the right panels of Figs. 12.6 and 12.7, we report the time evolution of the number of vapor bubbles exceeding the critical size. The correspondence with the snapshots on the left part of the figure is marked by the letters a, b, c, d.

The dynamics of the system can be divided into three main stages: during the first one the number of bubble increases almost linearly with time (with a constant rate); when the system is populated enough, the second stage consists of the expansion-coalescence dynamics when the nuclei increase their size up to the equilibrium radius

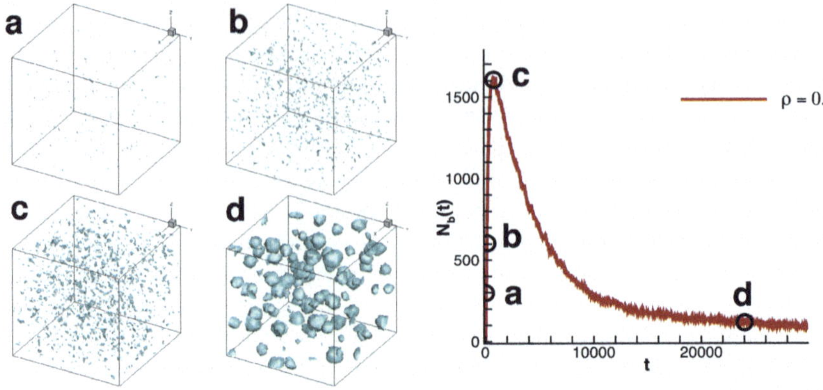

Fig. 12.6 Left panel: snapshots during the nucleation process in the thermodynamic condition $\rho^* = 0.46$ $\theta^* = 1.25$, the snapshots are taken at time $t^* = 300$, $t^* = 900$, $t^* = 2000$, $t^* = 24{,}000$. Right panel: number of stable bubble detected in the bulk versus time for the aforementioned thermodynamic condition

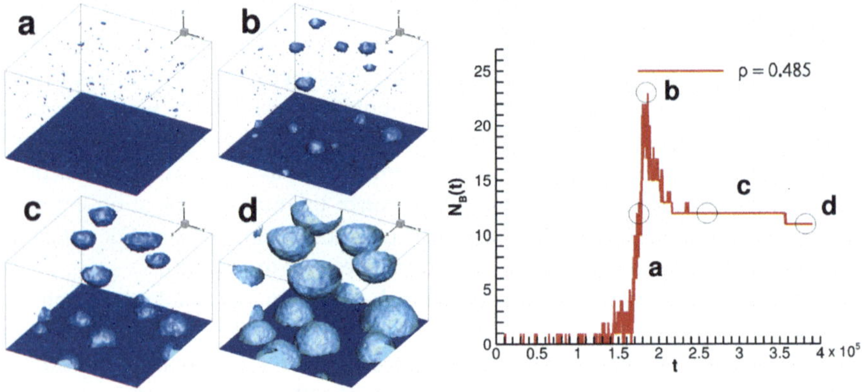

Fig. 12.7 Left panel: snapshots during the nucleation process in the thermodynamic condition $\rho^* = 0.485$ $\theta^* = 1.25$, the snapshots are taken at time $t^* = 180{,}000$, $t^* = 1900$, $t^* = 280{,}000$, $t^* = 360{,}000$. Right panel: number of stable bubble detected on the walls versus time for the aforementioned thermodynamic condition

and some of them coalesce with neighboring bubbles. Furthermore, the smallest nuclei start collapsing due to liquid compression. During the third stage, the system reaches a more stable thermodynamic condition, in which a small number of stable vapor bubbles are in equilibrium with the surrounding compressed liquid. The analysis of the first stage of the dynamics, when the number of bubbles increases, gives access to the nucleation rate J^*, representing the number of bubbles formed per unit time and per unit volume/area for homogeneous/heterogeneous cases, respectively. The nucleation rate is here calculated as the slope of the linear fit of the initial part of the curves in Figs. 12.6 and 12.7, as suggested in [8]. The homogeneous nucleation

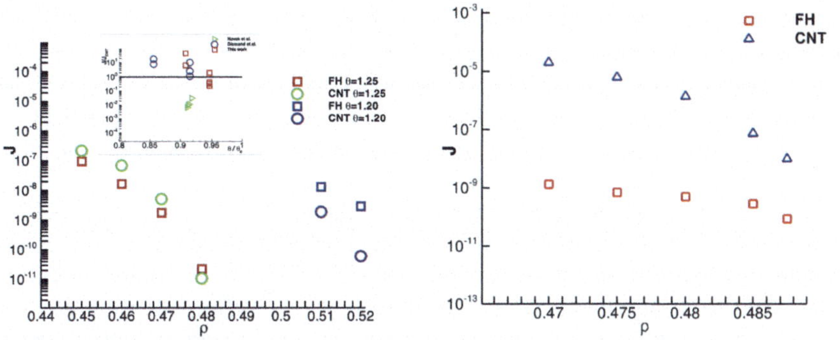

Fig. 12.8 Left panel: homogeneous nucleation rate comparison between fluctuating hydrodynamics numerical results, Blander and Katz rate predictions at different metastable conditions. In the inset, a comparison with MD simulations [8, 55] are reported. Right panel: heterogeneous nucleation rate comparison between fluctuating hydrodynamics numerical results, Blander and Katz rate predictions at different metastable conditions

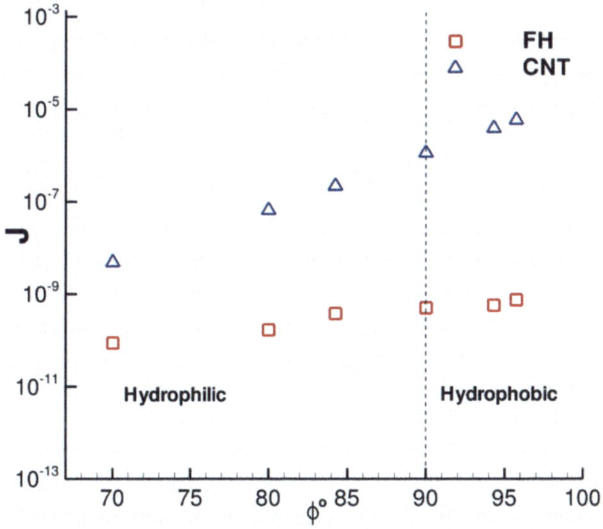

Fig. 12.9 Nucleation rate comparison between fluctuating hydrodynamics numerical results, Blander and Katz rate predictions at different contact angles ϕ

rate is normalized with the system volume, while the heterogeneous one is normalized with the wall areas. The calculated nucleation rates at different metastable conditions are compared in Fig. 12.8 with the theoretical predictions given by the Blander and Katz CNT formulation [56]. The rates calculated with our numerical simulations are smaller than predicted by classical theory. The discrepancy is significant when considering heterogeneous case. As explained in [12], this discrepancy is mostly related to the overcrowding of bubbles in the closed system where the constrain on mass, energy and volume play a crucial role. Apparently, the theoretically predicted rates are more precise as the system gets less populated. In the homogeneous case, the agreement with the classical theory is better, and the results are in line with MD simulations [8, 55] (see the inset in Fig. 12.8). Unfortunately, to the best of our knowledge, quantitative MD results of heterogeneous bubble nucleation are not available in the literature.

In Fig. 12.9, for a specific thermodynamic condition $\theta^* = 1.25$ and $\rho^* = 0.48$, we report a comparison between heterogeneous vapor bubble nucleation rate as evaluated by fluctuating hydrodynamics simulations and CNT, for different wall wettabilities. As expected from arguments based on the height of the free energy barrier, the rate shows a monotonic trend with the contact angle ϕ. In particular, nucleation is incentivized by hydrophobic surfaces.

12.5 Summary

In this work, we have discussed the continuum picture of liquid–vapor phase change. The foundation of most of the continuum techniques to address the complex problem of nucleation is the classical nucleation theory (CNT). Based on simple energetic arguments, CNT applies macroscopic thermodynamics informations to microscopic objects (vapor embryos on nanometric size) providing an energy landscape $\Omega(V)$ as a function of bubble volume (the reaction coordinates of the phase change process). Together with statistical mechanics arguments, CNT also provides the vapor bubble nucleation rate (the number of stable bubbles per unit time and volume). Despite the strong assumptions of the theory, CNT leads to an accurate description of vapor nucleation toward the saturation conditions. However, it fails when approaching spinodal limits. In fact, in the latter case, the typical liquid–vapor interface thickness is of the same order as the bubble radius, making the sharp-interface assumption too strong. Advanced theories like density functional theory can correct some CNT mispredictions, with a more accurate estimation of nucleation path. In any case, all these theories are quasi-static, and they deal with the nucleation of a single bubble. This sheds the light on the importance of adopting mesoscale approaches able to follow the phase change starting from the vapor bubble inception up to hydrodynamics motion. In the present work, the derivation of a promising mesoscale model was reviewed and discussed. The approach is based on a diffuse interface description of the liquid–vapor thermodynamics coupled with Navier–Stokes hydrodynamics. In order to take into account the stochastic nature of nucleation, thermal fluctuations have to be considered. For the latter purpose, we discussed the fluctuating hydrodynamics theory and its coupling with capillary fluids. The model consists of a set of stochastic partial differential equations (Capillary Landau–Lifshitz–Navier–Stokes equation CLLNS) having as a deterministic part the capillary Navier–Stokes equations and reproducing the Einstein-Boltzmann probability distribution for the fluctuating fields. In silico experiments of nucleation with CLLNS equations were discussed both for homogeneous and heterogeneous cases. The accuracy of the results together with the cheap computational cost of the method encourages its extension to more complex conditions of nucleation; for instance in the presence of mean flows, complex geometries as well as multispecies fluids.

References

1. Antonio Tinti, Alberto Giacomello, Yaroslav Grosu, and Carlo Massimo Casciola. Intrusion and extrusion of water in hydrophobic nanopores. *Proceedings of the National Academy of Sciences*, 114(48):E10266–E10273, 2017.
2. E Cramer and W Lauterborn. Acoustic cavitation noise spectra. In *Mechanics and Physics of Bubbles in Liquids*, pages 209–214. Springer, 1982.
3. Agostino Occhicone, Giorgia Sinibaldi, Norbert Danz, Carlo Massimo Casciola, and Francesco Michelotti. Cavitation bubble wall pressure measurement by an electromagnetic surface wave enhanced pump-probe configuration. *Applied Physics Letters*, 114(13):134101, 2019.
4. Chiara Scognamiglio, Francesco Magaletti, Yaroslava Izmaylov, Mirko Gallo, Carlo Massimo Casciola, and Xavier Noblin. The detailed acoustic signature of a micro-confined cavitation bubble. *Soft matter*, 2018.
5. G Sinibaldi, A Occhicone, F Alves Pereira, D Caprini, L Marino, F Michelotti, and CM Casciola. Laser induced cavitation: Plasma generation and breakdown shockwave. *Physics of Fluids*, 31(10):103302, 2019.
6. Dimo Kashchiev. *Nucleation*. Elsevier, 2000.
7. David W Oxtoby and R Evans. Nonclassical nucleation theory for the gas–liquid transition. *The Journal of chemical physics*, 89(12):7521–7530, 1988.
8. Jürg Diemand, Raymond Angélil, Kyoko K Tanaka, and Hidekazu Tanaka. Direct simulations of homogeneous bubble nucleation: Agreement with classical nucleation theory and no local hot spots. *Physical review E*, 90(5):052407, 2014.
9. Alberto Giacomello, Simone Meloni, Marcus Müller, and Carlo Massimo Casciola. Mechanism of the cassie-wenzel transition via the atomistic and continuum string methods. *The Journal of chemical physics*, 142(10):104701, 2015.
10. Mirko Gallo, Francesco Magaletti, and Carlo Massimo Casciola. Fluctuating hydrodynamics as a tool to investigate nucleation of cavitation bubbles. *International Journal of Computational Methods and Experimental Measurements*, 6(2):345–357, 2017.
11. Mirko Gallo, Francesco Magaletti, and Carlo Massimo Casciola. Thermally activated vapor bubble nucleation: The landau-lifshitz–van der waals approach. *Phys. Rev. Fluids*, 3:053604, May 2018.
12. Mirko Gallo, Francesco Magaletti, and Carlo Massimo Casciola. Heterogeneous bubble nucleation dynamics. *Journal of Fluid Mechanics*, 906, 2020.
13. Mirko Gallo, Francesco Magaletti, Davide Cocco, and Carlo Massimo Casciola. Nucleation and growth dynamics of vapour bubbles. *Journal of Fluid Mechanics*, 883, 2020.
14. Pablo G Debenedetti. *Metastable liquids: concepts and principles*. Princeton University Press, 1996.
15. John C Angus and Cliff C Hayman. Low-pressure, metastable growth of diamond and "diamondlike" phases. *Science*, 241(4868):913–921, 1988.
16. H_M Strong and RE Hanneman. Crystallization of diamond and graphite. *The Journal of Chemical Physics*, 46(9):3668–3676, 1967.
17. Emily B Moore and Valeria Molinero. Structural transformation in supercooled water controls the crystallization rate of ice. *Nature*, 479(7374):506, 2011.
18. Daniel M Murphy and Thomas Koop. Review of the vapour pressures of ice and supercooled water for atmospheric applications. *Quarterly Journal of the Royal Meteorological Society: A journal of the atmospheric sciences, applied meteorology and physical oceanography*, 131(608):1539–1565, 2005.

19. Sylvain Deville, Eric Maire, Guillaume Bernard-Granger, Audrey Lasalle, Agnès Bogner, Catherine Gauthier, Jérôme Leloup, and Christian Guizard. Metastable and unstable cellular solidification of colloidal suspensions. *Nature materials*, 8(12):966, 2009.
20. Amy E Larsen and David G Grier. Like-charge attractions in metastable colloidal crystallites. *Nature*, 385(6613):230, 1997.
21. Joachim Heierli. Solitary fracture waves in metastable snow stratifications. *Journal of Geophysical Research: Earth Surface*, 110(F2), 2005.
22. J-P Bouchaud, ME Cates, J Ravi Prakash, and SF Edwards. Hysteresis and metastability in a continuum sandpile model. *Physical review letters*, 74(11):1982, 1995.
23. J Karl Johnson, John A Zollweg, and Keith E Gubbins. The lennard-jones equation of state revisited. *Molecular Physics*, 78(3):591–618, 1993.
24. Hendrik Anthony Kramers. Brownian motion in a field of force and the diffusion model of chemical reactions. *Physica*, 7(4):284–304, 1940.
25. Christopher E Brennen. *Cavitation and bubble dynamics*. Cambridge University Press, 2013.
26. CA Ward, WR Johnson, RD Venter, S Ho, TW Forest, and WD Fraser. Heterogeneous bubble nucleation and conditions for growth in a liquid–gas system of constant mass and volume. *Journal of Applied Physics*, 54(4):1833–1843, 1983.
27. Georg Menzl, Miguel A Gonzalez, Philipp Geiger, Frédéric Caupin, José LF Abascal, Chantal Valeriani, and Christoph Dellago. Molecular mechanism for cavitation in water under tension. *Proceedings of the National Academy of Sciences*, 113(48):13582–13587, 2016.
28. James F Lutsko. Density functional theory of inhomogeneous liquids. ii. a fundamental measure approach. *The Journal of chemical physics*, 128(18):184711, 2008.
29. James F Lutsko and Miguel A Durán-Olivencia. A two-parameter extension of classical nucleation theory. *Journal of Physics: Condensed Matter*, 27(23):235101, 2015.
30. Raymond Angélil, Jürg Diemand, Kyoko K Tanaka, and Hidekazu Tanaka. Bubble evolution and properties in homogeneous nucleation simulations. *Physical review E*, 90(6):063301, 2014.
31. Rosalind J Allen, Daan Frenkel, and Pieter Rein ten Wolde. Simulating rare events in equilibrium or nonequilibrium stochastic systems. *The Journal of chemical physics*, 124(2):024102, 2006.
32. Rosalind J Allen, Chantal Valeriani, and Pieter Rein ten Wolde. Forward flux sampling for rare event simulations. *Journal of physics: Condensed matter*, 21(46):463102, 2009.
33. Peter G Bolhuis, David Chandler, Christoph Dellago, and Phillip L Geissler. Transition path sampling: Throwing ropes over rough mountain passes, in the dark. *Annual review of physical chemistry*, 53(1):291–318, 2002.
34. Christoph Dellago and Peter G Bolhuis. Transition path sampling and other advanced simulation techniques for rare events. In *Advanced Computer Simulation Approaches for Soft Matter Sciences III*, pages 167–233. Springer, 2009.
35. Sara Marchio, Simone Meloni, Alberto Giacomello, and Carlo Massimo Casciola. Wetting and recovery of nano-patterned surfaces beyond the classical picture. *Nanoscale*, 2019.
36. James F Lutsko. Density functional theory of inhomogeneous liquids. iv. squared-gradient approximation and classical nucleation theory. *The Journal of chemical physics*, 134(16):164501, 2011.
37. F Magaletti, L Marino, and CM Casciola. Shock wave formation in the collapse of a vapor nanobubble. *Physical Review Letters*, 114(6):064501, 2015.
38. Francesco Magaletti, Mirko Gallo, Luca Marino, and Carlo Massimo Casciola. Dynamics of a vapor nanobubble collapsing near a solid boundary. In *Journal of Physics: Conference Series*, volume 656(1), page 012012. IOP Publishing, 2015.
39. Francesco Magaletti, Luca Marino, and Carlo Massimo Casciola. Diffuse interface modeling of a radial vapor bubble collapse. In *Journal of Physics: Conference Series*, volume 656, page 012028. IOP Publishing, 2015.
40. Ronald Forrest Fox and George E Uhlenbeck. Contributions to non-equilibrium thermodynamics. i. theory of hydrodynamical fluctuations. *Physics of Fluids (1958–1988)*, 13(8):1893–1902, 1970.
41. LD Landau and EM Lifshitz. Statistical physics, vol. 5. *Course of theoretical physics*, 30, 1980.

42. Francesco Magaletti, Anastasios Georgoulas, and Marco Marengo. Unraveling low nucleation temperatures in pool boiling through fluctuating hydrodynamics simulations. *International Journal of Multiphase Flow*, page 103356, 2020.
43. James F Lutsko. Density functional theory of inhomogeneous liquids. i. the liquid-vapor interface in lennard-jones fluids. *The Journal of chemical physics*, 127(5):054701, 2007.
44. DM Anderson, GB McFadden, and AA Wheeler. Diffuse-interface methods in fluid mechanics. *Annual Review of Fluid Mechanics*, 30(1):139–165, 1998.
45. Francesco Dell'Isola, Henri Gouin, Pierre Seppecher, et al. Radius and surface tension of microscopic bubbles by second gradient theory. *Comptes Rendus de l Académie des Sciences-Series IIB-Mechanics*, 320, 1995.
46. D Jamet, O Lebaigue, N Coutris, and JM Delhaye. The second gradient method for the direct numerical simulation of liquid–vapor flows with phase change. *Journal of Computational Physics*, 169(2):624–651, 2001.
47. Weinan E, Weiqing Ren, and Eric Vanden-Eijnden. Simplified and improved string method for computing the minimum energy paths in barrier-crossing events. *Journal of Chemical Physics*, 126(16):164103, 2007.
48. Weiqing Ren. Wetting transition on patterned surfaces: transition states and energy barriers. *Langmuir*, 30(10):2879–2885, 2014.
49. Albert Einstein. Theorie der opaleszenz von homogenen flüssigkeiten und flüssigkeitsgemischen in der nähe des kritischen zustandes. *Annalen der Physik*, 338(16):1275–1298, 1910.
50. Albert Einstein. *Investigations on the Theory of the Brownian Movement*. Courier Corporation, 1956.
51. Rouslan L Stratonovich. *Nonlinear nonequilibrium thermodynamics I: linear and nonlinear fluctuation-dissipation theorems*, volume 57. Springer Science & Business Media, 2012.
52. Sybren Ruurds De Groot and Peter Mazur. *Non-equilibrium thermodynamics*. Courier Dover Publications, 2013.
53. Francesco Magaletti, Mirko Gallo, Luca Marino, and Carlo Massimo Casciola. Shock-induced collapse of a vapor nanobubble near solid boundaries. *International Journal of Multiphase Flow*, 84:34–45, 2016.
54. Giuseppe Da Prato. *Kolmogorov equations for stochastic PDEs*. Birkhäuser, 2012.
55. Brian R. Novak, Edward J. Maginn, and Mark J. McCready. Comparison of heterogeneous and homogeneous bubble nucleation using molecular simulations. *Phys. Rev. B*, 75:085413, 2007.
56. Milton Blander and Joseph L Katz. Bubble nucleation in liquids. *AIChE Journal*, 21(5):833–848, 1975.

Correction to: Vapor Nucleation in Metastable Liquids The Continuum Description

Mirko Gallo, Francesco Magaletti, Dario Abbondanza,
and Carlo Massimo Casciola

Correction to:
Chapter 12 in: M. Marengo and J. De Coninck (eds.), *The Surface Wettability Effect on Phase Change*, https://doi.org/10.1007/978-3-030-82992-6_12

The original version of this book was inadvertently published without including a chapter 12. The chapter 12 titled "Vapor Nucleation in Metastable Liquids The Continuum Description" has been added at the end of the book.

The updated version of this chapter can be found at
https://doi.org/10.1007/978-3-030-82992-6_12

M. Marengo and J. De Coninck (eds.), *The Surface Wettability Effect on Phase Change*, https://doi.org/10.1007/978-3-030-82992-6_13

The manufacturer's authorised representative in the EU is Springer Nature Customer Service Centre GmbH, Europaplatz 3, 69115 Heidelberg, Germany. If you have any concerns regarding our products, please contact ProductSafety@springernature.com

Printed and bound by CPI Group (UK) Ltd, Croydon, CR0 4YY
30/09/2026
02225694-0013